T0310838

Physics of Electronic Materials

Principles and Applications

Adopting a uniquely pedagogical approach, this comprehensive textbook on the quantum mechanics of semiconductor materials and devices focuses on the materials, components and devices themselves whilst incorporating a substantial amount of fundamental physics related to condensed matter theory and quantum mechanics.

Written primarily for advanced undergraduate students in physics and engineering, this book can also be used as a supporting text for introductory quantum mechanics courses, and will be of interest to anyone interested in how electronic devices function at a fundamental level.

Complete with numerous exercises, and with all the necessary mathematics and physics included in appendices, this book guides the reader seamlessly through the principles of quantum mechanics and the quantum theory of metals and semiconductors, before describing in detail how devices are exploited within electric circuits and in the hardware of computers: for example as amplifiers, switches and transistors.

Jørgen Rammer is Guest Professor in Physics at Lund University, Sweden. He has published extensively in the field of quantum transport theory and semiconductor physics.

Physics of Electronic Materials

Principles and Applications

JØRGEN RAMMER

Lund University

CAMBRIDGE
UNIVERSITY PRESS

University Printing House, Cambridge CB2 8BS, United Kingdom

One Liberty Plaza, 20th Floor, New York, NY 10006, USA

477 Williamstown Road, Port Melbourne, VIC 3207, Australia

314-321, 3rd Floor, Plot 3, Splendor Forum, Jasola District Centre, New Delhi - 110025, India

79 Anson Road, #06-04/06, Singapore 079906

Cambridge University Press is part of the University of Cambridge.

It furthers the University's mission by disseminating knowledge in the pursuit of education, learning and research at the highest international levels of excellence.

www.cambridge.org
Information on this title: www.cambridge.org/9781107084940
DOI: 10.1017/9781316027165

First published 2017

A catalogue record for this publication is available from the British Library

ISBN 978-1-107-08494-0 Hardback

Contents

Preface

Electronic devices play a crucial role in today's societies and in the physical sciences where they originated. Contemplating that, in just a few decades, technology guiding electrons and photons has emerged that makes possible oral and visual communication between people on opposite sides of the planet is truly a triumph of science and technology. Not to mention that, equipped with a computer with access to the internet, one can instantly access a wealth of human knowledge. The physical principles providing the understanding of the functioning of present-day electronic devices should therefore be of interest not only to physicists, electrical engineers and material scientists, but also to anyone with a general interest in how the *wired* world around us functions. Present-day information technology is based on the physical properties of semiconductors, in particular the functioning of the transistor. The intention of this book is to take the reader from the principles of quantum mechanics through the quantum theory of metals and semiconductors all the way to how devices are used to perform their duties in electric circuits: for example, functioning as amplifiers, switches and in the hardware of computers. The mechanics of arithmetic and logical operations are discussed, and it is shown how electronic devices in the present-day CMOS (complementary metal–oxide–semiconductor) technology can carry out arithmetic calculations and logic operations in computers.

We presently live not only with an extraordinary information processing technology, but also in a new type of age of creation. We are not restricted to building semiconductor structures from materials provided by their immediate availability in nature and an additional purification procedure. Today, artificial human-made semiconductor structures can be built atom by atom, allowing the construction of devices with new physical properties, such as for example the blue quantum well laser. The physical principles governing these so-called heterostructures, which are revolutionizing electronics and optoelectronics, are presented. The book presents a self-contained elementary introduction to semiconductor technology. It also describes new vistas such as the role of mesoscopic physics in materials science, and presents some of the recent advances in nanotechnology. Not only is the theory of devices presented, but also their functioning in basic electric circuits is described.

In order to understand the physics of the solid states of matter, such as, for example, the behavior of electrons in semiconductors, a knowledge of quantum mechanics is essential. The book therefore starts with a discussion of the basic properties of the solutions of the Schrödinger equation. For the interested reader, it is shown in Appendix A how to arrive at the Schrödinger equation from the superposition and correspondence principles. In Chapter 2, quantum tunneling is studied using a wave packet analysis, and the tunneling probabilities for single- and double-barrier structures are obtained.

In Chapter 3, the Sommerfeld model of a metal is introduced and its characteristics determined. It is then used to discuss a tunnel junction, and the results of Chapter 2 are used to give an elementary understanding of the scanning tunneling microscope. In Chapter 4, the Gaussian wave packet description of a metal containing impurities is introduced and used to understand basic conduction properties, the standard model of a conductor. Basic electronic circuit theory is discussed in Chapter 5.

In Chapter 6, the chemical bond that lumps atoms into molecules and semiconductors is discussed, and in Chapter 7 the energy band structure of a particle in a one-dimensional periodic potential is established. Chapter 8 considers motion in a one-dimensional periodic potential, Zener tunneling and Bloch wave dynamics, and the physics of three-dimensional crystalline solids is considered in Chapter 9.

In Chapter 10, the fundamentals of semiconductor physics are considered, namely the construction of n- and p-type semiconductors by doping with donor and acceptor impurities, respectively. The quantum dynamics of holes in semiconductors is presented, leading to the discussion in Chapter 11 of basic p–n junction device functioning and its numerous applications. In Chapter 12, the production methods of heterostructures and their physical description are presented, and their new electronic and optoelectronic possibilities discussed. In Chapter 13 transport through mesoscopic systems, such as double-barrier resonant tunneling nanostructures and quantum point contacts are presented. Their physics and applications in electronics are discussed. Finally, in Chapter 14 the implementation of arithmetic and logic in mechanical electronic machines is presented, and the functioning of a computer discussed.

The book is intended to be sufficiently broad to serve as a text for a one- or two-semester undergraduate course on semiconductor physics, mesoscopic and nanophysics and solid-state transport theory. It is also hoped that the book can serve as a useful reference book for introductory courses on quantum mechanics. The book is self-contained to the extent that it should be useful for students with basic mathematical skills to read it on their own. The basic physics needed to understand the content of the chapters is relegated to appendices, where it can be consulted if needed. A number of exercises (with solutions – indeed, always the case when the result is later used in the main text) has been provided in order to aid self-instruction.

It is a pleasure to thank the Division of Mathematical Physics at Lund University for its hospitality. I am grateful to Stephan Köhler for providing figures.

Jørgen Rammer
Lund, Sweden
August 2016

1 Quantum Mechanics

According to quantum mechanics, at a given moment in time t, a system of particles is described by a probability amplitude function, a complex-valued function $\psi(\mathbf{x}_1, \mathbf{x}_2, \ldots; t)$ of the (point) particle positions. The probability amplitude, which is also referred to as the wave function, has the significance that its absolute square $|\psi(\mathbf{x}_1, \mathbf{x}_2, \ldots; t)|^2$ at the time in question determines the probability for the event: *particles at the specified positions*. The description in terms of the wave function is complete, i.e. the dynamics of the particles is determined by a differential equation that is first order in time. The rate of change in time of the wave function is specified in terms of the wave function at the time in question and an operator, the Schrödinger equation,

$$i\hbar \frac{\partial \psi(\mathbf{x}_1, \mathbf{x}_2, \ldots; t)}{\partial t} = \hat{H}\psi(\mathbf{x}_1, \mathbf{x}_2, \ldots; t). \tag{1.1}$$

The linear operator $\hat{H}$ is called the Hamiltonian and is ultimately determined by experimental knowledge. The symbol $\hbar$ is a constant of nature, and as such also empirically determined. A "hat" has been introduced to signify that a quantity is an operator, i.e. it operates on a function, thereby turning it into another function, and standard notation for the result of an operator operating on a function has been used, $\hat{H}\psi(\mathbf{x}, t) \equiv (\hat{H}\psi)(\mathbf{x}, t)$. As an example, for a free particle of mass m, the Hamiltonian is the spatial differential operator

$$\hat{H}_0 = -\frac{\hbar^2}{2m} \frac{\partial^2}{\partial \mathbf{x}^2}, \tag{1.2}$$

where

$$\frac{\partial^2}{\partial \mathbf{x}^2} = \frac{\partial^2}{\partial x^2} + \frac{\partial^2}{\partial y^2} + \frac{\partial^2}{\partial z^2}. \tag{1.3}$$

Various notations for the Laplacian or Laplace operator will be employed:

$$\frac{\partial^2}{\partial \mathbf{x}^2} = \nabla_{\mathbf{x}}^2 = \nabla_{\mathbf{x}} \cdot \nabla_{\mathbf{x}} = \Delta_{\mathbf{x}}, \qquad \nabla_{\mathbf{x}} \equiv \frac{\partial}{\partial \mathbf{x}}. \tag{1.4}$$

In accordance with its probabilistic interpretation, at any moment in time, a proper wave function satisfies the normalization condition

$$\int d\mathbf{x}_1 \cdots \int d\mathbf{x}_N \, |\psi(\mathbf{x}_1, \mathbf{x}_2, \ldots, \mathbf{x}_N; t)|^2 = 1, \tag{1.5}$$

since the particles are assumed with certainty to be somewhere in space.

The free particle Schrödinger equation is analogous to a linear wave equation, and has the plane wave solutions

$$\psi_{\mathbf{k}}(\mathbf{x}, t) = A\, e^{i\mathbf{k}\cdot\mathbf{x} - i(\hbar\mathbf{k}^2/2m)t} \tag{1.6}$$

specified in terms of a wave vector $\mathbf{k}$. These are unnormalized solutions from which normalized solutions, i.e. wave packets, can be obtained by superposing solutions of different wave vectors (as constructed in Chapter 2).

We could at this point get right on with solving the Schrödinger equation for the physical problems of interest, having the Hamiltonian handed to us through the knowledge obtained by our forefathers.[1] A reader inclined to such a "tell me like it is" approach can jump directly to Chapter 2 and use the Schrödinger equation to study quantum tunneling or to the following chapters studying the properties of metals and semiconductors. However, such an "I believe in the Schrödinger equation and all its consequences" approach does not present the quantum mechanical concepts in the most instructive way. Our intuition is built on our direct experience with large objects, and there is no way in which we can directly experience quantum behavior by our senses. In fact, quantum mechanics is at odds with common sense, and unintelligible in terms of the way the macroscopic world behaves as we perceive directly with our senses and understand by the empirically confirmed laws of classical physics. Foremost, we note that, in quantum mechanics, probability has entered in a fundamental way, i.e. *chance* is a feature of how the world works. In general, for given identical circumstances, it is impossible to predict what will happen in the future: quantum mechanics is probabilistic in nature. Quantum mechanics only provides the odds for different outcomes. We also observe the strange feature that, in contrast to any physical statement, a description in terms of *complex* numbers is demanded. For the interested reader, it is shown in Appendix A that the Schrödinger equation can be arrived at from a few basic principles.

1.1 Hamiltonian

Consider the Schrödinger equation for a single particle of mass m in a potential,

$$i\hbar\frac{\partial\psi(\mathbf{x}, t)}{\partial t} = \left(-\frac{\hbar^2}{2m}\frac{\partial^2}{\partial\mathbf{x}^2} + V(\mathbf{x}, t)\right)\psi(\mathbf{x}, t). \tag{1.7}$$

The Hamiltonian, specifying the Schrödinger equation, then consists of the Laplacian and a multiplication operator, the space- and time-dependent potential, $V(\mathbf{x}, t)$, multiplying the wave function,

$$\hat{H} \equiv \hat{H}(t) = -\frac{\hbar^2}{2m}\frac{\partial^2}{\partial\mathbf{x}^2} + V(\mathbf{x}, t). \tag{1.8}$$

The real scalar potential, $V(\mathbf{x}, t)$, describes the fact that the particle is not free, but at different locations experiences different environments, which in addition can be changing

[1] In the same vein, you have probably solved Newton's equation having the expression for the gravitational force being *handed* to you.

in time. As shown in Appendix A, the potential is the potential energy the particle has according to classical mechanics, and the gradient of the potential equals the classical force, $\mathbf{F}(\mathbf{x},t) = -\nabla V(\mathbf{x},t)$.

That normalization at one instant of time,

$$\int d\mathbf{x}\, |\psi(\mathbf{x},t)|^2 = 1, \tag{1.9}$$

guarantees it at all times is a defining property of a Hamiltonian. If a function is normalized, it vanishes spatially at infinity in order for the normalization integral to be finite. For arbitrary normalized functions $\psi(\mathbf{x},t)$ and $\phi(\mathbf{x},t)$, two partial integrations, where the boundary terms at infinity vanish, transfer the Laplacian to the other function,

$$\int d\mathbf{x}\, \phi^*(\mathbf{x}) \left(\frac{\partial^2 \psi(\mathbf{x})}{\partial \mathbf{x}^2} \right) = \int d\mathbf{x}\, \psi(\mathbf{x}) \left(\frac{\partial^2 \phi(\mathbf{x})}{\partial \mathbf{x}^2} \right)^*, \tag{1.10}$$

and we have used that differentiation and complex conjugation are interchangeable operations. Since the potential is a real function, it can trivially be moved as a factor under the complex conjugation, and the Hamiltonian, Eq. (1.8), is seen to have the property

$$\int d\mathbf{x}\, \phi^*(\mathbf{x},t)\, \hat{H}\psi(\mathbf{x},t) = \int d\mathbf{x}\, \psi(\mathbf{x},t)(\hat{H}\phi(\mathbf{x},t))^*. \tag{1.11}$$

An operator having the property (1.11) is called a hermitian operator.

If $\psi(\mathbf{x},t)$ is a solution of the Schrödinger equation (1.7), then it follows that

$$\begin{aligned} \frac{d}{dt}\int d\mathbf{x}\, |\psi(\mathbf{x},t)|^2 &= \int d\mathbf{x} \left(\psi^*(\mathbf{x},t)\, \frac{\partial \psi(\mathbf{x},t)}{\partial t} + \psi(\mathbf{x},t)\, \frac{\partial \psi^*(\mathbf{x},t)}{\partial t} \right) \\ &= \frac{1}{i\hbar} \left(\left(\int d\mathbf{x}\, \psi^*(\mathbf{x},t)\, \hat{H}\psi(\mathbf{x},t) - \int d\mathbf{x}\, \psi(\mathbf{x},t)\hat{H}^*\psi^*(\mathbf{x},t) \right) \right. \\ &= 0, \end{aligned} \tag{1.12}$$

the last equality following from Eq. (1.11).[2] Therefore, if at one instant the wave function is normalized, the Schrödinger dynamics guarantees that it stays normalized at all times as a consequence of the Hamiltonian being hermitian.

1.2 Free Propagator

The fundamental solution of a Schrödinger equation, the propagator of the particle, is the solution that specifies the time evolution of an arbitrary wave function. The expression for

[2] For the considered case of a scalar potential, the Hamiltonian is real, $\hat{H}^* = \hat{H}$. For the case of a vector potential, this property is lost, but the Hamiltonian is still hermitian as discussed in Exercise A.2 on page 345.

the propagator for a free particle will be obtained here by solving the Schrödinger equation. For this we employ Fourier transformation (discussed in Appendix C).

Consider first, for simplicity, the one-dimensional case. Given a wave function, $\psi(x,t)$, the Fourier-transformed function with respect to the spatial variable is

$$\psi(k,t) = \frac{1}{2\pi}\int_{-\infty}^{\infty} dx\, e^{-ixk}\psi(x,t) \tag{1.13}$$

and the inverse Fourier transformation is

$$\psi(x,t) = \int_{-\infty}^{\infty} dk\, e^{ikx}\psi(k,t). \tag{1.14}$$

Inserting the Fourier expansion of the wave function, Eq. (1.14), into the free particle Schrödinger equation,

$$i\hbar\frac{\partial\psi(x,t)}{\partial t} = -\frac{\hbar^2}{2m}\frac{\partial^2\psi(x,t)}{\partial x^2}, \tag{1.15}$$

gives the equation

$$\int_{-\infty}^{\infty} dk\, e^{ixk}\left(i\hbar\frac{\partial\psi(k,t)}{\partial t} - \frac{\hbar^2k^2}{2m}\psi(k,t)\right) = 0. \tag{1.16}$$

The Fourier representation is unique (the Fourier transform of the zero function is the zero function), and the Fourier transform of the wave function thus satisfies the first-order differential equation

$$i\hbar\frac{\partial\psi(k,t)}{\partial t} = \frac{\hbar^2k^2}{2m}\psi(k,t). \tag{1.17}$$

The task of solving Eq. (1.17) thus amounts to finding the function that, when differentiated with respect to time once, gives back the same function multiplied by a number, here the imaginary number $-i\hbar k^2/2m$. This is the defining mark of the exponential function, and the solution of Eq. (1.17) is

$$\psi(k,t) = a_k\, e^{-i(\hbar k^2/2m)t}, \tag{1.18}$$

where a_k according to Eq. (1.13) is determined by

$$a_k = e^{i(\hbar k^2/2m)t}\psi(k,t) = e^{i(\hbar k^2/2m)t}\frac{1}{2\pi}\int_{-\infty}^{\infty} dx\, e^{-ixk}\psi(x,t), \tag{1.19}$$

the formula being valid for arbitrary time t.

The general solution of the free particle Schrödinger equation, Eq. (1.14), is thus a superposition of the functions given by Eq. (1.6), the plane wave solutions themselves corresponding to the choice $a_k = A\delta(k-k_0)$.

Assuming the wave function at time t', $\psi(x,t')$, is known, the Fourier transform, $\psi(k,t')$, at the same instant is known, and thereby the prefactor

$$a_k = e^{i(\hbar k^2/2m)t'}\psi(k,t'). \tag{1.20}$$

Inserting this expression into Eq. (1.18), the Fourier transform is then specified at all times by

$$\psi(k,t) = \psi(k,t')\, e^{-i(\hbar k^2/2m)(t-t')}, \tag{1.21}$$

where

$$\psi(k,t') = \frac{1}{2\pi}\int_{-\infty}^{\infty} dx\, e^{-ixk}\psi(x,t'). \tag{1.22}$$

The Fourier transform of a free particle wave function, Eq. (1.21), thus has a simple time dependence: an exponential with a phase varying linearly in time.

Inserting the expression (1.21) into Eq. (1.14) gives

$$\begin{aligned}\psi(x,t) &= \int_{-\infty}^{\infty} dk\, e^{ikx}\psi(k,t')\, e^{-i(\hbar k^2/2m)(t-t')} \\ &= \int_{-\infty}^{\infty} dk\, e^{ikx} e^{-i(\hbar k^2/2m)(t-t')}\frac{1}{2\pi}\int_{-\infty}^{\infty} dx'\, e^{-ix'k}\psi(x',t'),\end{aligned} \tag{1.23}$$

the last equality following from Eq. (1.22). Interchanging the order of the integrations gives

$$\psi(x,t) = \int_{-\infty}^{\infty} dx'\, K_0(x,t;x',t')\,\psi(x',t'), \tag{1.24}$$

where

$$K_0(x,t;x',t') = \frac{1}{2\pi}\int_{-\infty}^{\infty} dk\, e^{ik(x-x')-i(\hbar k^2/2m)(t-t')}. \tag{1.25}$$

Since $\psi(x,t)$ is a solution of the free particle Schrödinger equation, so is the kernel, K_0, according to Eq. (1.24). The Gaussian integral, Eq. (1.24), is performed by completing the square (see Appendix B), giving the expression

$$K_0(x,t;x',t') = \sqrt{\frac{m}{2\pi i\hbar(t-t')}}\ \exp\left(\frac{i}{\hbar}\frac{m(x-x')^2}{2(t-t')}\right). \tag{1.26}$$

According to Eq. (1.24), the kernel, K_0, propagates the arbitrary initial state, and is therefore called the propagator; here for the case of a free particle, it is the free propagator. Free quantum dynamics, initially described by the Schrödinger *differential* equation, has thus been inverted to be described by an integral equation.[3]

In three spatial dimensions, we simply have to do three-fold exactly the same calculation as above due to the multiplicative character of Fourier transformation:

$$\exp(-i\mathbf{x}\cdot\mathbf{k}) = \exp(-ik_1x - ik_2y - ik_3z) = \exp(-ik_1x)\exp(-ik_2y)\exp(-ik_3z).$$

[3] Propagation of a wave function by a kernel is generally valid: the fundamental equation of quantum dynamics arrived at in Appendix A, Eq. (A.12). The propagator is the basic quantum mechanical quantity, the quantum concept surviving even in relativistic quantum theory.

The general expression for the free particle propagator is therefore (here specified for d spatial dimensions)

$$K_0(\mathbf{x},t;\mathbf{x}',t') = \left(\frac{m}{2\pi\hbar i(t-t')}\right)^{d/2} \exp\left(\frac{im}{2\hbar}\frac{(\mathbf{x}-\mathbf{x}')^2}{t-t'}\right). \tag{1.27}$$

Exercise 1.1 Show by explicit differentiation that the free propagator $K_0(x,t;x',t')$ satisfies the free particle Schrödinger equation.

Exercise 1.2 Show that the plane wave solutions of the free particle Schrödinger equation (1.6), though non-normalized, are propagated by the free propagator, Eq. (1.27), i.e.

$$e^{i\mathbf{k}\cdot\mathbf{x}-i(\hbar\mathbf{k}^2/2m)t} = \int_{-\infty}^{\infty} d\mathbf{x}'\; K_0(\mathbf{x},t;\mathbf{x}',0)\, e^{i\mathbf{k}\cdot\mathbf{x}'}. \tag{1.28}$$

Exercise 1.3 Consider the free evolution of the Gaussian wave packet (consider the one-dimensional case for simplicity) which at time $t=0$ is centered around position $x=0$ and has a width δx (the parameters δx and p are real numbers),

$$\psi_{0p}(x,t=0) = \left(\frac{1}{2\pi\delta x^2}\right)^{1/4} \exp\left(-\frac{x^2}{4\delta x^2}+\frac{i}{\hbar}px\right). \tag{1.29}$$

Show that this initial state is normalized. Obtain, by using Eq. (1.24), the wave function and probability density at times $t>0$.

Solution

The integration to be performed is Gaussian and obtained by completing the square

$$\begin{aligned}\psi_{0p}(x,t) &= \int_{-\infty}^{\infty} dx'\; K_0(x,t;x',0)\,\psi_{0p}(x',0)\\ &= \left(\frac{\delta x^2}{2\pi\delta x_t^4}\right)^{1/4} \exp\left(\frac{im}{2\hbar t}x^2\right)\exp\left(-\frac{im\delta x^2}{2\hbar t}\frac{(x-x_0(t))^2}{\delta x_t^2}\right),\end{aligned} \tag{1.30}$$

where

$$\delta x_t^2 = \delta x^2\left(1+\frac{i\hbar t}{2m\delta x^2}\right), \qquad x_0(t) = \frac{p}{m}t. \tag{1.31}$$

The probability density at time t then becomes

$$P_{0p}(x,t) = |\psi_{0p}(x,t)|^2 = \sqrt{\frac{1}{2\pi\Delta x_t^2}}\exp\left(-\frac{(x-x_0(t))^2}{2\Delta x_t^2}\right), \tag{1.32}$$

where

$$\Delta x_t^2 = \frac{|\delta x_t^2|^2}{\delta x^2} = \delta x^2\left(1+\left(\frac{\hbar t}{2m\delta x^2}\right)^2\right). \tag{1.33}$$

As time goes by, the probability density keeps its Gaussian shape, but the profile does not propagate rigidly in space, instead experiencing wave packet spreading: its height decreasing and width increasing in concordance with the constraint of normalization of the probability distribution. The center of the Gaussian wave packet, $x_0(t) = pt/m$, moves with the constant velocity, p/m, the velocity with which a free particle of mass m and momentum p moves according to Newton's equation. Knowing the free propagator, we could turn to a discussion of the concept of *momentum*, or equivalently velocity, in quantum mechanics. However, this is postponed until needed, the details being relegated to Appendix G.

1.3 Probability Current

In view of the conservation of probability, Eq. (1.12), a diminishing probability in time in some region of space means that probability has streamed out of that volume. The Schrödinger equation identifies the probability current density describing this dynamics. The time derivative of the probability density, $P(\mathbf{x}, t) = |\psi(\mathbf{x}, t)|^2 = \psi(\mathbf{x}, t)\,\psi^*(\mathbf{x}, t)$, where $*$ denotes complex conjugation,

$$\frac{\partial P(\mathbf{x},t)}{\partial t} = \psi^*(\mathbf{x},t)\,\frac{\partial \psi(\mathbf{x},t)}{\partial t} + \psi(\mathbf{x},t)\,\frac{\partial \psi^*(\mathbf{x},t)}{\partial t} \tag{1.34}$$

becomes, according to the Schrödinger equation,

$$\frac{\partial P(\mathbf{x},t)}{\partial t} = \frac{1}{i\hbar}[\psi^*(\mathbf{x},t)\hat{H}(t)\psi(\mathbf{x},t) - \psi(\mathbf{x},t)(\hat{H}(t))^*\psi^*(\mathbf{x},t)]. \tag{1.35}$$

Consider the case of a particle in a scalar potential, i.e. described by the Hamiltonian in Eq. (1.8). The two terms in Eq. (1.35) containing the potential cancel each other, leaving

$$\begin{aligned}\frac{\partial P(\mathbf{x},t)}{\partial t} &= \frac{i\hbar}{2m}(\psi^*(\mathbf{x},t)\,\Delta_{\mathbf{x}}\psi(\mathbf{x},t) - \psi(\mathbf{x},t)\Delta_{\mathbf{x}}\psi^*(\mathbf{x},t)) \\ &= -\frac{\hbar}{2im}\nabla_{\mathbf{x}}\cdot(\psi^*(\mathbf{x},t)\nabla_{\mathbf{x}}\psi(\mathbf{x},t) - \psi(\mathbf{x},t)\nabla_{\mathbf{x}}\psi^*(\mathbf{x},t)),\end{aligned} \tag{1.36}$$

where the last equality is a trivial rewriting, in terms of the *divergence*, as the additional two generated terms cancel each other. Introducing the vector field (which we note is real, $\mathbf{S}^* = \mathbf{S}$)

$$\begin{aligned}\mathbf{S}(\mathbf{x},t) &= \frac{\hbar}{2im}(\psi^*(\mathbf{x},t)\,\nabla_{\mathbf{x}}\psi(\mathbf{x},t) - \psi(\mathbf{x},t)\nabla_{\mathbf{x}}\psi^*(\mathbf{x},t)) \\ &= \frac{\hbar}{2im}\,\psi^*(\mathbf{x},t)\nabla_{\mathbf{x}}\psi(\mathbf{x},t) + \text{c.c.}\end{aligned} \tag{1.37}$$

for a particle in the state described by the wave function $\psi(\mathbf{x}, t)$, Eq. (1.36) takes the form of a continuity equation,

$$\frac{\partial P(\mathbf{x},t)}{\partial t} + \nabla_{\mathbf{x}} \cdot \mathbf{S}(\mathbf{x},t) = 0. \tag{1.38}$$

Integrating Eq. (1.38) over volume Ω, and using Gauss's theorem from vector calculus turns the divergence term into a surface integral[4]

$$\frac{d}{dt}\int_{\Omega} d\mathbf{x}\, P(\mathbf{x},t) = \int_{\Omega} d\mathbf{x}\, \frac{\partial P(\mathbf{x},t)}{\partial t} = -\int_{\Omega} d\mathbf{x}\, \nabla \cdot \mathbf{S}(\mathbf{x},t) = -\int_{S} d\mathbf{s} \cdot \mathbf{S}(\mathbf{x},t), \tag{1.39}$$

where S is the surface enclosing the volume Ω, and $d\mathbf{s} = ds\,\mathbf{n}$ is the directed surface element, where $\mathbf{n}$ is a unit vector normal to the surface area element, ds, directed outward from the enclosed volume. Equation (1.39) expresses the fact that the change in time of the probability for the particle to be in volume Ω at the time in question is expressed in terms of the net flow of the vector field $\mathbf{S}(\mathbf{x},t)$ through the surface enclosing the volume. The vector field $\mathbf{S}(\mathbf{x},t)$ therefore has the meaning of a probability current density or flux, i.e. $\Delta s\,\mathbf{n} \cdot \mathbf{S}(\mathbf{x},t)$ is the probability per unit time that the particle at time t will pass through the small surface at position $\mathbf{x}$ with area Δs. If $\mathbf{n} \cdot \mathbf{S}$ is positive, the flow is in the direction specified by $\mathbf{n}$, and if negative the flow is opposite, i.e. out- or in-flow. Since $|\psi(\mathbf{x},t)|^2$ is a probability, $\mathbf{S}(\mathbf{x},t)$ is also an average quantity, the average particle current density. According to Eq. (1.36), the probability current density is generated by the kinetic energy part of the Hamiltonian, the potential only entering implicitly through the wave function, a solution of the Schrödinger equation where the potential is present.[5] The probability current density is a measurable quantity. Just as the probability statements of the wave function are obtainable by repeated measurements, so is the probability current density by measuring the particle flux in terms of, for example, the blackening of photographic emulsions constituting the surface elements of interest.

Even for non-normalizable wave functions, such as the plane wave equation (1.6) corresponding to a state of a free particle, Eq. (1.39) renders $|\psi|^2$ a measure of the relative probability density for bounded spatial regions.

Writing the wave function in terms of its modulus and phase, two real functions, $\psi = |\psi|e^{i\Lambda}$, the expression in Eq. (1.37) becomes

$$\mathbf{S}(\mathbf{x},t) = \frac{\hbar}{m}\,|\psi(\mathbf{x},t)|^2\,\nabla_{\mathbf{x}}\Lambda(\mathbf{x},t). \tag{1.40}$$

1.4 Stationary States and Energy

For an isolated system, the Hamiltonian being time-independent, the Schrödinger equation is now shown to have solutions describing situations where all physical properties are

[4] If unfamiliar with Gauss's theorem, then first do the calculation in one spatial dimension where the boundary terms appear immediately by the fundamental theorem of integration. Then the three-dimensional case where Ω is a box corresponds to doing the partial integration three times.

[5] In the case of a vector potential, its influence will in addition enter explicitly in the probability current density as discussed in Exercise A.2 on page 345.

independent of time, a stationary state of the system. In particular, the probability density is time-independent in a stationary state, and, for a single particle $|\psi(\mathbf{x},t)|^2 = P(\mathbf{x})$, nothing happens in the whereabouts of the particle as time passes. A stationary state of the particle must thus be described by a function whose time dependence can only occur in the phase,

$$\psi(\mathbf{x},t) = \sqrt{P(\mathbf{x})}\, e^{i\Lambda(\mathbf{x},t)}, \tag{1.41}$$

where so far Λ can be an arbitrary but real function. The probability current density in a stationary state is, according to Eq. (1.40),

$$\mathbf{S}(\mathbf{x},t) = \frac{\hbar}{m} P(\mathbf{x}) \nabla_{\mathbf{x}} \Lambda(\mathbf{x},t). \tag{1.42}$$

For a stationary state where no physical properties are to change in time, the probability current density must also be time-independent, which then requires that the gradient of $\Lambda(\mathbf{x},t)$ be independent of time, and $\Lambda(\mathbf{x},t)$ therefore has the form (the minus sign just for convenience)

$$\Lambda(\mathbf{x},t) = \varphi(\mathbf{x}) - f(t), \tag{1.43}$$

where φ and f are real functions in view of Λ being real.[6] The wave function of a stationary state is thus the product of independent spatial and temporal parts,

$$\psi(\mathbf{x},t) = \sqrt{P(\mathbf{x})}\, e^{i\varphi(\mathbf{x})} e^{-if(t)}. \tag{1.44}$$

The time derivative is

$$i\hbar \frac{\partial \psi(\mathbf{x},t)}{\partial t} = \hbar \dot{f}(t)\, \psi(\mathbf{x},t) \tag{1.45}$$

and a stationary state function is a solution of the Schrödinger equation only if (dividing out the overall factor $e^{if(t)}$)

$$\hbar \dot{f}(t) \sqrt{P(\mathbf{x})}\, e^{i\varphi(\mathbf{x})} = \hat{H} \sqrt{P(\mathbf{x})}\, e^{i\varphi(\mathbf{x})}. \tag{1.46}$$

Since the right-hand side of the equation is independent of time, the time derivative of $f(t)$ must be a constant, and therefore

$$f(t) = r_1 t + r_2, \tag{1.47}$$

where r_1 and r_2 are real numbers. According to Eq. (1.46), a stationary state is thus described by a wave function of the form[7]

$$\psi_E(\mathbf{x},t) = \psi_E(\mathbf{x})\, e^{-(i/\hbar)Et}, \tag{1.48}$$

where E is a real number, $E \equiv r_1 \hbar$, and $\psi_E(\mathbf{x}) = \sqrt{P(\mathbf{x})}\, e^{i\varphi(\mathbf{x})}$ is determined by

$$\hat{H}\psi_E(\,\mathbf{x}) = E\psi_E(\mathbf{x}). \tag{1.49}$$

[6] The other option, $\Lambda(\mathbf{x},t) = -f(t)g(\mathbf{x}) + \varphi(\mathbf{x})$, $\nabla_{\mathbf{x}}\, g(\mathbf{x}) = \mathbf{0}$, reduces to the former case as the constraint on $g(\mathbf{x})$ demands it to be constant in space, $g(\mathbf{x}) = c$.

[7] In view of its physical interpretation, a wave function is not uniquely defined, but can always be subjected to a phase factor change, $\psi \to \psi\, e^{i\varphi}$, where the phase, φ, can be any real number. The state of a physical system is thus properly represented by a so-called ray, the (equivalence) class of wave functions $e^{i\varphi}\psi$, differing only by an overall phase factor of modulus one (with respect to which the observable quantities, $|\psi|^2$ and $\mathbf{S}$, are invariant).

The spatial part of the wave function for a stationary state is a solution of the so-called time-independent Schrödinger equation, an eigenfunction of the Hamiltonian corresponding to an eigenvalue E.

The real number E determines the time dependence of the stationary state at all times. Stationary states are the only solutions of the Schrödinger equation whose time dependence is a phase factor with a phase linear in time and characterized by a single real number, and we shall call this conserved quantity, or constant of motion, the *energy* of the particle in the stationary state in question. A particle in a stationary state would, if isolated, stay in this state of definite energy forever.

The energy eigenfunctions and eigenvalues for a free particle were already encountered in Eq. (1.6). The kinetic energy is related to the wave vector, $\mathbf{k}$, according to $E = \hbar^2\mathbf{k}^2/2m$. We note that, if momentum $\mathbf{p} = \hbar\mathbf{k}$ (the de Broglie relation) is associated with the wave vector of a plane wave, the energy dispersion is identical to that of a classical particle, $E = \mathbf{p}^2/2m$.

For a stationary state, the probability density is constant in time, and since then $\nabla\cdot\mathbf{S} = 0$, the net flow into any volume vanishes, i.e. in-flow equals out-flow. For the stationary free particle state, Eq. (1.6), the probability current density at each point in space equals

$$\mathbf{S} = \frac{\hbar\mathbf{k}}{m}\,|A|^2 = \frac{1}{\hbar}\frac{\partial E(\mathbf{k})}{\partial\mathbf{k}}\,|A|^2 = \frac{\partial E(\mathbf{p})}{\partial\mathbf{p}}\,|A|^2. \tag{1.50}$$

A solution of a time-independent Schrödinger equation that is real (up to the usual overall phase factor) has in the corresponding stationary state, according to Eq. (1.40) or (1.37), a probability current density that vanishes everywhere.

Shifting a Hamiltonian by a real constant V_0, $\hat{H} \to \hat{H} + V_0$, has no physical consequences as it is equivalent to subjecting the wave function to the phase transformation $\psi(x,t) \to \psi(x,t)\exp(-iV_0t/\hbar)$. If $\psi(x,t)$ is a solution to the Schrödinger equation described by the Hamiltonian $\hat{H}$, then $\psi(x,t)\exp(-iV_0t/\hbar)$ is a solution to the Schrödinger equation described by the Hamiltonian $\hat{H} + V_0$. The physical observables, the probability density and current density, are identical for either of the wave functions, as the two wave functions describe the same physical situation. For a stationary state, the change corresponds to shifting the energy values by the amount V_0, i.e. shifting the zero level from which energy is measured. The energy value of a quantum state is thus only defined modulo a constant, i.e. only an energy *difference* has physical significance in quantum mechanics.

Exercise 1.4 Show that, as a consequence of the Hamiltonian being a hermitian operator, the energy eigenvalue in Eq. (1.49) must be real.

Exercise 1.5 Show that, for a particle in a bounded potential, $V(\mathbf{x}) \geq V_{\text{min}}$, the energy spectrum is bounded from below, $E \geq V_{\text{min}}$.

Exercise 1.6 Show that $\nabla \cdot S = 0$ for a stationary state also follows directly from the time-independent Schrödinger equation.

1.5 Energy Eigenfunctions

The set of energy eigenvalues, i.e. the real values E for which a corresponding eigenfunction satisfies Eq. (1.49), is called the energy spectrum of the particle. As discussed in Appendix E, a characteristic feature of quantum mechanics is that the energy values of a spatially confined particle are quantized, i.e. the particle can only take on discrete energies E_n, $n = 1, 2, \ldots$, with corresponding normalizable energy eigenfunctions ψ_n. This is a general feature of confined waves and a well-known phenomenon in other contexts such as the vibrations of, for example, a guitar string, where the vibrations of the string (small enough to ensure linearity and thereby the validity of the string wave equation) are built from the fundamental and its higher harmonics, the Fourier series representation as discussed in Appendix C. If wave functions are only confined for energies below a certain value, then the energy spectrum has in addition a continuous part, corresponding to non-normalizable wave functions (as discussed in Appendix E). Different potentials can thus lead to either only a discrete spectrum or only a continuous spectrum, such as is the case of a free particle, or a combination of both.

1.5.1 Orthogonality

An important property of energy eigenfunctions,

$$\hat{H}\psi_n(\mathbf{x}) = E_n\psi_n(\mathbf{x}) \quad \text{and} \quad \hat{H}\psi_m(\mathbf{x}) = E_m\psi_m(\mathbf{x}), \tag{1.51}$$

corresponding to different energy values, $E_n \neq E_m$, is that they are so-called orthogonal. In the following, a system with only a discrete spectrum is considered (the continuous case is discussed in Exercise G.1 on page 376). Since the wave functions then are normalizable, it is now shown that they can be chosen orthonormal, i.e. the overlaps of energy eigenfunctions satisfy

$$\int d\mathbf{x}\ \psi_n^*(\mathbf{x})\psi_m(\mathbf{x}) = \delta_{nm}, \tag{1.52}$$

where δ_{nm} denotes the Kronecker symbol, the function, or unit matrix,

$$\delta_{nm} = \begin{cases} 1, & \text{for } n = m, \\ 0, & \text{for } n \neq m. \end{cases} \tag{1.53}$$

The equality

$$\int d\mathbf{x}\ \psi_n^*(\mathbf{x})(\hat{H}\psi_m(\mathbf{x})) = E_m \int d\mathbf{x}\ \psi_n^*(\mathbf{x})\,\psi_m(\mathbf{x}) \tag{1.54}$$

follows immediately from Eq. (1.51), and recalling the hermitian property of the Hamiltonian, Eq. (1.11), gives

$$\int d\mathbf{x}\ \psi_n^*(\mathbf{x})(\hat{H}\psi_m(\mathbf{x})) = \int d\mathbf{x}\ \psi_m(\mathbf{x})(\hat{H}\psi_n(\mathbf{x}))^* = E_n \int d\mathbf{x}\ \psi_n^*(\mathbf{x})\,\psi_m(\mathbf{x}), \tag{1.55}$$

the last equality following from Eq. (1.51) and the fact that an energy eigenvalue is real. Subtracting the two different expressions for the *Hamiltonian wave function overlap* integral gives

$$(E_n - E_m) \int d\mathbf{x}\; \psi_n^*(\mathbf{x})\, \psi_m(\mathbf{x}) = 0 \tag{1.56}$$

and thereby Eq. (1.52) for normalized energy eigenfunctions.

1.5.2 Completeness

A superposition of the energy eigenfunctions

$$\psi(\mathbf{x}) = \sum_{n=1}^{\infty} a_n \psi_n(\mathbf{x}) \quad \text{and} \quad \hat{H}\psi_n(\mathbf{x}) = E_n \psi_n(\mathbf{x}) \tag{1.57}$$

has, by the orthogonality of the functions, coefficients given by

$$a_n = \int d\mathbf{x}\; \psi_n^*(\mathbf{x})\, \psi(\mathbf{x}). \tag{1.58}$$

Inserting these expressions into Eq. (1.57) gives

$$\psi(\mathbf{x}) = \sum_{n=1}^{\infty} \psi_n(\mathbf{x}) \int d\mathbf{x}'\; \psi_n^*(\mathbf{x}')\, \psi(\mathbf{x}') = \int d\mathbf{x}' \left(\sum_{n=1}^{\infty} \psi_n(\mathbf{x})\, \psi_n^*(\mathbf{x}') \right) \psi(\mathbf{x}'), \tag{1.59}$$

where the last equality follows from interchanging integration and summation. Since different choices of sets of constants a_n in Eq. (1.57) generate a large class of functions for which Eq. (1.59) must be valid, it is suggestive to conclude, in view of the definition of the Dirac delta function (see Appendix B), that

$$\sum_{n=1}^{\infty} \psi_n(\mathbf{x})\, \psi_n^*(\mathbf{x}') = \delta(\mathbf{x} - \mathbf{x}'). \tag{1.60}$$

Equation (1.60) expresses the completeness of energy eigenfunctions: any normalizable function can be expressed as an expansion on the energy eigenfunctions, i.e. is of the form in Eq. (1.57). This follows from Eq. (1.60) as one immediately works one's way back to Eq. (1.57) for an arbitrary wave function $\psi(\mathbf{x})$. Different Hamiltonians have different eigenfunctions, but each set will be complete. The completeness relation is shortly shown to follow from the basic property of a propagator, Eq. (A.13).

In the case of a free particle, the plane waves are the energy eigenfunctions, and their completeness for this continuum energy spectrum case is, as shown in Appendix B, Eq. (B.25),

$$\int \frac{d\mathbf{k}}{(2\pi)^3}\, e^{i\mathbf{k}\cdot(\mathbf{x}-\mathbf{x}')} = \delta(\mathbf{x} - \mathbf{x}'). \tag{1.61}$$

Equation (1.61) expresses the fact that any wave function can be expanded on the set of eigenfunctions for the free particle Hamiltonian, the plane waves. In mathematical terms, a wave function has a Fourier expansion.

A superposition of the stationary state wave functions

$$\psi(\mathbf{x},t)=\sum_{n=1}^{\infty}a_n\psi_n(\mathbf{x},t)=\sum_{n=1}^{\infty}a_n\psi_n(\mathbf{x})\,e^{-(i/\hbar)E_nt} \tag{1.62}$$

is, due to the linearity of the Hamiltonian, a solution of the Schrödinger equation, where for all times t orthonormality gives

$$a_n\,e^{-(i/\hbar)E_nt}=\int d\mathbf{x}\;\psi_n^*(\mathbf{x})\,\psi(\mathbf{x},t). \tag{1.63}$$

Employing Eq. (1.63) for time t', Eq. (1.62) can therefore be rewritten as

$$\psi(\mathbf{x},t)=\sum_{n=1}^{\infty}e^{-(i/\hbar)E_n(t-t')}\psi_n(\mathbf{x})\int d\mathbf{x}'\;\psi_n^*(\mathbf{x}')\,\psi(\mathbf{x}',t') \tag{1.64}$$

and interchanging summation and integration gives

$$\psi(\mathbf{x},t)=\int d\mathbf{x}'\left(\sum_{n=1}^{\infty}e^{-(i/\hbar)E_n(t-t')}\psi_n(\mathbf{x})\,\psi_n^*(\mathbf{x}')\right)\psi(\mathbf{x}',t') \tag{1.65}$$

and thereby the spectral representation of the propagator (see Eq. (A.12))

$$K(\mathbf{x},t;\mathbf{x}',t')=\sum_{n=1}\psi_n(\mathbf{x})\,\psi_n^*(\mathbf{x}')\,e^{-(i/\hbar)E_n(t-t')}=\sum_{n=1}\psi_n(\mathbf{x},t)\,\psi_n^*(\mathbf{x}',t'). \tag{1.66}$$

According to Eq. (1.25), the free particle propagator has the spectral representation in terms of the plane waves (for the three-dimensional case)

$$K_0(\mathbf{x},t;\mathbf{x}',t')=\int\frac{d\mathbf{k}}{(2\pi)^3}\,e^{i\mathbf{k}\cdot(\mathbf{x}-\mathbf{x}')}\,e^{-(i/\hbar)\epsilon_{\mathbf{k}}(t-t')}=\int\frac{d\mathbf{k}}{(2\pi)^3}\,\psi_{\mathbf{k}}(\mathbf{x},t)\,\psi_{\mathbf{k}}^*(\mathbf{x}',t'), \tag{1.67}$$

the equivalent of Eq. (1.66) for the case of a continuous energy spectrum, $\epsilon_{\mathbf{k}}=\hbar^2\mathbf{k}^2/2m$.

The completeness of the energy eigenfunctions, Eq. (1.60), can be arrived at from the fundamental dynamic equation (A.12). For a stationary state, the fundamental dynamic equation (A.12) reads

$$\psi_E(\mathbf{x})=e^{(i/\hbar)E(t-t')}\int d\mathbf{x}'\;K(\mathbf{x},t;\mathbf{x}',t')\,\psi_E(\mathbf{x}'). \tag{1.68}$$

For each energy eigenfunction and arbitrary spans of time, the propagator thus projects out the energy eigenfunction in question. This orthogonality projection is executed by the spectral representation of the propagator

$$K(\mathbf{x},t;\mathbf{x}',t')=\sum_{E}\psi_E(\mathbf{x})\,\psi_E^*(\mathbf{x}')\,e^{-(i/\hbar)E(t-t')}, \tag{1.69}$$

where the summation symbolizes summation and integration over discrete and continuous parts of the energy spectrum, respectively. The completeness relation now follows from the initial condition for the propagator, Eq. (A.13). Once the energies and energy eigenfunctions are known, the solutions of the *time-independent* Schrödinger equation, the propagator and thereby the quantum dynamics is known.

1.6 Summary

The quantum description of a system of N particles is provided by the wave function, $\psi(\mathbf{x}_1, \mathbf{x}_2, \ldots, \mathbf{x}_N, t)$, the probability amplitude for the basic event of particle locations at a time in question. The quantum dynamics of the system is determined by the Schrödinger equation

$$i\hbar \frac{\partial \psi(\mathbf{x}_1, \mathbf{x}_2, \ldots, \mathbf{x}_N, t)}{\partial t} = \hat{H}\psi(\mathbf{x}_1, \mathbf{x}_2, \ldots, \mathbf{x}_N, t), \tag{1.70}$$

where the Hamiltonian for the case of, say, a two-particle interaction, such as Coulomb interaction, is given by

$$\hat{H} = \sum_{i=1}^{N} \left(-\frac{\hbar^2}{2m_i} \frac{\partial^2}{\partial \mathbf{x}_i^2} \right) + \frac{1}{2} \sum_{i \neq j} V(\mathbf{x}_i, \mathbf{x}_j). \tag{1.71}$$

2 Quantum Tunneling

A most striking phenomenon is that of quantum tunneling, where a particle traverses a region of space that classically is energetically forbidden.[1] A process that is *strictly* forbidden in classical mechanics is possible in quantum mechanics. Tunneling is the mechanism behind mono- and covalent bonding, binding atoms into molecules and further into crystalline solids such as semiconductors, as discussed in Section 6.4. Tunneling results in a wide variety of phenomena, ranging from radioactive decay and fusion to controlled single electron tunneling in ultra-small tunnel junctions. Important practical applications of tunneling, such as the scanning tunneling microscope and electronic devices based on double-barrier resonant tunneling, are discussed in Chapters 3 and 13.

2.1 Wave Packet Analysis

To understand tunneling, it is instructive to consider the wave packet analysis in one dimension. In this scattering approach, a particle is launched toward a potential barrier and the solution of the Schrödinger equation containing all relevant information about the long-time behavior of the particle is obtained.

2.1.1 Wave Packet Construction

Consider a particle of mass m in one spatial dimension experiencing a non-constant potential only in a region of finite extent where the potential can be arbitrary. Since physical properties are unchanged by shifting the Hamiltonian by a constant (recall Section 1.4), the potential in the free regions can be chosen to be zero:

$$V(x) = \begin{cases} 0, & x \leq x_l, \\ V(x), & x_l \leq x \leq x_r, \\ 0, & x \geq x_r. \end{cases} \tag{2.1}$$

An example of such a potential barrier is depicted in Figure 2.1. The asymmetric case, where the potential has different constant values in the force-free regions, is presented in Exercise 2.2.

[1] The escape of an alpha particle (a ^{4}He nucleus, the tightest bound system of protons and neutrons) from a nucleus was coined *tunneling* by G. Gamow, as if the particle had escaped through a tunnel in the potential wall.

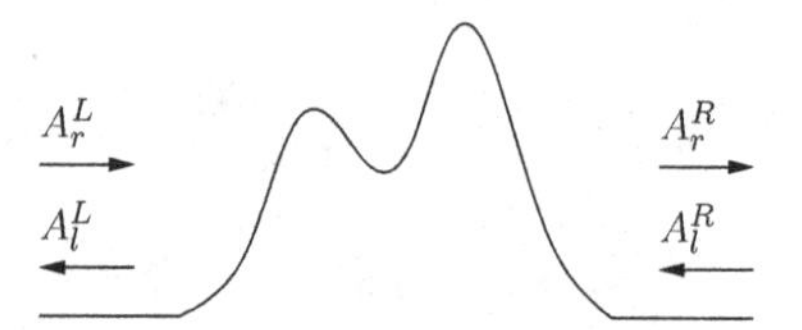

Figure 2.1 Potential barrier.

The solution of the time-independent Schrödinger equation,

$$-\frac{\hbar^2}{2m}\frac{d^2\psi(x)}{dx^2} + V(x)\,\psi(x) = E\,\psi(x), \tag{2.2}$$

is sought, E being an arbitrary energy value, $E \geq 0$.

In the left and right regions the particle is free, and the time-independent Schrödinger equation becomes

$$-\frac{\hbar^2}{2m}\frac{d^2\psi(x)}{dx^2} = E\,\psi(x). \tag{2.3}$$

The task of solving Eq. (2.3) thus amounts to finding the function which, when differentiated twice, gives back the same function multiplied by a negative real number,[2] $-2mE/\hbar^2$. The exponential function of a purely imaginary argument provides the desired outcome, and the twice differentiation allows for either overall sign of the argument. The general solution of Eq. (2.3) for the energy eigenvalue $E > 0$ is thus the sum, or superposition, of the two independent plane wave solutions,

$$\psi_E(x) = A_r\, e^{ikx} + A_l\, e^{-ikx}, \tag{2.4}$$

where the wave number $k \equiv k(E) = \sqrt{2mE}/\hbar$, corresponding to energy value $E = \hbar^2k^2/2m$, has been introduced. The potential, or here rather its absence, determines the *type* of wave function, and the energy eigenvalue specifies the shape of the solution, the wave length of the spatial oscillations of the energy eigenfunction.

The general solution of the time-independent Schrödinger equation (2.2), being a solution of a linear second-order differential equation, is a superposition of two independent solutions, and the solution corresponding to energy E is in the three regions

$$\psi_E(x) = \begin{cases} A_r^L(E)\, e^{ik(E)x} + A_l^L(E)\, e^{-ik(E)x}, & x \leq x_l, \\ A_1(E)\,\psi_1(x) + A_2(E)\,\psi_2(x), & x_l \leq x \leq x_r, \\ A_r^R(E)\, e^{ikx} + A_l^R(E)\, e^{-ikx}, & x \geq x_r, \end{cases} \tag{2.5}$$

where ψ_1 and ψ_2 are independent solutions in the barrier region, $x_l \leq x \leq x_r$.

[2] A particle in the state with zero kinetic energy, $E = 0$, as we shall see, has zero probability for traversing from one free region of the potential to the other, i.e. cannot tunnel.

The stationary state wave function corresponding to energy E is then

$$\psi_E(x,t) = \psi_E(x)\, e^{-(i/\hbar)Et}$$
$$= \begin{cases} (A_r^L(E)\, e^{ikx} + A_l^L(E)\, e^{-ikx})\, e^{-(i/\hbar)Et}, & x \leq x_l, \\ (A_1(E)\, \psi_1(x) + A_2(E)\, \psi_2(x))\, e^{-(i/\hbar)Et}, & x_l \leq x \leq x_r, \\ (A_r^R(E)\, e^{ikx} + A_l^R(E)\, e^{-ikx})\, e^{-(i/\hbar)Et}, & x \geq x_r, \end{cases} \tag{2.6}$$

and, as for any stationary state, its physical properties are time-independent. This wave function is therefore a far cry from a state corresponding to a particle emitted from a source and then approaching the spatial region where the potential is present. To obtain a wave function describing such a scenario, a localized wave function that moves toward the barrier must be constructed. This is obtained by superposing stationary solutions. A general solution of the Schrödinger equation

$$i\hbar \frac{\partial \psi(x,t)}{\partial t} = \left(-\frac{\hbar^2}{2m} \frac{\partial^2}{\partial x^2} + V(x) \right) \psi(x,t) \tag{2.7}$$

is, due to the linearity of the Schrödinger equation, obtained by a superposition of energy eigenstates,

$$\psi(x,t) = \int_0^\infty dE\; g(E)\, \psi_E(x)\, e^{-(i/\hbar)Et}. \tag{2.8}$$

Choosing the so far arbitrary weight function g suitably, a description of the quantum dynamics for the scattering situation is obtained.

2.1.2 Wave Packets in Free Regions

We first show how, in a free region, the solution of the Schrödinger equation

$$\psi_g(x,t) = \int_0^\infty dE\; g(E - E_0)[A_r(E)\, e^{ik(E)x} + A_l(E)\, e^{-ik(E)x}]\, e^{-(i/\hbar)Et} \tag{2.9}$$

can be turned into wave packets by choosing the weight function g to be localized around a definite energy value. The argument of the weight function has been shifted so that, choosing $g(E)$ to be peaked around the value $E = 0$, only the states near the arbitrary energy $E_0 > 0$ contribute to the superposition in Eq. (2.9). The so far arbitrary coefficients $A_{l,r}(E)$ are now chosen as continuous functions of energy. Assuming $g(E)$ to be non-vanishing only in a tiny region, the smoothly varying amplitude functions can be assumed constant over this arbitrarily small energy range, and taken outside the integral in Eq. (2.9) upon inserting E_0 as argument. The constructed, arbitrarily good, approximate solution of the free particle Schrödinger equation thus has the form

$$\psi_g(x,t) = A_r(E_0)\, \phi^+(x,t) + A_l(E_0)\, \phi^-(x,t), \tag{2.10}$$

where

$$\phi^{\pm}(x,t) = \int_0^\infty dE\; g(E-E_0)\, e^{-(i/\hbar)Et \pm ik(E)x}$$

$$= e^{-(i/\hbar)E_0 t \pm ik(E_0)x} \int_{-E_0}^{\infty} dE\; g(E)\, e^{-(i/\hbar)Et \pm ix(k(E_0+E)-k(E_0))}. \quad (2.11)$$

The exponentially oscillating prefactor in time comes from shifting the integration variable, $E \to E - E_0$, and the spatially dependent phase is simply added and subtracted for later convenience.

The integral in Eq. (2.11) cannot be evaluated exactly due to the energy dispersion, the square root dependence of the wave vector on the energy, but the wave function can with arbitrary accuracy be obtained for the physical situation of interest. The peaked weight function, $g(E)$, limits the region of integration in Eq. (2.11) to values in the narrow range $|E| \sim \Delta E$, and choosing $\Delta E \ll E_0$, the peakedness of the weight function allows a lowest-order Taylor expansion giving

$$k(E_0+E) \simeq k(E_0) + k'(E_0)E = k_0 + \frac{k_0}{2E_0}E = k_0 + \frac{m}{\hbar^2 k_0}E, \qquad k_0 \equiv k(E_0). \quad (2.12)$$

Introducing the magnitude of the velocity that the particle would have classically corresponding to energy E_0, $mv_0^2/2 \equiv E_0$, or equivalently in terms of the wave vector $v_0 = \hbar k_0/m$, we obtain

$$k(E_0+E) \simeq k_0 + \frac{E}{\hbar v_0} \quad (2.13)$$

and

$$\phi^{\pm}(x,t) = e^{-(i/\hbar)E_0 t \pm ik_0 x} \int_{-E_0}^{\infty} dE\; g(E)\, e^{-(i/\hbar)E(t \pm x/v_0)}$$

$$= e^{-(i/\hbar)E_0 t \pm ik_0 x}\, \Phi(v_0 t \pm x), \quad (2.14)$$

which is an envelope function, Φ, multiplied by the wave functions for the two stationary states corresponding to energy E_0, plane waves oscillating in time with frequency $\omega_0 = E_0/\hbar$ and spatially oscillating with wave length $\lambda_0 = 2\pi/k_0$. Since g is peaked, the integration can be extended to minus infinity, and choosing g real and even makes Φ real and even.

Once the function g is chosen such that the envelope function, Φ, is of finite extent, the ensuing calculations can be performed. However it can be convenient to have explicit expressions to look at, and a Gaussian weight is therefore employed,

$$g(E) \equiv \left(\frac{2}{\pi \Delta E^2}\right)^{1/4} \exp\left(-\frac{E^2}{\Delta E^2}\right). \quad (2.15)$$

Extending the energy integration in Eq. (2.14) to minus infinity incurs only an exponentially small error, determined by the parameter $\Delta E/E_0$, and the integral is now a simple Gaussian and

$$\phi^{\pm}(x,t) = e^{-(i/\hbar)E_0 t \pm i k_0 x}\phi_{\pm}(x,t), \tag{2.16}$$

where the Gaussian integral is obtained by completing the square (see Appendix B) giving

$$\begin{aligned}\phi_{\pm}(x,t) &= \left(\frac{2}{\pi\Delta E^2}\right)^{1/4} \int_{-\infty}^{\infty} dE\, e^{-E^2/\Delta E^2} e^{-(i/\hbar)E[t \mp mx/\hbar k_0]} \\ &= (2\pi\Delta E^2)^{1/4} \exp\left(-\frac{(x \mp v_0 t)^2}{(2\hbar v_0/\Delta E)^2}\right).\end{aligned} \tag{2.17}$$

As a consequence of $g(E)$ being peaked on the scale ΔE, the functions $\phi_{\pm}(x,t)$ are spatial envelope functions, exponentially fast spatially decaying Gaussian functions. They are bell-shaped and vanish outside a spatial range of size $\Delta x = \hbar v_0/\Delta E$ centered around the positions $x = \mp v_0 t$, respectively. The functions $\phi_{\pm}(x,t)$ are left- and right-moving rigid envelope functions, moving with velocities $\pm v_0$, i.e. in the positive and negative x directions, respectively (explaining the subscripts on the corresponding amplitudes, $A_{l,r}$, as referring to left- and right-moving).

Thus, the superposition of energy states gives the wave function in free space,

$$\psi_{g(E_0)}(x,t) = A_r(E_0)\, e^{-(i/\hbar)E_0 t + i k_0 x}\phi_+(x,t) + A_l(E_0)\, e^{-(i/\hbar)E_0 t - i k_0 x}\phi_-(x,t), \tag{2.18}$$

a function that is spatially localized at two different moving locations $x = \mp v_0 t$, respectively.[3] The subscript of the constructed solution indicates that only energy states in a small energy range, ΔE, around the energy E_0 are superimposed.

2.1.3 Wave Packet in Barrier Region

At large negative times, the constructed wave function in the free regions is, according to Eq. (2.18), only non-zero far away from the barrier potential, which by choice of origin is located near $x = 0$. By continuity, the wave function is expected to be exponentially small in the barrier region at such times. As the wave packet approaches the barrier, at times $t \sim 0$ it becomes non-vanishing in the barrier region, but at large positive times the wave function is again expected to vanish in the barrier region, as suggested by Eq. (2.18). The time scale after which the particle's probability distribution in the region of the potential vanishes is readily obtained for the case of a square barrier of height V. For energies of the particle below the barrier, $E_0 < V$, the tunneling regime of interest, the superposition solution of the Schrödinger equation is in the barrier region, $x_l < x < x_r$,

$$\begin{aligned}\psi_{g(E_0)}(x,t) &= \left(\frac{2}{\pi\Delta E^2}\right)^{1/4} \\ &\times \int_0^{\infty} dE\, e^{-(E-E_0)^2/\Delta E^2}[A_1(E)\, e^{-\kappa(E)x} + A_2(E)\, e^{\kappa(E)x}]\, e^{-(i/\hbar)Et},\end{aligned} \tag{2.19}$$

[3] The locations of the wave packets can of course be changed by inserting phase factors $\exp[ix_0(k(E_0+E)-k_0)]$ into the terms in Eq. (2.9). However, the option of changing the (launching) position of a wave packet is irrelevant for the scattering problem we shall consider.

where $\kappa = \kappa(E) = \sqrt{2m(V-E)}/\hbar$ (the square barrier is discussed in detail in Section 2.5). Owing to the peaked character of the weight function, κ can be Taylor-expanded to lowest order around the energy value E_0, and the resulting Gaussian integral gives, with $\kappa_0 \equiv \kappa(E_0)$,

$$\psi_{g(E_0)}(x,t) = A_1(E_0)\, e^{-(i/\hbar)E_0 t - \kappa_0 x} B_+(x,t) + A_2(E_0)\, e^{-(i/\hbar)E_0 t + \kappa_0 x} B_-(x,t), \qquad (2.20)$$

where (corresponding to the substitution $imx/\hbar^2 k_0 \to mx/\hbar^2 \kappa_0$ in Eq. (2.17))

$$B_\pm(x,t) = (2\pi \Delta E^2)^{1/4} \exp\left[\frac{\Delta E^2}{4}\left(-\frac{i}{\hbar}t \pm \frac{m}{\hbar^2 \kappa_0}x\right)^2\right]$$
$$= (2\pi \Delta E^2)^{1/4} \exp\left[\mp i \frac{m\Delta E^2}{2\hbar^3 \kappa_0} t x\right] \exp\left[\left(\frac{mx\Delta E}{2\hbar^2 \kappa_0}\right)^2\right] \exp\left[-\left(\frac{\Delta E\, t}{2\hbar}\right)^2\right]. \qquad (2.21)$$

The spatial variable is limited, $x_l \leq x \leq x_r$, and the factor determining the magnitude of the wave function in the barrier region as a function of time is the exponential damping in time factor. Thus at times $t \sim 0$ the wave function is non-zero in the barrier region, whereas at large positive or negative times, $|t| \gg \hbar/\Delta E$, the wave function in the barrier region is seen to be vanishingly small. The estimate is valid also for a spatially varying potential since the analysis is local. For locations in the barrier region, the wave function of the form in Eq. (2.19) can be used with the corresponding value of the potential.

2.1.4 Asymptotic Wave Packets

The free regions are separated by the region of the potential, and in different regions the superpositions are done with the corresponding amplitudes dictated by the solution, Eq. (2.5), of the time-independent Schrödinger equation. For times where the distance $v_0|t|$ is much larger than the extension of the envelopes, Δx, the functions $\phi_\pm$ are, according to Eq. (2.18), only non-vanishing in their respective small bell-shaped regions far away from the potential region. At large negative times, $\phi_+(x,t)$ is only non-vanishing close to $x = v_0 t = -v_0|t|$, and $\phi_-(x,t)$ is only non-vanishing close to $x = -v_0 t = v_0|t|$. For large negative times, $t \ll -\hbar/\Delta E$, the constructed arbitrarily accurate solution to the Schrödinger equation built from energy states close to energy value E_0 is thus

$$\psi_{g(E_0)}(x,t) \simeq \begin{cases} A_r^L(E_0)\, e^{-(i/\hbar)E_0 t + ik_0 x} \phi_+(x,t), & x < x_l, \\ 0, & x_l < x < x_r, \\ A_l^R(E_0)\, e^{-(i/\hbar)E_0 t - ik_0 x} \phi_-(x,t), & x > x_r, \end{cases} \qquad (2.22)$$

corresponding to the particle in a superposition of two incoming wave packets: one with envelope ϕ_+ approaching the barrier from the left, and one with envelope ϕ_- approaching the barrier from the right.

At large positive times, $t \gg \hbar/\Delta E$, the envelopes have swapped sides: $\phi_+(x,t)$ is only non-vanishing close to $x = v_0 t = v_0|t|$, and $\phi_-(x,t)$ is only non-vanishing close to $x =$

$-v_0 t = -v_0|t|$. For large positive times, $t \gg \hbar/\Delta E$, the constructed arbitrarily accurate solution to the Schrödinger equation is thus

$$\psi_{g(E_0)}(x,t) \simeq \begin{cases} A_l^L(E_0)\, e^{-(i/\hbar)E_0 t - ik_0 x}\phi_-(x,t), & x < x_l, \\ 0, & x_l < x < x_r, \\ A_r^R(E_0)\, e^{-(i/\hbar)E_0 t + ik_0 x}\phi_+(x,t), & x > x_r, \end{cases} \tag{2.23}$$

corresponding to a superposition of two outgoing wave packets: one with envelope ϕ_+ moving away from the barrier in the right region, and one with envelope ϕ_- moving away from the barrier on the left side.

2.2 Scattering Scenario

Consider the scattering problem of an incoming particle from the left. This initial condition corresponds to letting the incoming wave amplitude on the right vanish, $A_l^R(E_0) = 0$, in the general solution (2.22). Then, at large negative times, $t \ll -\hbar/\Delta E$, there is only a wave to the left of the barrier, representing the incoming particle,

$$\psi_{g(E_0)}(x,t) \simeq \begin{cases} A_r^L(E_0)\, e^{-(i/\hbar)E_0 t + ik_0 x}\phi_+(x,t), & x < x_l, \\ 0, & x_l < x < x_r, \\ 0, & x > x_r. \end{cases} \tag{2.24}$$

This wave packet is centered at $x = v_0 t$, its envelope moving to the right with velocity v_0 as depicted in Figure 2.2. The plane wave inside the envelope function has a large number of spatial oscillations, on the order of $\Delta x/\lambda_0 = E_0/\pi\Delta E \gg 1$.

Once the wave packet hits the barrier, the shape of the envelope function suffers contortions, and becomes wildly oscillating while penetrating into the classically forbidden region. However, at large positive times, $t \gg \hbar/\Delta E$, as shown in Section 2.1.3, the wave function again vanishes in the barrier region, and according to Eq. (2.23) the state of the particle is represented by two outgoing wave packets as depicted in Figure 2.3.

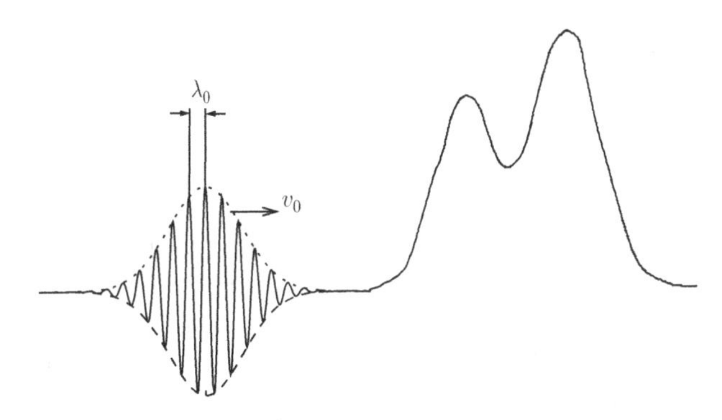

Figure 2.2 Real or imaginary part of the incoming wave packet.

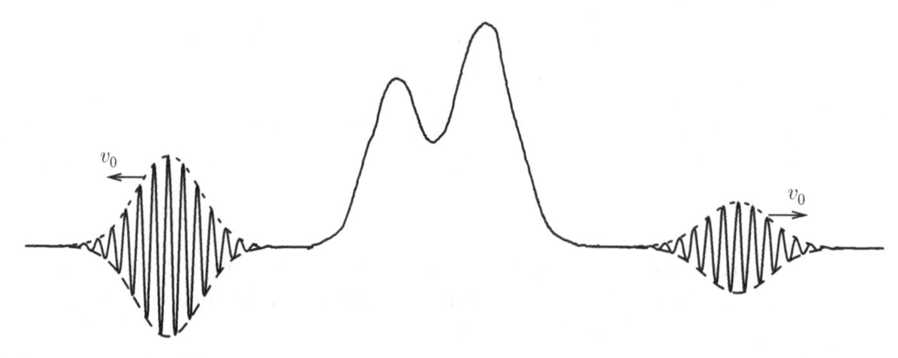

Figure 2.3 Real or imaginary part of the scattered wave packets.

Thus, in the far past and future, i.e. for times $|t| \gg \hbar/\Delta E$, the solution to the scattering problem is specified by the wave function

$$\psi_{g(E_0)}(x,t) \simeq \begin{cases} [A_r^L(E_0)\, e^{ik_0x}\phi_+(x,t) + A_l^L(E_0)\, e^{-ik_0x}\phi_-(x,t)]\, e^{-(i/\hbar)E_0t}, & x < x_l, \\ 0, & x_l < x < x_r, \\ A_r^R(E_0)\, e^{ik_0x}\phi_+(x,t)\, e^{-(i/\hbar)E_0t}, & x > x_r, \end{cases} \tag{2.25}$$

representing at large negative times an incoming wave packet from the left, as depicted in Figure 2.2, and at large positive times the scattered wave, two outgoing wave packets, as depicted in Figure 2.3. We stress again that at large times, $t_f \gg \hbar/\Delta E$, the constructed outgoing waves approximate with arbitrary accuracy the exact solution of the Schrödinger equation.

2.3 Tunneling Probability

Of interest is the probability for a particle described by the incoming wave packet, Eq. (2.24), corresponding to energy E_0, to end up on the right-hand side of the barrier at large times, $t \gg \hbar/\Delta E$. This transmission probability, $T(E_0)$, obtained by adding up the probabilities for particle location on the right, is

$$T(E_0) = \int_{x_r}^{\infty} dx\, |\psi_{g(E_0)}(x, t \gg \hbar/\Delta E)|^2 \simeq |A_r^R(E_0)|^2 \int_{-\infty}^{\infty} dx\, [\phi_+(x, t \gg \hbar/\Delta E)]^2, \tag{2.26}$$

as the only contribution to the integral to the right of x_r, according to Eq. (2.23) or (2.25), is from the outgoing wave packet on the right. The last approximate equality is of arbitrary exponential accuracy as the change in the lower integration limit extends the range of integration into locations where the function ϕ_+ is exponentially vanishing. Mathematically, the equality sign is only exact for the asymptotic states where t approaches infinity. However, from a physical standpoint, $\exp(-10^{10})$ is indistinguishable from zero.

The neglect of dispersion, i.e. the approximation of only keeping the first-order term in E in Eq. (2.12), leading to the missing feature of wave packet spreading, is seen to be of no consequence. Once the transmitted wave has propagated far enough away from the barrier, i.e. $t \gg \hbar/\Delta$, the moment of time chosen to evaluate the transmission probability is irrelevant, wave spreading or not. The envelope integral has the same value as the integral of the exact solution as dictated by the conservation of probability ensured by the Schrödinger equation.

At large negative times, where only an incoming wave packet to the left of the barrier is present, normalization determines the magnitude of the incoming wave packet amplitude,

$$1 = \int_{-\infty}^{\infty} dx\, |\psi_{g(E_0)}(x, t \ll -\hbar/\Delta E)|^2 \simeq |A_r^L(E_0)|^2 \int_{-\infty}^{x_l} dx\, [\phi_+(x, t \ll -\hbar/\Delta E)]^2. \quad (2.27)$$

The two integrals in Eqs. (2.26) and (2.27) give identical values (recall their definition in Eq. (2.17)): both integrands are bell-shaped functions with the same area under them. The transmission probability for a particle in a wave packet state corresponding to energy E_0, the transmission coefficient at energy E_0, is thus

$$T(E_0) = \frac{|A_r^R(E_0)|^2}{|A_r^L(E_0)|^2}. \quad (2.28)$$

Similarly, the probability that the particle is reflected, the reflection coefficient $R(E_0)$, is obtained by adding up the probabilities for particle location on the left side of the barrier at large times, which according to Eq. (2.23) or (2.25) becomes

$$R(E_0) = \int_{-\infty}^{x_l} dx\, |\psi_{g(E_0)}(x, t \gg \hbar/\Delta E)|^2 \simeq |A_l^L(E_0)|^2 \int_{-\infty}^{\infty} dx\, [\phi_-(x, t \gg \hbar/\Delta E)]^2. \quad (2.29)$$

Since the areas under $\phi_-^2(x, t \gg \hbar/\Delta E)$ and $\phi_+^2(x, t \ll -\hbar/\Delta E)$ are equal, and the latter is given by Eq. (2.27), the reflection coefficient becomes[4]

$$R(E_0) = \frac{|A_l^L(E_0)|^2}{|A_r^L(E_0)|^2}. \quad (2.30)$$

The normalization of the wave function at a time $t \gg \hbar/\Delta E$ gives

$$\begin{aligned} 1 &= \int_{-\infty}^{\infty} dx\, |\psi_{g(E_0)}(x, t \gg \hbar/\Delta E)|^2 \\ &\simeq \int_{-\infty}^{x_l} dx\, |\psi_{g(E_0)}(x, t \gg \hbar/\Delta E)|^2 + \int_{x_r}^{\infty} dx\, |\psi_{g(E_0)}(x, t \gg \hbar/\Delta E)|^2 \end{aligned} \quad (2.31)$$

and according to Eq. (2.23)

[4] The reflection and transmission probabilities for a particle incoming from the right are for the considered symmetric case obtained by everywhere substituting $L \leftrightarrow R$ and $l \leftrightarrow r$.

$$1 = \int_{-\infty}^{x_l} dx \, |A_l^L(E_0)|^2 [\phi_-(x, t \gg \hbar/\Delta E)]^2 + \int_{x_r}^{\infty} dx \, |A_r^R(E_0)|^2 [\phi_+(x, t \ll -\hbar/\Delta E)]^2$$

$$= [\, |A_l^L(E_0)|^2 + |A_r^R(E_0)|^2] \int_{-\infty}^{\infty} dx \, [\phi_-(x, t \gg \hbar/\Delta E)]^2$$

$$= \frac{|A_l^L(E_0)|^2 + |A_r^R(E_0)|^2}{|A_r^L(E_0)|^2}, \tag{2.32}$$

where the last two equalities follow from the previously noted identity of the involved integrals. Equation (2.32) is the consequence of the probability conservation property of the Schrödinger equation, and according to Eqs. (2.30) and (2.28) becomes the statement that the probabilities for reflection and transmission add up to one,

$$T(E_0) + R(E_0) = 1. \tag{2.33}$$

The particle is with certainty either transmitted or reflected.

Exercise 2.1 Evaluate the Gaussian integrals

$$\int_{-\infty}^{\infty} dx \, [\phi_+(x, t \ll -\hbar/\Delta E)]^2 = \frac{2\pi\hbar^2}{m} k_0 = 2\pi\hbar v_0 = \int_{-\infty}^{\infty} dx \, [\phi_+(x, t \gg \hbar/\Delta E)]^2. \tag{2.34}$$

Exercise 2.2 Consider the case where the constant potentials in the two force-free regions are different, i.e. an asymmetric barrier, $V_r > 0$,

$$V(x) = \begin{cases} 0, & x \leq x_l, \\ V(x), & x_l \leq x \leq x_r, \\ V_r, & x \geq x_r. \end{cases} \tag{2.35}$$

Repeat the wave packet analysis and show that the transmission coefficient from left to right corresponding to energy E_0 is given by

$$T_{l\to r}(E_0) = \begin{cases} 0, & 0 \leq E_0 \leq V_r, \\ \dfrac{k_r(E_0)}{k_0} \dfrac{|A_r^R(E_0)|^2}{|A_r^L(E_0)|^2} = \dfrac{v}{v_0} \dfrac{|A_r^R(E_0)|^2}{|A_r^L(E_0)|^2}, & E_0 \geq V_r, \end{cases} \tag{2.36}$$

where $v_0 = \sqrt{2mE_0}/\hbar$ and $v = \sqrt{2m(E_0 - V_r)}/\hbar$ are the wave packet envelope velocities in the left and right regions, respectively. Note that the only change relative to the previous discussion is that, to the right of x_r, the solution is obtained by the change $k \to k_r(E) \equiv \sqrt{2m(E - V_r)}/\hbar$ in Eq. (2.5), thereby invalidating Eq. (2.34). Knowledge of the potential levels in the free regions is contained in the wave vectors for these regions, reflecting the corresponding amount of kinetic energy.

Show that the reflection coefficient is given by

$$R_{l\to l}(E_0) = \begin{cases} 1, & 0 \le E_0 \le V_r, \\ \dfrac{|A_l^L(E_0)|^2}{|A_r^L(E_0)|^2} = 1 - T_{l\to r}(E_0), & E_0 \ge V_r. \end{cases} \tag{2.37}$$

Consider now a particle incoming from the right, its total energy of course exceeding its potential energy, $E_0 > V_r$. Show that the transmission coefficient is

$$T_{l\leftarrow r}(E_0) = \frac{k_0}{k_r(E_0)} \frac{|A_l^L(E_0)|^2}{|A_l^R(E_0)|^2} = \frac{v_0}{v} \frac{|A_l^L(E_0)|^2}{|A_l^R(E_0)|^2} \tag{2.38}$$

and the reflection coefficient is

$$R_{r\leftarrow r}(E_0) = \frac{|A_r^R(E_0)|^2}{|A_l^R(E_0)|^2} = 1 - T_{l\leftarrow r}(E_0). \tag{2.39}$$

2.4 Stationary Limit

The wave packet analysis reveals that the information needed to obtain the transmission and reflection probabilities for a particle scattered by a potential is in fact contained in the solution of the time-*independent* Schrödinger equation. The plane waves corresponding to the scattering of an incoming particle from the left (considering first the symmetric case),

$$\psi_{\text{inc}}(x) = A_r^L(E)\, e^{ik(E)x}, \quad \psi_{\text{r}}(x) = A_l^L(E)\, e^{-ik(E)x}, \quad \psi_{\text{t}}(x) = A_r^R(E)\, e^{ik(E)x}, \tag{2.40}$$

have, according to Eq. (1.37), associated right- and left-directed probability currents in respective regions

$$S_{\text{inc}} = \frac{\hbar k(E)}{m} |A_r^L(E)|^2, \quad S_{\text{r}} = -\frac{\hbar k(E)}{m} |A_l^L(E)|^2, \quad S_{\text{t}} = \frac{\hbar k(E)}{m} |A_r^R(E)|^2. \tag{2.41}$$

The transmission and reflection coefficients are according to Eqs. (2.28) and (2.30) the ratios of the currents[5]

$$T(E) = \frac{S_{\text{t}}}{S_{\text{inc}}}, \quad R(E) = \frac{-S_{\text{r}}}{S_{\text{inc}}}. \tag{2.42}$$

According to Eq. (2.33)

$$S_{\text{inc}} + S_{\text{r}} = S_{\text{t}}. \tag{2.43}$$

This relationship also follows from the fact that in any stationary state the current is, according to the continuity equation (1.38), conserved, $dS/dx = 0$, and the probability current has the same value everywhere. Indeed, calculating the probability current for the state in Eq. (2.6), here considering the case $A_l^R = 0$, and using current conservation, Eq. (2.43)

[5] The linearity of quantum mechanics makes overall normalization irrelevant for obtaining the scattering probabilities; the incoming wave amplitude can be chosen to have unit value, $A_r^L(E) = 1$.

is obtained. The sum of the two oppositely directed currents on the left of the barrier equals the current on the right, which also equals the constant current inside the barrier.

The stationary situation can be viewed as the limit where the energy of the wave packet state becomes ever more precise, i.e. ΔE is made smaller. In the limit $\Delta E \to 0$, the envelope functions become infinitely elongated, $\Delta x = \hbar v_0/\Delta E$, and the wave packets turn into the stationary state, Eq. (2.6), corresponding to the incoming, transmitted and reflected parts of the solution to the time-independent Schrödinger equation.

From Eq. (2.40) it follows that the transmission coefficient can be expressed in terms of the wave functions of the transmitted and incoming waves at, for example, their respective barrier boundaries,

$$T(E) = \frac{|\psi_t(x_r)|^2}{|\psi_{\rm inc}(x_l)|^2}. \tag{2.44}$$

Exercise 2.3 Show that for the stationary state specified by the wave function in Eq. (2.6), the probability currents in the free regions on the left and right of the barrier are given by

$$S_L(x) = \frac{\hbar k}{m}\left(|A_r^L|^2 - |A_l^L|^2\right) \tag{2.45}$$

and

$$S_R(x) = \frac{\hbar k}{m}\left(|A_r^R|^2 - |A_l^R|^2\right). \tag{2.46}$$

According to the continuity equation, the current is independent of the position since the probability density is constant, and the two expressions are therefore equal, generalizing the scattering scenario case of Eq. (2.43).

Exercise 2.4 Consider the asymmetric case of Exercise 2.2 on page 24, where the constant potentials in the two force-free regions are different. Show that the reflection and transmission properties, irrespective of whether the particle is incoming from the left or right, can be expressed as the ratio of the currents,

$$T(E) = \frac{S_t}{S_{\rm inc}}, \quad R(E) = -\frac{S_r}{S_{\rm inc}}. \tag{2.47}$$

2.5 Square Barrier

Consider the case of a single scatterer in the form of a square barrier of height V,

$$V(x) = \begin{cases} 0, & x < x_l, \\ V, & x_l < x < x_r, \\ 0, & x > x_r, \end{cases} \tag{2.48}$$

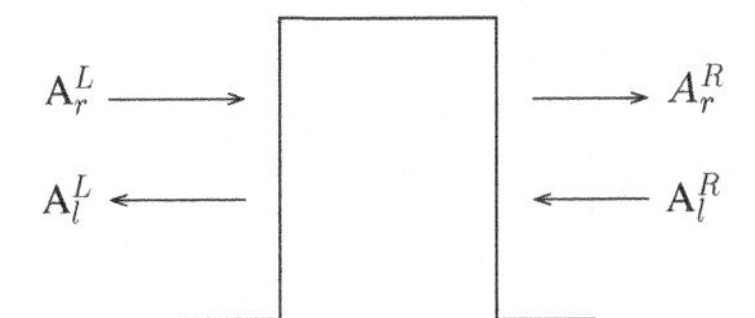

Figure 2.4 Amplitudes for left- and right-moving waves.

the potential depicted in Figure 2.4. This example is not only of academic interest because it is realizable by semiconductor technology, as discussed in Chapter 12.

For the *under-barrier* case, $0 < E < V$, the solution to the time-independent Schrödinger equation corresponding to energy E is

$$\psi_E(x) = \begin{cases} A_r^L\, e^{ikx} + A_l^L\, e^{-ikx}, & x \in L,\ x < x_l, \\ A_r^B\, e^{-\kappa x} + A_l^B\, e^{\kappa x}, & x \in B,\ x_l < x < x_r, \\ A_r^R\, e^{ikx} + A_l^R\, e^{-ikx}, & x \in R,\ x > x_r, \end{cases} \tag{2.49}$$

where $k = k(E) = \hbar^{-1}\sqrt{2mE}$ and $\kappa = \kappa(E) = \hbar^{-1}\sqrt{2m(V-E)}$. The *over-barrier* case, $E > V$, corresponds to the substitution

$$\kappa(E) \to -ik_V(E) = -i\hbar^{-1}\sqrt{2m(E-V)}, \tag{2.50}$$

since in that case the solution in the barrier region is also a plane wave, just oscillating with a longer wave length compared to the waves in the free regions, as the potential energy in the barrier region is subtracted from the total energy, making the kinetic energy, and therefore the wave vector, smaller.

The wave function of given energy, in the free regions, is specified by the amplitudes, as illustrated in Figure 2.4.

2.5.1 Transfer Matrix

As discussed in Appendix E, solutions to the time-independent Schrödinger equation for a particle in a stepwise constant potential are wave functions with continuous derivative everywhere. Matching by continuity the wave function expressions, Eq. (2.49), approaching the left edge of the square barrier,

$$\psi(x_l - 0) = \psi(x_l + 0), \tag{2.51}$$

requires

$$A_r^L\, e^{ikx_l} + A_l^L\, e^{-ikx_l} = A_r^B\, e^{-\kappa x_l} + A_l^B\, e^{\kappa x_l}, \tag{2.52}$$

and continuity of the derivative

$$\psi'(x_l - 0) = \psi'(x_l + 0) \tag{2.53}$$

demands

$$ikA_r^L\, e^{ikx_l} - ikA_l^L\, e^{-ikx_l} = -\kappa A_r^B\, e^{-\kappa x_l} + \kappa A_l^B\, e^{\kappa x_l}. \tag{2.54}$$

The two equations can be combined into the matrix equation

$$\begin{pmatrix} e^{ikx_l} & e^{-ikx_l} \\ ik\,e^{ikx_l} & -ik\,e^{-ikx_l} \end{pmatrix} \begin{pmatrix} A_r^L \\ A_l^L \end{pmatrix} = \begin{pmatrix} e^{-\kappa x_l} & e^{\kappa x_l} \\ -\kappa\,e^{-\kappa x_l} & \kappa\,e^{\kappa x_l} \end{pmatrix} \begin{pmatrix} A_r^B \\ A_l^B \end{pmatrix} \tag{2.55}$$

or introducing shorter notation

$$\mathbf{K}(x_l)\mathbf{A}^L = \mathcal{K}(x_l)\mathbf{A}^B, \tag{2.56}$$

leaving the k and κ dependence, i.e. the dependence on the energy of the particle, implicit through the labeling of the matrices, $\mathbf{K}$ and $\mathcal{K}$.

Similarly, matching at the right barrier edge gives (simply letting $x_l \to x_r$ and $L \to R$ in Eq. (2.55))

$$\mathbf{K}(x_r)\mathbf{A}^R = \mathcal{K}(x_r)\mathbf{A}^B, \tag{2.57}$$

where

$$\mathcal{K}(x_r) = \begin{pmatrix} e^{-\kappa x_r} & e^{\kappa x_r} \\ -\kappa\,e^{-\kappa x_r} & \kappa\,e^{\kappa x_r} \end{pmatrix} \tag{2.58}$$

and

$$\mathbf{K}(x_r) = \begin{pmatrix} e^{ikx_r} & e^{-ikx_r} \\ ik\,e^{ikx_r} & -ik\,e^{-ikx_r} \end{pmatrix}. \tag{2.59}$$

The determinants of the matrices are non-zero, $\det \mathbf{K}(x_{l,r}) = -2ik(E)$ and $\det \mathcal{K}(x_{l,r}) = 2\kappa(E)$, and the matrices can therefore be inverted (barring the cases $E = V$ and $E = 0$, where there are not two independent solutions, cases dealt with separately). Rewriting Eqs. (2.57) and (2.56) in the form

$$\mathbf{A}^R = [\mathbf{K}(x_r)]^{-1}\mathcal{K}(x_r)\mathbf{A}^B, \quad \mathbf{A}^B = [\mathcal{K}(x_l)]^{-1}\mathbf{K}(x_l)\mathbf{A}^L, \tag{2.60}$$

the amplitudes on the two sides of the barrier are related according to

$$\mathbf{A}^R = [\mathbf{K}(x_r)]^{-1}\mathcal{K}(x_r)[\mathcal{K}(x_l)]^{-1}\mathbf{K}(x_l)\mathbf{A}^L \equiv \mathcal{T}(x_r, x_l)\,\mathbf{A}^L, \tag{2.61}$$

where the transfer matrix $\mathcal{T}$, relating the amplitudes on the right and left sides of the barrier, has been introduced. A transfer matrix relates the amplitudes on the right and left sides for an arbitrary potential, the only change being that the $\mathcal{K}$ matrices are changed (the general discussion is presented in Appendix F).

Since the parameters $k(E)$ and $\kappa(E)$ depend continuously on the energy, so does the transfer matrix. Thus if, say, $A_{l,r}^L(E)$ are chosen to be continuous, this automatically leaves the other coefficients also as continuous. In contrast, any choice of discontinuity in, say, $\mathbf{A}^L$ is proportionally transferred to $\mathbf{A}^R$, thus leaving the transmission and reflection probabilities unchanged. The choice of continuous amplitude coefficients was thus just for convenience.

Choosing a set of amplitudes on, say, the left side of the barrier, the transfer matrix produces the correct amplitudes on the right side. The linear relationship between the amplitudes on the two sides of the barrier reflects the linearity of quantum mechanics, *viz.* that the time-independent Schrödinger equation is a linear equation. The linearity implies a rigidity of energy eigenfunctions. For example, if $A_r^L(E) = 0 = A_l^L(E)$, then the linear relationship between amplitudes implies that all amplitudes vanish, resulting in a vanishing function, i.e. not a proper wave function (a wave function only vanishes at points, its nodes). At least one of the amplitudes, $A_r^L(E)$ or $A_l^L(E)$, must thus be non-vanishing, and it then follows from Eq. (2.56) that both the coefficients in the barrier region are non-vanishing. Since in the finite barrier region both the independent solutions are at our disposal, the matching of the wave function can be achieved for any energy, i.e. the time-independent Schrödinger equation for a barrier has solutions for any energy value $E \geq 0$.

Introducing the matrix elements of the transfer matrix

$$\mathcal{T} = \begin{pmatrix} T_{11} & T_{12} \\ T_{21} & T_{22} \end{pmatrix}, \tag{2.62}$$

the amplitudes on the two sides of the barrier are related according to

$$\begin{pmatrix} A_r^R \\ A_l^R \end{pmatrix} = \begin{pmatrix} T_{11} & T_{12} \\ T_{21} & T_{22} \end{pmatrix} \begin{pmatrix} A_r^L \\ A_l^L \end{pmatrix}. \tag{2.63}$$

2.5.2 Square Barrier Transfer Matrix

The transfer matrix for a square barrier can now be evaluated by brute force. Remember the rule for inverting a two-by-two matrix,

$$\begin{pmatrix} \cdots & \cdots \\ \cdots & \cdots \end{pmatrix}^{-1} = \frac{1}{\det} \begin{pmatrix} \searrow & - \\ - & \nwarrow \end{pmatrix}, \tag{2.64}$$

i.e. interchange the diagonal elements, multiply the off-diagonal elements by minus one and divide by the determinant (the matrix is thus only invertible if its determinant is different from zero). From the inversion formula, Eq. (2.64), it immediately follows that the determinant of the inverse matrix is the reciprocal of the determinant of the original matrix, $\det M^{-1} = 1/\det M$.

Matrix multiplication is associative, and to obtain the transfer matrix we start by evaluating

$$\begin{aligned} \mathcal{K}(x_r)\,[\mathcal{K}(x_l)]^{-1} &= \begin{pmatrix} e^{-\kappa x_r} & e^{\kappa x_r} \\ -\kappa\, e^{-\kappa x_r} & \kappa\, e^{\kappa x_r} \end{pmatrix} \frac{1}{2\kappa} \begin{pmatrix} \kappa\, e^{\kappa x_l} & -e^{\kappa x_l} \\ \kappa\, e^{-\kappa x_l} & e^{-\kappa x_l} \end{pmatrix} \\ &= \frac{1}{\kappa} \begin{pmatrix} \kappa \cosh s_B & \sinh s_B \\ \kappa^2 \sinh s_B & \kappa \cosh s_B \end{pmatrix}, \end{aligned} \tag{2.65}$$

where we have introduced the shorthand $s_B \equiv \kappa(x_r - x_l) \equiv \kappa a_B$, i.e. a_B is the barrier width.[6]

[6] For $E = V$, where $\kappa = 0$, the matrix $\mathcal{K}(x_l^B)$ cannot be inverted. However, the formula for the transfer matrix, Eq. (2.61), is still correct upon substitution of the unit matrix, $\mathcal{K}(x_r^B)[\mathcal{K}(x_l^B)]^{-1} \to \mathbf{1}$, as seen by using the expressions for the matrices $\mathcal{K}(x_r^B)$ and $\mathcal{K}(x_l^B)$ in Eqs. (2.55) and (2.57) for the case $\kappa = 0$. Note that this detour is in fact not needed since the expressions for the transfer matrix elements have the correct $\kappa \to 0$ limit by continuity.

The transfer matrix now has the form

$$\begin{pmatrix} T_{11} & T_{12} \\ T_{21} & T_{22} \end{pmatrix} = \frac{1}{-2ik\kappa} \begin{pmatrix} -ik\, e^{-ikx_r} & -e^{-ikx_r} \\ -ik\, e^{ikx_r} & e^{ikx_r} \end{pmatrix} \begin{pmatrix} M_{11} & M_{12} \\ M_{21} & M_{22} \end{pmatrix}, \tag{2.66}$$

where

$$\begin{pmatrix} M_{11} & M_{12} \\ M_{21} & M_{22} \end{pmatrix} = \begin{pmatrix} \kappa \cosh s_B & \sinh s_B \\ \kappa^2 \sinh s_B & \kappa \cosh s_B \end{pmatrix} \begin{pmatrix} e^{ikx_l} & e^{-ikx_l} \\ ik\, e^{ikx_l} & -ik\, e^{-ikx_l} \end{pmatrix} \tag{2.67}$$

giving

$$\begin{pmatrix} M_{11} & M_{12} \\ M_{21} & M_{22} \end{pmatrix} = \begin{pmatrix} e^{ikx_l}(\kappa \cosh s_B + ik \sinh s_B) & e^{-ikx_l}(\kappa \cosh s_B - ik \sinh s_B) \\ \kappa e^{ikx_l}(\kappa \sinh s_B + ik \cosh s_B) & \kappa e^{-ikx_l}(\kappa \sinh s_B - ik \cosh s_B) \end{pmatrix}. \tag{2.68}$$

The transfer matrix elements can now be read off, and for the under-barrier case, $E < V$ (the energy dependence of κ and k is suppressed), we have

$$T_{11} = -\frac{e^{-ik(x_r - x_l)}}{2ik\kappa} [(k^2 - \kappa^2) \sinh s_B - 2ik\kappa \cosh s_B] = T_{22}^* \tag{2.69}$$

and

$$T_{12} = \frac{e^{-ik(x_r + x_l)}}{2ik\kappa} (k^2 + \kappa^2) \sinh s_B = T_{21}^*. \tag{2.70}$$

For the case of a square barrier, according to Eqs. (2.69) and (2.70), the components of the transfer matrix are related. These symmetry properties are not accidental for the square barrier, but are indeed valid for arbitrary potentials as a consequence of time reversal symmetry and current conservation, as shown in Appendix F, where it is also shown that the determinant of a transfer matrix equals one. Such symmetry constraints are of general use in guiding one to correct analytical results by back-tracking calculational errors.

Exercise 2.5 As an example of checking the correctness of the expressions for the transfer matrix elements, show that they (in a highly non-trivial way) lead to a determinant equal to one:

$$\det \mathcal{T} = T_{11}T_{22} - T_{21}T_{12} = |T_{11}|^2 - |T_{12}|^2 = \cosh^2 \kappa a_B - \sinh^2 \kappa a_B = 1. \tag{2.71}$$

That the determinant of the transfer matrix for a square barrier equals one also follows from Eq. (2.65),[7]

$$\det(\mathcal{K}(x_r)\, [\mathcal{K}(x_l)]^{-1}) = \cosh^2 s_B - \sinh^2 s_B = 1, \tag{2.72}$$

[7] This *intermediate* conclusion is valid for an arbitrary potential. The matrices in general are the Wronski matrices for the two independent solutions in the barrier region evaluated at $x = x_l$ and $x = x_r$, respectively. The conclusion follows since the determinant, the Wronskian, is independent of position, as generally shown in Appendix E.

and the fact that the determinant of a product of matrices is the product of their determinants, as the two other matrices (or Wronskians) in Eq. (2.61), according to $\det K(x_{l,r}) = -2ik(E)$, also has canceling determinants by being each other's inverse.

For the over-barrier case, $E > V$, the result for the transfer matrix is according to Eq. (2.50) obtained by the substitution $\kappa(E) \to -ik_V(E) = -i\hbar^{-1}\sqrt{2m(E-V)}$. The matrix elements of the transfer matrix is in that case (as $\cosh(\pm iu) = \cos u$ and $\sinh(\pm iu) = \pm i \sin u$, u real)

$$T_{11} = e^{-ik(x_r^B - x_l^B)} \left(i\frac{k^2 + k_V^2}{2kk_V} \sin(k_V a_B) + \cos(k_V a_B) \right) = T_{22}^* \tag{2.73}$$

and

$$T_{12} = -i\frac{e^{-ik(x_r^B + x_l^B)}}{2kk_V}(k^2 - k_V^2)\sin(k_V a_B) = T_{21}^*. \tag{2.74}$$

Exercise 2.6 Show that for $E = V$ ($\kappa(E = V) = 0 = k_V(E = V)$), the expressions in Eqs. (2.73) and (2.74) have the same limit as those in Eqs. (2.69) and (2.70).

Exercise 2.7 Consider the limit of the square barrier where $a_B \to 0$ and $V \to \infty$, their product being constant, $a_B V = w$. The square barrier potential then becomes the delta potential $V(x) = w\delta(x)$. Show that the 11-component of the transfer matrix becomes

$$T_{11}(E) = 1 - i\frac{mw}{\hbar^2 k} = 1 - i\frac{w}{\hbar\sqrt{2E/m}}. \tag{2.75}$$

2.6 Transmission Coefficient

Consider an incoming particle from the left scattering off an arbitrary potential. The physical boundary condition on the right is therefore $A_l^R = 0$; there is no incoming wave on the right. The amplitudes for the scattering situation is thus as depicted in Figure 2.5.

Figure 2.5 Amplitudes for the case of an incoming particle from the left.

Inverting the relationship in Eq. (2.63) gives

$$\begin{pmatrix} A_r^L \\ A_l^L \end{pmatrix} = \begin{pmatrix} T_{22} & -T_{12} \\ -T_{21} & T_{11} \end{pmatrix} \begin{pmatrix} A_r^R \\ 0 \end{pmatrix} \tag{2.76}$$

and

$$A_r^L = T_{22} A_r^R. \tag{2.77}$$

The 22-component describes the magnitude and phase relationship between the transmitted and incoming wave amplitudes. In this section it is assumed that the constant potentials in the free regions have the same value (otherwise consult Exercise 2.2 on page 24 for amendments). The transmission coefficient, Eq. (2.28), for a particle of energy E is therefore

$$T(E) = \frac{1}{|T_{22}(E)|^2} = \frac{1}{|T_{11}(E)|^2}. \tag{2.78}$$

Using the fact that the determinant of a transfer matrix equals one, $|T_{11}|^2 - |T_{12}|^2 = 1$, the transmission coefficient can alternatively be expressed in terms of the off-diagonal element,

$$T(E) = \frac{1}{1 + |T_{12}(E)|^2}, \tag{2.79}$$

explicitly expressing the probability constraint, $0 \leq T(E) \leq 1$.

For energies below the barrier height of an arbitrary potential, $E_0 < V_{\max}$, where classical mechanics leads to the particle being reflected with certainty, a quantum particle can make its way through the barrier and end up on the other side with finite probability, the hallmark of quantum dynamics.[8] The transmission probability for $E_0 < V_{\max}$ is referred to as the *tunneling* probability.

The diagonal elements of the transfer matrix do not depend on the relative location of the barrier, depending only on the barrier width, and the transmission coefficient is independent of the barrier location, as the translational invariance of space dictates (recall, for example, the explicit expression for a square barrier).

The scattering of a particle incoming from the right is specified by having no incoming wave from the left, $A_r^L = 0$, corresponding to having amplitudes as depicted in Figure 2.6.

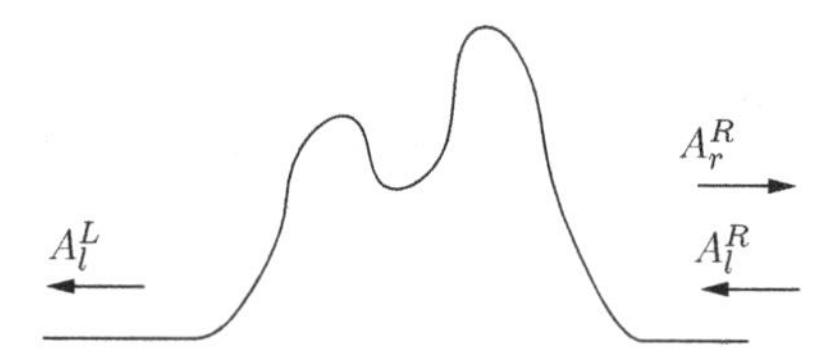

Figure 2.6 Scattering amplitudes for particle incoming from the right.

[8] In classical physics, the process is strictly forbidden. As a free particle enters the region of the potential, its kinetic energy, E_K, is converted into potential energy. At the location x_t for which $V(x_t) = E_K$, at that instant, t, it comes to rest. It is then pushed by the force, $F(x_t) = -V'(x_t)$, back into the region from which it came, eventually ending up moving with the same magnitude of velocity in the free region, but in the opposite direction. The particle is totally reflected.

The transfer matrix equation

$$\begin{pmatrix} A_r^R \\ A_l^R \end{pmatrix} = \begin{pmatrix} T_{11} & T_{12} \\ T_{21} & T_{22} \end{pmatrix} \begin{pmatrix} 0 \\ A_l^L \end{pmatrix} \tag{2.80}$$

then gives for the transmission coefficient

$$T_{l\leftarrow r}(E) = \frac{|A_l^L(E)|^2}{|A_l^R(E)|^2} = \frac{1}{|T_{22}(E)|^2} = T_{l\rightarrow r}(E). \tag{2.81}$$

The transmission coefficient from right to left equals the transmission coefficient for transmission from left to right for the considered symmetric case (regarding the asymmetric case, consult Exercise 2.2 on page 24).

It can seem excessive for a single square barrier to introduce the transfer matrix instead of just solving the matching equations directly and obtain the transmission probability. However, a potential can consist of several barriers, and the total transfer matrix can be expressed in terms of the transfer matrices for the individual barriers. In such cases, the transfer matrix is a convenient tool. In Section 2.8 the double barrier is studied, and in Section 7.6 the transfer matrix is used to analyze the motion of a particle in a periodic potential, the infinite repetition of the same barrier.

2.7 Square Barrier

For a square barrier, the under-barrier transmission coefficient, $E < V$, or tunneling probability, becomes according to Eqs. (2.79) and (2.70)

$$T(E) = \frac{1}{1 + ((k^2+\kappa^2)^2/4k^2\kappa^2)\sinh^2\kappa a_B}, \quad k = \frac{\sqrt{2mE}}{\hbar}, \quad \kappa = \frac{\sqrt{2m(V-E)}}{\hbar}. \tag{2.82}$$

For the over-barrier transmission coefficient, $E > V$, the transfer matrix is obtained by the substitution $\kappa(E) \rightarrow -ik_V(E) = -i\hbar^{-1}\sqrt{2m(E-V)}$ (recall Eq. (2.50)). Since $\sinh(-iu) = -i\sin u$ for u real, the transmission probability for $E > V$ becomes, according to Eq. (2.82),

$$T(E) = \frac{1}{1 + ((k^2-k_V^2)^2/4k^2k_V^2)\sin^2 k_V a_B}, \tag{2.83}$$

where

$$k_V = \frac{\sqrt{2m(E-V)}}{\hbar}. \tag{2.84}$$

Thus for a square barrier the transmission coefficient as a function of the energy E of the particle has the form

$$T(E) = \begin{cases} \left(1 + \dfrac{V^2}{4E(V-E)} \sinh^2\left(a_B \dfrac{\sqrt{2m(V-E)}}{\hbar}\right)\right)^{-1}, & E \leq V, \\ \left(1 + \dfrac{V^2}{4E(E-V)} \sin^2\left(a_B \dfrac{\sqrt{2m(E-V)}}{\hbar}\right)\right)^{-1}, & E \geq V. \end{cases} \tag{2.85}$$

The result should be contrasted with the classical case, where an incoming particle is either totally reflected, $T_{\rm cl}(E < V) = 0$, or transmitted with certainty, $T_{\rm cl}(E > V) = 1$, depending on the energy of the particle being smaller or larger than the height of the barrier.

When the energy of the incoming particle equals the barrier height, $E = V$, where $\kappa(E = V) = 0$, the two expressions in Eq. (2.85) should be equal since the transfer matrix elements are continuous functions of the energy, and indeed both give the value

$$T(E = V) = \frac{1}{1 + ma_B^2 V/2\hbar^2}. \tag{2.86}$$

At the threshold, $E = V$, for total classical transmission, $T_{\rm cl}(E = V^+) = 1$, the tunneling probability is thus less than one, as specified by $ma_B^2 V/2\hbar^2$, the strength of the barrier.

For a graphical plot, it is useful to express the transmission coefficient in terms of the dimensionless variable $\epsilon \equiv E/V$,

$$T(\epsilon) = \begin{cases} \left(1 + \dfrac{1}{4\epsilon(1-\epsilon)} \sinh^2\left(\epsilon_0 \sqrt{1-\epsilon}\right)\right)^{-1}, & \epsilon \leq 1, \\ \left(1 + \dfrac{1}{4\epsilon(1-\epsilon)} \sin^2\left(\epsilon_0 \sqrt{\epsilon - 1}\right)\right)^{-1}, & \epsilon \geq 1, \end{cases} \tag{2.87}$$

and the dimensionless barrier strength

$$\epsilon_0 = \sqrt{\frac{2ma_B^2 V}{\hbar^2}}. \tag{2.88}$$

The transmission coefficient for a square barrier as a function of the dimensionless energy variable is shown in Figure 2.7 for two values of the barrier strength.

For a strong barrier, i.e. wide and/or high barrier, the dimensionless barrier strength is large, $\epsilon_0 \gg 1$, and tunneling is strongly suppressed for energies below the barrier height. For energies above the barrier height, $\epsilon > 1$, there are then, due to the discontinuity in the potential step, pronounced oscillations in the transmission probability. Introducing the wave length for the waves in the barrier region, $\lambda_V \equiv 2\pi/k_V$, total transmission, i.e. the absence of reflection, occurs according to Eq. (2.83) when a multiple of such half wave lengths equals the barrier width, $n\lambda_V/2 = a_B$, $n = 1, 2, \ldots$.

For a strong barrier, i.e. $\kappa(E)a_B \gg 1$, $\sinh^2 \kappa(E)a_B \simeq e^{2\kappa(E)a_B}$ is very large and dominates the denominator in Eq. (2.82), and the tunneling probability becomes

$$T(E) \simeq \frac{4E(V-E)}{V^2} e^{-2a_B\kappa(E)} = \frac{4E(V-E)}{V^2} e^{-2\sqrt{(2ma_B^2/\hbar^2)(V-E)}}, \tag{2.89}$$

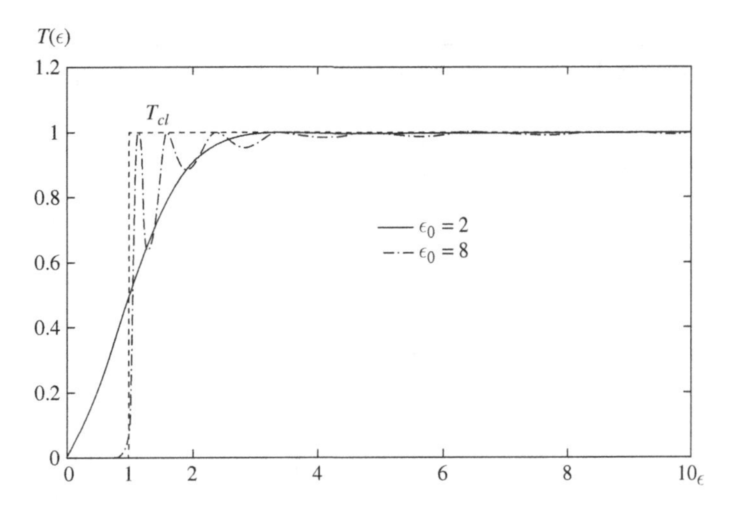

Figure 2.7 Transmission coefficient for two different square barriers.

i.e. the tunneling probability is exponentially small for a strong barrier.

The wave function in the barrier region, $x_l < x < x_r$, is a superposition of right and left decaying exponential functions,

$$\psi^B(x) = \psi_r^B(x) + \psi_l^B(x) = A_r^B\, e^{-\kappa x} + A_l^B\, e^{\kappa x}. \tag{2.90}$$

Considering the scenario of an incoming particle from the left, the barrier amplitudes are related (recall Eq. (2.60)) to the amplitude or wave function on the right side of the barrier according to

$$A_r^B = \frac{(\kappa - ik)\, e^{\kappa x_r}}{2\kappa} A_r^R\, e^{ikx_r} = \frac{(\kappa - ik)\, e^{\kappa x_r}}{2\kappa} \psi_R(x_r) \tag{2.91}$$

and

$$A_l^B = \frac{(\kappa + ik)\, e^{-\kappa x_r}}{2\kappa} A_r^R\, e^{ikx_r} = \frac{(\kappa + ik)\, e^{-\kappa x_r}}{2\kappa} \psi_R(x_r). \tag{2.92}$$

The relative magnitude decay of the right decaying barrier function is thus

$$\frac{\min |\psi_r^B(x)|}{\max |\psi_r^B(x)|} = \frac{|\psi_r^B(x_r)|}{|\psi_r^B(x_l)|} = \frac{\psi_r^B(x_r)}{\psi_r^B(x_l)} = \frac{e^{-\kappa x_r}}{e^{-\kappa x_l}} = e^{-\kappa a_B}. \tag{2.93}$$

In the weak tunneling regime, $T \propto \exp(-2\kappa a_B)$, the tunneling probability can thus be expressed in terms of the relative *under-barrier* decay at distance $x = a_B$, the measure of the amount that an incoming plane wave penetrates into a confining barrier, as further discussed in Section 6.1.

Introducing the expression for the penetration depth of the wave function (squared) into the barrier region, $d(E) \equiv 1/2\kappa(E) = \hbar/\sqrt{8m(V-E)}$, and noting that the exponential damping factor in Eq. (2.89) is the dominating factor, the tunneling probability is approximately (note the prefactor is one for $E = V/2$)

$$T(E) \simeq e^{-a_B/d(E)}, \quad d(E) < a_B. \tag{2.94}$$

Exercise 2.8 Verify the properties for the left decaying exponential wave function in the barrier region,

$$\frac{\min|\psi_l^B(x)|}{\max|\psi_l^B(x)|} = \frac{|\psi_l^B(x_l)|}{|\psi_l^B(x_r)|} = \frac{\psi_l^B(x_l)}{\psi_l^B(x_r)} = \frac{e^{\kappa x_l}}{e^{\kappa x_r}} = e^{-\kappa a_B}, \tag{2.95}$$

and that the left and right decaying parts of the barrier wave function at $x = x_r$ have equal modulus, $|\psi_r^B(x_r)| = |\psi_l^B(x_r)|$.

Exercise 2.9 Consider the potential well of constant depth $V > 0$,

$$V(x) = \begin{cases} 0, & x < x_l^B, \\ -V, & x_l^B < x < x_r^B, \\ 0, & x > x_r^B. \end{cases} \tag{2.96}$$

Obtain the reflection probability. Hint: Note that the result is obtainable from the square barrier result by the substitution $\kappa \to -ik_B$, $k_B(E) = \sqrt{2m(E+V)}/\hbar$.

Exercise 2.10 Use the result of Exercise 2.7 on page 31 to obtain the transmission probability for the delta potential $V(x) = w\delta(x)$ (see also the discussion in Appendix E)

$$T(E) = \frac{1}{1 + mw^2/2\hbar^2 E}. \tag{2.97}$$

2.8 Double Barrier

Consider a potential consisting of three free regions with arbitrary potentials in between. A simple example of such a situation is depicted in Figure 2.8.

The linearity of quantum mechanics turns the stacking of barriers into matrix multiplication of the corresponding transfer matrices. For the considered case of two barriers, a transfer matrix, $\mathcal{T}^{(1)}$, according to Eq. (2.61) relates the amplitudes on the left of the left barrier to those on the right of the left barrier, i.e. the amplitudes in the middle region. These are in turn related to the amplitudes on the right of the right barrier through a second

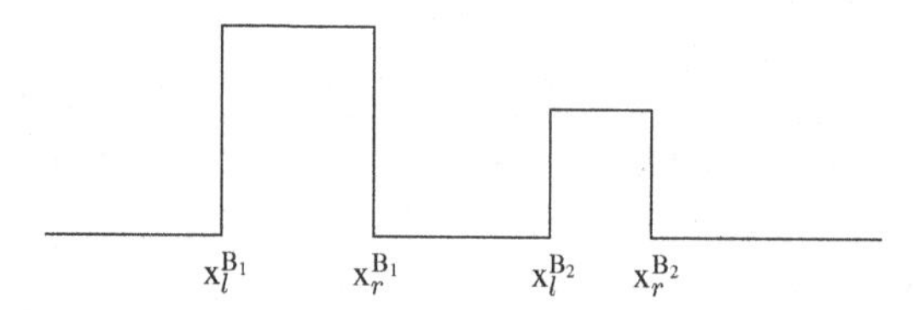

Figure 2.8 Double square barrier potential.

transfer matrix $\mathcal{T}^{(2)}$. The total transfer matrix, $\mathcal{T}^{(12)}$, for the double-barrier structure is therefore the product of the transfer matrices for the two barriers,

$$\mathcal{T}^{(12)} = \mathcal{T}^{(2)}\mathcal{T}^{(1)}, \tag{2.98}$$

expressing the linear relationship between amplitudes on the right and left of the double barrier. Adding additional barriers is thus mechanized into the routine of matrix multiplication.

The transmission coefficient is according to Eq. (2.78) the inverse of the absolute square of the 22-component of the transfer matrix $\mathcal{T}^{(12)}$, giving for an arbitrary double barrier

$$T_{DB}(E) = \frac{1}{|T_{22}^{(12)}(E)|^2} = \frac{1}{|T_{21}^{(2)}(E)\,T_{12}^{(1)}(E) + T_{22}^{(2)}(E)\,T_{22}^{(1)}(E)|^2}. \tag{2.99}$$

Consider now the case where the two barriers are identical square barriers of width a_B, separated by distance W, the double barrier depicted in Figure 2.9.

In that case, simplifications occur since the only difference between the transfer matrices for the two barriers is their reference to different spatial locations. Since the diagonal elements of the transfer matrix for a square barrier, according to Eq. (2.69), do not depend on the relative location of the barrier, depending only on the barrier width, they are identical for the two identical square barriers, i.e. $T_{22}^{(2)}(E) = T_{22}^{(1)}(E)$. In contrast, the off-diagonal transfer matrix elements sense the barrier location, and since $x_{r,l}^{B_2} = x_{r,l}^{B_1} + W + a_B$ according to Figure 2.9, the off-diagonal transfer matrix elements for the two barriers are, according to Eq. (2.70), phase shifted relative to each other according to

$$T_{21}^{(2)} = e^{2ik(W+a_B)}T_{21}^{(1)}. \tag{2.100}$$

The 22-component of the transfer matrix for the double barrier consisting of two identical square barriers is therefore

$$\begin{aligned} T_{22}^{(12)}(E) &= T_{21}^{(2)}(E)\,T_{12}^{(1)}(E) + T_{22}^{(2)}(E)\,T_{22}^{(1)}(E) \\ &= e^{2ik(W+a_B)}|T_{12}^{(1)}(E)|^2 + (T_{22}^{(1)}(E))^2 \end{aligned} \tag{2.101}$$

and the transmission coefficient becomes

$$T_{DSB}(E) = \frac{1}{|T_{22}^{(12)}(E)|^2} = \frac{1}{e^{2ik(W+a_B)}|T_{12}^{(1)}(E)|^2 + (T_{22}^{(1)}(E))^2}. \tag{2.102}$$

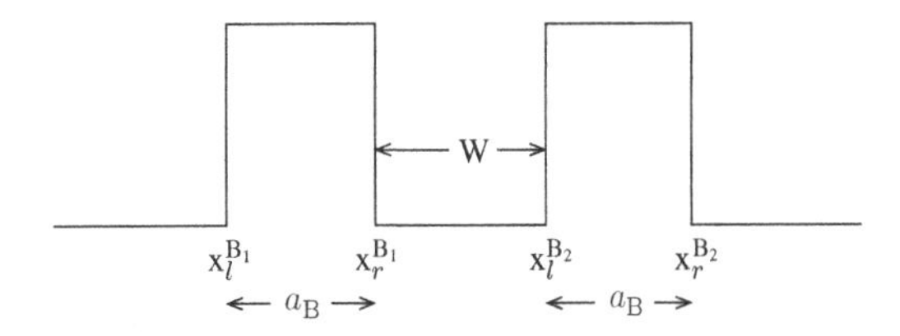

Figure 2.9 Two identical square barriers.

Exercise 2.11 Insert into Eq. (2.102) the expressions for the transfer matrix for the single barrier, Eqs. (2.69) and (2.70), to obtain for the double barrier the tunneling probability, $E < V$,

$$T_{DSB}(E) = \frac{1}{\left| e^{2ikW} \frac{(k^2+\kappa^2)^2}{4k^2\kappa^2} \sinh^2 s_B + \left(\cosh s_B + i \frac{\kappa^2 - k^2}{2k\kappa} \sinh s_B \right)^2 \right|^2}. \tag{2.103}$$

A check of the correctness of Eq. (2.103) is provided by setting the barrier separation parameter equal to zero, $W = 0$, turning the double barrier into a single square barrier of width $2a_B$. Show that Eq. (2.103) for $W = 0$ reduces to

$$T^{DSB}_{W=0}(E) = \frac{1}{\left| \cosh^2 s_B + \sinh^2 s_B + i \frac{\kappa^2 - k^2}{k\kappa} \cosh s_B \sinh s_B \right|^2}$$
$$= \frac{1}{(\cosh^2 s_B + \sinh^2 s_B)^2 + \frac{(\kappa^2 - k^2)^2}{k^2\kappa^2} \cosh^2 s_B \sinh^2 s_B}, \tag{2.104}$$

which upon using the identities $\cosh^2 x + \sinh^2 x = \cosh 2x$ and $2\cosh x \sinh x = \sinh 2x$ becomes

$$T^{DSB}_{W=0}(E) = \frac{1}{1 + \frac{(\kappa^2 + k^2)^2}{4k^2\kappa^2} \sinh^2 2\kappa a_B} = T^B_{2a_B}(E) \tag{2.105}$$

the one-barrier formula, Eq. (2.82), for a square barrier that is twice as wide.

Using the fact that the determinant of a transfer matrix equals one, $|T^{(1)}_{12}|^2 = |T^{(1)}_{22}|^2 - 1$, the transmission coefficient for the identical double square barrier, Eq. (2.102), can be rewritten as

$$T_{DSB}(E) = \frac{1}{\left| e^{2ik(W+a_B)}(|T^{(1)}_{22}(E)|^2 - 1) + (T^{(1)}_{22}(E))^2 \right|^2}. \tag{2.106}$$

Since $|T^{(1)}_{22}(E)|^2 = 1/T_B(E)$, where $T_B(E)$ is the transmission coefficient for the single barrier,

$$|T^{(1)}_{22}(E)|^2 - 1 = \frac{1}{T_B(E)} - 1 = \frac{1 - T_B(E)}{T_B(E)} = \frac{R_B(E)}{T_B(E)}, \tag{2.107}$$

where R_B denotes the reflection coefficients for the single barrier. The double identical square barrier transmission coefficient, Eq. (2.106), then takes the form, with $T^{(1)}_{22}(E) \equiv |T^{(1)}_{22}(E)|\, e^{i\phi(E)} = e^{i\phi(E)}/\sqrt{T_B(E)}$,

$$T_{DSB}(E) = \frac{1}{\left|\frac{R_B(E)}{T_B(E)} e^{2ik(W+a_B)} + \left(\frac{e^{i\phi(E)}}{\sqrt{T_B(E)}}\right)^2\right|^2} \tag{2.108}$$

or in the more transparent form

$$T_{DSB}(E) = \frac{T_B(E)\, T_B(E)}{\left|1 + R_B(E)\, e^{2i[k(E)(W+a_B)-\phi(E)]}\right|^2}. \tag{2.109}$$

For a single strong barrier, the tunneling probability is, according to Eq. (2.89), exponentially small. For a double barrier, however, this need not be the case for all energies. Rewriting Eq. (2.109) solely in terms of the one-barrier reflection coefficient, the transmission coefficient, Eq. (2.109), becomes

$$T_{DSB}(E) = \frac{(1 - R_B(E))^2}{\left|1 + R_B(E)\, e^{2i[k(E)(W+a_B)-\phi(E)]}\right|^2} \tag{2.110}$$

and the transmission coefficient for the double barrier is seen to equal unity for energies satisfying the condition

$$k(E)(W + a_B) - \phi(E) = (2n + 1)\frac{\pi}{2} \tag{2.111}$$

for any $n = 0, \pm 1, \pm 2, \ldots$. Energy values for which the double-barrier transmission coefficient equals one are called resonant energies. A wave packet initially localized mainly in between the two barriers will for such *energies* remain localized for a long time in the well region.

In order to show that, for the case of identical square barriers, at special energy values, the double barrier can have a tunneling probability equal to one, $T_{DSB}(E_s < V) = 1$, consider the transmission coefficient expression in Eq. (2.103). Let us analyze the transmission coefficient for the special energy of half the barrier height, $E = V/2$. Then $k = \kappa$ and Eq. (2.103) reduces to

$$T_{DSB}(E = V/2) = \frac{1}{\left|\cosh^2 s_B + e^{2ikW} \sinh^2 s_B\right|^2}. \tag{2.112}$$

When the well width and the wave vector are related such that $k(E = V/2)W = n\pi/2$ for any $n = 1, 2, \ldots$, the phase factor in front of $\sinh^2 s_B$ equals minus one, and the transmission is thus perfect, $T(E = V/2) = 1$. Depending on the parameters, there can be several resonance energies for which, quantum mechanically, the particle can pass right through the two barriers as if they were not there.

The study of a double barrier consisting of identical square barriers is not just of academic interest. As shown in Chapter 12, such a potential can be experienced by an electron in a gallium aluminum arsenide (GaAlAs) heterostructure, where the potential barriers are due to tiny layers of AlAs semiconductor material sandwiched in between GaAs. As discussed in Chapter 13, the current in such a device is determined by its transmission coefficient, and the double-barrier structure will exhibit interesting electronic properties such as negative differential resistance as discussed in Section 13.1.

For an off-resonance energy and small single-barrier tunneling probability, Eq. (2.109) for the tunneling probability for the double barrier becomes approximately the product of the tunneling probability for each of the barriers. In this approximation, Eq. (2.109) can therefore be used to obtain approximately the tunneling probability for a smooth single-barrier potential. A smoothly varying potential can be approximated by square barriers of barrier width Δx_i and different heights $V(x_i)$, each barrier's tunneling probability being, according to Eq. (2.94), approximately given by

$$T_i(E) \simeq \exp\left(-\frac{\Delta x_i}{d_i}\right) = \exp\left(-\frac{1}{\hbar}\sqrt{8m(V(x_i)-E)}\,\Delta x_i\right), \qquad (2.113)$$

$$d_i = \frac{\hbar}{\sqrt{8m(V(x_i)-E)}}.$$

The total transmission probability for the sequence of barriers is approximately the product of the individual transmission coefficients for the local square barriers, and the tunneling probability for a smooth potential is approximately

$$T(E) \simeq \prod_i T_i(E) = \prod_i \exp\left(-\frac{1}{\hbar}\sqrt{8m(V(x_i)-E)}\,\Delta x_i\right)$$

$$= \exp\left(-\frac{1}{\hbar}\int_{x_1}^{x_2} dx\,\sqrt{8m(V(x)-E)}\right). \qquad (2.114)$$

This is Gamow's tunneling formula, where x_1 and x_2 are the classical turning points, $V(x_{1,2}) = E$, as the over-barrier potential parts have their transmission coefficients approximated by one.

2.9 Summary

The wave packet analysis of the scattering of a particle showed that, once the time-*independent* Schrödinger equation is solved, the information of interest for the dynamical scattering problem is known. Namely, the transmission and reflection probabilities, $T(E)$ and $R(E)$, can be extracted from the amplitudes and wave vectors of the plane wave solutions in the free regions. To each energy eigenvalue are associated two opposite currents, a left- and right-directed probability current. An incoming plane wave leads to transmitted and reflected currents given by

$$S_t = T(E)S_{inc} \quad \text{and} \quad S_r = -R(E)S_{inc} \qquad (2.115)$$

and thereby a constant current everywhere of magnitude S_t.

We noted that the asymptotic outgoing scattering state is the arbitrarily good approximation to the exact solution of the Schrödinger equation, and for the physical quantities of interest, the reflection and transmission probabilities, the phenomenon of wave packet spreading has no bearing on calculating the probabilities for the various outcomes of the scattering process.

Only a few potential barriers allow for exact calculation of the tunneling probability, but all exhibit the same qualitative feature of being exponentially sensitive to barrier width and height, $T \propto e^{-2\kappa a_B}$, in the weak tunneling regime. The probability for tunneling is then small and specified by the exponential of the ratio of the barrier width and the penetration length of the wave function into the barrier region.

3 Standard Metal Model

Solid-state objects, such as metals, are assemblies of a huge number of atoms sticking together, and from a first principles point of view their description constitutes an unsolvable many-body problem, as the number of particles, nuclei and electrons, is astronomical (on the order of 10^{23} per cm^3). Electron or X-ray diffraction experiments reveal the grainy character of a metal. Charge is separated spatially into two: ions, the nuclei and closed shell electrons (so-called core electrons, tightly bound in their almost unperturbed atomic states); and valence electrons, which will have their density spread almost equally throughout the metal, thereby providing the glue to counteract the repulsive Coulomb force between the ions. The ions are solidified in a regular manner, forming a crystal, as discussed in Section 6.6. The spread-out valence electrons are called conduction electrons, since, in response to an applied electric field, they will conduct an electric current.

One of the earliest applications of quantum theory to solids was Sommerfeld's (1928) free electron model of the conduction electrons in a metal. Despite its neglect of the crystal periodic potential, it captures many essential features of the physics of metals, even quantitatively for alkali metals.

3.1 Sommerfeld Model

In the Sommerfeld model, the positively charged ionic lattice of a metal is represented by a smeared-out fixed background charge, keeping the system overall charge neutral. In view of the huge number of conduction electrons, the net effect of their Coulomb interaction is assumed to be replaceable by a mean field potential representing the effect of all the other conduction electrons. In a constant charge background, the mean field potential is the same at every point, i.e. a constant potential (expressing the translation invariance of the model), which for convenience is chosen to be zero. The Sommerfeld model is thus the independent particle model, an assembly of N non-interacting electrons confined to a box.

3.1.1 Energy States

The Sommerfeld model is characterized by the kinetic energy Hamiltonian for N particles,

$$\hat{H}_0 = -\sum_{i=1}^{N} \frac{\hbar^2}{2m} \Delta_i, \tag{3.1}$$

and the all important quantum feature that electrons are fermions.

An energy eigenstate of a single electron in a box of volume Ω is specified by its momentum and spin quantum numbers (see the discussion in Appendices H and I),

$$\psi_{\mathbf{p},s_z}(\mathbf{x}, s_z') = \psi_{\mathbf{p}}(\mathbf{x})\, \chi_{s_z}(s_z') = \frac{1}{\sqrt{\Omega}} e^{(i/\hbar)\mathbf{p}\cdot\mathbf{x}} \chi_{s_z}(s_z') = \frac{1}{\sqrt{\Omega}} e^{(i/\hbar)\mathbf{p}\cdot\mathbf{x}} \delta_{s_z,s_z'}, \tag{3.2}$$

where both spin states correspond to the same energy, $\epsilon(\mathbf{p}) = \mathbf{p}^2/2m$. According to the periodic boundary condition, only the subset of discrete momentum values are allowed, assuming for simplicity $\Omega = L^3$,

$$\mathbf{p_n} = \frac{2\pi\hbar}{L}\mathbf{n}, \quad \mathbf{n} = (n_x, n_y, n_z), \quad n_x, n_y, n_z = 0, \pm 1, \pm 2, \ldots. \tag{3.3}$$

The quantum numbers specifying the different energy states can thus be visualized as a lattice of allowed momentum (or wave vector) values.

Electrons are fermions, and any wave function representing a state of the N electrons must be antisymmetric upon interchange of any pair of electrons, as discussed in Appendix I. The energy eigenstates of the Hamiltonian, Eq. (3.1), for a system of N identical fermions, are thus the antisymmetric N-particle functions specified by the occupied momentum and spin states $\mathbf{p}_1, s_1^z; \ldots; \mathbf{p}_N, s_N^z$, i.e.

$$\begin{aligned} &\psi_{\mathbf{p}_1,s_1^z;\ldots;\mathbf{p}_N,s_N^z}(\mathbf{x}_1, s_1^{z\prime}; \ldots; \mathbf{x}_N, s_N^{z\prime}) \\ &= \frac{1}{\sqrt{N!}} \sum_P (-1)^{\zeta_P} \psi_{\mathbf{p}_{P(1)},s_{P(1)}^z}(\mathbf{x}_1, s_1^{z\prime}) \cdots \psi_{\mathbf{p}_{P(N)},s_{P(N)}^z}(\mathbf{x}_N, s_N^{z\prime}). \end{aligned} \tag{3.4}$$

The sum is over all permutations of the N particles, where ζ_P counts the number of transpositions in the permutation P, and gives the sign of the permutation

$$(-1)^{\zeta_P} = \prod_{1 \le i < j \le N} \frac{j-i}{P_j - P_i}, \tag{3.5}$$

i.e. a minus sign appears for an odd number of particle interchanges, corresponding to an odd number of transpositions in the permutation P. The single-particle wave functions, Eq. (3.2), being normalized, have as a consequence that the above N-particle wave function is normalized, as there are $N!$ different permutations of N particles.

The state represented by the wave function in Eq. (3.4) is an energy eigenstate of the free N-particle Hamiltonian,

$$\hat{H}_0 \psi_{\mathbf{p}_1,s_1^z;\ldots;\mathbf{p}_N,s_N^z}(\mathbf{x}_1, \ldots, s_N^{z\prime}) = \left(\sum_{i=1}^{N} \epsilon(\mathbf{p}_i) \right) \psi_{\mathbf{p}_1,s_1^z;\ldots;\mathbf{p}_N,s_N^z}(\mathbf{x}_1, \ldots, s_N^{z\prime}), \tag{3.6}$$

each occupied single-particle state contributing its energy to the total energy.

The antisymmetric basis wave function, Eq. (3.4), can be expressed as a determinant, the Slater determinant,

$$\psi_{\mathbf{p}_1,s_1^z;\ldots;\mathbf{p}_N,s_N^z}(\mathbf{x}_1,s_1^{z\prime};\ldots;\mathbf{x}_N,s_N^{z\prime})$$

$$= \frac{1}{\sqrt{N!}} \begin{vmatrix} \psi_{\mathbf{p}_1,s_1^z}(\mathbf{x}_1,s_1^{z\prime}) & \psi_{\mathbf{p}_1,s_1^z}(\mathbf{x}_2,s_2^{z\prime}) & \ldots & \psi_{\mathbf{p}_1,s_1^z}(\mathbf{x}_N,s_N^{z\prime}) \\ \psi_{\mathbf{p}_2,s_2^z}(\mathbf{x}_1,s_1^{z\prime}) & \psi_{\mathbf{p}_2,s_2^z}(\mathbf{x}_2,s_2^{z\prime}) & \ldots & \psi_{\mathbf{p}_2,s_2^z}(\mathbf{x}_N,s_N^{z\prime}) \\ \vdots & \vdots & \ddots & \vdots \\ \psi_{\mathbf{p}_N,s_N^z}(\mathbf{x}_1,s_1^{z\prime}) & \psi_{\mathbf{p}_N,s_N^z}(\mathbf{x}_N,s_N^{z\prime}) & \ldots & \psi_{\mathbf{p}_N,s_N^z}(\mathbf{x}_N,s_N^{z\prime}) \end{vmatrix}. \tag{3.7}$$

All the single-particle states should be different, since the determinant vanishes if two rows are identical, corresponding to two elements in the set $\{\mathbf{p}_i, s_i^z\}_{i=1,\ldots,N}$ being equal. Only one electron can occupy a given one-particle state in accordance with the exclusion principle (as also discussed in Appendix I). In an N-particle state, Eq. (3.7), a single-particle state, labeled by the quantum numbers $(\mathbf{p}_i, s_i^z)$, can thus be either unoccupied, so absent from the expression, or singly occupied in one of the two possible spin states or doubly occupied when both spin states are present.

Exercise 3.1 Show that the probability density that one of the particles is located at position $\mathbf{x}$ for an antisymmetric wave function is

$$P(\mathbf{x}) = \frac{1}{N}\int d\mathbf{x}_2 \ldots d\mathbf{x}_N \; |\psi_{\mathbf{p}_1,\ldots,\mathbf{p}_N}(\mathbf{x},\mathbf{x}_2,\ldots,\mathbf{x}_N)|^2 + \cdots$$

$$+ \frac{1}{N}\int d\mathbf{x}_1 \ldots d\mathbf{x}_{N-1} \; |\psi_{\mathbf{p}_1,\ldots,\mathbf{p}_N}(\mathbf{x}_1,\mathbf{x}_2,\ldots,\mathbf{x})|^2$$

$$= \frac{1}{N}\sum_{i=1}^{N} |\psi_{\mathbf{p}_i}(\mathbf{x})|^2 = \frac{1}{\Omega}. \tag{3.8}$$

Hint: To begin with, the spin degree of freedom is neglected, i.e. consider the state in Eq. (3.4) with no spin.

The average density of the particles is

$$n(\mathbf{x}) = NP(\mathbf{x}) = \sum_{i=1}^{N} |\psi_{\mathbf{p}_i}(\mathbf{x})|^2 = \frac{N}{\Omega}. \tag{3.9}$$

Next include the spin degree of freedom and show that the result is only changed by the feature that a momentum label can appear twice if both spin states are occupied.

Exercise 3.2 Show that the average current density for the state considered in the previous exercise is (see Eq. (A.50))

$$\mathbf{S}(\mathbf{x}) = \sum_{i=1}^{N} \frac{\mathbf{p}_i}{m} |\psi_{\mathbf{p}_i}(\mathbf{x})|^2 = \frac{1}{\Omega}\sum_{i=1}^{N} \frac{\mathbf{p}_i}{m}. \tag{3.10}$$

3.1.2 Ground State

The ground state for N independent electrons is obtained in accordance with the exclusion principle: filling up higher-energy single-particle states until all N electrons are accommodated. In Figure 3.1 is depicted the ground state for the case $N = 30$ (in the two-dimensional case for visual simplicity), each momentum state accommodating one spin-up and one spin-down electron.

The spacing between the allowed momentum values in the momentum lattice is, according to Eq. (3.3), very small, $2\pi\hbar/L$, where L is the linear size of the box confining the electrons, $L = \Omega^{1/3}$. Since the number of conduction electrons in a metal is huge, the ground state of the Sommerfeld model is therefore described by having all momentum, and spin-up and spin-down, states within a sphere occupied, and all states outside the sphere unoccupied. The ground state for a system of independent fermions is depicted in Figure 3.2, the filled Fermi sphere.

The radius of the Fermi sphere, the highest momentum value of an occupied single-particle state in the ground state, is called the Fermi momentum and denoted by p_F.

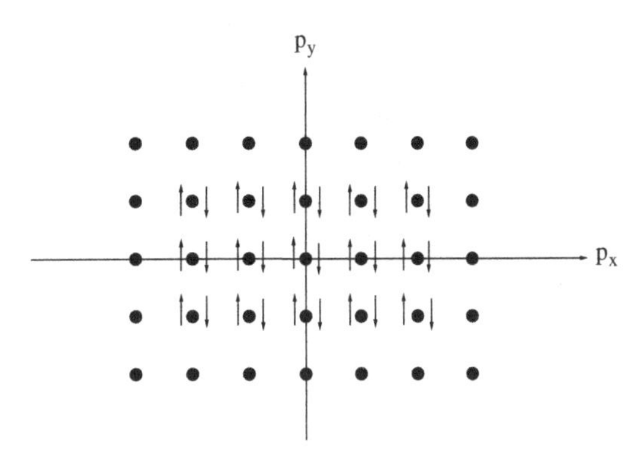

Figure 3.1 Ground-state configuration for 30 independent fermions.

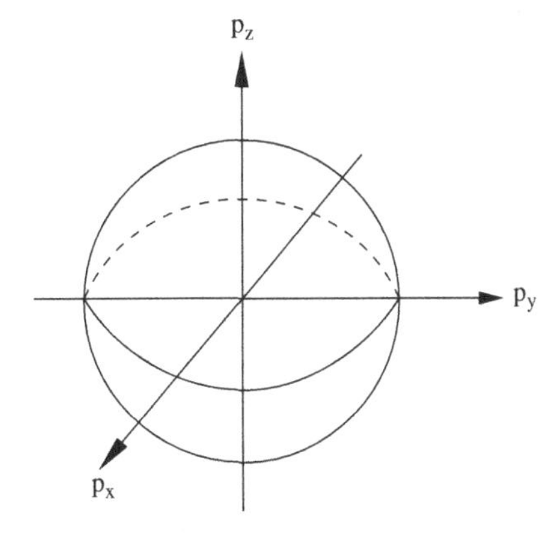

Figure 3.2 Fermi sphere.

Since the lattice constant of the momentum lattice of allowed states is $2\pi\hbar/L$, one momentum state can be accommodated in each tiny momentum volume of size $\Delta\mathbf{p} = (2\pi\hbar)^3/L^3 = (2\pi\hbar)^3/\Omega$. Thus two particles can be accommodated in each $(2\pi\hbar)^3/\Omega$ volume in momentum space owing to the spin degree of freedom. The volume of the Fermi sphere is $4\pi p_F^3/3$, and the number of momentum state boxes it contains is therefore $4\pi p_F^3/3\Delta\mathbf{p}$. Because of the spin degeneracy, the number of occupied boxes must be equal to $N/2$, i.e.

$$2\frac{\frac{4}{3}\pi p_F^3}{(2\pi\hbar/L)^3} = N \tag{3.11}$$

or

$$p_F = \hbar(3\pi^2 n)^{1/3}, \tag{3.12}$$

where n is the conduction electron density, $n = N/\Omega$.

The energy surface separating occupied and unoccupied single-particle states in the ground state is called the Fermi surface, and the Fermi energy, the energy value on the Fermi surface, is

$$\epsilon_F = \frac{p_F^2}{2m} = \frac{\hbar^2(3\pi^2 n)^{2/3}}{2m}, \tag{3.13}$$

i.e. determined by the electron density. The Fermi energy is the energy needed to add a particle to the ground state of the Fermi gas.

The energy distribution function describing the ground state is the step function, $f_{T=0}(\epsilon_{\mathbf{p}}) = \theta(\epsilon_F - \epsilon_{\mathbf{p}})$; all levels below the Fermi energy are with certainty occupied, and all levels above are with certainty empty.

Exercise 3.3 Show that, in two spatial dimensions, the Fermi momentum is related to the density, $n = N/L^2$, according to

$$p_F = \hbar\sqrt{2\pi n}. \tag{3.14}$$

Show that, in one spatial dimension, the Fermi momentum is related to the density, $n = N/L$, according to

$$p_F = \tfrac{1}{2}\pi\hbar n. \tag{3.15}$$

Exercise 3.4 Estimate the order of magnitude of ϵ_F, p_F, $v_F = p_F/m$, $k_F = p_F/\hbar$ and $\lambda_F = 2\pi/k_F$. Hint: See Section 3.4.

3.1.3 Density of States

In the following, we encounter summations over the momentum lattice

$$\Sigma = \sum_{\mathbf{p_n}} F(\mathbf{p_n}), \tag{3.16}$$

where $\mathbf{p_n}$ is specified in Eq. (3.3). An example of such a summation is the ground-state energy

$$E_G = 2 \sum_{\epsilon_{\mathbf{p_n}} \leq \epsilon_F} \epsilon_{\mathbf{p_n}} = 2 \sum_{\mathbf{p_n}} \epsilon_{\mathbf{p_n}} \, \theta(\epsilon_F - \epsilon_{\mathbf{p_n}}), \tag{3.17}$$

the factor of 2 being due to the spin degeneracy of single-particle energy states.

For sufficiently large volume, the size of the momentum lattice cells, $\Delta\mathbf{p} = (2\pi\hbar)^3/\Omega$, becomes infinitesimal, and the sum turns into the integral

$$\frac{1}{\Omega} \sum_{\mathbf{p_n}} F(\mathbf{p_n}) = \sum_{\mathbf{p_n}} \frac{\Delta\mathbf{p}}{(2\pi\hbar)^3} F(\mathbf{p_n}) = \int \frac{d\mathbf{p}}{(2\pi\hbar)^3} F(\mathbf{p}). \tag{3.18}$$

In spherical coordinates, the integral becomes

$$\int \frac{d\mathbf{p}}{(2\pi\hbar)^3} F(\mathbf{p}) = \frac{1}{(2\pi\hbar)^3} \int_0^{2\pi} d\varphi \int_0^{\pi} d\theta \, \sin\theta \int_0^{\infty} dp \, p^2 F(\varphi, \theta, p). \tag{3.19}$$

Integration over the azimuthal and polar angle directions is equivalent to integration over the unit sphere area, and letting $\hat{\mathbf{p}}$ denote a unit vector the integral is rewritten

$$\int \frac{d\mathbf{p}}{(2\pi\hbar)^3} F(\mathbf{p}) = \frac{4\pi}{(2\pi\hbar)^3} \int \frac{d\hat{\mathbf{p}}}{4\pi} \int_0^{\infty} dp \, p^2 F(\hat{\mathbf{p}}, p), \tag{3.20}$$

where the area of the unit sphere, 4π, has been introduced for normalization. When the integrand is a spherically symmetric function, i.e. independent of momentum direction $\hat{\mathbf{p}}$, it can be expressed in terms of the energy variable, $F(\epsilon_p)$, $\epsilon_{\mathbf{p}} = \epsilon_p = p^2/2m$, and the integral reduces to

$$\int \frac{d\mathbf{p}}{(2\pi\hbar)^3} F(\mathbf{p}) = \frac{4\pi}{(2\pi\hbar)^3} \int_0^{\infty} dp \, p^2 F(\epsilon_p). \tag{3.21}$$

Changing from momentum magnitude variable to energy variable gives

$$d\epsilon_p = \frac{p}{m} dp, \quad p^2 \, dp = mp \frac{p}{m} dp = mp \, d\epsilon_p = \sqrt{2m^3 \epsilon_p} \, d\epsilon_p \tag{3.22}$$

and the momentum magnitude integral has been turned into an energy integration,

$$\int \frac{d\mathbf{p}}{(2\pi\hbar)^3} F(\epsilon_{\mathbf{p}}) = \int_0^{\infty} d\epsilon_p \, N(\epsilon_p) F(\epsilon_p), \tag{3.23}$$

where

$$N(\epsilon) = \frac{\sqrt{2m^3 \epsilon}}{2\pi^2 \hbar^3}. \tag{3.24}$$

The function (3.24) is an important quantity, whose meaning can be established as follows. In Figure 3.3 is depicted the cross-section of two concentric spherical surfaces with radii of $p = \sqrt{2m\epsilon}$ and $p + \Delta p = \sqrt{2m(\epsilon + \Delta\epsilon)}$, assumed close in value.

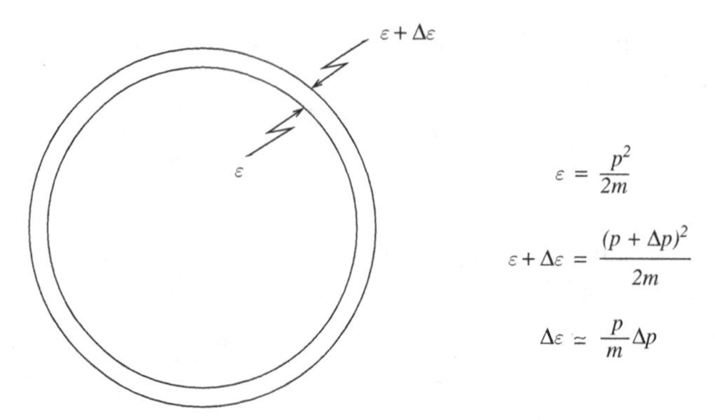

Figure 3.3 Cross-section of two concentric spherical surfaces of radii $p = \sqrt{2m\epsilon}$ and $p + \Delta p = \sqrt{2m(\epsilon + \Delta\epsilon)}$.

Counting the number of momentum states, $N(\epsilon, \epsilon + \Delta\epsilon)$, in between these two constant energy surfaces specified by the energies $\epsilon = p^2/2m$ and $\epsilon + \Delta\epsilon = (p + \Delta p)^2/2m$ ($\Delta\epsilon = p\Delta p/m$) gives (using # as shorthand for number)

$$\begin{aligned} N(\epsilon, \epsilon + \Delta\epsilon) &\equiv \text{\# of momentum states between spheres} \\ &= \frac{\frac{4}{3}\pi[(p + \Delta p)^3 - p^3]}{(2\pi\hbar/L)^3} \simeq \frac{\Omega p^2}{2\pi^2\hbar^3}\,\Delta p \\ &= \frac{mp}{2\pi^2\hbar^3}\,\Omega\Delta\epsilon = N(\epsilon)\Omega\Delta\epsilon, \end{aligned} \tag{3.25}$$

and $N(\epsilon)$ is identified as the number of momentum states per unit energy per unit volume,

$$N(\epsilon) = \frac{N(\epsilon, \epsilon + \Delta\epsilon)}{\Delta\epsilon\Omega}, \tag{3.26}$$

which equals the number of free particle energy levels per unit energy per unit volume per spin, the density of states or energy levels per spin.[1]

The density of states of a metal at the Fermi energy, $N(\epsilon_{\mathrm{F}})$, can be determined by measuring the specific heat at low temperatures, as discussed in Appendix K.

Exercise 3.5 Show that the free particle density of states is the same irrespective of the use of periodic boundary conditions or box boundary conditions, the latter assuming that a wave function vanishes on the box boundary.

Exercise 3.6 Show that the density of states per spin is specified by the counting formula for energy levels,

$$N_d(\epsilon) = \frac{1}{L^d}\sum_{\mathbf{p}} \delta(\epsilon - \epsilon_{\mathbf{p}}), \tag{3.27}$$

[1] Recall that, according to Appendix M, one need not rely on the trick of a quantization volume and its boundary condition to count the density of energy levels. The density of states can, for the infinite volume case, be obtained from the expression for the propagator.

either by direct calculation or using

$$N(\epsilon, \epsilon + \Delta\epsilon) = \sum_{\mathbf{p}} (\theta(\epsilon + \Delta\epsilon - \epsilon_{\mathbf{p}}) - \theta(\epsilon - \epsilon_{\mathbf{p}})). \tag{3.28}$$

Exercise 3.7 Show that the ratio of the density and the density of states at the Fermi energy in three spatial dimensions is given by

$$\frac{n}{N(\epsilon_F)} = \frac{4}{3}\epsilon_F. \tag{3.29}$$

Exercise 3.8 Show that, in one and two spatial dimensions, the free particle density of states per spin is

$$N_d(\epsilon) = \begin{cases} \sqrt{\dfrac{m}{2\pi^2\hbar^2\epsilon}}, & d = 1, \\ \dfrac{m}{2\pi\hbar^2}, & d = 2. \end{cases} \tag{3.30}$$

In Figure 3.4 is displayed the density of states for relevant spatial dimensions.

Returning to the ground-state energy, Eq. (3.17), we first note that, since each single-particle state has an energy that is a fraction of the Fermi energy, the ground-state energy takes the form $E_G \propto N\epsilon_F$, where the proportionality constant is less than one. To determine the constant, the ground-state energy, Eq. (3.17), or

$$E_G = 2\Omega \int \frac{d\mathbf{p}}{(2\pi\hbar)^3}\, \epsilon_{\mathbf{p}}\, \theta(\epsilon_F - \epsilon_{\mathbf{p}}) \tag{3.31}$$

is expressed in terms of the density of states according to Eq. (3.23), and the integral simply evaluated, $N(\epsilon) = C\epsilon^{1/2}$,

$$E_G = 2\Omega \int_0^{\epsilon_F} d\epsilon\, N(\epsilon)\epsilon = 2\Omega C \int_0^{\epsilon_F} d\epsilon\, \epsilon^{3/2} = 2\Omega C\,\frac{2}{5}\left[\epsilon^{5/2}\right]_0^{\epsilon_F} = 2\Omega\,\frac{2}{5}\left[\epsilon^2 N(\epsilon)\right]_0^{\epsilon_F}, \tag{3.32}$$

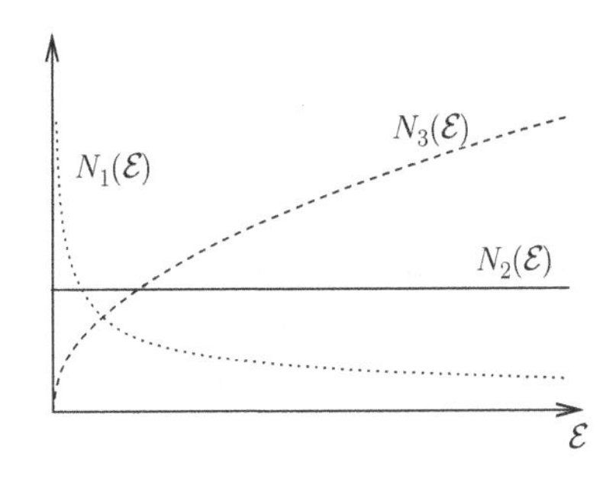

Figure 3.4 Free particle density of states in one, two and three spatial dimensions.

giving

$$E_G = 2\Omega \frac{2}{5} N(\epsilon_F)\, \epsilon_F^2, \tag{3.33}$$

which according to Eq. (3.29) equals

$$E_G = \frac{3}{5} N \epsilon_F. \tag{3.34}$$

Owing to the exclusion principle's demand that fermions must occupy different states, a Fermi gas has a non-zero ground-state energy.

At zero temperature, a classical ideal gas exerts no force on its confining walls, the particles simply lying still – the pressure of the gas (and its energy) is zero. In contrast, the ground-state energy of a Fermi gas depends, according to Eq. (3.34), on the volume of the box to which the electrons are confined, $E_G \propto p_F^2 \propto n^{2/3} \propto \Omega^{-2/3}$. The gas has lower energy if less confined. Therefore, even at zero temperature, the free Fermi gas exerts a non-zero pressure on its confining walls, the degeneracy pressure,

$$P_G = -\left(\frac{\partial E_G}{\partial \Omega}\right)_N = \frac{2}{3}\frac{E_G}{\Omega} = \frac{2}{5} n \epsilon_F. \tag{3.35}$$

The exclusion principle prevents identical fermions of the same spin from being lumped too closely together, and they need a force to keep them lumped together at all. For the case of the electrons in a metal, this force is the Coulomb force due to the positive charge of the ions of the metal.

3.2 Confinement and Contact Potential

The net effect of the forces from the charges making up a metal is to confine the electrons. The electrons experience a confining potential, i.e. a potential jump at the surface of the metal, which is a quantity that is complicated to calculate analytically with good precision due to electron–electron Coulomb interaction. At a surface of a metal, a layer of electronic charge spills out from the fixed positive background charge, leaving behind a layer of average positive charge – a dipole layer is formed, giving rise to part of the confining potential. The width of the charge inhomogeneity is of atomic size, as follows from the discussion of screening in Section 4.6. A spatial cross-sectional view of the ground state of the electrons thus gives a *bath tub* or Fermi sea picture of a metal, where in the ground state all energy states up to the Fermi energy are filled, as depicted in Figure 3.5.

The binding energy, E_B, of the electrons, also called the work function of the metal, can for example be measured through the photoelectric effect, the lowest photon energy required to eject an electron from the metal.

Two different metals separated from each other have their own different Fermi energies, as depicted in Figure 3.6. In the Sommerfeld model, different Fermi energies correspond to different electron densities of the metals.

Upon being brought into contact, there will be a net flow of electrons from the metal with the larger Fermi energy. Equilibration happens through thermal excitation over the barrier

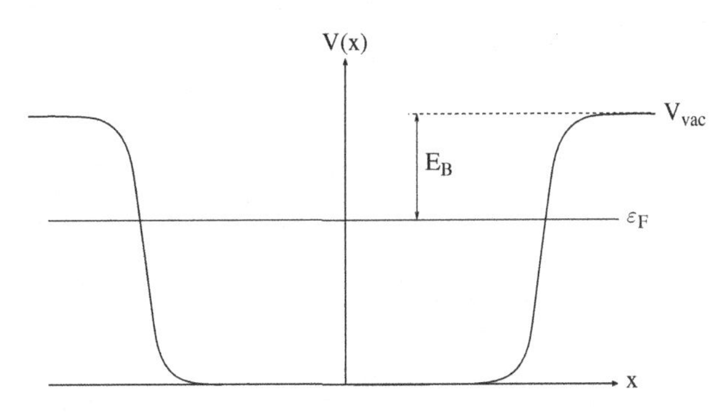

Figure 3.5 Cross-sectional view of confinement potential.

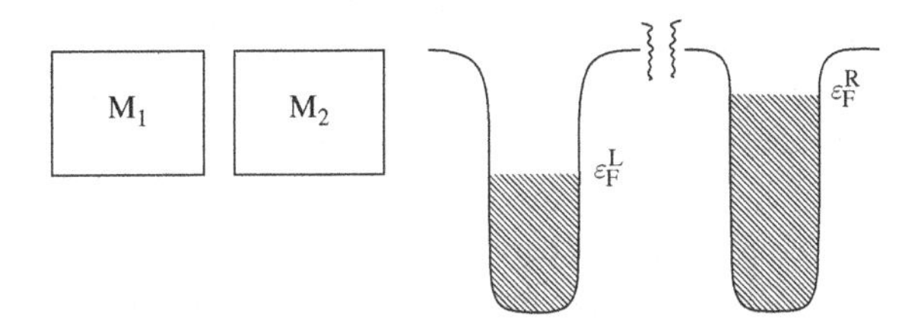

Figure 3.6 Cross-sectional view of two metals with different Fermi energies.

potential or for the considered case of zero temperature by the mechanism of quantum tunneling provided the metals are close enough. The net result will be that electrons are transferred to the metal of lower Fermi energy. In equilibrium, the electric field inside a metal vanishes, since the large number of free electrons will move until this state of affairs is quickly reached. The electrostatic potential is then constant throughout the metal, its surface being an equipotential surface corresponding to this value. Consequently, the divergence of the electric field vanishes, and the charge densities in the bulk of the metals vanish according to Gauss's law: the bulk electron charge density is equal in magnitude but opposite to the charge density of the smeared-out background charge. An excess charge on a metal thus resides on its surface (in reality, in a sheet only a few atomic layers thick). The metal of lower Fermi energy thus has a negative surface charge due to the excess of electrons, and the other metal a deficiency of electrons, resulting in a positive surface charge. These surface charges reside mainly on the opposing surfaces of the metals, as plus and minus charges attract each other, the arrangement forming a capacitor.[2]

An electric field thus exists in between the two metals, specified by an electrostatic potential difference $\Delta\phi < 0$, the electrostatic potential of the negatively charged metal being lower than that of the positively charged one. An electron in the bulk of the negatively charged metal with a certain kinetic energy therefore has the additional positive potential

[2] Such a bimetallic junction is a primitive battery. Connecting the metals by a wire leads to a current as the system discharges.

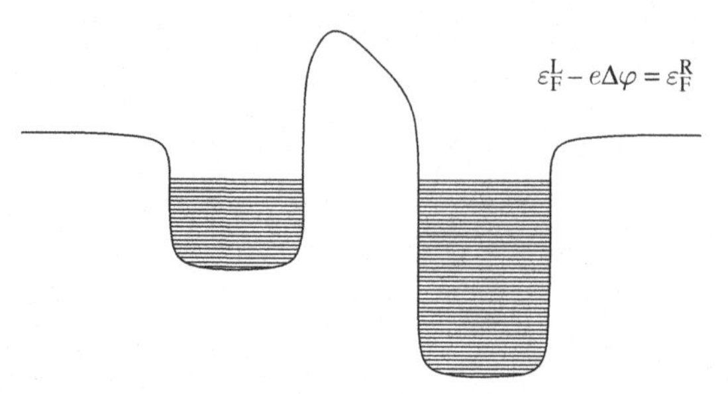

Figure 3.7 Electrochemical equilibration between two metals.

energy $-e\Delta\phi$ relative to the kinetic energy levels in the other metal.[3] A relative raising of the electron energy levels occurs with a corresponding tilt of the electron potential energy profile in between the metals. The equilibrium state of two metals in contact is reached when no net current is flowing, corresponding to the electrochemical potentials in the two metals being equal, $\epsilon_F^L - e\Delta\phi = \epsilon_F^R$; the electrochemical potential is constant throughout a system in equilibrium, as depicted in Figure 3.7.

When two conductors are in close enough contact to give an appreciable probability for tunneling through the barrier constituted by the confining potential in the vacuum region, an electronic device has been constructed. When a battery is connected to the two metals, their electrostatic potentials are made different, i.e. the electrochemical potentials in the two metals are held at different values, and a net tunneling current will flow. The constructed tunneling device, a metal–vacuum–metal junction, is further discussed in Section 3.5, the scanning tunneling microscope.

3.3 Tunnel Junction

Two metals separated by a thin insulating layer constitute a tunnel junction. Such a device may be created by lumping aluminum atoms together in a vacuum chamber by spraying them onto a substrate, and then doing the lumping in the presence of air so that a layer of the insulator aluminum oxide is formed on top of the aluminum, and then again growing aluminum on top of the insulator. A tunnel junction is schematically depicted in Figure 3.8.

In the insulator region there are no plane wave energy states available for the electrons, and the insulator constitutes a barrier for the electrons. The barrier can to a good approximation be taken as a square potential barrier, separating the two Fermi seas of electrons in the metals, if not for tunneling. The height of the barrier is determined by the band gap of the insulator.[4]

[3] We succumb to the convention of denoting the charge of the electron by $-e$, thus e denotes the unit of electric charge, $e \simeq 1.602 \times 10^{-19}$ C.

[4] A discussion of how stepwise constant potentials arise as an approximate description is given in Chapter 12. The equal energy wave functions in the barrier region are the mid-gap decaying *Bloch* states, as discussed in Chapter 7, joining to the wave function of the metallic states in the electrodes. In the potential barrier case, they are the imaginary wave vector solutions in the barrier region, Eq. (2.49).

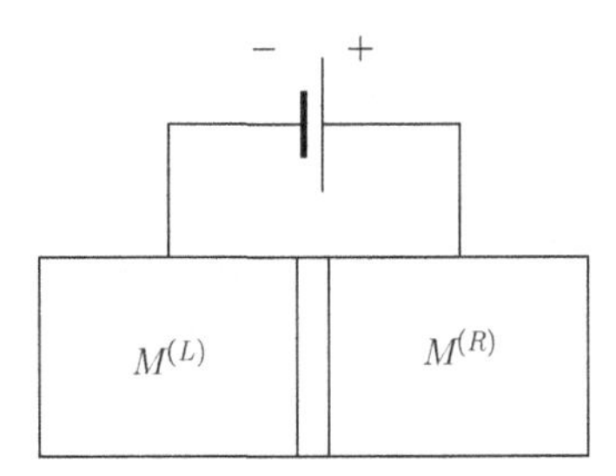

Figure 3.8 Tunnel junction consisting of two metals separated by an insulator.

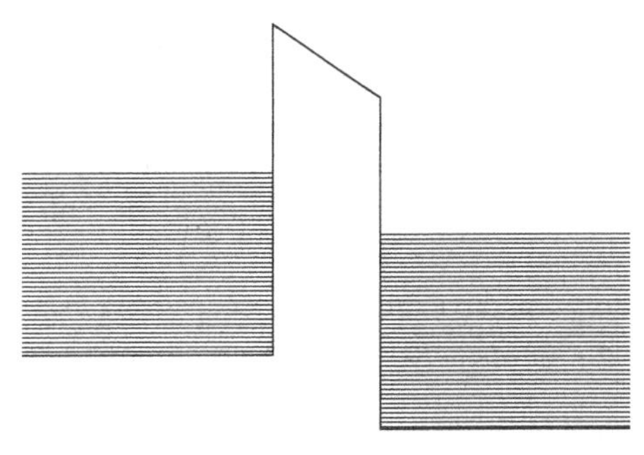

Figure 3.9 Elevated Fermi sea due to a battery voltage.

When a battery of voltage V is connected to the tunnel junction, as depicted in Figure 3.8, the two metals are given different electrostatic potentials. The metal attached to the negative battery electrode gets electrons injected, resulting in a surface charge of electrons on the left metal, and an equal deficit of electrons appears on the right metal. The two surface charge densities of opposite sign reside on the opposing metal–insulator surfaces, thereby forming a capacitor. Owing to the electrostatic potential difference, the barrier, instead of being square, will get a tilt, as illustrated in Figure 3.9.[5]

The battery raises the electrochemical potential in the negative electrode relative to that of the other one by the amount eV. The electron energy levels in the negatively charged metal are thereby raised by the amount eV, $V > 0$, relative to the levels in the positive electrode.

In equilibrium, left to itself, no net current flows through the junction. By attaching a battery to the two metals, the junction is kept in a non-equilibrium steady state due to the different electrochemical potentials and the scattering taking place at the barrier. A net flow of electricity through the junction occurs, an excess of electrons tunneling from left to right, each tunneled electron being replenished by the battery.

Of interest is the average current through the biased junction. For simplicity, consider the one-dimensional case.[6] Then current density equals current, and an electron in a free particle stationary state with kinetic energy ϵ has associated probability fluxes

[5] If the voltage applied to the junction is small compared to the barrier height, $eV \ll V_B$, the charging effect can be neglected. If the surface charges are not negligible, the real junction should be modeled as an ideal tunnel junction in parallel with a capacitor.

[6] In higher spatial dimensions, as discussed in Chapter 13, modifications to the one-dimensional current expression occurs as particles are not only transferred with normal incidence. Furthermore, the two electrodes need

$$S(\epsilon) = \pm v_\epsilon, \qquad v_\epsilon = \sqrt{\frac{2\epsilon}{m}} = \frac{\hbar|k|}{m} = \frac{|p|}{m}, \tag{3.36}$$

where certainty of occupation of the right-flowing current state, $A_r^L = 1$, corresponds to the plus sign, and the certainty of occupation of the left-flowing current state, $A_l^R = 1$, corresponds to the minus sign. As discussed in Chapter 2, an incoming flux is partly transmitted and partly reflected due to the barrier, and the magnitude of the current is $T(\epsilon)S(\epsilon)$. The improper dimension of the current in Eq. (3.36) is readily remedied by noting that the magnitude of, say, the right-flowing current carried by energy states in a small energy region $\Delta\epsilon$ is[7]

$$\mathcal{S}(\epsilon)\,\Delta\epsilon = 2 \times \frac{1}{2} N_1(\epsilon)\, S(\epsilon)\, T(\epsilon)\, \Delta\epsilon, \tag{3.37}$$

where the factor of 2 accounts for spin, and only half the density of states, $N_1(\epsilon)$, appears since only energy states corresponding to current direction toward the barrier should be counted.

Assuming zero temperature, the occupation of states with left-flowing current (incoming from the right) is specified by the distribution of occupied states,

$$f^R(\epsilon) = \theta(\epsilon_F - \epsilon) = \begin{cases} 1, & 0 < \epsilon < \epsilon_F, \\ 0, & \epsilon > \epsilon_F, \end{cases} \tag{3.38}$$

and the occupation of states with right-flowing current (incoming from the left) is elevated and specified by the distribution function

$$f^L(\epsilon) = \begin{cases} 1, & eV < \epsilon < \epsilon_F + eV, \\ 0, & \text{otherwise}. \end{cases} \tag{3.39}$$

The left- and right-flowing currents in the states in the energy range $eV < \epsilon < \epsilon_F$ cancel each other, and a net current of electrons thus flows from left to right due to the states in the energy range $\epsilon_F < \epsilon < \epsilon_F + eV$ of right-flowing current. Choosing the electric current direction through the device to be from right to left (see the general discussion in Section 5.1), the electric current through the tunnel junction is

$$I = 2e \int_0^\infty d\epsilon\, \frac{1}{2} N_1(\epsilon)\, v_\epsilon\, T(\epsilon)\, G(\epsilon), \tag{3.40}$$

where $G(\epsilon)$ is the gate function, which equals one for ϵ in the interval $[\epsilon_F, \epsilon_F + eV]$ and vanishes outside,

$$G(\epsilon) = \begin{cases} 1, & \epsilon_F < \epsilon < \epsilon_F + eV, \\ 0, & \text{otherwise}. \end{cases} \tag{3.41}$$

not be of the same material, and the product of the density of states for the two materials enters the tunnel current formula.

[7] The discussion can equally well be phrased in terms of momentum states, for which a momentum summation appearing with a quantization *volume* would lead to the density of states as shown in Section 13.1.

The gate function restricts the integration to the voltage band at the Fermi energy, and the current becomes

$$I = \frac{2e}{2\pi\hbar} \int_{\epsilon_F}^{\epsilon_F+eV} d\epsilon\, T(\epsilon) = \frac{e}{\pi\hbar}\, eVT(\epsilon_m) \tag{3.42}$$

(note that $N_1(\epsilon) = 1/\pi\hbar v_\epsilon$ according to Eq. (3.30)), where ϵ_m is the mean value point of the integral, $\epsilon_F < \epsilon_m < \epsilon_F + eV$. Since $eV \ll \epsilon_F$, the tunnel current is expressed in terms of the conductance as

$$I = G_t V, \qquad G_t = \frac{e^2}{\pi\hbar}\, T(\epsilon_F). \tag{3.43}$$

The tunnel conductance is thus governed by the tunneling probability of an electron at the Fermi energy. This observation is sufficient for the basic understanding of the tunneling microscope, as discussed in Section 3.5.

The above discussion of a tunnel junction operating at zero temperature could at first sight seem academic. However, as discussed in the next section, a metal at room temperature behaves almost identically to that of a metal at zero temperature. The room temperature thermal distribution of electrons in a metal is almost identical to the zero temperature step function.

3.4 Fermi–Dirac Distribution

The physical characteristics of the Sommerfeld model are, according to the discussion in Section 3.1.2, solely determined by the density of the conduction electrons, which can be roughly estimated as follows. The size of an atom is on the order of 10^{-8} cm.[8] In a metal the atoms are lumped together, but since their electron clouds, due to the exclusion principle, cannot overlap, the atomic size is a lower limit for the distance between the nuclei in a metal, setting this distance, say, at an order of magnitude larger, on the order of 10^{-7} cm. For the case of just one valence electron from each atom free to roam the metal, such as for alkali metals, there is thus a density of 10^{21} electrons per cm^3. This gives, according to Eq. (3.13), for the characteristic energy of an *alkali* electron gas, a Fermi energy on the order of one electronvolt, $\epsilon_F \sim 1\,\text{eV}$ (other metals reach up to an order of magnitude higher, $\epsilon_F \sim 10\,\text{eV}$). Introducing the characteristic temperature of a Fermi gas, $kT_F = \epsilon_F$, gives typically for the Fermi temperature of a metal $T_F \sim 10^4$ K. The characteristic temperature of the electron gas is thus huge compared to room temperature. In fact, at temperatures reaching the Fermi temperature, the metal has long since ceased to exist, having melted.

At zero temperature, all energy levels up to the Fermi energy are filled and the energy distribution function of the electrons is the step function, $f_{T=0}(\epsilon_{\mathbf{p}}) = \theta(\epsilon_F - \epsilon_{\mathbf{p}})$. Since

[8] The Schrödinger equation for the hydrogen atom is specified by Planck's constant, the electron mass and the squared electron charge, or in SI units $e_0^2 \equiv e^2/(4\pi\epsilon_0)$, and the characteristic length of the problem is according to dimensional analysis given by $a = \hbar^\alpha m^\beta e_0^\gamma$, which can only become a length if $\beta = -1$ and $\alpha = -\gamma = 2$, making a equal to the Bohr radius, $a \simeq 0.5$ Å.

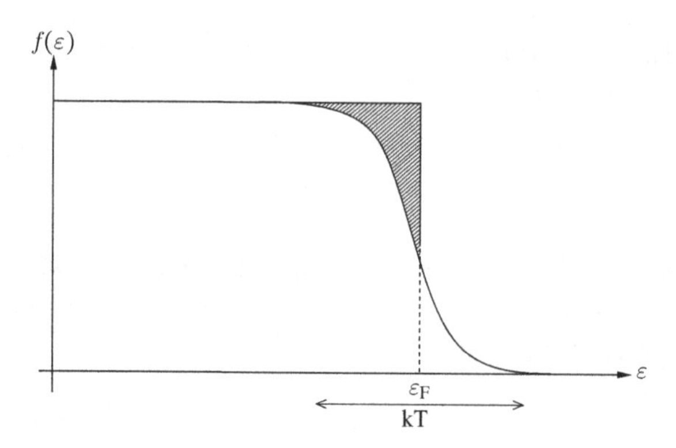

Figure 3.10 The Fermi function.

the Fermi temperature is two orders of magnitude larger than room temperature, the electrons in a metal at room temperature are only slightly excited out of the Fermi sphere. The distribution function deviates only slightly from the ground-state distribution, and the electron gas is degenerate. As shown in Appendix K, the probability that the energy state $\epsilon_{\mathbf{p}}$ for a Fermi gas in thermal equilibrium at temperature T is occupied is given by the Fermi function

$$f(\epsilon_{\mathbf{p}}) = \frac{1}{e^{(\epsilon_{\mathbf{p}}-\epsilon_{\mathrm{F}})/kT} + 1}. \tag{3.44}$$

The Fermi function is depicted in Figure 3.10. At finite temperatures a small fraction of the electrons in the Fermi sea are excited above the Fermi level, and the probability distribution differs from the zero temperature distribution only in a thin shell of size kT around the Fermi energy.

The derivative of the Fermi function is peaked at the Fermi energy,

$$-\frac{\partial f(\epsilon)}{\partial \epsilon} = \frac{1}{4kT}\frac{1}{\cosh^2((\epsilon - \epsilon_{\mathrm{F}})/2kT)}, \tag{3.45}$$

and has the narrow width kT, i.e. has the shape shown in Figure 3.11.

We note that the derivative of the Fermi function becomes progressively peaked as the temperature approaches zero, and in accordance with Appendix B becomes the Dirac delta function at zero temperature,

$$-\frac{\partial f_{T=0}(\epsilon)}{\partial \epsilon} = \delta(\epsilon - \epsilon_{\mathrm{F}}). \tag{3.46}$$

The derivative of the step function is a delta function, Eq. (B.22); the derivative of a discontinuous unit step gives a delta spike.

Returning to the tunnel junction, now assumed at room temperature, the electron distribution functions will be slightly changed due to thermal excitation. The electric current through the tunnel junction is then (for a detailed discussion see Section 13.1)

$$I = e\int_0^\infty d\epsilon\; N_1(\epsilon)\, v_\epsilon (f^L(\epsilon) - f^R(\epsilon)). \tag{3.47}$$

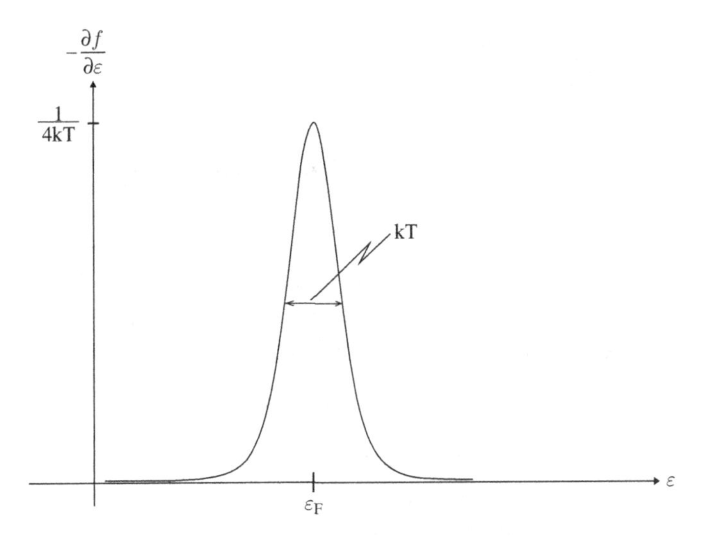

Figure 3.11 The derivative of minus the Fermi function.

Since the Fermi level on the left is raised by eV relative to the Fermi level on right,

$$f^L(\epsilon) = f^R(\epsilon - eV), \tag{3.48}$$

the average electric current through the junction at finite temperature is

$$I = \frac{e}{\pi\hbar} \int_0^\infty d\epsilon \; T(\epsilon)(f(\epsilon - eV) - f(\epsilon)). \tag{3.49}$$

The difference of the distribution functions is a gate function slightly smoothed by the temperature, and if the tunneling probability varies smoothly on the scale of the voltage and temperature, then

$$\begin{aligned} I &\simeq \frac{e}{\pi\hbar} T(\epsilon_F) \int_0^\infty d\epsilon \; (f(\epsilon - eV) - f(\epsilon)) \\ &\simeq \frac{e}{\pi\hbar} T(\epsilon_F) \int_{-\infty}^\infty d\epsilon \; (f(\epsilon - eV) - f(\epsilon)) = \frac{e}{\pi\hbar} T(\epsilon_F)\, eV. \end{aligned} \tag{3.50}$$

Or, to lowest order in the voltage, Eq. (3.49) becomes

$$I = \frac{e}{\pi\hbar} eV \int_0^\infty d\epsilon \; T(\epsilon) \left(-\frac{\partial f(\epsilon)}{\partial \epsilon}\right) \simeq \frac{e}{\pi\hbar} eV\, T(\epsilon_F) \int_0^\infty d\epsilon \left(-\frac{\partial f(\epsilon)}{\partial \epsilon}\right), \tag{3.51}$$

and upon using (corrections being exponentially small in the parameter T/T_F)

$$\int_0^\infty d\epsilon \left(-\frac{\partial f(\epsilon)}{\partial \epsilon}\right) \simeq \int_{-\infty}^\infty d\epsilon \left(-\frac{\partial f(\epsilon)}{\partial \epsilon}\right) = 1, \tag{3.52}$$

the finite temperature tunnel conductance in fact equals the zero temperature result of Eq. (3.43).

3.5 Tunneling Microscope

The tunneling phenomenon can be used to construct a *microscope* that is sensitive enough to *see* atoms. For a strong barrier, the tunneling probability is (recall the estimate (2.89) for a square barrier or Gamow's formula (2.114))

$$T(E) \propto e^{-2a_B\sqrt{(2m/\hbar^2)(V_B-E)}}, \tag{3.53}$$

where V_B is the height of the barrier. The exponential sensitivity to the barrier width, a_B, is utilized in a scanning tunneling microscope (STM). A sharply pointed piece of metal, an atomically sharp tip, is brought close to a conducting surface, and by applying a voltage a tunnel current flows.

The tunneling probability determines the tip-to-surface distance dependence of the tunneling current. Investigating the change in current due to a change in tip position, consider varying the tip-to-surface distance from 2 to 1 Å (1 Å = 10^{-1} nm = 10^{-10} m). The tunneling current according to Eq. (3.43) is proportional to the tunneling probability at the Fermi energy, and the current ratio becomes

$$\begin{aligned}\frac{I(a_B = 2\,\text{Å})}{I(a_B = 1\,\text{Å})} = \frac{T(a_B = 2\,\text{Å})}{T(a_B = 1\,\text{Å})} &= \exp\left\{-2(a_B(2) - a_B(1))\sqrt{\frac{2m}{\hbar^2}(V_B - \epsilon_F)}\right\}\\ &= \exp\left\{-\frac{2}{a_0}(2-1)\,\text{Å}\sqrt{\frac{V_B - \epsilon_F}{13.6\,\text{eV}}}\right\},\end{aligned} \tag{3.54}$$

where, in the last equality, the ratio is expressed in atomic units: the Bohr radius $a_0 \equiv 4\pi\epsilon_0\hbar^2/m_e e^2 \simeq 0.5$ Å, and the binding energy of the hydrogen atom, $E_R \simeq 13.6$ eV ($1\,\text{eV} \simeq 1.602 \times 10^{-19}$ J). The current ratio is thus determined by the binding energy of the electrons in the tip, $V_B - \epsilon_F$, the energy needed to take an electron at the Fermi energy over the confining barrier. An STM tip typically has a binding energy on the order of three electronvolts, 3 eV. Changing the tip-to-surface distance from 2 to 1 Å, the current thus increases by a factor 10, an order of magnitude that is easily detected. Using a scanning tunneling microscope, one can therefore detect tip position changes of sub-ångström order, which is precisely the size of atoms, and one can thus identify atoms placed on a surface, and bumps in the atomic structure of a surface.

Like moving a fingertip across an object to reveal its roughness, the tip is moved up and down sub-ångström distances across a conducting surface to scan its atomic composition. In the constant current mode operation of an STM, the distance of the tip to the conducting surface is adjusted, controlled by feedback electronics, such that the current is held fixed, as sketched in Figure 3.12. The tip topography across a sample is thus recorded.

A nanometer is a very short distance, and the way in which the tip is moved these incredibly small distances cannot be done by traditional mechanical means, but is achieved by

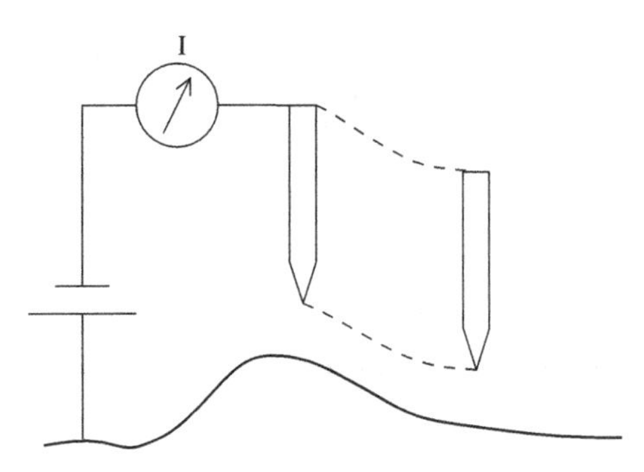

Figure 3.12 Schematics of the constant current mode operation.

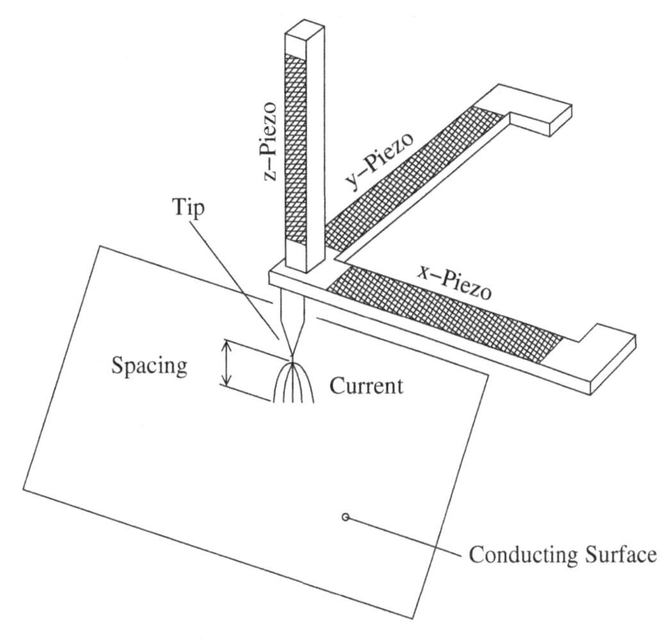

Figure 3.13 STM sensing surface structure operated by a piezo robot.

utilizing the piezoelectric effect (Figure 3.13). If two opposite sides of a crystal with lack of inversion symmetry are squeezed, changing its size in that direction a little, a voltage difference appears. A piezoelectric crystal also displays the opposite effect, whereby applying a voltage to the opposite sides expands or contracts the crystal. In this way, one changes the tip position in the sub-nanometer range. The typical resolutions of an STM are 0.1 nm in the vertical direction and nanometer resolution in the lateral directions. The sensitive monitoring and data collection is computer controlled.

Of vital importance for the operation of the tunneling microscope is elimination of vibrations, because tip motion of just 0.5 Å will invalidate data interpretation. This is achieved by having strong springs attached between the tip mounting and the substrate on which the sample is mounted, and a series of soft springs connecting the latter to the surrounding casing.

Besides being proportional to the tunneling probability, the tunneling current is dependent on the density of electron states in the tip and sample,

$$I \propto \int_0^{eV} d\epsilon \, N_S(\epsilon_F + \epsilon - eV) N_T(\epsilon_F + \epsilon). \tag{3.55}$$

The tunneling current thus measures the sample electron density at the surface, i.e. the derivative of the current–voltage characteristic is proportional to the local density of states,

$$\frac{dI}{dV} \propto N_S(\epsilon_F + \epsilon - eV). \tag{3.56}$$

The STM provides the density of electrons at different points on a surface. For a metal surface, the STM image is typically featureless, reflecting the almost uniform electron density. It was thus semiconductor surfaces that gave the technique its first successes, such as determining the reconstructed (111)-surface of silicon, elusive to other resolution techniques.

To investigate the atomic structure of non-conducting surfaces (insulators), one can use an atomic force microscope (AFM). In this case the probe is, for example, a diamond tip mounted on a thin gold foil, and the operation mode can be that of constant force between the almost touching tip and the sample surface, a constant position of the foil indicating a constant force. The foil, fixed at one end, is so thin that it responds as a cantilever to the small change in molecular forces as the tip scans a sample surface. The sample position is adjustable by a piezoelectric element on which it is mounted. The cantilever deflections can be measured by an STM tip behind the AFM tip. Keeping the STM current constant keeps the force on the AFM tip constant, and the signals controlling the piezoelectric elements show how the position of the sample has to be adjusted to achieve this. In this way the non-conducting surface is mapped with atomic precision. The cantilever deflections can alternatively be measured by a laser beam bouncing off the foil.

Tunneling microscopes are standard equipment for studying materials and even biological samples with atomic precision. Tools the size of a hand, tunneling microscopes were key instruments opening up the field of nanoscience and furthering its progress: imaging, measuring and manipulating matter at the nanoscale.

3.6 Occupation Number Representation

The many-body system considered in the Sommerfeld model consists of independent fermions, and its energy states are specified by the occupied single-particle energy levels, labeled in the following generally by λ corresponding to energy values ϵ_λ. The extent to which a state is occupied is, due to the exclusion principle, a binary question. Any N-particle energy state of the system is specified by stating the occupation numbers of the energy levels, which for a fermion system will be a string, $\{N_\lambda\}$, consisting of an infinite sequence of "1"s and "0"s, specifying by "1" that a level is occupied and by "0" that a level is unoccupied. For example,

$$(1, 0, 1, \ldots) \equiv \{N_\lambda\} \tag{3.57}$$

specifies that energy level ϵ_1 is occupied, level ϵ_2 is unoccupied, level ϵ_3 is occupied, etc. This occupation number representation is more flexible than the notation used in Appendix K.

In thermal equilibrium, the probability that the system is in the N-particle state specified by $\{N_\lambda\}$ is according to Eq. (K.2)

$$P(\{N_\lambda\}) = Z^{-1} \exp\left(-\frac{1}{kT}\sum_{i=1}^{\infty} \epsilon_{\lambda_i} N_{\lambda_i}\right), \tag{3.58}$$

where the normalization factor, the partition function, Eq. (K.4), is given by

$$Z = \sum_{\{N_\lambda\}}{}' \exp\left(-\frac{1}{kT}\sum_{i=1}^{\infty} \epsilon_{\lambda_i} N_{\lambda_i}\right). \tag{3.59}$$

The prime indicates that the summation over occupation number strings, $\{N_\lambda\}$, respects the constraint of a finite number of particles, i.e. summation is only over the set of occupation numbers, $\{N_\lambda\}$, for which

$$\sum_{i=1}^{\infty} N_{\lambda_i} = N. \tag{3.60}$$

A non-equilibrium state is, in the occupation number representation, specified by an occupation number probability distribution, $P(\{N_\lambda\})$, attributing to a string, $\{N_\lambda\}$, the probability that the state of the N particles is described by the corresponding many-body state. Like any probability distribution, it satisfies a normalization condition,

$$\sum_{\{N_\lambda\}}{}' P(\{N_\lambda\}) = 1 \tag{3.61}$$

but can otherwise be arbitrary (except for the restriction, due to the exclusion principle, that a state is either occupied or unoccupied, which is embodied in the definition of a fermionic string, Eq. (3.57)).

Multiplying the normalization condition, Eq. (3.61), by the number of particles, N, and using Eq. (3.60) gives for the total number of particles

$$N = \sum_{\{N_\lambda\}}{}' \left(\sum_{\lambda=1}^{\infty} N_\lambda\right) P(\{N_\lambda\}). \tag{3.62}$$

The two summations can be interchanged because

$$N = \sum_{\lambda=1}^{\infty} \sum_{\{N_\lambda\}}{}' N_\lambda P(\{N_\lambda\}) \tag{3.63}$$

gives rise to the same sum of terms. Since N_λ is either 0 or 1 for any λ, only configurations where $N_\lambda = 1$ are counted in the λ sum, and the number of particles is given by the expression

$$N = \sum_{\lambda=1}^{\infty} \sum_{\substack{\{N_\lambda\} \\ N_\lambda=1}}' 1 \times P(\{N_\lambda\}). \tag{3.64}$$

The terms in the λ summation are seen to be the mean occupation of λ levels, i.e.

$$N = \sum_{\lambda=1}^{\infty} f_\lambda, \quad f_\lambda = \sum_{\substack{\{N_\lambda\} \\ N_\lambda=1}}' P(\{N_\lambda\}), \tag{3.65}$$

where f_λ is the probability that energy level λ is occupied, i.e. the average value of occupation (see the equivalent discussion for the equilibrium case, Eq. (K.7)). The number of particles, N, equals the sum of the occupation probabilities of the levels, the normalization condition on the energy level distribution function, f_λ, expressing particle number conservation. A general requirement of a fermion energy distribution function, f_λ, is its respect of the exclusion principle, i.e. the average occupation of a level satisfies $0 \leq f_\lambda \leq 1$.

3.6.1 Density

The density for a given occupation of energy levels is (recall Exercise 3.1 on page 44)

$$n_{\{N_\lambda\}}(\mathbf{x}) = \sum_{\lambda_i} |\psi_{\lambda_i}(\mathbf{x})|^2 = \sum_{\lambda=1}^{\infty} |\psi_\lambda(\mathbf{x})|^2 N_\lambda. \tag{3.66}$$

For a given distribution of the states, $P(\{N_\lambda\})$, the average density is

$$n(\mathbf{x}) = \sum_{\{N_\lambda\}}' n_{\{N_\lambda\}}(\mathbf{x}) P(\{N_\lambda\}) = \sum_{\{N_\lambda\}}' \left(\sum_{\lambda=1}^{\infty} |\psi_\lambda(\mathbf{x})|^2 N_\lambda \right) P(\{N_\lambda\}). \tag{3.67}$$

Interchanging the two summations gives

$$n(\mathbf{x}) = \sum_{\lambda=1}^{\infty} \left(|\psi_\lambda(\mathbf{x})|^2 \sum_{\{N_\lambda\}}' N_\lambda P(\{N_\lambda\}) \right) = \sum_{\lambda=1}^{\infty} |\psi_\lambda(\mathbf{x})|^2 f_\lambda. \tag{3.68}$$

For the electron gas, where the energy levels are specified by a momentum and spin label, the electron density becomes, according to Eq. (3.65) for an arbitrary state,

$$n = \frac{1}{\Omega} \sum_{\mathbf{p},\, s_z=\pm 1} f(\mathbf{p}, s_z) = \sum_{s_z=\pm 1} \int \frac{d\mathbf{p}}{(2\pi\hbar)^3} f(\mathbf{p}, s_z). \tag{3.69}$$

In thermal equilibrium, the distribution function is the Fermi function, Eq. (3.44), and Eq. (3.69) thus relates the chemical potential, μ, to the electron density according to (the factor of 2 accounting for the spin degeneracy)

$$n = 2 \int \frac{d\mathbf{p}}{(2\pi\hbar)^3} \frac{1}{e^{(\epsilon_{\mathbf{p}}-\mu)/kT} + 1} \simeq 2 \int_0^{\infty} d\epsilon\, N(\epsilon) \frac{1}{e^{(\epsilon-\epsilon_F)/kT} + 1}, \tag{3.70}$$

where in the last equality we have introduced the fact that, for a degenerate electron gas, the chemical potential is effectively equal to the Fermi energy, as shown in Appendix K.

Performing a partial integration in Eq. (3.70) introduces the derivative of the Fermi function, a function that is strongly peaked at the Fermi energy, and using this property the integral is immediately evaluated in the standard (seemingly, approximate) way but in fact providing the exact relationship, Eq. (3.29).

3.6.2 Current Density

Our interest lies in non-equilibrium states, in particular the steady current states encountered in electronic devices due to applied voltages. The probability current density for a particle in an energy eigenstate labeled by λ is, according to Eq. (1.37),

$$\mathbf{S}_\lambda(\mathbf{x}) = \frac{\hbar}{2mi}[\psi_\lambda^*(\mathbf{x})\nabla_\mathbf{x}\psi_\lambda(\mathbf{x}) - \psi_\lambda(\mathbf{x})\nabla_\mathbf{x}\psi_\lambda^*(\mathbf{x})] \tag{3.71}$$

and the average electric current density for a particle in that state is

$$\mathbf{j}_\lambda(\mathbf{x}) = q\,\mathbf{S}_\lambda(\mathbf{x}), \tag{3.72}$$

where q denotes the charge of the particle.

For an arbitrary N-body basis state, Eq. (3.7), the average current density according to Eq. (3.10) is the sum of the single-particle currents. In the occupation number representation, the electric probability current density in an antisymmetrized N-body energy state is thus specified by

$$\mathbf{j}_{\{N_\lambda\}}(\mathbf{x}) = q\sum_\lambda{}' \mathbf{S}_\lambda(\mathbf{x})N_\lambda, \tag{3.73}$$

where the prime indicates that exactly N levels are occupied, and $N_\lambda = 1$ or $N_\lambda = 0$ specifies whether level λ is occupied or not.

For a mixture of states, such as for example the thermal state, the system is not in a definite N-body state, but has a probability, $P(\{N_\lambda\})$, for being in the state specified by the occupation numbers. The mixture statistical average then gives for the average electric current

$$\mathbf{j}(\mathbf{x}) = \sum_{\{N_\lambda\}}{}' \left(q\sum_\lambda \mathbf{S}_\lambda(\mathbf{x})\,N_\lambda\right) P(\{N_\lambda\}). \tag{3.74}$$

Only configurations where $N_\lambda = 1$ contribute to the λ summation, and the average current density can upon interchange of summations be rewritten as

$$\mathbf{j}(\mathbf{x}) = q\sum_{\lambda=1}^{\infty} \mathbf{S}_\lambda(\mathbf{x}) \cdot \sum_{\substack{\{N_\lambda\}\\ N_\lambda=1}}{}' P(\{N_\lambda\}) = q\sum_{\lambda=1}^{\infty} \mathbf{S}_\lambda(\mathbf{x}) f_\lambda. \tag{3.75}$$

The average current density is the sum of the current densities for *all* single-particle energy states weighted by the probability that the state is occupied.

In the Sommerfeld model, the current density in a single-particle energy state labeled by $\mathbf{p}$ is

$$\mathbf{S}_\mathbf{p}(\mathbf{x}) = \frac{1}{\Omega}\frac{\mathbf{p}}{m} = \frac{1}{\Omega}\frac{\partial\epsilon(\mathbf{p})}{\partial\mathbf{p}}, \tag{3.76}$$

where Ω is the volume of the system, and the average charge current density becomes (the factor of 2 accounting for spin degeneracy, $\mathbf{v_p} \equiv \mathbf{p}/m$),

$$\mathbf{j} = \frac{-2e}{\Omega} \sum_{\mathbf{p}} \frac{\mathbf{p}}{m} f(\mathbf{p}) = -2e \int \frac{d\mathbf{p}}{(2\pi\hbar)^3} \, \mathbf{v_p} \, f(\mathbf{p}). \tag{3.77}$$

In thermal equilibrium, the average current density vanishes since the two states of opposite momentum are equally populated, as they have the same energy, $\epsilon_{-\mathbf{p}} = \epsilon_{\mathbf{p}}$, and $\mathbf{v}_{-\mathbf{p}} = -\mathbf{v_p}$ as guaranteed by time reversal symmetry.

3.7 Summary

The Sommerfeld model provides the standard model of metallic behavior, describing qualitatively the properties of metals. The reason for the success of the Sommerfeld model, despite its complete neglect of the ionic lattice and Coulomb interaction, is the crucial role of the quantum statistics of the electrons, the exclusion principle, a simplifying feature whose consequences will be taken advantage of also in the following chapters.

The Sommerfeld model was also able to describe the current-carrying state of a tunnel junction. However, the use of macroscopically extended exact energy states is not useful when we want to understand transport in semiconductor devices where internal electric potentials can vary rapidly. For such situations, a local description is needed, as the exact energy states are not available. In the next chapter we introduce such a description, the wave packet description, and apply it first to understand the properties of a metallic conductor before progressing to the semiconductor case.

4 Standard Conductor Model

Unavoidable defects, such as foreign atoms, in a metal will, due to screening, give rise to short-range potentials that scatter electrons elastically into states with different velocity directions. Including such impurity atoms in the Sommerfeld model leads to a simple but adequate description of a metal. Owing to elastic scattering of the conduction electrons off impurities, an electric field is needed to sustain an electric current, i.e. the system exhibits resistance. The wave packet description, already employed in the study of tunneling and scattering, is now developed for the Sommerfeld metal, providing understanding of the main transport properties of metals. However, the wave packet description is now needed for an assembly of electrons, a many-body system of fermions whose wave function is subject to the exclusion principle.

4.1 Gaussian States

An electron in a Gaussian wave packet state with a definite spin, s_z, is described by the wave function

$$\psi_{\mathbf{x}_i,\mathbf{p}_i,s_z}(\mathbf{x}, s'_z) = \psi_{\mathbf{x}_i,\mathbf{p}_i}(\mathbf{x})\, \chi_{s_z}(s'_z) \tag{4.1}$$

with the Gaussian spatial part

$$\psi_{\mathbf{x}_i,\mathbf{p}_i}(\mathbf{x}) = \left(\frac{1}{2\pi\delta x^2}\right)^{3/4} \exp\left(-\frac{(\mathbf{x}-\mathbf{x}_i)^2}{4\delta x^2} + \frac{i}{\hbar}\mathbf{p}_i \cdot \mathbf{x}\right). \tag{4.2}$$

The position and momentum labels, $(\mathbf{x}_i, \mathbf{p}_i)$, of the normalized Gaussian function equal the average position and momentum of the particle (see Appendix M). The real parameter, δx, is according to Eq. (M.9) the position variance in the Gaussian state.

The utility of the Gaussian functions rests on their completeness, since they can therefore be used as a basis set of functions to describe electron states. As discussed in Appendix M, the set of Gaussian wave packets is an over-complete set of states, which is pruned to completeness by reducing the $(\mathbf{x}_i, \mathbf{p}_i)$ parameters to reside on a lattice of points in phase space where the size of the unit cell is

$$\Delta\mathbf{x}_c\, \Delta\mathbf{p}_c = (2\pi\hbar)^3. \tag{4.3}$$

To each such unit cell of the phase space lattice is thus associated one spatial Gaussian wave packet state with corresponding average position and momentum as specified by the lattice point $(\mathbf{x}_i, \mathbf{p}_i)$ in question. Owing to the spin degeneracy, each such phase space volume, or equivalently each lattice point state in phase space, can thus accommodate two electrons. The Gaussian state is a minimal uncertainty state, i.e. the momentum variance is $\delta p = \hbar/2\delta x$. Since $\Delta x_c\, \Delta p_c = 2\pi\hbar > \hbar/2 = \delta p\, \delta x$, the variances can be chosen such that $\delta x < \Delta x_c$ and $\delta p < \Delta p_c$, and an electron in a Gaussian state is mainly localized in its corresponding spatial cell and immediate surrounding.

Electrons are fermions, and a wave function representing a state of the N electrons must be antisymmetric upon interchange of pairs of electron position and spin coordinates, i.e. interchange of the indistinguishable electrons as discussed in Appendix I. The basis set of Gaussian N-particle electron states is thus the antisymmetric wave functions (using notation $(\mathbf{x}, \mathbf{p}, s^z)_i \equiv (\mathbf{x}_i, \mathbf{p}_i, s^z_i)$)

$$\begin{aligned}
&\psi^{AS}_{(\mathbf{x},\mathbf{p},s^z)_1;\ldots;(\mathbf{x},\mathbf{p},s^z)_N}(\mathbf{x}'_1, s^{z\prime}_1; \ldots; \mathbf{x}'_N, s^{z\prime}_N) \\
&= \frac{1}{\sqrt{N!}} \sum_P (-1)^{\xi_P}\, \psi_{(\mathbf{x},\mathbf{p},s^z)_{P(1)}}(\mathbf{x}'_1, s^{z\prime}_1) \ldots \psi_{(\mathbf{x},\mathbf{p},s^z)_{P(N)}}(\mathbf{x}'_N, s^{z\prime}_N) \\
&\equiv \frac{1}{\sqrt{N!}} \begin{vmatrix}
\psi_{(\mathbf{x},\mathbf{p},s^z)_1}(\mathbf{x}'_1, s^{z\prime}_1) & \psi_{(\mathbf{x},\mathbf{p},s^z)_1}(\mathbf{x}'_2, s^{z\prime}_2) & \cdots & \psi_{(\mathbf{x},\mathbf{p},s^z)_1}(\mathbf{x}'_N, s^{z\prime}_N) \\
\psi_{(\mathbf{x},\mathbf{p},s^z)_2}(\mathbf{x}'_1, s^{z\prime}_1) & \psi_{(\mathbf{x},\mathbf{p},s^z)_2}(\mathbf{x}'_2, s^{z\prime}_2) & \cdots & \psi_{(\mathbf{x},\mathbf{p},s^z)_2}(\mathbf{x}'_N, s^{z\prime}_N) \\
\vdots & \vdots & \ddots & \vdots \\
\psi_{(\mathbf{x},\mathbf{p},s^z)_N}(\mathbf{x}'_1, s^{z\prime}_1) & \psi_{(\mathbf{x},\mathbf{p},s^z)_N}(\mathbf{x}'_2, s^{z\prime}_2) & \cdots & \psi_{(\mathbf{x},\mathbf{p},s^z)_N}(\mathbf{x}'_N, s^{z\prime}_N)
\end{vmatrix}.
\end{aligned} \tag{4.4}$$

These are Slater determinants of Gaussian wave packets where one particle is in Gaussian state $(\mathbf{x}_1, \mathbf{p}_1, s^z_1)$, one is in state $(\mathbf{x}_2, \mathbf{p}_2, s^z_2)$, etc. (recall the analogous discussion in Section 3.1.1). All the single-particle states are different because, if two rows are identical, the determinant vanishes. Only one electron of given spin thus occupies a spatial one-particle Gaussian state in accordance with the exclusion principle. Owing to the spin degree of freedom, any single-particle Gaussian state in the N-particle basis state, Eq. (4.4), can thus be either unoccupied, so absent from the expression, or singly occupied in one of the two possible spin states or doubly occupied when both spin states are present. For a Fermi gas enclosed in a volume Ω, the position labeling of a Gaussian state, $\mathbf{x}_i$, being the average position in that state, is of course restricted to that volume, i.e. the lattice positions $\mathbf{x}_i$ are all located inside the volume of the system, whereas lattice momentum values, $\mathbf{p}_i$, are unlimited.

4.2 Gaussian Many-Body States

Consider first the wave packet description of the one-dimensional electron gas, N electrons on a line of length L. In that case a picture of the cell division of phase space, *viz.* boxes of phase space area $\Delta x_c\, \Delta p_c = 2\pi\hbar$, is depicted in Figure 4.1. A point in each of the boxes is chosen as the phase space lattice point labeling the Gaussian wave packet state. The boxes

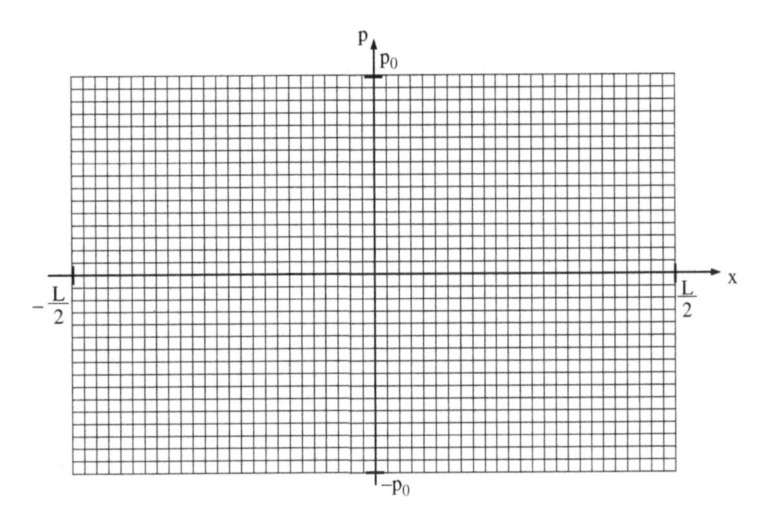

Figure 4.1 Phase space cells of associated Gaussian wave packets.

stretch in the x direction the length of the sample, from $-L/2$ to $L/2$, and are unbounded in the momentum direction.

The average kinetic energy of a single particle in a Gaussian state is for the one-dimensional case (see Eq. (M.19))

$$-\frac{\hbar^2}{2m}\int dx\, \psi^*_{x_ip_i}(x)\,\frac{d^2}{dx^2}\,\psi_{x_ip_i}(x) = \left(\frac{\delta p^2}{2m} + \epsilon(p_i)\right), \tag{4.5}$$

where $\epsilon(p_i) = p_i^2/2m$, and $\delta p = \hbar/2\delta x$ is the momentum variance of the particle in the Gaussian state. Shifting the single-particle Hamiltonian by the constant amount $-\delta p^2/2m$ leaves the physics unchanged, as it just changes the zero level from which energy is measured. A particle in the Gaussian state labeled by x_i and p_i can thus be attributed the average kinetic energy $\epsilon(p_i) = p_i^2/2m$, equal to the kinetic energy of a free particle with definite momentum p_i.

Consider first the non-antisymmetrized N-particle product of Gaussian functions

$$\psi^{\text{NAS}}_{(x,p,s^z)_1;\ldots;(x,p,s^z)_N}(x_1', s_1^{z\prime};\ldots;x_N', s_N^{z\prime}) = \psi_{(x,p,s^z)_1}(x_1', s_1^{z\prime})\cdots\psi_{(x,p,s^z)_N}(x_N', s_N^{z\prime}). \tag{4.6}$$

The state is normalized as a consequence of the Gaussian functions being normalized. The average of the kinetic energy operator in this state is

$$\begin{aligned}&\langle\psi^{\text{NAS}}_{(x,p,s^z)_1;\ldots;(x,p,s^z)_N}|\hat{H}_{\text{KE}}|\,\psi^{\text{NAS}}_{(x,p,s^z)_1;\ldots;(x,p,s^z)_N}\rangle\\ &\quad\equiv \sum_{s_1^{z\prime},\ldots,s_N^{z\prime}}\int dx_1'\ldots dx_N'\,\psi^{*\,\text{NAS}}_{(x,p,s^z)_1;\ldots;(x,p,s^z)_N}(x_1', s_1^{z\prime};\ldots;x_N', s_N^{z\prime})\\ &\quad\times \hat{H}_{\text{KE}}\,\psi^{\text{NAS}}_{(x,p,s^z)_1;\ldots;(x,p,s^z)_N}(x_1', s_1^{z\prime};\ldots;x_N', s_N^{z\prime}),\end{aligned} \tag{4.7}$$

where the total kinetic energy operator for the N-particle system is the sum of the single-particle kinetic energy Hamiltonians,

$$\hat{H}_{\text{KE}} = -\frac{\hbar^2}{2m} \sum_{n=1}^{N} \frac{\partial^2}{\partial x_n'^2}. \tag{4.8}$$

Any term in the expression (4.7) thus contains one factor that is a single-particle Hamiltonian average kinetic energy term and the remaining factors that are just normalization integrals of Gaussian functions, the spin summations, due to the spin parts of the wave function being Kronecker functions, also just producing factors of 1 (see Appendix M). As a result, the sum of single-particle average kinetic energies, up to an additive constant, of the Gaussian states in question is produced,

$$\langle \psi^{\text{NAS}}_{(x,p,s^z)_1;\ldots;(x,p,s^z)_N} | \hat{H}_{\text{KE}} | \psi^{\text{NAS}}_{(x,p,s^z)_1;\ldots;(x,p,s^z)_N} \rangle = \sum_{i=1}^{N} \left(\epsilon(p_i) + \frac{\delta p^2}{2m} \right). \tag{4.9}$$

Next, consider the average of the kinetic energy operator in the Gaussian many-body basis state represented by the antisymmetrized fermionic N-particle function in Eq. (4.4),

$$\langle \psi^{\text{AS}}_{(x,p,s^z)_1;\ldots;(x,p,s^z)_N} | \hat{H}_{\text{KE}} | \psi^{\text{AS}}_{(x,p,s^z)_1;\ldots;(x,p,s^z)_N} \rangle.$$

It takes the form of a double sum over all permutations accompanied by the factor $1/N!$. There are therefore $N!$ permutations, which gives the expression in Eq. (4.9) multiplied by $1/N!$, i.e. the term in Eq. (4.9) is recovered (see Appendix M for explicit cases). In addition to the sum of average single-particle energies, the average of the kinetic energy operator contains many more terms corresponding to the additional permutations. The additional terms consist of one factor of Hamiltonian overlap multiplied by factors of overlaps of Gaussian states (the former being expressible in terms of the latter according to Exercise M.4 on page 410). This is to be contrasted with the average kinetic energy for the state built by energy eigenstates, Eq. (3.7), where these terms vanish by orthogonality, and only the sum of energy eigenvalues survives. For the present case, only the spin parts of the wave functions are exactly orthogonal (see Eq. (M.42)).

We note that for any sum of one-body operators, such as the kinetic energy, only one antisymmetrization operation is needed as

$$\begin{aligned} &\langle \psi^{\text{AS}}_{(x,p,s^z)_1;\ldots;(x,p,s^z)_N} | \hat{H}_{\text{KE}} | \psi^{\text{AS}}_{(x,p,s^z)_1;\ldots;(x,p,s^z)_N} \rangle \\ &= \sqrt{N!}\, \langle \psi^{\text{NAS}}_{(x,p,s^z)_1;\ldots;(x,p,s^z)_N} | \hat{H}_{\text{KE}} | \psi^{\text{AS}}_{(x,p,s^z)_1;\ldots;(x,p,s^z)_N} \rangle \\ &= \sum_P (-1)^{\xi_P} \langle \psi^{\text{NAS}}_{(x,p,s^z)_1;\ldots;(x,p,s^z)_N} | \hat{H}_{\text{KE}} | \psi^{\text{NAS}}_{(x,p,s^z)_{P(1)};\ldots;(x,p,s^z)_{P(N)}} \rangle. \end{aligned} \tag{4.10}$$

The case of the unit operator gives the normalization integral

$$\begin{aligned} &\langle \psi^{\text{AS}}_{(x,p,s^z)_1;\ldots;(x,p,s^z)_N} \mid \psi^{\text{AS}}_{(x,p,s^z)_1;\ldots;(x,p,s^z)_N} \rangle \\ &= \sum_P (-1)^{\xi_P} \sum_{s_1^{z\prime},\ldots,s_N^{z\prime}} \int d\mathbf{x}_1' \cdots \int d\mathbf{x}_N'\, \psi^*_{(x,p,s^z)_1}(\mathbf{x}_1', s_1^{z\prime})\, \psi_{(x,p,s^z)_{P_1}}(\mathbf{x}_1', s_1^{z\prime}) \cdots \\ &\cdots \psi^*_{(x,p,s^z)_N}(\mathbf{x}_N', s_N^{z\prime})\, \psi_{(x,p,s^z)_{P_N}}(\mathbf{x}_N', s_N^{z\prime}) \end{aligned}$$

$$= \det\langle x_i, p_i, s_i^z \mid x_j, p_j, s_j^z \rangle, \quad i, j = 1, \ldots, N. \tag{4.11}$$

The diagonal term gives the value one, since the single-particle Gaussian functions are normalized, and the many other terms give various products of overlaps with alternating signs.

The spatial overlap of Gaussian functions corresponding to lattice points labeled by (x_1, p_1) and (x_2, p_2), respectively, is according to Eq. (M.25) given by

$$\langle x_1, p_1 \mid x_2, p_2 \rangle = \exp\left(-\frac{(x_1 - x_2)^2}{8\delta x^2} - \frac{(p_1 - p_2)^2}{8\delta p^2} - \frac{i}{2\hbar}(p_1 - p_2)(x_1 + x_2) \right). \tag{4.12}$$

The largest overlaps for different Gaussian states are thus for states of the same position labeling and occupying neighboring momentum cells, or for states of the same momentum labeling and neighboring spatial cells. If the momentum values are identical, $p_2 = p_1$, the lower bound on the overlap occurs for neighboring spatial cells, $x_2 = x_1 \mp \Delta x_c$, the spatial overlap being

$$\langle x_1, p_1 \mid x_2, p_1 \rangle = \exp\left(-\frac{\Delta x_c^2}{8\delta x^2} \right). \tag{4.13}$$

We are free to choose the position variance such that $\delta x < \Delta x_c$ and the overlap is small, the exponential of a negative number being less than one. If the position labels are identical, $x_2 = x_1$, and the momentum cells adjacent, $p_2 = p_1 \pm \Delta p_c$, the spatial overlap is

$$\langle x_1, p_1 \mid x_1, p_2 \rangle = \exp\left(-\frac{\Delta p_c^2}{8\delta p^2} \mp \frac{i}{\hbar} x_1 \Delta p_c \right). \tag{4.14}$$

We are still free to choose the momentum variance such that $\delta p < \Delta p_c$ since the constraint $2\pi\hbar = \Delta x_c \, \Delta p_c > \delta x \, \delta p = \hbar/2$ is not violated, and the overlap, Eq. (4.14), is small. Say choosing $\Delta x_c = \sqrt{4\pi}\, \delta x$, whereby $\Delta p_c = \sqrt{4\pi}\, \delta p$, gives

$$\exp\left(-\frac{\Delta x_c^2}{8\delta x^2} \right) = \exp\left(-\frac{\Delta p_c^2}{8\delta p^2} \right) = e^{-\pi/2} \simeq e^{-1.57} \simeq \frac{2}{10}. \tag{4.15}$$

The overlap of the next nearest terms, where both x and p labels differ, is

$$e^{-\pi}\, e^{-\pi/2} = (e^{-\pi/2})^3 \simeq \frac{8}{10^3}. \tag{4.16}$$

Even the largest possible overlaps are thus much less than one, and most of the Gaussian overlaps are completely negligible since they are far apart in either position or momentum labels. Most of the terms in the expression for the average of the kinetic energy operator are therefore completely negligible, being proportional to a product of a large number of factors much smaller than one. However, there will always be a multitude of terms of a Hamiltonian overlap multiplied by only one factor of Gaussian overlap. The average of the kinetic energy operator in the antisymmetric N-particle state, Eq. (4.4),

$$\langle \psi^{AS}_{(x,p)_1, s_1^z; \ldots; (x,p)_N, s_N^z} | \hat{H}_{\mathrm{KE}} | \psi^{AS}_{(x,p)_1, s_1^z; \ldots; (x,p)_N, s_N^z} \rangle = \sum_{i=1}^{N} \left(\epsilon(p_i) + \frac{\delta p^2}{2m} \right) + \cdots, \tag{4.17}$$

thus has correction terms which, as shown in Appendix M, can be expressed as corrections to the sum of single-particle energy terms multiplied by at least two wave function overlaps. Since the overlap terms are smaller than the main term, it is tempting to simply ignore them (as if the Gaussian states are orthogonal independent particle states). In view of the many overlap terms, their smallness and alternating sign suggests this, though this is an approximation not easily justified in the present scheme. However, since interest is only in deviations from equilibrium, this approximation turns out to provide correct results.

For now it suffices to note that the state for which the average of the kinetic energy operator has the smallest value is obtained by filling up Gaussian states of lowest momentum labels until all electrons are accommodated. We shall refer to this state as the Gaussian *ground* state, and though its usability as a reference state at the moment is not so clear, we start by finding the value of the largest momentum label of occupied Gaussian states.

4.3 Gaussian Ground State

The N-body state built from Gaussian states of lowest average kinetic energy, the Gaussian *ground* state, is dictated by the exclusion principle: fill up successively states of lowest average energy until all N fermions are accommodated. We thus fill up the Gaussian states with increasing momentum labels in the lattice depicted in Figure 4.1, recalling that the average energy is independent of the position labeling of a Gaussian state. Each quantized phase space area accommodates a spin-up and spin-down electron. Gaussian states with momentum labels $\pm p$ have the same average energy, reflecting the equal energy of an electron moving in either direction of the line. The two rows of boxes of different spatial positions above and below the phase space momentum value $p = 0$ contain the states with lowest momentum values. These two rows are filled first, and then the higher pair of rows of states of equal momentum magnitude are successively filled until the huge number of electrons is accommodated. Let us denote the maximum momentum label reached by p_0. The area of the square in phase space of filled Gaussian lattice states is thus $2p_0L$ as depicted in Figure 4.1. The largest momentum value, p_0, of an occupied Gaussian lattice state is determined by counting. The number of boxes in the x direction is $L/\Delta x_c$, and the number of occupied boxes in the p direction is by definition of p_0 equal to $2p_0/\Delta p_c$. The number of occupied phase space boxes corresponding to the state of lowest average value of the kinetic energy operator is therefore $(L/\Delta x_c)(2p_0/\Delta p_c)$. Because of the spin degeneracy, the number of occupied boxes equals half the number of particles, i.e.

$$\frac{N}{2} = \frac{L}{\Delta x_c}\frac{2p_0}{\Delta p_c} = \frac{2p_0L}{2\pi\hbar}. \tag{4.18}$$

The uppermost occupied Gaussian wave packet states thus have momentum value of magnitude $p_0 = \pi\hbar n/2$, where $n = N/L$, i.e. the electron density. The value of p_0 is identical to the Fermi momentum for the considered one-dimensional case, i.e. $p_0 = p_F$ (recall Exercise 3.3 on page 46).

Another way of specifying the Gaussian ground state is to view the filling process vertically in Figure 4.1, noting that, for each interval in real space, Δx_c, Gaussian states with momentum magnitudes $|\pm p_i|$ up to the Fermi momentum are filled. Each spatial interval, Δx_c, thus has associated with it a local Fermi sea (or column) of Gaussian states where the momentum spacing between the Gaussian states is Δp_c. The number of Gaussian states, or corresponding electrons, associated with interval Δx_c is

$$N_{\Delta x_c} = 2\,\frac{2p_0}{\Delta p_c} = 2\,\frac{\pi\hbar n}{\Delta p_c}, \tag{4.19}$$

the first factor of 2 being due to the spin degeneracy, and the corresponding density is

$$n_{\Delta x_c} = \frac{N_{\Delta x_c}}{\Delta x_c} = \frac{2\pi\hbar n}{\Delta x_c\,\Delta p_c} = n, \tag{4.20}$$

i.e. the density of electrons in the local Fermi sea of Gaussian states equals the density, n, of electrons in the system.

Choosing $\Delta x_c = L/\sqrt{N}$ gives $\Delta p_c = 4p_F/\sqrt{N}$, i.e. a p spacing much smaller than the Fermi momentum, $\Delta p_c \ll p_F$. Choosing $\delta x = \Delta x_c/\sqrt{4\pi}$ (and thereby $\delta p = \Delta p_c/\sqrt{4\pi}$), then for $\Delta x_c = L/\sqrt{N} = \lambda_F\sqrt{N}/4$ we have $\delta x = \lambda_F\sqrt{N}/4\sqrt{4\pi}$, i.e. $\delta x \gg \lambda_F$, and for Gaussian states with momentum labeling on the order of the Fermi momentum (or higher), many plane wave oscillations are executed under the Gaussian envelope.

In three spatial dimensions, the filling of Gaussian phase space states for the Gaussian *ground* state involves, in real space, covering the volume Ω of the sample with cells of volume $\Delta\mathbf{x}_c$. For each such real space cell, or corresponding position label $\mathbf{x}_i$, momentum states of increasing average kinetic energy are filled, i.e. momentum lattice states inside a sphere, the radius of which we denote p_0. The number of space boxes is $\Omega/\Delta\mathbf{x}_c$ and for each of these there are $4\pi p_0^3/3\Delta\mathbf{p}_c$ filled momentum boxes. The number of filled phase space boxes should equal half the number of electrons due to the spin degeneracy, i.e.

$$\frac{N}{2} = \frac{\Omega}{\Delta\mathbf{x}_c}\,\frac{\frac{4}{3}\pi p_0^3}{\Delta\mathbf{p}_c} = \frac{\Omega}{(2\pi\hbar)^3}\,\frac{4\pi}{3}\,p_0^3, \tag{4.21}$$

giving

$$p_0 = \hbar(3\pi^2 n)^{1/3}, \tag{4.22}$$

where $n = N/\Omega$ is the electron density, i.e. p_0 is equal to the Fermi momentum, $p_0 = p_F$. The picture that has evolved is thus that each spatial volume of the sample, $\Delta\mathbf{x}_c$, has associated with it a local Fermi sea (or sphere) of Gaussian wave packet states of different lattice momenta, their spacing being Δp_c.

Since (again using # as shorthand for number)

$$\sum_i \theta(\epsilon_F - \epsilon(\mathbf{p}_i)) = \#\text{ of }\mathbf{p}\text{-space cells in the local Fermi sphere}$$

$$= \frac{\frac{4}{3}\pi p_F^3}{\Delta\mathbf{p}_c}, \tag{4.23}$$

the sum over all occupied phase space cells gives

$$\sum_c \frac{\Delta \mathbf{x}_c \, \Delta \mathbf{p}_c}{(2\pi\hbar)^3} \theta(\epsilon_F - \epsilon(\mathbf{p}_c)) = \Omega \frac{\Delta \mathbf{p}_c}{(2\pi\hbar)^3} \sum_i \theta(\epsilon_F - \epsilon(\mathbf{p}_i)) = \Omega \frac{\frac{4}{3}\pi p_F^3}{(2\pi\hbar)^3}, \tag{4.24}$$

as each spatial cell gives the same contribution, and the space cell volumes, $\Delta \mathbf{x}_c$, add up to the system volume Ω. According to the expression for the Fermi momentum, Eq. (4.22), Eq. (4.24) takes the form

$$\sum_c \frac{\Delta \mathbf{x}_c \, \Delta \mathbf{p}_c}{(2\pi\hbar)^3} \theta(\epsilon_F - \epsilon(\mathbf{p}_c)) = \frac{\Omega}{2} n. \tag{4.25}$$

The counting of occupied states in the Gaussian ground state can therefore be rephrased in terms of the density of electrons, n, as the density is equal to the sum over occupied phase space cells according to ($\Delta \mathbf{p}_i \equiv \Delta \mathbf{p}_c$)

$$n = \frac{2}{\Omega} \sum_c \frac{\Delta \mathbf{x}_c \, \Delta \mathbf{p}_c}{(2\pi\hbar)^3} \theta(\epsilon_F - \epsilon(\mathbf{p}_c)) = 2 \sum_i \frac{\Delta \mathbf{p}_i}{(2\pi\hbar)^3} \theta(\epsilon_F - \epsilon(\mathbf{p}_i)). \tag{4.26}$$

In any cell volume $\Delta \mathbf{x}_c$ there is the number, $N_{\Delta \mathbf{x}_c}$, of electrons occupying the local Fermi sea of Gaussian states with different momentum and spin-up and spin-down values,

$$N_{\Delta \mathbf{x}_c} = 2 \, \frac{\frac{4}{3}\pi p_F^3}{\Delta \mathbf{p}_c}. \tag{4.27}$$

The density of Gaussian states, or corresponding electrons, associated with volume $\Delta \mathbf{x}_c$ is thus

$$n_{\Delta \mathbf{x}_c} = \frac{N_{\Delta x_c}}{\Delta \mathbf{x}_c} = \frac{\frac{8}{3}\pi p_F^3}{\Delta \mathbf{x}_c \, \Delta \mathbf{p}_c} = \frac{\frac{8}{3}\pi p_F^3}{(2\pi\hbar)^3} = \frac{\frac{8}{3}\pi\hbar^3(3\pi^2 n)}{(2\pi\hbar)^3} = n, \tag{4.28}$$

i.e. the density of electrons in the local Fermi sea of Gaussian states in any volume, $\Delta \mathbf{x}_c$, equals the density, n, of electrons in the system.

Treating the Gaussian states as independent particle states, i.e. forming the quantity

$$n(\mathbf{x}) = 2 \sum_l |\psi_{\mathbf{x}_l \mathbf{p}_l}(\mathbf{x})|^2, \tag{4.29}$$

where the sum is over the states occupied in the Gaussian ground state, it indeed equals the density of the electron gas. Owing to the localized shape of the Gaussian function, the main contribution from the summation for given $\mathbf{x}$ is from the spatial cells near $\mathbf{x}$. In the bulk of the sample, this summation gives an identical result, and $n(\mathbf{x})$ is the same in each cell, $n(\mathbf{x}) = \bar{n}$, showing a negligible variation within each small cell. In the limit of a large sample volume, $\Omega \gg \Delta \mathbf{x}_c$, a constant bulk density can thus be ascribed to each space cell volume $\Delta \mathbf{x}_c$ given by

$$\begin{aligned} \bar{n} &\equiv \frac{1}{\Delta \mathbf{x}_c} \int_{\Delta \mathbf{x}_c} d\mathbf{x} \; n(\mathbf{x}) = \frac{1}{\Omega} \int_{\Omega} d\mathbf{x} \; n(\mathbf{x}) \\ &= \frac{2}{\Omega} \int_{\Omega} d\mathbf{x} \sum_l |\psi_{\mathbf{x}_l \mathbf{p}_l}(\mathbf{x})|^2 = \frac{2}{\Omega} \sum_l 1, \end{aligned} \tag{4.30}$$

where in the last equality the normalization of the Gaussian functions has been used. The summation equals the number of cells in the Gaussian ground state and

$$\bar{n} = \frac{2}{\Omega} \frac{\Omega}{\Delta \mathbf{x}_c} \frac{\frac{4}{3}\pi p_F^3}{\Delta \mathbf{p}_c} = \frac{\frac{8}{3}\pi p_F^3}{(2\pi\hbar)^3} = \frac{\frac{8}{3}\pi\hbar^3(3\pi^2 n)}{(2\pi\hbar)^3} = n, \tag{4.31}$$

indeed equal to the density of the electron gas.

Exercise 4.1 Show that the Gaussian Fermi momentum equals the plane wave Fermi momentum also for the two-dimensional case.

4.4 Gaussian Density of States

Next we want to obtain, for given spin, the density of Gaussian states (GS) with average energy in a small energy interval $\Delta\epsilon$. Consider first the one-dimensional case. Introduce the number, $N_1(\epsilon, \epsilon+\Delta\epsilon)$, of Gaussian states with average energies between ϵ and $\epsilon+\Delta\epsilon$, i.e.

$$\begin{aligned} N_1(\epsilon, \epsilon+\Delta\epsilon) &\equiv \#\text{ of GS with average energy between } \epsilon \text{ and } \epsilon+\Delta\epsilon \\ &= \#\text{ of GS with momentum labels between } p \text{ and } p+\Delta p \\ &\quad + \#\text{ of GS with momentum labels between } -p \text{ and } -p-\Delta p, \end{aligned} \tag{4.32}$$

where $\epsilon = p^2/2m$ and $\epsilon + \Delta\epsilon = (p+\Delta p)^2/2m$, and therefore $\Delta p = m\Delta\epsilon/p$. Thus

$$N_1(\epsilon, \epsilon+\Delta\epsilon) = 2 \times (\#\text{ of GS with momentum labels between } p \text{ and } p+\Delta p), \tag{4.33}$$

the factor of 2 counting the two momentum values $\pm p$ corresponding to the same energy value, $\epsilon(\pm p) = \epsilon(p) = \epsilon$. For each spatial cell Δx_c, there are $\Delta p/\Delta p_c$ Gaussian states with momentum values between p and $p+\Delta p$, and therefore $2\Delta p/\Delta p_c$ Gaussian states with energies between ϵ and $\epsilon+\Delta\epsilon$. Since there are $L/\Delta x_c$ spatial cells, Eq. (4.33) becomes

$$N_1(\epsilon, \epsilon+\Delta\epsilon) = \frac{2\Delta p}{\Delta p_c} \frac{L}{\Delta x_c}. \tag{4.34}$$

The density of Gaussian states per unit length of the system and per unit energy is thus

$$N_1(\epsilon) = \frac{N_1(\epsilon, \epsilon+\Delta\epsilon)}{\Delta\epsilon\, L} = \frac{2\Delta p}{\Delta\epsilon} \frac{1}{\Delta x_c\, \Delta p_c} = \frac{2\Delta p}{\Delta\epsilon} \frac{1}{2\pi\hbar}. \tag{4.35}$$

Since $\Delta p/\Delta\epsilon = m/p = 1/v_\epsilon$, where $v_\epsilon = \sqrt{2\epsilon/m}$, the density of Gaussian states is thus

$$N_1(\epsilon) = \frac{1}{\pi\hbar v_\epsilon}. \tag{4.36}$$

The density of states obtained by counting Gaussian wave packet states is thus the same as counting exact energy eigenstates, Eq. (3.30), the conclusion also reached when the two Fermi momenta were seen to be equal.

Despite the different ways of counting states, the density of plane wave energy (momentum) states (PW) equals the density of Gaussian states (GS) for the following reason. Consider the energy interval $\Delta\epsilon$ with its corresponding associated doublet momentum interval Δp. Momentum values for plane wave states are macroscopically close, their separation being $\Delta p_L = 2\pi\hbar/L$, and

$$\text{\# of PW with energy in interval } \Delta\epsilon = \frac{2\Delta p}{\Delta p_L} = \frac{L}{\pi\hbar v_\epsilon}\,\Delta\epsilon. \tag{4.37}$$

The Gaussian momentum labels are less dense, their separation being the distance between neighboring momentum cells, $\Delta p_c = 2\pi\hbar/\Delta x_c$, but the Gaussian momentum states are replicated $L/\Delta x_c$ times, corresponding to the different spatial cells, so that

$$\text{\# of GS with average energy in interval } \Delta\epsilon = \frac{2\Delta p}{\Delta p_c}\,\frac{L}{\Delta x_c} = \frac{L}{\pi\hbar v_\epsilon}\,\Delta\epsilon. \tag{4.38}$$

The macroscopic degeneracy in location of Gaussian states thus compensates to give exactly the same density of states as for plane wave states.

Next consider the three-dimensional case. For each spatial cell volume $\Delta\mathbf{x}_c$, momentum labeled Gaussian states with average energies between ϵ and $\epsilon + \Delta\epsilon$ are located between the corresponding spherical energy surfaces in the momentum directions of phase space. The radii of the two close spheres, p and $p + \Delta p$, are specified by the energies according to $\epsilon = p^2/2m$ and $\epsilon + \Delta\epsilon = (p + \Delta p)^2/2m$. The number of Gaussian states between the spheres is the volume between the spheres divided by $\Delta\mathbf{p}_c$. Since there are $\Omega/\Delta\mathbf{x}_c$ spatial cells, the number, $N(\epsilon, \epsilon + \Delta\epsilon)$, of Gaussian lattice states with average energies between ϵ and $\epsilon + \Delta\epsilon$ is

$$\begin{aligned} N(\epsilon, \epsilon + \Delta\epsilon) &\equiv \text{\# of GS with average energy between } \epsilon \text{ and } \epsilon + \Delta\epsilon \\ &= \frac{\Omega}{\Delta\mathbf{x}_c}\,\frac{\frac{4}{3}\pi[(p + \Delta p)^3 - p^3]}{\Delta\mathbf{p}_c} \\ &= \frac{4\pi p^2 \Delta p\,\Omega}{\Delta\mathbf{x}_c\,\Delta\mathbf{p}_c} = \frac{4\pi p^2 \Delta p\,\Omega}{(2\pi\hbar)^3}. \end{aligned} \tag{4.39}$$

The density of Gaussian states per unit volume of the system per unit energy is thus ($\Delta p/\Delta\epsilon = m/p = m/\sqrt{2m\epsilon}$)

$$N(\epsilon) = \frac{N(\epsilon, \epsilon + \Delta\epsilon)}{\Delta\epsilon\,\Omega} = \frac{mp}{2\pi^2\hbar^3} = \frac{m\sqrt{2m\epsilon}}{2\pi^2\hbar^3}, \tag{4.40}$$

i.e. the same density of states as for the plane wave states.

Exercise 4.2 Obtain the Gaussian density of states for the two-dimensional case.

4.5 Equilibrium State

The usefulness of the Gaussian *ground* state is not at this point clear, but it will be shown to serve as reference state. We would like to obtain a description of thermal equilibrium in terms of the Gaussian states. However, from the statistical argument in Appendix K for obtaining the Fermi–Dirac distribution, energy eigenstates are called for, which the Gaussian states are not. To the rescue comes the degeneracy of the electron gas. In the discussion of the tunnel junction in Section 3.3, we realized that, for the physical quantity of interest, the current, only the energy states close to the Fermi surface effectively contribute. Similarly, the room temperature thermal excitation of the electron gas leads to changes in the occupation of energy levels only in a thin layer, of size kT, around the Fermi surface as $kT \ll \epsilon_F$. Most of the states in the Fermi sea are inert to external perturbations due to the exclusion principle blocking transitions to occupied states. This is a general feature, and also the non-equilibrium states encountered in transport phenomena have the feature that the states contributing to a physical quantity can be interpreted as stemming only from energy states close to the Fermi surface.

Consider a Gaussian wave packet labeled by $(\mathbf{x}_i, \mathbf{p}_i)$, where $|\mathbf{p}_i| \sim p_F$. Such a Gaussian state is approximately an energy (and momentum) eigenstate according to Appendix M if the spatially dependent term in Eq. (M.18) is negligible compared to the momentum label dependent term. Since the spatial extension of the Gaussian function is on the order of its variance, δx, this requires $1/\delta x \ll p_i/\hbar$, or equivalently $\delta p \ll p_i$ (in which case the irrelevant constant factor term on the right-hand side of Eq. (M.18) is similarly negligible). Since $\delta p \ll p_F$ (recall Section 4.3), Gaussian states with momentum labels of magnitude the Fermi momentum (or higher), $|\mathbf{p}_i| \sim p_F$, can thus be considered approximate eigenstates of the kinetic energy Hamiltonian,

$$-\frac{\hbar^2}{2m}\nabla_{\mathbf{x}}^2\, \psi_{\mathbf{x}_i,\mathbf{p}_i}(\mathbf{x}) \simeq \epsilon(\mathbf{p}_i)\, \psi_{\mathbf{x}_i,\mathbf{p}_i}(\mathbf{x}). \tag{4.41}$$

In each cell volume $\Delta\mathbf{x}_c$, the Gaussian states near the Fermi surface can thus be considered energy eigenstates. For such an approximate energy eigenstate, the criterion $\delta p \ll p_F$ converts in real space into the criterion $\lambda_F \ll \delta x$. The Gaussian function thus executes many oscillations under the envelope. Inside most of the space volume $\Delta\mathbf{x}_c$, the Gaussian state approximates a plane wave corresponding to the momentum label in question. The density contribution from the plane waves in an energy interval near the Fermi energy is thus equal to the contribution of the Gaussian wave packets in the different cells, as discussed in Section 4.3.

The energy weight counting argument in Appendix K leading to the Fermi–Dirac distribution is governed by the energy eigenvalues of the Hamiltonian. However, for energy intervals near the Fermi surface, the counting is indifferent to using the energies associated with the plane wave exact energy states or the Gaussian approximate energy eigenstates, since their energy values and density of states are the same. A plane wave is substituted by chopped-up plane waves, the Gaussian wave packets. When using the Gaussian states, the counting is done for the states in each cell volume $\Delta\mathbf{x}_c$. In thermal equilibrium, the counting argument in Appendix K thus associates the Gaussian state labeled by $(\mathbf{x},\mathbf{p})$, $|\mathbf{p}| \sim p_{\mathrm{F}}$, with the probability of being occupied, as specified by the Fermi function

$$f(\mathbf{x},\mathbf{p}) = \frac{1}{e^{(\epsilon(\mathbf{p})-\epsilon_{\mathrm{F}})/kT}+1} = f(\epsilon(\mathbf{p})), \tag{4.42}$$

the thermal Fermi distribution for Gaussian states.

Summing the distribution function in Eq. (4.42) over all phase space cells gives

$$\frac{2}{\Omega}\sum_c f(\mathbf{x}_c,\mathbf{p}_c) = \frac{2}{\Omega}\sum_c \frac{\Delta\mathbf{x}_c\,\Delta\mathbf{p}_c}{(2\pi\hbar)^3}\,\frac{1}{e^{(\epsilon(\mathbf{p}_c)-\epsilon_{\mathrm{F}})/kT}+1} = 2\int_0^\infty d\epsilon\; N(\epsilon)f(\epsilon) = n, \tag{4.43}$$

where the phase space cell summation according to Eq. (M.37) can be turned into the energy integration, with $N(\epsilon)$ being the density of states, and therefore according to Eq. (3.70) gives the density. Each spatial cell, $\Delta\mathbf{x}_c$, gives the same contribution, and since the sum of space cell volumes, $\Delta\mathbf{x}_c$, adds up to the system volume Ω, Eq. (4.43) can be rewritten as

$$\frac{2}{\Omega}\sum_c f(\mathbf{x}_c,\mathbf{p}_c) = 2\sum_{\mathbf{p}_i} \frac{\Delta\mathbf{p}_c}{(2\pi\hbar)^3}\,\frac{1}{e^{(\epsilon(\mathbf{p}_i)-\epsilon_{\mathrm{F}})/kT}+1}. \tag{4.44}$$

We would like to turn the summation into an integral; thus we write the formula for the density as

$$n = 2\int \frac{d\mathbf{p}}{(2\pi\hbar)^3}\,\frac{1}{e^{(\epsilon(\mathbf{p})-\epsilon_{\mathrm{F}})/kT}+1}, \tag{4.45}$$

even when interpreted in terms of occupied Gaussian states. However, such a conversion is only valid for momentum values much larger than the linear size, Δp_c, of the momentum volume $\Delta\mathbf{p}_c$, $|\mathbf{p}| \gg \Delta p_c$, such that $\Delta\mathbf{p}_c$ can be considered infinitesimal. However, we can safely write, both integrals producing the density (recall Eq. (4.26)),

$$0 = 2\int \frac{d\mathbf{p}}{(2\pi\hbar)^3}\left(\frac{1}{e^{(\epsilon(\mathbf{p})-\epsilon_{\mathrm{F}})/kT}+1} - \theta(\epsilon_{\mathrm{F}}-\epsilon(\mathbf{p}))\right) \tag{4.46}$$

and interpret the distribution functions as denoting occupation of Gaussian wave packet states in a given spatial cell, since the integrand is only non-zero in the thin thermal layer around the Fermi surface. Equation (4.45) is retained but with the proviso that a subtraction procedure must be employed in order that the interpretation is only applied to Gaussian states near the Fermi surface. Such a scenario always appears naturally, an example of which is discussed in the next section.

If the particles experience an applied potential, $V(\mathbf{x})$, which varies slowly on the scale of the position variance of the Gaussian state, the Gaussian wave packet satisfies with high accuracy

$$V(\mathbf{x})\,\psi_{\mathbf{x}_i,\mathbf{p}_i}(\mathbf{x}) \simeq V(\mathbf{x}_i)\,\psi_{\mathbf{x}_i,\mathbf{p}_i}(\mathbf{x}), \tag{4.47}$$

i.e. the Gaussian state, $\psi_{\mathbf{x}_i,\mathbf{p}_i}$ ($|\mathbf{p}_i| \sim p_F$), is an approximate solution of the time-independent Schrödinger equation for the particle in the potential corresponding to the energy eigenvalue $E(\mathbf{x}_i, \mathbf{p}_i) = \epsilon(\mathbf{p}_i) + V(\mathbf{x}_i)$ as

$$\left(-\frac{\hbar^2}{2m}\nabla_{\mathbf{x}}^2 + V(\mathbf{x})\right)\psi_{\mathbf{x}_i,\mathbf{p}_i}(\mathbf{x}) \simeq \left(\epsilon(\mathbf{p}_i) + V(\mathbf{x}_i)\right)\psi_{\mathbf{x}_i,\mathbf{p}_i}(\mathbf{x}). \tag{4.48}$$

The counting argument in Appendix K then gives that the probability that the Gaussian state labeled by $(\mathbf{x}_i, \mathbf{p}_i)$ is occupied is

$$f(\mathbf{x}_i, \mathbf{p}_i) = \frac{1}{e^{(\epsilon(\mathbf{p}_i)+V(\mathbf{x}_i)-\epsilon_F)/kT} + 1}, \tag{4.49}$$

the occupation of $\psi_{\mathbf{x}_i,\mathbf{p}_i}(\mathbf{x})$ as given by the local Fermi function.

Treating the Gaussian states as independent particle states (recall Eq. (4.29)) thermally populated according to the distribution function (4.49), the sum

$$n(\mathbf{x}) \equiv 2\sum_{\mathbf{x}_i,\mathbf{p}_i} |\psi_{\mathbf{x}_i\mathbf{p}_i}(\mathbf{x})|^2 f(\mathbf{x}_i, \mathbf{p}_i) \tag{4.50}$$

gives the density of the electron gas. Owing to the localized shape of the Gaussian function, the main contribution to the $\mathbf{x}_i$ summation for given $\mathbf{x}$ is from the spatial cells near $\mathbf{x}$. Since the potential is slowly varying on the scale of the cell size, these contributions have the same thermal weight,

$$n(\mathbf{x}) = 2\sum_{\mathbf{p}_i} f(\mathbf{x}, \mathbf{p}_i) \sum_{\mathbf{x}_i} |\psi_{\mathbf{x}_i\mathbf{p}_i}(\mathbf{x})|^2. \tag{4.51}$$

In the bulk of the sample, the spatial cell summation gives an identical result and

$$\sum_{\mathbf{x}_i} |\psi_{\mathbf{x}_i\mathbf{p}_i}(\mathbf{x})|^2 = \frac{1}{\Omega}\int_{\Omega} d\mathbf{x} \sum_{\mathbf{x}_i} |\psi_{\mathbf{x}_i\mathbf{p}_i}(\mathbf{x})|^2 = \frac{1}{\Omega}\sum_{\mathbf{x}_i} 1 = \frac{1}{\Omega}\frac{\Omega}{\Delta\mathbf{x}_c}, \tag{4.52}$$

and the expression (4.50) reduces to

$$n(\mathbf{x}) = 2\sum_{\mathbf{p}_i} \frac{\Delta\mathbf{p}_i}{(2\pi\hbar)^3} f(\mathbf{x}, \mathbf{p}_i), \tag{4.53}$$

the density specified by the population of Gaussian states.

The thermal equilibrium density thus becomes in the presence of a potential

$$n(\mathbf{x}) = 2\int \frac{d\mathbf{p}}{(2\pi\hbar)^3} \frac{1}{e^{(\epsilon(\mathbf{p})+V(\mathbf{x})-\epsilon_F)/kT} + 1}, \tag{4.54}$$

of course, again with the proviso that the formula by a subtraction procedure is only interpreted in terms of the Gaussian states that are approximate energy states (approximating for this part of the energy spectrum the contribution from the exact energy eigenstates). That such a scenario naturally presents itself we turn to next.

4.6 Impurity Atoms and Screening

A metal, like any large object, always has imperfections, for example foreign atoms substituting for atoms of the metal, referred to as impurities. To be specific, consider a magnesium atom, Mg, substituted for a copper atom in an otherwise pure copper crystal. Magnesium is a divalent atom, two valence electrons, whereas copper, Cu, is monovalent, one valence electron. The substitution of the impurity atom thus results in an extra ionic charge e ($e \equiv |e|$) at the impurity site and one extra conduction electron. In a metal, conduction electrons are free to move and electrons will be attracted to the positive impurity point charge, resulting in a non-uniform average electronic charge distribution, $-en(\mathbf{r}) = -en_e - e\delta n(\mathbf{r})$, where n_e is the constant electron density in the Sommerfeld model of pure copper. Describing the metal and impurity in the Sommerfeld model, the ionic charge density is $en_i + e\delta(\mathbf{r})$, where the last term is the extra impurity ionic point charge, as the location of the impurity is chosen as the origin of the coordinate system. The charge neutrality in the Sommerfeld model of the pure copper crystal reads $-en_e + en_i = 0$, since copper is monovalent.

4.6.1 Mean Field Approximation

Since electric fields in a metal are small due to the many free conduction electrons, charge fluctuations are strongly suppressed and a mean field description is adopted, i.e. an electron is ascribed an average potential energy representing the Coulomb interaction with the myriad of other conduction electrons. The mean field potential energy of an electron at position $\mathbf{r}$ is $V(\mathbf{r}) = -e\varphi(\mathbf{r})$, where $\varphi(\mathbf{r})$ is the electrostatic potential due to both the Coulomb field from the fixed impurity point charge and the potential created by the adjusted electronic charge distribution and fixed ionic charge. The mean field is an average treatment of the mutual Coulomb interaction between the electrons, accounting for the many-body interaction implicitly through its influence on the mean density of the electrons. In the present case, the mean field is not translation-invariant as in the Sommerfeld model, but now determined by the inhomogeneity created by the presence of the impurity point charge. In terms of the electrostatic potential, Gauss's law, $\nabla \cdot \mathbf{E} = \rho/\epsilon_0$, turns into Poisson's equation for the charge distribution in question,

$$-\Delta\varphi(\mathbf{r}) = \frac{-e}{\epsilon_0}(n_e(\mathbf{r}) - n_i - \delta(\mathbf{r})). \tag{4.55}$$

Employing the charge neutrality condition, $-en_e + en_i = 0$, Eq. (4.55) becomes

$$-\Delta_{\mathbf{r}}\,\varphi(\mathbf{r}) = \frac{-e}{\epsilon_0}(\delta n(\mathbf{r}) - \delta(\mathbf{r})), \tag{4.56}$$

where $\delta n(\mathbf{r}) \equiv n_e(\mathbf{r}) - n_e$ is the deviation from the constant electron density of the Sommerfeld model.

Assuming the mean field is slowly varying, the quasi-classical description is valid where an electron is ascribed a Gaussian wave packet specified by its average position and momentum, $(\mathbf{r}, \mathbf{p})$. The corresponding energy for an electron in that state is

$E(\mathbf{r},\mathbf{p}) = \epsilon(\mathbf{p}) - e\varphi(\mathbf{r})$, where $\epsilon(\mathbf{p})$ is the kinetic energy of the electron, which in the Sommerfeld model has quadratic dispersion, $\epsilon(\mathbf{p}) = \mathbf{p}^2/2m$.[1]

The metal is in thermal equilibrium and the distribution of the electrons in the Gaussian states labeled by $(\mathbf{r},\mathbf{p})$ is governed by the exclusion principle as specified by the local Fermi function

$$f_\varphi(\mathbf{r},\mathbf{p}) = \frac{1}{e^{(\epsilon(\mathbf{p})-e\varphi(\mathbf{r})-\epsilon_F)/kT}+1}. \tag{4.57}$$

The change in electron density due to the introduction of the impurity atom is, according to Eq. (4.45),

$$\delta n(\mathbf{r}) = n_e(\mathbf{r}) - n_e = 2\int \frac{d\mathbf{p}}{(2\pi\hbar)^3}\,(f_\varphi(\mathbf{r},\mathbf{p}) - f_0(\mathbf{p})), \tag{4.58}$$

where $f_0(\mathbf{p})$ is the Fermi function for Gaussian states in the absence of the impurity atom, $f_0 = f_{\varphi=0}$. The potential energy is small compared to the Fermi energy, $|e\varphi(\mathbf{r})| \ll \epsilon_F$, since electric fields in a metal are weak due to the many free electrons. The contribution to the momentum integral is then only from the region near the Fermi surface, and Eqs. (4.57) and (4.45) are thus only used for momentum values near the Fermi surface. A subtraction term has occurred naturally. The local density (deviation) can thus be thought of as stemming from a local Fermi sea of electrons.

Taylor-expanding the distribution function, f_φ, in terms of the mean field potential gives

$$\delta n(\mathbf{r}) = -2e\varphi(\mathbf{r})\int \frac{d\mathbf{p}}{(2\pi\hbar)^3}\frac{\partial f(\epsilon(\mathbf{p}))}{\partial\epsilon(\mathbf{p})} = -2e\varphi(\mathbf{r})\int_0^\infty d\epsilon\; N(\epsilon)\,\frac{\partial f(\epsilon)}{\partial\epsilon}. \tag{4.59}$$

The integral is immediately evaluated using the properties of the Fermi function discussed in the previous chapter[2]

$$\delta n(\mathbf{r}) = 2eN_0\varphi(\mathbf{r}) \tag{4.60}$$

where N_0 denotes the density of states at the Fermi energy, $N_0 \equiv N(\epsilon_F) = \sqrt{2m^3\epsilon_F}/2\pi^2\hbar^3$, the explicit expression for the density of states referring to the free particle quadratic dispersion.

Inserting the expression (4.60) for the electron density into the Poisson equation (4.56) gives the equation determining the mean field impurity potential,

$$\Delta\varphi(\mathbf{r}) = \frac{e}{\epsilon_0}(2eN_0\,\varphi(\mathbf{r}) - \delta(\mathbf{r})). \tag{4.61}$$

[1] However, as we shall soon see, the result of the calculation is independent of the dispersion relation, as only the property of the derivative of the Fermi function being peaked at the Fermi surface is important. The result is thus also valid for Bloch electrons.

[2] Equation (4.60) contains the general message for a conductor: if an electric field, $\mathbf{E}(\mathbf{r}) = -\nabla\varphi(\mathbf{r})$, is present in a conductor, an inhomogeneous charge distribution is present, their relationship being obtained by taking the gradient of Eq. (4.60),

$$\nabla n(\mathbf{r}) \simeq -2eN_0\,\mathbf{E}(\mathbf{r}).$$

Conversely, an unwanted charge distribution can impair the penetration of an electric field, as field lines end up on charges.

4.6.2 Screened Impurity Potential

Fourier analysis is a general way to solve differential equations, as already encountered in Section 1.2. For the problem at hand, a needed inversion of the Fourier transformation invokes the residue theorem, and, for the unfamiliar reader, Eq. (4.61) is also solved by elementary means in Appendix O, where the problem is simplified by invoking the symmetry of the problem.

Introducing the Fourier transform

$$\varphi(\mathbf{r}) = \int \frac{d\mathbf{k}}{(2\pi)^3}\, e^{i\mathbf{k}\cdot\mathbf{r}} \varphi(\mathbf{k}) \tag{4.62}$$

and recalling the Fourier representation of the delta function, Eq. (B.25), Fourier-transforming Eq. (4.61) gives the equation for the Fourier transform,

$$-\mathbf{k}^2 \varphi(\mathbf{k}) = \frac{e}{\epsilon_0}(2eN_0\varphi(\mathbf{k}) - 1), \tag{4.63}$$

with the solution

$$\varphi(\mathbf{k}) = \frac{e/\epsilon_0}{k^2 + \kappa_s^2}, \tag{4.64}$$

where $\kappa_s^2 = 2e^2N_0/\epsilon_0$. The solution of the differential equation (4.61) has thus been transformed to performing the Fourier transform[3]

$$\varphi(\mathbf{r}) = \int \frac{d\mathbf{k}}{(2\pi)^3}\, e^{i\mathbf{k}\cdot\mathbf{r}} \frac{e/\epsilon_0}{k^2 + \kappa_s^2}. \tag{4.65}$$

Changing to spherical coordinates and choosing the polar $\mathbf{k}$ axis parallel to $\mathbf{r}$ gives

$$\varphi(\mathbf{r}) = \int_0^\infty \frac{dk\, k^2}{(2\pi)^3} \frac{e/\epsilon_0}{k^2 + \kappa_s^2} \int_0^{2\pi} d\phi \int_0^{\pi} (-d\cos\theta)\, e^{ikr\cos\theta}, \tag{4.66}$$

where the azimuthal integration is trivial and the polar angle integration is made simple by a substitution of variable, $u \equiv \cos\theta$,

$$\varphi(\mathbf{r}) = 2\pi \int_0^\infty \frac{dk\, k^2}{(2\pi)^3} \frac{e/\epsilon_0}{k^2 + \kappa_s^2} \int_{-1}^{1} du\, e^{ikru}. \tag{4.67}$$

Performing the angular integrations thus gives

$$\varphi(\mathbf{r}) = \frac{e}{\epsilon_0 (2\pi)^2 ir} \int_0^\infty dk\, \frac{k}{k^2 + \kappa_s^2}(e^{ikr} - e^{-ikr}). \tag{4.68}$$

[3] If one checks that the Fourier transform of e^{-cr}/r is $4\pi/(k^2 + c^2)$, an elementary integral to perform, the following evaluation is superfluous by Fourier's inversion theorem.

Shifting the integration variable in the second term, $k \to -k$, gives

$$\varphi(\mathbf{r}) = \frac{e}{\epsilon_0 (2\pi)^2 ir} \int_{-\infty}^{\infty} dk \, \frac{k}{k^2 + \kappa_s^2} e^{ikr}. \tag{4.69}$$

The integral can be evaluated by the method of residues, and the shifting of the contour should, by the convergence of the integrand, be into the upper complex k plane. By the partial fraction decomposition

$$\frac{k}{k^2 + \kappa_s^2} = \frac{1}{2}\left(\frac{1}{k + i\kappa_s} + \frac{1}{k - i\kappa_s}\right), \tag{4.70}$$

the function is seen to have a single simple pole in the upper half-plane at $k = i\kappa_s$, and the k integral gives $2\pi i \exp(-\kappa_s r)/2$. Thus the impurity mean field potential becomes[4]

$$\varphi(\mathbf{r}) = \frac{e}{4\pi\epsilon_0} \frac{e^{-|\mathbf{r}|/\lambda_{TF}}}{|\mathbf{r}|}, \tag{4.71}$$

where the Thomas–Fermi screening length

$$\lambda_{TF} \equiv \kappa_s^{-1} = (\epsilon_0 / 2N_0 e^2)^{1/2} = \sqrt{\frac{\epsilon_0}{me^2}}\, \pi\hbar (3\pi^2 n)^{-1/6} = \sqrt{\frac{\lambda_F a_0}{8}} \tag{4.72}$$

can be expressed in terms of the electron density, n, or in terms of the Fermi wave length and the Bohr radius, $a_0 = (4\pi\epsilon_0\hbar^2)/me^2$. The bare Coulomb potential of the impurity charge is, according to Eq. (4.71), screened. An electron further away from the impurity than the screening length does not sense the presence of the impurity.

To estimate the screening length, we recall that the ions in a metal are mainly bound by the conduction electrons, the core electrons producing much smaller binding. The energy of a conduction electron, being bound in the metal, is negative, $E_{kin} + E_{pot} < 0$, the average kinetic energy being on the order of the Fermi energy, $E_{kin} \sim p_F^2/2m$. The average potential energy of a conduction electron attracted to an ion, $e^2/4\pi\epsilon_0\langle r\rangle$, cannot be much smaller than the average kinetic energy, since that would be in the direction of favoring the crystal to dissociate, i.e. form separated charge neutral atoms. We therefore have the estimate for the strength of Coulomb attraction by the ions

$$\frac{e^2}{4\pi\epsilon_0 \langle r\rangle} \sim \frac{p_F^2}{2m}, \tag{4.73}$$

where $\langle r\rangle$ is the average distance between electrons and ions, or equivalently the average distance between the conduction electrons, i.e. $\langle r\rangle = 1/n^{1/3}$. The estimate can be expressed in terms of the Bohr radius according to

$$k_F^2 \sim \frac{n^{1/3}}{a_0}. \tag{4.74}$$

[4] All coefficients in Eq. (4.63) are rotation-invariant in **k**-space, and thus lead to $\varphi(\mathbf{k}) = \varphi(|\mathbf{k}|)$, being rotation-invariant, and consequently also the potential in real space is rotation-invariant, $\varphi(\mathbf{r}) = \varphi(|\mathbf{r}|)$. This can also be deduced directly from Eq. (4.61), the Laplacian and the delta function being rotation-invariant. Once $\varphi(\mathbf{r}) = \varphi(|\mathbf{r}|)$ is realized, as a consequence of the isotropy of the model, the potential is obtainable by elementary means, as done in Appendix O.

In the Sommerfeld model, the Fermi wave vector is $k_F = (3\pi^2 n)^{1/3}$, i.e. Eq. (4.74) becomes $k_F a_0 \sim 1$, or the statement $n^{-1/3} \sim a_0$, that the distance between the ions is roughly on the order of the size of an atom. Noting that the Thomas–Fermi screening length can be expressed as $\lambda_{TF} = (\pi a_0/4k_F)^{1/2}$, we obtain $\lambda_{TF} \sim a_0$, i.e. the screening length is also on the order of the interatomic distance. The impurity is screened at atomic distance, i.e. screening in a metal is *perfect*.

The phenomenon of screening provides the reason why the effects of electron properties due to Coulomb interaction are suppressed. Although the bare average Coulomb energy between electrons, separated by a distance on the order of the Bohr radius, is $e^2/4\pi\epsilon_0 a_0 \simeq 30\,\text{eV}$, and thus much larger than the Fermi energy, the characteristic kinetic energy of an electron, the strong Coulomb force is screened and becomes effectively weak. Instead of infinite range, the Coulomb force is cut off at atomic distance, $\lambda_{TF} \sim a_0$. The Sommerfeld model neglecting Coulomb interaction, dealing with independent particles, thereby becomes a meaningful approximation. There are small corrections due to Coulomb interaction, such as a small correction of the electron mass. These tiny effects are described by Landau's phenomenological Fermi liquid theory, but are of no consequence for electronics.

The excess electron density profile, being proportional to the screened Coulomb potential, Eq. (4.60), is healing on the same length as the potential,

$$\delta n(\mathbf{r}) = \frac{\kappa_s^2}{4\pi} \frac{e^{-\kappa_s|\mathbf{r}|}}{|\mathbf{r}|} = \frac{e^2 N_0}{2\pi\epsilon_0} \frac{e^{-r/\lambda_{TF}}}{r}, \tag{4.75}$$

and the electronic density is healed to the undisturbed density at atomic distances. The density of the electron gas adjusts such that most of the electrons do not sense the impurity charge; the general feature of screening is that an external charge, or any charge fluctuation in the electron gas for that matter, is suppressed beyond the screening length due to the mobile charges in the conductor. Screening allows charge inhomogeneities of almost only atomic size extension.

Exercise 4.3 Show that the expression for the density profile can be rewritten as

$$\delta n(\mathbf{r}) = n_e \frac{3\pi}{4(k_F\lambda_{TF})^2} \frac{e^{-|\mathbf{r}|/\lambda_{TF}}}{k_F r} \tag{4.76}$$

and thereby

$$n_e(\mathbf{r}) = n_e \left(1 + \frac{3\pi}{4(k_F\lambda_{TF})^2} \frac{e^{-|\mathbf{r}|/\lambda_{TF}}}{k_F r}\right). \tag{4.77}$$

Integrate the density over the volume of the system.

Electrons accumulate in the region where the potential energy

$$V(\mathbf{r}) = -e\varphi(\mathbf{r}) = -\frac{e^2}{4\pi\epsilon_0} \frac{e^{-|\mathbf{r}|/\lambda_{TF}}}{|\mathbf{r}|} \tag{4.78}$$

is low, i.e. near the positive impurity charge according to $\delta n(\mathbf{r}) = -2N_0V(\mathbf{r})$, expressing the fact that, at positions of high potential energy, the electron density is depleted, and vice versa.

When $r = \lambda_F^3/16\pi^2\lambda_{TF}^2 = \lambda_F^2/2\pi^2 a_0$, the potential energy equals the Fermi energy, $|e\varphi(\mathbf{r})| = \epsilon_F$, and at distances shorter than this the Thomas–Fermi theory could fail quantitatively. The distance r for which $|e\varphi(\mathbf{r})| \sim \epsilon_F$ is of atomic size, and the calculation is thus consistent throughout the bulk of the metal.

Generally, even a purified metal contains a huge number of impurities, and we have arrived at the conclusion that the electrons in a disordered metal are moving freely through space in Gaussian wave packets in between being elastically scattered into other momentum states by the atomic short-range potentials due to impurity atoms. The resulting electron dynamics is discussed in Section 4.8.

4.6.3 Electrochemical Equilibration

The Thomas–Fermi relation, Eq. (4.60), between electron density and potential energy has a thermodynamic interpretation, *viz.* as equilibration of the electrochemical potential. The electrochemical potential, μ_{ec}, is the energy required to add an electron to the system. In the absence of the impurity charge, the electrochemical potential is simply the chemical potential, i.e. the Fermi energy of the electron gas. In response to the potential (or force) due to the impurity charge, the electrons will move until equilibrium is reached where the electrochemical potential is constant throughout the system, as otherwise electrons could still be pushed to places of lower energy.[5] The constant value attained is equal to the chemical potential far away from the local disturbance since the system there is unchanged. The Thomas–Fermi theory is the mean field approximation where the electrochemical potential is the sum of the local chemical potential, $\mu(\mathbf{r})$, and the work done by the mean field in the process of bringing an electron to the position in question, the potential energy $V(\mathbf{r})$ or $-e\varphi(\mathbf{r})$. Electrochemical equilibration thus gives for the electrochemical potential

$$\mu_{ec} = \mu(\mathbf{r}) - e\varphi(\mathbf{r}) = \mu(|\mathbf{r}| = \infty). \tag{4.79}$$

In the electron gas, the chemical potential is the Fermi energy, and Eq. (4.79) becomes

$$\epsilon_F(n + \delta n(\mathbf{r})) - e\varphi(\mathbf{r}) = \epsilon_F(n). \tag{4.80}$$

The Thomas–Fermi model assumes small deviations from the equilibrium electron density and a Taylor expansion gives

$$\epsilon_F(n + \delta n(\mathbf{r})) \simeq \epsilon_F(n) + \epsilon_F'(n)\,\delta n(\mathbf{r}), \tag{4.81}$$

and since the Fermi energy is specified by the density according to Eq. (3.13)

$$\epsilon_F(n) \propto p_F^2 \propto n^{2/3}, \tag{4.82}$$

[5] The current vanishes in thermal equilibrium as witnessed by the spherical symmetry of the distribution function, Eq. (4.57). However, this is due to two canceling electron currents. The diffusion current due to the electron density gradient $\mathbf{s}_D = -D_0\nabla n(\mathbf{r}) = 2eN_0D_0\mathbf{E}(\mathbf{r})$, where the last equality follows from Eq. (4.60), is canceled by the current driven by the electric field, $\mathbf{s}_E = \sigma_0\mathbf{E}/-e$, since conductivity and diffusion constant for the electron gas are related according to $\sigma_0 = 2e^2N_0D_0$.

the density derivative gives

$$\epsilon'_{\mathrm{F}}(n) = \frac{d\epsilon_{\mathrm{F}}(n)}{dn} = \frac{2}{3}\frac{\epsilon_{\mathrm{F}}}{n} = \frac{1}{2N_0(\epsilon_{\mathrm{F}})}, \tag{4.83}$$

the last equality following from Eq. (3.29), whereupon the relationship (4.60) follows.

4.7 Wave Packet Dynamics

The free motion of a particle in a Gaussian wave packet state can be evaluated exactly as discussed in Exercise 1.3 on page 6 and in Appendix N. Here, for generality, the dynamics of less specified wave packets is discussed first. The general solution of the free particle Schrödinger equation is a superposition, $\epsilon(\mathbf{p}) = \mathbf{p}^2/2m$,

$$\psi(\mathbf{x}, t) = \int d\mathbf{p}\; g(\mathbf{p})\, e^{(i/\hbar)(\mathbf{p}\cdot\mathbf{x} - \epsilon(\mathbf{p})t)} . \tag{4.84}$$

Choosing the weight function, g, peaked around the momentum value $\mathbf{p}_0$, the energy $\epsilon(\mathbf{p})$ can be Taylor-expanded around that value, and only the lowest-order term needs to be kept,

$$\begin{aligned}\psi(\mathbf{x}, t) &\simeq e^{(i/\hbar)(\mathbf{p}_0\cdot\mathbf{x} - \epsilon(\mathbf{p}_0)t)} \int \frac{d\mathbf{p}}{(2\pi\hbar)^3}\, g(\mathbf{p})\, e^{(i/\hbar)(\mathbf{p}-\mathbf{p}_0)\cdot(\mathbf{x} - \nabla_{\mathbf{p}}\epsilon(\mathbf{p}_0)t)} \\ &= e^{(i/\hbar)(\mathbf{p}_0\cdot\mathbf{x} - \epsilon(\mathbf{p}_0)t)} F(\mathbf{x} - t\nabla_{\mathbf{p}}\epsilon(\mathbf{p}_0)).\end{aligned} \tag{4.85}$$

The constructed wave packet has, as already encountered in Chapter 2, an envelope, F, which moves rigidly along with the group velocity (the velocity with which the group of plane waves with slightly different wave vectors moves)[6]

$$\mathbf{v}(\mathbf{p}) = \nabla_{\mathbf{p}}\epsilon(\mathbf{p}) = \frac{\partial \epsilon(\mathbf{p})}{\partial \mathbf{p}}, \tag{4.86}$$

which equals the velocity of a free particle of momentum $\mathbf{p}$ according to classical mechanics.

To have explicit expressions, the Gaussian weight is chosen,

$$g_{\mathbf{x}_0\mathbf{p}_0}(\mathbf{p}) = N \exp\left(-\frac{(\mathbf{p} - \mathbf{p}_0)^2}{4\delta p^2} - \frac{i}{\hbar}(\mathbf{p} - \mathbf{p}_0)\cdot \mathbf{x}_0 \right), \tag{4.87}$$

[6] The group velocity can be contrasted with the velocity with which a plane wave front, say peak or trough, corresponding to wave vector $\mathbf{k} = \mathbf{p}/\hbar$, moves, the so-called phase velocity, which has the magnitude $v_{\mathrm{ph}} = \epsilon/p = v_{\mathrm{g}}/2$ (for the quadratic dispersion in question) directed along $\mathbf{p}$. Riding along the wave packet with the group velocity, the wave fronts move relative to the envelope (opposite to the direction of the group velocity $\mathbf{v}_{\mathrm{g}}$) due to the difference between the group and phase velocities. The group and phase velocities are only identical if the waves are dispersion-free, i.e. if the energy depends linearly on the wave vector, as is the case for electromagnetic waves. Owing to the linear dispersion of sound waves, we can communicate by speech at classroom distances.

corresponding to the state in Eq. (4.84) at time $t = 0$ being the Gaussian state centered around $\mathbf{x}_0$ and $\mathbf{p}_0$, Eq. (4.2). The normalized envelope function in Eq. (4.85) then becomes

$$F(\mathbf{x} - t\nabla_{\mathbf{p}}\epsilon(\mathbf{p}_0)) = \left(\frac{1}{2\pi\delta x^2}\right)^{3/4} \exp\left(-\frac{(\mathbf{x} - \mathbf{x}_0 - t\nabla_{\mathbf{p}}\epsilon(\mathbf{p}_0))^2}{4\delta x^2}\right). \tag{4.88}$$

The average position value is

$$\langle \mathbf{x}(t) \rangle = \int d\mathbf{x}\, \mathbf{x}\, |\psi(\mathbf{x}, t)|^2 = \mathbf{x}_0 + \mathbf{v}(\mathbf{p}_0)\, t, \tag{4.89}$$

where $\mathbf{x}_0 \equiv \langle \mathbf{x}(t = 0) \rangle$. For a free particle, the center of the Gaussian wave packet moves with the velocity $\mathbf{v}(\mathbf{p}_0)$ at all times but the envelope function is broadened in time (as noted in Appendix N), a feature left out in the above short-time approximation. The wave packet description is the closest quantum mechanics gets to classical mechanic's trajectory motion; the particle is attributed simultaneous position and momentum with accuracy set by Heisenberg's uncertainty relation.

Since the envelope equation (4.88) is real, the probability current corresponding to wave function (4.85) is

$$\begin{aligned} S_{\mathbf{x}_0\mathbf{p}_0}(\mathbf{x}, t) &\simeq \frac{\mathbf{p}_0}{m} F^2(\mathbf{x} - t\nabla_{\mathbf{p}}\epsilon(\mathbf{p}_0)) \\ &= \frac{\partial \epsilon(\mathbf{p}_0)}{\partial \mathbf{p}} \left(\frac{1}{2\pi\delta x^2}\right)^{3/2} \exp\left(-\frac{(\mathbf{x} - \mathbf{x}_0 - t\nabla_{\mathbf{p}}\epsilon(\mathbf{p}_0))^2}{2\delta x^2}\right). \end{aligned} \tag{4.90}$$

The transport properties of the Sommerfeld metal are then determined by the scattering of the Gaussian wave packets (as discussed in Section 4.8). The evolved quasi-classical picture of Gaussian states describing the motion of electrons in a metal is thus similar to that provided by classical physics, except for the feature that, instead of classical particles moving on straight lines in between scattering, quantum particles in Gaussian wave packet states have their centers move as dictated by classical mechanics.

As discussed in Section 4.8.3, the sum of the probability currents of occupied Gaussian states gives the average current density, and a simple picture of the electron gas dynamics therefore emerges: the electrons can be thought of as occupying different Gaussian states as dictated by the physical situation in question, each such adding its contribution to the average current density. Each cell volume in real space, $\Delta\mathbf{x}_c$, thus contains electrons occupying Gaussian states with different momenta describing electrons moving in all directions as depicted in Figure 4.2, the picture emulating that of classical physics.

4.8 Drude Transport Theory

A conductor, such as a metal in its normal state, exhibits resistance due to its imperfections: foreign atoms substituting for atoms of the crystal, vacancies due to missing atoms, dislocations in the crystal, grain boundaries. These imperfections lead to scattering of conduction electrons, as does the interaction of electrons with the thermal vibrations of the

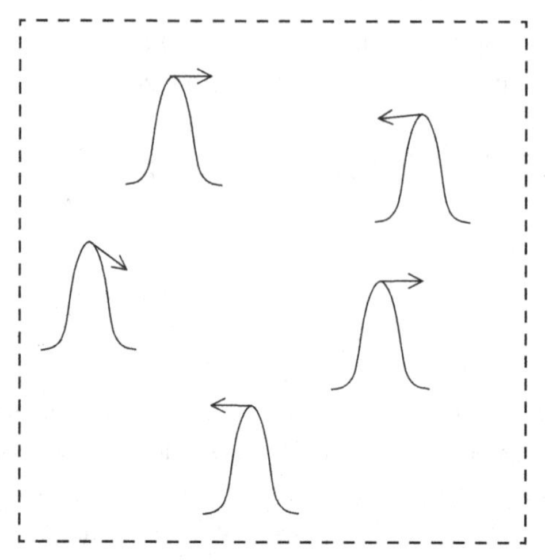

Figure 4.2 Each spatial volume contains electrons in wave packet states moving in all directions.

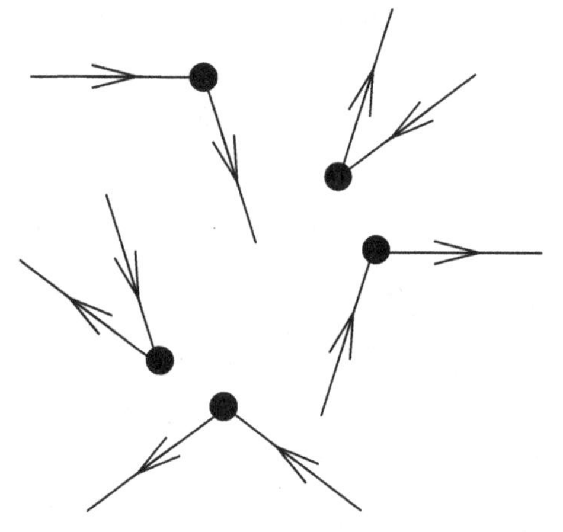

Figure 4.3 Myriad of impurity scatterings of electrons.

ionic lattice and Coulomb interaction between the electrons. To understand the main effect of scattering on the electrical properties of a conductor, it is sufficient to consider the scattering by impurities, characterized, as discussed in Section 4.6, by a short-range impurity potential, Eq. (4.78). The scattering off the short-range impurity potentials in the presence of an external force leads to a dynamics of the electrons that is like that of classical particles bouncing off hard spheres, except that the scattering is a quantum mechanical process that gives probabilities for scattering into all Gaussian states of the same energy.[7]

In a piece of metal there is an astronomically large number of electrons bouncing off the many impurities, as illustrated in Figure 4.3. In any short time interval, many electrons are scattered, and a statistical theory is called for: it is the average velocity of the multitude

[7] Recall the discussion of scattering off a potential barrier in Chapter 2, where in the considered one-dimensional case only two energy-conserving channels are present, transmission and reflection.

of electrons that is of interest. The simplest such theory is due to Drude, and asserts that an electron will scatter off an impurity with probability $\Delta t/\tau$, in any short time interval Δt, τ being the phenomenological parameter thus specifying the theory.[8] In view of the myriad of random scatterings taking place in each time interval, the electrons that *do* scatter have their total velocity averaged to zero, as the averaging over all the random directions in Figure 4.3 suggests. The electrons thus only pick up an average velocity if a force is present.

4.8.1 Drift Velocity

The fraction of electrons, $\Delta t/\tau$, that *do* suffer a scattering in time interval Δt contribute to the average momentum at time $t + \Delta t$ only if there is a force present, as any average initial momentum is lost by randomization due to the scattering. The amount by which their average momentum is changed is the fraction of such electrons multiplied by their change in average momentum in time interval Δt, $\mathbf{F}(t)\Delta t$, that is, $\mathbf{F}(t)\Delta t(\Delta t/\tau)$. Denoting the average momentum of the electrons at time t by $\mathbf{P}(t)$, the fraction, $(1 - \Delta t/\tau)$, of electrons that *do not* suffer a collision in time interval Δt will conserve their fraction of this initial average momentum, $(1 - \Delta t/\tau)\mathbf{P}(t)$, and pick up an additional momentum due to the force given by $(1 - \Delta t/\tau)\mathbf{F}(t)\Delta t$. The average momentum at time $t + \Delta t$ is thus

$$\mathbf{P}(t + \Delta t) = (1 - \Delta t/\tau)(\mathbf{P}(t) + \mathbf{F}(t)\Delta t) + \mathbf{F}(t)\frac{\Delta t^2}{\tau}. \tag{4.91}$$

Moving the term $\mathbf{P}(t)$ to the left-hand side of the equation and dividing by Δt gives the equation satisfied by the average momentum of the electrons:

$$\frac{d\mathbf{P}(t)}{dt} = -\frac{\mathbf{P}(t)}{\tau} + \mathbf{F}(t). \tag{4.92}$$

The impurity scattering thus gives rise to a friction force.[9]

Introducing the average velocity, $\mathbf{V} = \mathbf{P}/m$, where m is the electron mass, the equation for the average velocity of the electrons is

$$m\frac{d\mathbf{V}(t)}{dt} = -\frac{m}{\tau}\mathbf{V}(t) + \mathbf{F}(t). \tag{4.93}$$

Multiplying Eq. (4.93) by $\exp(t/\tau)$, the resulting equation can be rewritten as

$$\frac{d}{dt}(e^{t/\tau}\mathbf{V}(t)) = \frac{1}{m}\, e^{t/\tau}\mathbf{F}(t), \tag{4.94}$$

[8] Drude's theory (1900), obviously classical, assumed (erroneously) that the ions of a perfect lattice were the cause of scattering. Such an assumption has a problem with the low resistance of metals. Drude's theory precedes the advent of quantum mechanics according to which an electron can move without any change of velocity through a perfect lattice, as discussed in Chapter 8, resistance thus being due to deviations from a perfect lattice. Drude, of course, used the classical thermal velocity for an electron (which is much lower than the Fermi velocity), thus not making his estimate of metallic resistance in complete disagreement with experiment. The success of Drude's theory is a testimony to the power and delusion of phenomenological predictions: although the microscopic mechanism initially conceived is completely erroneous, reinterpreting the phenomenological parameter leaves empirical confirmation correct.

[9] We refer the reader interested in a derivation of the Drude theory from kinetic theory to chapter five of Rammer, *Quantum Transport Theory* [1].

which can immediately be integrated from time t_0 to t giving

$$\mathbf{V}(t) = \mathbf{V}(t_0)\, e^{(t_0-t)/\tau} + \frac{1}{m}\int_{t_0}^{t} dt'\; e^{-(t-t')/\tau}\mathbf{F}(t'), \tag{4.95}$$

where $\mathbf{V}_0$ is the average velocity at the arbitrary time t_0.

In the case of a time-independent force, $\mathbf{F}$, the solution to Eq. (4.95) gives for the average velocity, assuming the force is switched on at time $t_0 = 0$,

$$\mathbf{V}(t) = \frac{\tau}{m}(1 - e^{-t/\tau})\mathbf{F} + \mathbf{V}_0\, e^{-t/\tau}. \tag{4.96}$$

The average velocity behavior contains two transient effects due to the collisions with impurities: the loss of memory of the initial velocity direction, and a delay in full effectiveness of the force. The transients are determined by the time scale τ, which according to Exercise 4.4 on page 93 is extremely short. The stationary solution of Eq. (4.93), $\dot{\mathbf{V}} = \mathbf{0}$, is thus almost immediately reached, giving for the average or drift velocity of the electrons in the steady state

$$\mathbf{v}_{\mathrm{d}} = \mathbf{V}(t \gg \tau) = \frac{\tau}{m}\,\mathbf{F} = \mu\,\mathbf{F}, \tag{4.97}$$

where μ is called the mobility of the particle. Since each electron is affected identically by the force, each electron attains this drift velocity.

In the absence of a force, Eq. (4.96) becomes

$$\mathbf{V}(t) = \mathbf{V}_0\, e^{-t/\tau}, \tag{4.98}$$

stating that the memory of an initial average velocity, $\mathbf{V}_0$, is lost due to collisions on the time scale τ. The appearance of exponential damping is worth interpretation, its appearance being a general feature of physical processes happening at a constant rate. Recall the product expansion of the exponential function, $t = N\Delta t$,

$$e^{-t/\tau} = \lim_{N\to\infty}\left(1 - \frac{\Delta t}{\tau}\right)^N, \tag{4.99}$$

which follows from the binomial expansion

$$\left(1 - \frac{\Delta t}{\tau}\right)^N = \sum_{n=0}^{N}\binom{N}{n} 1^{N-n}\left(-\frac{\Delta t}{\tau}\right)^n \tag{4.100}$$

since in the limit of large N, $N \gg n$,

$$\binom{N}{n} = \frac{N!}{n!\,(N-n)!} = \frac{N(N-1)\cdots(N-n+1)}{n!} \simeq \frac{N^n}{n!} \tag{4.101}$$

and thus

$$\lim_{N\to\infty}\left(1 - \frac{\Delta t}{\tau}\right)^N = \sum_{n=0}^{\infty}\frac{1}{n!}\left(\frac{-N\Delta t}{\tau}\right)^n = e^{-t/\tau}. \tag{4.102}$$

According to the input assumption of the Drude theory, $(1 - \Delta t/\tau)^N$ is the probability that an electron is not scattered in time interval $N\Delta t = t$. The exponential factor, $\exp(-t/\tau)$, thus has the following interpretation: the probability an electron does not suffer a collision in time interval t, so the average particle velocity survives, $1/\tau$ being the survival rate. The probability for not being scattered thus decays exponentially in time. For a short time interval, this probability becomes $\exp(-\Delta t/\tau) \simeq 1 - \Delta t/\tau$, the input assumption of the Drude theory.

4.8.2 Conductivity

The average electric current density of the electron gas is

$$\mathbf{j}(t) = -en\mathbf{V}(t), \tag{4.103}$$

where n is the density of the electrons.

The case of interest is the force due to an applied electric field, $\mathbf{F} = -e\mathbf{E}$. Consider first the case of a static electric field, in which case, according to Eq. (4.97), the steady current density flows,

$$\mathbf{j} = -en\mathbf{v}_\mathrm{d} = \frac{ne^2\tau}{m}\,\mathbf{E}, \tag{4.104}$$

i.e. the electron gas has the d.c. conductivity

$$\sigma_0 = \frac{ne^2\tau}{m}, \qquad \mathbf{j} = \sigma_0\,\mathbf{E}. \tag{4.105}$$

Impurity scattering causes current degradation, an average current only being present if a driving force is present. The current density response to the electric field is linear in the Drude model, and the conductivity tensor is diagonal due to the isotropy of the model. Since the density of the electrons in a metal is fixed, the conductivity is a direct measure of the scattering or collision time τ.

4.8.3 Shifted Local Fermi Sea

The current density can also be calculated explicitly taking into account that the electrons are in thermal equilibrium, i.e. there is a thermally distributed occupation of Gaussian states. The degeneracy of the electron gas then explains the absent temperature dependence of the conductivity when the scattering is dominated by elastic scattering. A Gaussian wave packet state is characterized by its average momentum (or velocity), and the average momentum of all the electrons is taken as the sum of average momenta of each occupied Gaussian state divided by the number of electrons as in Section 4.8.1. As discussed in Appendix N, the dynamics of a Gaussian state in the presence of a force is equivalent to that of classical mechanics: the average velocity is changed as dictated by Newton's equation. For a spatially homogeneous force, all electrons thus have their average velocity changed by the same amount.

The electric field accelerates the electrons and the collisions slow them down. In the steady state of constant average electron velocity, described by the stationary solution of Eq. (4.92), $\dot{\mathbf{P}} = \mathbf{0}$, the average momentum of the electron gas is given by $\mathbf{P} = -e\tau\mathbf{E}$, which is the gain achieved in momentum by each electron in a Gaussian state when the two

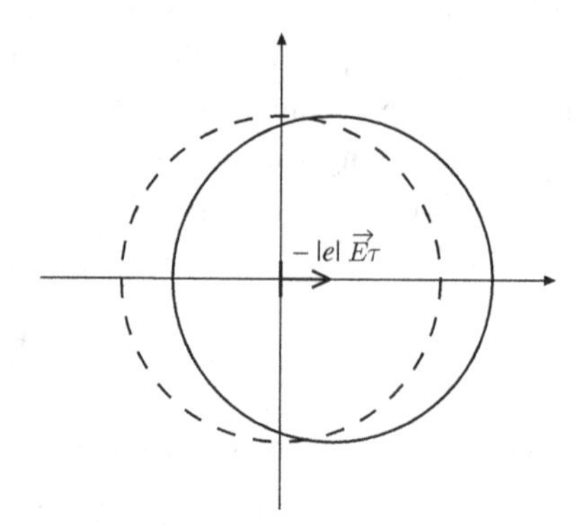

Figure 4.4 Shifted local Fermi sea.

opposing forces, external force and friction force, balance each other in the steady state. Owing to the electric field and collisions, an electron that in the absence of the electric field is in the Gaussian state labeled by momentum $\mathbf{p}$ is now shifted into state $\mathbf{p}-e\tau\mathbf{E} = \mathbf{p}+m\mathbf{v}_{\rm d}$. In the Gaussian wave packet description of the electron gas, in each space volume $\Delta\mathbf{x}_c$, the electric field shifts the local Fermi sea by the amount $-e\tau\mathbf{E} = \tau\mathbf{F}$, as depicted in Figure 4.4, resulting in an asymmetric distribution of populated Gaussian states.

The probability current density in the Gaussian state, Eq. (4.2), labeled by $(\mathbf{x}_i, \mathbf{p}_i)$, is (see Appendix M)

$$S_{\mathbf{x}_i\mathbf{p}_i}(\mathbf{x}) = \frac{\mathbf{p}_i}{m}\,|\psi_{\mathbf{x}_i,\mathbf{p}_i}(\mathbf{x})|^2 = \mathbf{v}_{\mathbf{p}_i}\,|\psi_{\mathbf{x},\mathbf{p}_i}(\mathbf{x}_i)|^2$$
$$= \mathbf{v}_{\mathbf{p}_i}\left(\frac{1}{2\pi\delta x^2}\right)^{3/2}\exp\left(-\frac{(\mathbf{x}-\mathbf{x}_i)^2}{2\delta x^2}\right), \qquad (4.106)$$

corresponding to the electric current density $\mathbf{j}_{\mathbf{x}_i\mathbf{p}_i}(\mathbf{x}) = -eS_{\mathbf{x}_i\mathbf{p}_i}(\mathbf{x})$.

In thermal equilibrium, the Gaussian state of high momentum label is populated with probability specified by the Fermi function, $f(\epsilon(\mathbf{p}))$, as discussed in Section 4.5. Owing to the shift in the electron distribution by the presence of the electric field, the Gaussian state contributing with velocity $\mathbf{v}_\mathbf{p}$ is populated as described by the Fermi function $f(\epsilon(\mathbf{p}+\tau e\mathbf{E})) = f(\epsilon(\mathbf{p}-m\mathbf{v}_{\rm d}))$.[10] The total electric current density can be viewed as the sum of contributions from the Gaussian states weighted according to their population,

$$\mathbf{j}(\mathbf{x}) = -2e\sum_{\mathbf{x}_i\mathbf{p}_i}\mathbf{v}_{\mathbf{p}_i}\,|\psi_{\mathbf{x}_i,\mathbf{p}_i}(\mathbf{x})|^2[f(\epsilon(\mathbf{p}_i+e\tau\mathbf{E}))-f(\epsilon(\mathbf{p}_i))], \qquad (4.107)$$

where the factor of 2 is due to spin degeneracy. The current density in the absence of the electric field, the equilibrium current, vanishes due to $\mathbf{v}_{-\mathbf{p}_i} = -\mathbf{v}_{\mathbf{p}_i}$ and $|\psi_{\mathbf{x}_i,-\mathbf{p}_i}(\mathbf{x})|^2 = |\psi_{\mathbf{x}_i,\mathbf{p}_i}(\mathbf{x})|^2$, and is subtracted such that, as it should be, attribution of probability is only needed for Gaussian states of momentum on the order of the Fermi momentum. Owing to the localized shape of the Gaussian wave function, the main contribution from the summation over $\mathbf{x}_i$ is from the spatial cells near $\mathbf{x}$. In the bulk of the sample, this summation gives

[10] The assumption of a thermal equilibrium distribution in the presence of the electric field assumes a heat sink as discussed in Section 4.9.

an identical result, and $\mathbf{j}(\mathbf{x})$ is the same in each cell, showing a negligible variation within each small cell (which is irrelevant since eventually interest lies in the current, the surface integrated current density). In the limit of a large sample volume, $\Omega \gg \Delta\mathbf{x}_c$, a constant bulk current density can thus be ascribed to each space cell volume $\Delta\mathbf{x}_c$ given by

$$\begin{aligned}\mathbf{j} &\equiv \frac{1}{\Delta\mathbf{x}_c}\int_{\Delta\mathbf{x}_c} d\mathbf{x}\,\mathbf{j}(\mathbf{x}) = \frac{1}{\Omega}\int_{\Omega} d\mathbf{x}\,\mathbf{j}(\mathbf{x}) \\ &= \frac{-2e}{\Omega}\int_{\Omega} d\mathbf{x}\sum_{\mathbf{x}_i\mathbf{p}_i}\mathbf{v}_{\mathbf{p}_i}\,|\psi_{\mathbf{x}_i,\mathbf{p}_i}(\mathbf{x})|^2[f(\epsilon(\mathbf{p}_i+e\tau\mathbf{E}))-f(\epsilon(\mathbf{p}_i))] \\ &= \frac{-2e}{\Omega}\sum_{\mathbf{x}_i\mathbf{p}_i}\frac{\Delta\mathbf{x}_c\Delta\mathbf{p}_c}{(2\pi\hbar)^3}\mathbf{v}_{\mathbf{p}_i}[f(\epsilon(\mathbf{p}_i+e\tau\mathbf{E}))-f(\epsilon(\mathbf{p}_i))],\end{aligned} \tag{4.108}$$

where in the last equality the normalization of the Gaussian functions has been used, and a factor of 1 inserted. The summation over $\mathbf{x}_i$, i.e. each associated spatial cell volume $\Delta\mathbf{x}_c$, sums up to the volume and the electric current density becomes

$$\mathbf{j} = -2e\sum_{\mathbf{p}_i}\frac{\Delta\mathbf{p}_c}{(2\pi\hbar)^3}\mathbf{v}_{\mathbf{p}_i}[f(\epsilon(\mathbf{p}_i+e\tau\mathbf{E}))-f(\epsilon(\mathbf{p}_i))]. \tag{4.109}$$

Since only a weak electric field is present in the metal, $\tau e|\mathbf{E}| \ll p_{\mathrm{F}}$, a Taylor expansion gives

$$\begin{aligned}\mathbf{j} &= -2e\sum_{\mathbf{p}_i}\frac{\Delta\mathbf{p}_c}{(2\pi\hbar)^3}\frac{\mathbf{p}_i}{m}\left(\frac{\partial f(\epsilon(\mathbf{p}_i))}{\partial\mathbf{p}_i}\cdot(e\tau\mathbf{E})\right) \\ &= -2e\sum_{\mathbf{p}_i}\frac{\Delta\mathbf{p}_c}{(2\pi\hbar)^3}\frac{\mathbf{p}_i}{m}\frac{e\tau\mathbf{E}\cdot\mathbf{p}_i}{m}\frac{\partial f(\epsilon(\mathbf{p}_i))}{\partial\epsilon(\mathbf{p}_i)}.\end{aligned} \tag{4.110}$$

The derivative of the Fermi function makes the integrand only non-vanishing near the Fermi surface, and since the size of momentum cells are small compared to the Fermi momentum, $\Delta p_c \ll p_{\mathrm{F}}$, the summation over momentum cells can be replaced by the integral

$$\begin{aligned}\mathbf{j} &= \frac{2e^2\tau}{m^2}\int\frac{d\mathbf{p}}{(2\pi\hbar)^3}(\mathbf{E}\cdot\mathbf{p})\mathbf{p}\left(-\frac{\partial f(\epsilon(\mathbf{p}))}{\partial\epsilon(\mathbf{p})}\right) \\ &= \frac{e^2\tau}{m^2}\mathbf{E}\int_0^{\pi} d\theta\,\sin\theta\cos^2\theta\int_0^{\infty} d\epsilon\,N(\epsilon)2m\epsilon\left(-\frac{\partial f(\epsilon)}{\partial\epsilon}\right),\end{aligned} \tag{4.111}$$

where in the last integral spherical coordinates have been introduced, the polar axis chosen along $\mathbf{E}$, and the momentum length integration shifted into energy integration, as already discussed in Section 3.1.3. The polar angle integral is elementary, giving $2/3$, and performing the energy integration is trivial in view of the peakedness of the derivative of the Fermi function, restricting the integration to the Fermi surface, giving

$$\mathbf{j} = \sigma\,\mathbf{E}, \qquad \sigma = \frac{2e^2}{3}N(\epsilon_{\mathrm{F}})v_{\mathrm{F}}^2\tau. \tag{4.112}$$

In the Gaussian wave packet description of the electron gas, the electric field shifts each local Fermi sea. The corresponding asymmetric distribution produces the net local current, Eq. (4.112), in each spatial cell, which in view of Eq. (3.29) is the same result for the current as in Eq. (4.104). The conductivity can thus be viewed as due to only states near the Fermi surface or as in Section 4.8.2 due to all of the electrons in a shifted Fermi sea.

4.9 Resistor

A conductor of length L has the voltage V between its two ends,

$$V \equiv \int_0^L d\mathbf{l} \cdot \mathbf{E}, \tag{4.113}$$

the direction of the line integral taken along the electric field. In the bulk of the conductor, the driving electric field is homogeneous, and the voltage is proportional to the length, $V = EL$. Denoting the cross-sectional area of the conductor by S, the current

$$I = \int_S d\mathbf{s} \cdot \mathbf{j} \tag{4.114}$$

becomes, according to Eq. (4.105) (assuming the current uniform or introducing an effective cross-sectional area),

$$I = \sigma_0 ES = \frac{S}{L} \sigma_0 V = GV \tag{4.115}$$

the proportionality factor being called the conductance of the conductor.

A conductor with a small conductance, a bad conductor, is referred to as a resistor. The inverse of the conductance, $R = 1/G$, is called the resistance of the resistor, and is the proportionality constant between voltage and current in Ohm's law,

$$V = RI, \qquad R = \frac{L}{S\sigma_0} = \frac{L}{S}\rho_0. \tag{4.116}$$

The inverse of the conductivity, $\rho_0 = 1/\sigma_0$, is called the resistivity. The resistance of a wire is proportional to the length of the wire and inversely proportional to its cross-sectional area. A simple model of a resistor is thus a very disordered conductor whose resistance is due to elastic scattering of the conduction electrons off impurities.

The electric field does work on the electrons. Since the dynamics of an electron in a Gaussian state is identical to that of classical dynamics, the change in kinetic energy of the electron in the Gaussian state is the same as that of a classical particle (see the discussion in Appendix N). Owing to the electric field, each electron thus gains energy per unit time, i.e. absorbs the power

$$P_1 = -e\mathbf{E} \cdot \mathbf{V} = \frac{e^2\tau}{m} \mathbf{E}^2 = \frac{e^2\tau}{m} \frac{V^2}{L^2}, \tag{4.117}$$

where the second equality follows from the steady-state solution of Eq. (4.104), and in the last equality the voltage has been introduced. The power absorbed by the N electrons is thus

$$P = \frac{Ne^2\tau}{m}\frac{V^2}{L^2}, \tag{4.118}$$

or introducing the density of the electron gas, $n = N/\Omega = N/LS$,

$$P = \sigma_0 \frac{S}{L} V^2 = \frac{V^2}{R} = IV = RI^2 \tag{4.119}$$

which is Joule's law. The assumption that the electron gas is described by a constant temperature, through a displaced equilibrium Fermi distribution, is thus only valid if the energy absorbed, Eq. (4.119), is taken out of the electron gas as it otherwise heats up. The energy absorbed by the electrons is transferred to the bulk of the system, the ionic lattice, by inelastic interaction of the electrons with the ions. In the absence of cooling to carry away the absorbed energy, Joule heating of the material takes place at the rate given in Eq. (4.119). Typically cooling is not perfect and one can feel that operating electronic devices heat up: their temperature rises as the power absorbed by the electrons is dissipated into irregular or thermal motion of the constituents of the material.

Exercise 4.4 For a copper wire, the resistivity is about $\rho_{\text{Cu}} \simeq 1.7 \times 10^{-8}\,\Omega\,\text{m}$. Use the Drude model to estimate the collision time τ for conduction electrons in a copper wire.

4.10 Frequency-Dependent Conductivity

An alternating voltage produces a time-dependent electric field, $\mathbf{E}(t)$, in a conductor. Assuming that the electric field vanishes in the far past and the initial average velocity of the electrons correspondingly vanishes, the average velocity at time t is, according to Eq. (4.95),

$$\mathbf{V}(t) = \frac{-e}{m}\int_{-\infty}^{t} dt'\; e^{-(t-t')/\tau}\mathbf{E}(t'). \tag{4.120}$$

The average current density is then

$$\mathbf{j}(t) = -en\mathbf{V}(t) = \frac{ne^2}{m}\int_{-\infty}^{t} dt'\; e^{-(t-t')/\tau}\mathbf{E}(t'). \tag{4.121}$$

The current density and electric field are linearly related. The linear response is the exact response in the Drude model, and it suffices to study the response to a single frequency,

$$\mathbf{E}(t) = \Re(\mathbf{E}(\omega)\,e^{-i\omega t}) = \tfrac{1}{2}(\mathbf{E}(\omega)\,e^{-i\omega t} + \mathbf{E}^*(\omega)\,e^{i\omega t}). \tag{4.122}$$

The response to an arbitrary time-dependent field can then be built by superposing fields of different frequencies. The integration in Eq. (4.121) is then elementary, giving

$$\mathbf{j}(t) = \Re\left(\frac{ne^2}{m}\,\frac{1}{(1/\tau) - i\omega}\,\mathbf{E}(\omega)\,e^{-i\omega t}\right) \tag{4.123}$$

or

$$\mathbf{j}(t) = \Re(\mathbf{j}(\omega)\,e^{-i\omega t}), \tag{4.124}$$

where the complex amplitude relationship

$$\mathbf{j}(\omega) = \sigma(\omega)\,\mathbf{E}(\omega) \tag{4.125}$$

is specified by the frequency-dependent complex conductivity

$$\sigma(\omega) = \frac{\sigma_0}{1 - i\omega\tau}, \tag{4.126}$$

which is the frequency-dependent Drude conductivity.

The frequency-dependent conductivity satisfies the sum rule

$$\int_{-\infty}^{\infty} d\omega\,\sigma(\omega) = 2\int_{0}^{\infty} d\omega\,\Re(\sigma(\omega)) = 2\int_{0}^{\infty} d\omega\,\frac{\sigma_0}{1 + (\omega\tau)^2} = \frac{\pi ne^2}{m}, \tag{4.127}$$

which is independent of τ. The sum rule for the conductivity thus allows for the determination of the density of charge carriers, here the electron density n.

4.11 Hall Effect

Consider the electrons experiencing spatially homogeneous electromagnetic fields so that the average force on the electrons is the Lorentz force, $\mathbf{F}(t) = -e\mathbf{E}(t) - e\mathbf{V}(t) \times \mathbf{B}(t)$, and the equation of motion for the average velocity is, according to Eq. (4.93),

$$m\dot{\mathbf{V}}(t) + \frac{m}{\tau}\,\mathbf{V}(t) = -e\mathbf{E}(t) + (-e\mathbf{V}(t) \times \mathbf{B}(t)). \tag{4.128}$$

In the case of time-independent electric and magnetic fields, the stationary solution of Eq. (4.128), $\dot{\mathbf{V}} = \mathbf{0}$, is determined by

$$\frac{m}{\tau}\,\mathbf{V} = -e\mathbf{E} - e\mathbf{V} \times \mathbf{B}. \tag{4.129}$$

Rewriting in matrix form

$$\frac{-e\tau}{m}\,\mathbf{E} = \left(\begin{pmatrix} 1 & 0 & 0 \\ 0 & 1 & 0 \\ 0 & 0 & 1 \end{pmatrix} + \frac{e\tau}{m}\begin{pmatrix} 0 & B_z & -B_y \\ -B_z & 0 & B_x \\ B_y & -B_x & 0 \end{pmatrix}\right)\begin{pmatrix} V_x \\ V_y \\ V_z \end{pmatrix}, \tag{4.130}$$

the force due to the electric field is related to the average velocity according to

$$-e\mathbf{E} = \boldsymbol{\mu}^{-1}\mathbf{V}, \tag{4.131}$$

where the inverse mobility tensor is seen to be (choosing the $\hat{\mathbf{z}}$ direction parallel to the magnetic field, $\mathbf{B} = B\hat{\mathbf{z}}$)

$$\boldsymbol{\mu}^{-1} = \frac{m}{\tau} \begin{pmatrix} 1 & \omega_c\tau & 0 \\ -\omega_c\tau & 1 & 0 \\ 0 & 0 & 1 \end{pmatrix}, \tag{4.132}$$

where $\omega_c = eB/m$ is the Larmor or cyclotron frequency. The average velocity response to a force, $\mathbf{V} = \boldsymbol{\mu}\mathbf{F}$, is thus specified by the mobility tensor

$$\boldsymbol{\mu} = \frac{\tau/m}{1 + (\omega_c\tau)^2} \begin{pmatrix} 1 & -\omega_c\tau & 0 \\ \omega_c\tau & 1 & 0 \\ 0 & 0 & 1 + (\omega_c\tau)^2 \end{pmatrix}, \tag{4.133}$$

inverting the matrix equation (4.132) by recalling the ease in inverting a 2×2 matrix, Eq. (2.64), the prefactor being equal to the inverse of the determinant.[11]

The current density formula, $\mathbf{j} = -en\mathbf{V}$, together with Eq. (4.131), gives the relationship between the conductivity and mobility tensors, $\boldsymbol{\sigma} = ne^2\boldsymbol{\mu}$, and thereby explicitly for the conductivity tensor

$$\boldsymbol{\sigma} = \frac{\sigma_0}{1 + (\omega_c\tau)^2} \begin{pmatrix} 1 & -\omega_c\tau & 0 \\ \omega_c\tau & 1 & 0 \\ 0 & 0 & 1 + (\omega_c\tau)^2 \end{pmatrix}. \tag{4.134}$$

The introduction of the mobility in the context of metals is superfluous since the density of electrons is fixed. However, in a semiconductor the density of charge carriers varies with temperature, and it is relevant to separate this effect from the effect of scattering.

We observe explicitly the Onsager symmetry relation (as demanded by time reversal symmetry)

$$\sigma_{\alpha\beta}(\mathbf{B}) = \sigma_{\beta\alpha}(-\mathbf{B}). \tag{4.135}$$

Inverting the conductivity tensor gives the resistivity tensor (recall the inverse mobility tensor, Eq. (4.132))

$$\boldsymbol{\rho} = [\boldsymbol{\sigma}]^{-1} = \rho_0 \begin{pmatrix} 1 & \omega_c\tau & 0 \\ -\omega_c\tau & 1 & 0 \\ 0 & 0 & 1 \end{pmatrix}, \qquad \rho_0 = \frac{1}{\sigma_0}. \tag{4.136}$$

We note that the considered isotropic model does not display magnetoresistance, i.e. ρ_{xx} does not depend on the magnetic field as a consequence of the isotropic dispersion, $\epsilon_{\mathbf{p}} = \epsilon(|\mathbf{p}|)$.

Assuming the electric field is along the x direction, $\mathbf{E} = E\hat{\mathbf{x}}$, the stationary value for the average velocity in an infinite system is, according to Eqs. (4.131) and (4.133),

$$\mathbf{V} = -e\boldsymbol{\mu}\mathbf{E} = \frac{1}{-en}\boldsymbol{\sigma}\mathbf{E}, \tag{4.137}$$

[11] The zz component of the mobility tensor should be independent of the magnetic field, $\mu_{zz} = \tau/m$, reflecting that if the electric field is chosen along the direction of the magnetic field (which defined the z direction), the average velocity of the electrons is parallel to the magnetic field and the resulting cross-product in the Lorentz force vanishes, $\mathbf{V} \times \mathbf{B} = \mathbf{0}$, i.e. the presence of the magnetic field has no effect in the case $\mathbf{B} \parallel \mathbf{E}$.

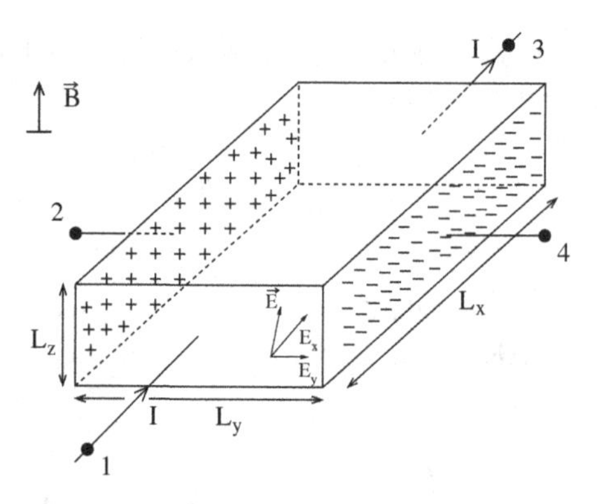

Figure 4.5 Hall bar measurement.

given by

$$\mathbf{V} = \frac{E}{-en}\begin{pmatrix}\sigma_{xx}\\ \sigma_{yx}\\ 0\end{pmatrix} = \frac{-e\tau/m}{1+(\omega_c\tau)^2}\,E\begin{pmatrix}1\\ \omega_c\tau\\ 0\end{pmatrix}. \tag{4.138}$$

We note that, in the limit $\omega_c\tau \gg 1$, the inequality $\sigma_{yx} \gg \sigma_{xx}$ is satisfied, signifying that the Hall current, proportional to V_y, is much larger than the current in the x direction. In fact, the current in the x direction is due to the *presence* of scattering as σ_{xx} vanishes in the limit $\tau \to \infty$, whereas the Hall current is finite in the absence of scattering.

A Hall bar is a finite rectangular piece of material where the voltage drop in the transverse (to the current) direction can be measured, as depicted in Figure 4.5.

When the indicated magnetic field is turned on, the conduction electrons are deflected from the transport current direction by the Lorentz force due to the magnetic field. Since the sample is finite, electrons pile up at one side and deficiency arises at the opposite side due to the charge neutrality of the metal. An electric field in the direction transverse to the current, say the y direction, thus develops in a Hall bar, in addition to the driving field, in say the x direction, E_x, or, equivalently, a Hall voltage appears in addition to the current driving voltage. Electron deflection takes place until the built-up electric field, E_y, cancels the force from the magnetic field, $-eV_xB$, i.e. $E_y = -V_xB$, or in terms of the current density $j_x = -enV_x$, $E_y = -j_xB/en$. The Hall coefficient then becomes

$$R_{\mathrm{H}} \equiv \frac{E_y}{Bj_x} = \frac{1}{-en}, \qquad \frac{d\rho_{yx}}{dB} = R_{\mathrm{H}}. \tag{4.139}$$

The transverse electric field, E_y, can also be determined by the equilibrium condition of a current only in the x direction, $j_y = 0$, giving (we can neglect the small fringe field due to the finite size of the sample, i.e. $E_z \simeq 0$), according to

$$j_y = \sigma_{xx}E_x + \sigma_{yx}E_y, \tag{4.140}$$

the relationship

$$\frac{E_y}{E_x} = -\frac{\sigma_{yx}}{\sigma_{yy}}, \tag{4.141}$$

whereby the Hall coefficient becomes

$$R_H = \frac{(-\sigma_{yx}/\sigma_{yy})E_x}{B(\sigma_{xx}E_x + \sigma_{xy}E_y)} = \frac{(-\sigma_{yx}/\sigma_{yy})E_x}{B\left(\sigma_{xx} - \dfrac{\sigma_{xy}\sigma_{yx}}{\sigma_{yy}}\right)E_x}. \tag{4.142}$$

Using the expressions in the conductivity tensor, Eq. (4.134), the expression in Eq. (4.139) is again arrived at.

The Hall voltage measured between terminals 2 and 4 in Figure 4.5 is

$$V_H = E_y L_y = L_y B j_x \frac{1}{-en}. \tag{4.143}$$

A Hall experiment, determining the slope of the Hall voltage as a function of the magnetic field strength, thus determines the density of the electron gas.[12]

The quasi-classical Drude theory is not valid in the high magnetic field limit where Landau quantization of the electron energy levels is important. In that regime the electron gas exhibits interesting effects such as Shubnikov–de Haas oscillations in the ρ_{xx} component, and the quantum Hall effect in two-dimensional electron gases.

4.12 Summary

The introduced wave packet description of the independent electron model of a metal was shown to describe the transport properties relevant for electronics, thereby providing the standard disordered metal model. Though states of an electron gas are described by anti-symmetric wave functions, the degeneracy of the electron gas allowed for a description of its properties in terms of Gaussian states whose electric current dynamics is identical to that of particles obeying classical mechanics. The local Gaussian state description will be essential when dynamics in semiconductors is considered where typically spatially varying potentials are present.

[12] A Hall measurement for a semiconductor reveals the sign of the charge carriers, i.e. distinguishes between electrons and holes.

5 Electric Circuit Theory

Electronics is about pushing electrons around inside matter in a controlled manner, creating flows of electric charge in circuits. This should be done to provide calculated voltages over components, thereby providing the right current flows, ensuring the right voltage drops over other components, etc. Designing the functioning of electronic circuits can be done without understanding the physics of the components. All that is needed is knowledge of the components' current–voltage relationships, their I–V characteristics: what is the current response through a component to an electric potential difference, or voltage, between its terminals, or vice versa, what is the needed current through a device to provide a certain voltage across its terminals. The art and craft of electronics then emerges from understanding electronic circuitry as represented by symbolic circuit diagrams. The main intention of this book is to provide the physical understanding of the I–V characteristics of electronic components, the physics of electronics, but electric circuits are discussed throughout the chapters, here basic circuit theory.

5.1 Wires and Currents

An electric circuit consists of electronic components connected by *wires*, thin filaments of good conducting material such as a clean metal. The resistance of a wire is assumed so small in comparison with the resistance of other components that the wires can be assumed perfect conductors, i.e. having zero resistivity or infinite conductivity. Inside a perfect conductor, the electric field vanishes, since the slightest field would create an infinite current, i.e. instantaneously nullifying it. The electric potential is thus the same in a wire, and no charge inhomogeneity exists in wires. In a *wire* an electric current thus flows without the presence of a voltage difference; the charged electrons flow like an incompressible liquid.

The electric current in a wire, or lead, is the amount of charge per unit time passing any cross-section of the wire. In circuit diagram notation, a wire is depicted as a directed line, indicated by an arrow, connecting two component terminals indicated by dots. In conjunction with the arrow stands the symbol I, the electric current in the wire (and out of and into the components it connects). The instantaneous average current through any cross-sectional surface, S, of a wire is given by the surface integral across S of the current density

$$I = \int_S d\mathbf{s} \cdot \mathbf{j}, \tag{5.1}$$

where $d\mathbf{s}$ denotes a local vector perpendicular to the surface, its length $|d\mathbf{s}|$ equal to the magnitude of its assigned small surface area and its direction specified by the wire arrow. If I is positive, it indicates that the amount I of positive charge is transferred per unit time through any cross-section of the wire, flowing in the direction indicated by the arrow. If I is negative, it indicates that an amount of positive charge per unit time is flowing in the direction opposite to the arrow. Often the charge flow is solely due to electrons, and a positive current then corresponds to the indicated amount of electron charge moving in the direction opposite to the arrow. The unit of charge in the SI system is called the *coulomb*, and denoted C. The elementary unit of charge e is $e \simeq 1.6 \times 10^{-19}$ C, i.e. the charge of the electron is approximately -1.6×10^{-19} C. The rate of flow of charge, the current, is measured in the unit *ampere*, one coulomb per second, and denoted A. If a current of 1.6 A flows in a metallic wire, it corresponds to the passing of an enormous number of electrons, almost 10^{19} electrons per second.

When talking about a current in an electric circuit, the average electric current is meant, and circuit design should of course be such that the inevitable fluctuations of the current should be small such that the average value is the only relevant quantity. In particular, fluctuations in current should never cause appreciable fluctuations of voltages over components, indeed never by a fluctuation change of sign, so as to provoke opposite current flows, causing failure of the intended circuit performance. Since the strong Coulomb force and the multitude of free charge carriers suppress charge and current fluctuations, and as the exclusion principle governing the quantum states of electrons by itself also provides rigidity of the charge flow, fluctuations in electronic systems are small, as discussed in Appendix L.

Exercise 5.1 Calculate the resistance of a copper wire of length 1 mm and transverse dimension 0.1 mm (recall Exercise 4.4).

5.2 Resistor

Electric components exhibit resistance. As discussed in Section 4.8, due to scattering of electrons, a piece of dirty metal, say, is a resistor, i.e. to push a current through the material requires an electric field inside it, pushing the carriers in the direction of the electric force in between scattering. The electric field is created by plus and minus charges piled up at the respective ends of the resistor, the deficiency and excess of electrons located in the region where the electric field changes from zero in the wires to the value in the resistor.

In a metal, only electrons conduct electric current, but other components can have charge carriers of either sign contributing to the current, as depicted in Figure 5.1. For example, the electrolyte inside a battery contains ions of both positive and negative charge. When later discussing the physics of certain types of semiconductors, we shall realize that, although the only physically moving parts are their electrons, it will be important to describe the

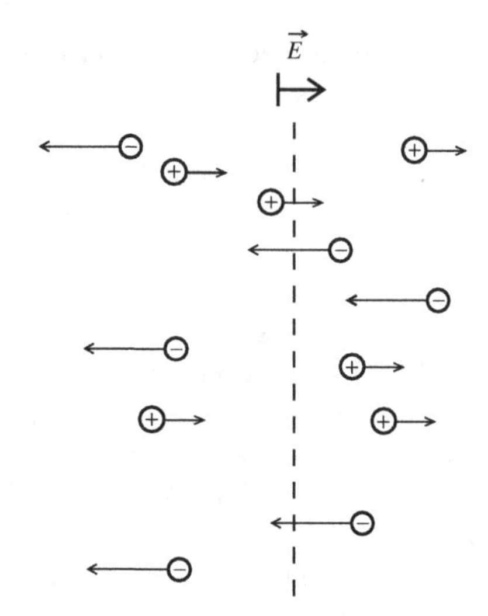

Figure 5.1 Flow of positive and negative charge carriers.

transport properties in terms of oppositely charged particles, so-called holes. In general, the electronic transport in semiconductors is due to both electrons and holes.

The electrons in a metallic resistor is a many-body system, and their dynamics is described by the Gaussian wave packet states occupied by the electrons, as discussed in Section 4.8. An electron in a Gaussian state moves as dictated by classical mechanics, and, if an electric field is present, an average electric current flows. The electric current is built by the drift velocity of each individual occupied electron wave packet state.[1]

As discussed in Section 4.8.1, in the presence of an electric field and scattering, a charge carrier in a resistor has a drift velocity, the field induced velocity of its associated Gaussian wave packet. Since different charge carriers can have different mass, their drift velocities can differ. The positive and negative charge carriers, though moving in opposite directions, have the same direction of their charge current densities, i.e. their charge current density vectors are parallel, $\mathbf{j}_{-} \parallel \mathbf{j}_{+}$, and the total electric current density is their sum, $\mathbf{j} = \mathbf{j}_{-} + \mathbf{j}_{+}$, and has the direction of the electric field. A positive electric current through a resistor thus flows from the high electric potential, or voltage, terminal to the low voltage terminal, irrespective of whether the current in the resistor is carried by positive charges or negative charges or both.

Consider a resistor, depicted as the black box in Figure 5.2, attached to perfectly conducting wires leading to the terminals labeled 1 and 2 just outside the resistor. The magnetic field inside a resistor created by its current can usually be assumed negligible, as is the magnetic field in the external region between the terminals of the resistor (if needed, physically achievable by having a shielding box enclosing the resistor). The electric field in the region of the resistor and the external region between the terminals is thus *curl*-free,

$$\nabla \times \mathbf{E} = \mathbf{0}, \qquad \mathbf{E}(\mathbf{r}) = -\nabla \phi(\mathbf{r}), \tag{5.2}$$

and the electric field can thus be expressed as the gradient of a scalar function ϕ, the electric potential. If the current is constant, ϕ is the electrostatic potential, and for time-dependent

[1] When learning circuit theory in terms of classical mechanics, it therefore gave a correct description.

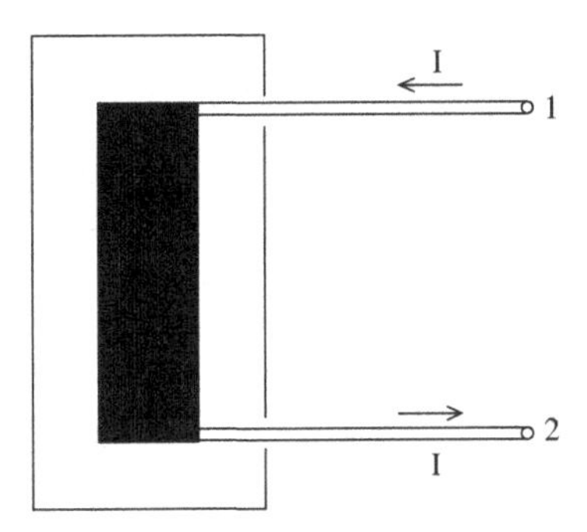

Figure 5.2 Resistor depiction.

currents it is the instantaneous electrostatic potential. The time dependence is irrelevant for the following argument and therefore suppressed. With the minus sign in the definition of the electric potential, $q\phi(\mathbf{r})$ equals *minus* the amount of work performed against the electric field needed to bring the charge q from a reference point to the location $\mathbf{r}$, i.e. a charge q thus has the potential energy, $q\phi(\mathbf{r})$, at location $\mathbf{r}$ with respect to the chosen reference point (usually where $\phi = 0$).

The flux of the curl of a vector field through a surface S equals, according to,

$$\int_S d\mathbf{s} \cdot \nabla \times \mathbf{E}(\mathbf{x}) = \int_C d\mathbf{l} \cdot \mathbf{E}(\mathbf{x}) \tag{5.3}$$

the line integral along the curve C enclosing the surface S (the direction specified by the direction of $d\mathbf{s}$ and the right-hand rule). Choosing the closed loop, C, to consist of a curve lying inside the resistor and the perfect conductors making up the wires and an external curve stretching from terminal 2 to terminal 1, the vanishing of the curl of the electric field gives that the line integral along the closed curve vanishes, i.e.

$$\oint d\mathbf{l} \cdot \mathbf{E} = 0. \tag{5.4}$$

Partitioning the curve into its various segments gives

$$\oint d\mathbf{l} \cdot \mathbf{E} = \int_{\substack{\text{inside}\\ \text{resistor}}} d\mathbf{l} \cdot \mathbf{E} + \int_{\substack{\text{inside}\\ \text{wires}}} d\mathbf{l} \cdot \mathbf{E} + \int_2^1 d\mathbf{l} \cdot \mathbf{E} \tag{5.5}$$

for the case where the direction of the curve is chosen in accordance with the wire arrows depicted in Figure 5.2. The wires are assumed perfect conductors where the electric field vanishes, so

$$\int_{\substack{\text{inside}\\ \text{resistor}}} d\mathbf{l} \cdot \mathbf{E} = -\int_2^1 d\mathbf{l} \cdot \mathbf{E} = \int_2^1 d\mathbf{l} \cdot \nabla\phi(\mathbf{r}) = \phi(1) - \phi(2) \equiv V, \tag{5.6}$$

where we have defined the voltage over the resistor, V, the difference in electric potential between the *in* and *out* terminals, $V = \phi_{\text{in}} - \phi_{\text{out}}$. For voltages of interest, the current is proportional to the small voltage, Ohm's law,

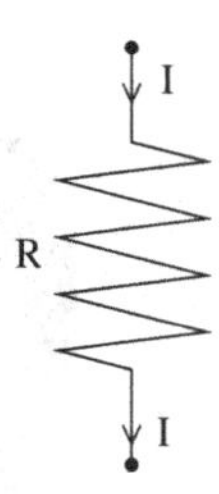

Figure 5.3 Electric circuit symbol for a resistor.

$$V = RI \tag{5.7}$$

where the positive proportionality constant is called the resistance of the resistor. The resistance depends on the resistor's geometry and the strength of scattering processes inside it. A simple model of a resistor was discussed in Section 4.8. The electric circuit symbol for a resistor is often chosen as a rectangular box or as depicted in Figure 5.3.

The unit of electric potential and thereby that of voltage is, in SI units, the *volt* (V). Charge times electric potential has the dimension of energy, which is measured in *joule* (J), defined as newton meter (N m), the *newton* (N) being the unit of force equal to kg m/s^2. A charge of 1 C at a potential of 1 V (relative to a reference point) has an electric potential energy of 1 J. The *ohm* (Ω) is the unit of electrical resistance, equaling one volt per ampere (V/A).

A resistor is typically made from powdered carbon mixed with insulating material such as ceramics, the mixture held together by resin and wrapped in a plastic covering with two metallic leads sticking out at opposite ends. Its resistance then depends on the fraction of good conducting material, the amount of carbon.

5.3 Voltage Divider

In an electric circuit, definite output voltages are needed at terminals in response to input voltages, and to achieve this resistors are put to essential use. Consider a voltage divider as depicted by the circuit symbols in Figure 5.4.

In Figure 5.4 we have introduced a circuit symbol, at the bottom, which we display separately in Figure 5.5 and is referred to as *ground* (or *earth* in the UK), but is in practice just a large good conductor, a charge reservoir, say a metal casing, so that its electric potential stays at the same constant value, which is taken as voltage reference, i.e. equal to zero, $V_{\text{ground}} = 0$. All terminals ending on *ground* are thus electrically joined.

Regarding the voltage divider in Figure 5.4, a voltage $V = V_+$ is distributed over two resistors in series, and by Ohm's law $V_1 = IR_1$, and the rest of the voltage is dropped over the other resistor, $V - V_1 = IR_2$, giving for the output voltage $V_{\text{out}} = V_1 = IR_1$ in terms of the input voltage, $V_{\text{in}} = V$,

$$V_{\text{out}} = V_{\text{in}} \frac{R_1}{R_1 + R_2}. \tag{5.8}$$

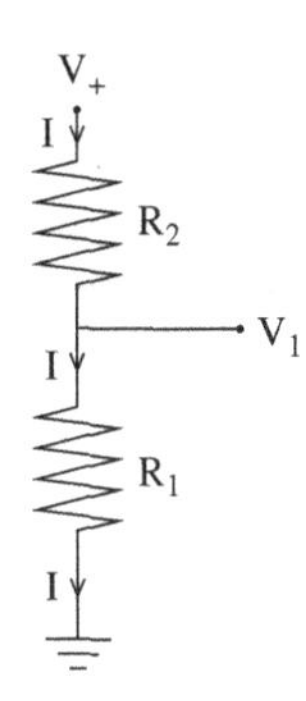

Figure 5.4 Voltage divider.

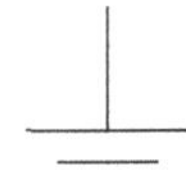

Figure 5.5 Electric circuit symbol for *ground*.

Thus, by choice of suitable resistors, the voltage on the output terminal can be tuned to any desired value.

5.4 Electric Power

Owing to interaction with the ionic lattice, inelastic scattering, energy is transferred from the electric agents of interest, say electrons, to the multitude of other degrees of freedom making up a component, typically an ionic lattice. The work done by the electric field on the electrons ends up through inelastic scattering as irregular thermal motion of the ionic crystal lattice, i.e. is dissipated to the bulk and the resistor heats up. The electrons in the Gaussian wave packet states are moving with a drift velocity specified by the electric field, as discussed in Section 4.8.1. Work is done by the electric field on each of the electrons and their average energy is changed as dictated by Newton's equation: force times particle distance moved, i.e. the amount of work done on each electron per unit time is the force times the drift velocity, Eq. (4.117) (recall the resistor discussion in Section 4.9). In a small time interval, Δt, by definition of current, charge $\Delta q = I\Delta t$ is transported through any cross-section of the resistor. In each cross-sectional volume, charge Δq enters and leaves, and the total work done by the electric field on the electrons in the component equals that of the charge Δq being transported in time interval Δt through the length of the component. Since the line integral of the electric field through the component equals the potential difference between its terminals, the voltage V, the work done is $\Delta W = \Delta q V = I\Delta t V$. The work provided by the voltage across the component is thus dissipated at the rate $P = \Delta W/\Delta t = \dot{W} = IV$. Heating of a component increases its temperature, and the thermal agitation of the ionic lattice increases electron scattering and thereby its resistance as a function of temperature. The ever-present heating in electric circuits is thus a nuisance that has to be countered by cooling, such as for example for computers and servers.

The above power dissipation argument is quite general even when current is due to both positive and negative charge carriers. When an electric current, I, is flowing through a device, the work performed per unit time, the electric power, is

$$P = \dot{W} = IV, \tag{5.9}$$

the power equals current multiplied by voltage. In other words, when charge per unit time, I, is transported between two terminals of voltage difference V, the voltage provides the electric power IV to the device between the terminals. The unit of power is called the *watt* (W), 1 W equaling one joule per second (W = J/s). A current of one ampere passing through a device with voltage difference one volt turns one watt of electrical energy into some other form of energy, here heat.

5.5 Capacitor

A capacitor is a device for storing charge. Consider for example two metal plates close to each other with attached wires brought out to terminals as depicted in Figure 5.6. The plates contain opposite charges, $\pm Q$, the surface charges residing mainly on opposing plate surfaces.

The plates and wires are assumed to be perfect conductors, and their actual imperfection due to scattering processes can be taken into account by assuming an ideal resistor in series. At any point in a capacitor plate and its wire terminal arm there is thus the same electric potential, but a different value in the two arms due to the charge of the capacitor.

Consider first any number of charged conductors separated by vacuum (or a dielectric). The electric field outside the conductors is determined by Gauss's law, $\nabla \cdot \mathbf{E} = \rho/\epsilon$. Recall Gauss's theorem,

$$\int_V d\mathbf{x}\, \nabla \cdot \mathbf{E} = \int_S d\mathbf{s} \cdot \mathbf{E}, \tag{5.10}$$

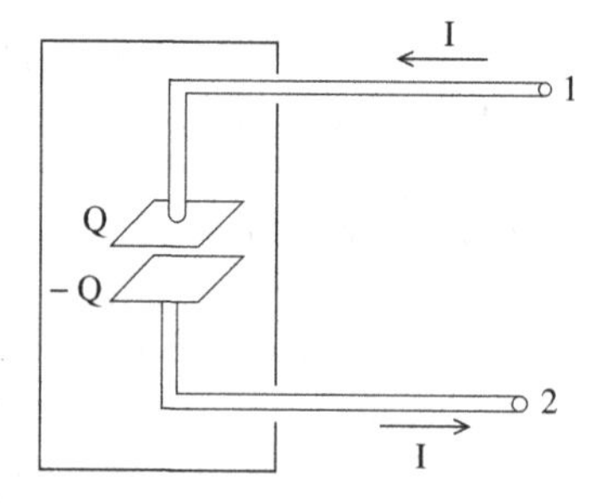

Figure 5.6 Parallel plate capacitor.

stating that the divergence of a vector field, $\mathbf{E}$, integrated over a volume V is equal to the flux of the vector field out through the surface S enclosing the volume. Apply Gauss's theorem to a small volume with top and bottom side areas ΔA parallel to the surface of one of the conductors, one side located inside the conductor, where the electric field vanishes, and the other side of the volume just outside the surface. The total charge inside this small volume is $\sigma_s \Delta A$, where σ_s is the local surface charge density, and the left-hand side of Eq. (5.10) becomes, according to Gauss's law, $\sigma_s \, \Delta A/\epsilon$. The conductor surface is an equipotential surface, and the electric field is therefore perpendicular to the surface and has only a flux through the outer surface equaling $E \Delta A$, and Eq. (5.10) and Gauss's law thus relate the electric field strength just outside the surface of the conductor to its surface charge density according to $E = \sigma_s/\epsilon$. The (normal component of the) electric field has a jump at the conductor surface determined by the surface charge density, a delta function in the coordinate perpendicular to the surface (in reality, of course, a charge density smooth on atomic distance).

Returning to the plate capacitor, the electric field lines in the middle, away from the plate ends, are straight parallel lines connecting the opposite charges of the plates, and at the ends of the plates, the field lines bulge out as the surface charge spills outside the inner surfaces. For an estimate, let us first neglect this effect, thus assuming uniform surface charge densities, $\pm\sigma_s$, on opposing plate areas, specifying the capacitor charge according to $Q = \sigma_s A$. The electric field lines are thus parallel and the flux of the electric field is the same everywhere between the capacitor plates, giving for the electric field strength $E = \sigma_s/\epsilon = Q/\epsilon A$, as a dielectric, with dielectric constant ϵ, is assumed to fill the space between the plates.[2] Denoting the distance between the plates as d, the electric potential difference between the plates, the voltage of the capacitor, is $V = Ed = Qd/\epsilon A$, or

$$Q = CV. \tag{5.11}$$

The proportionality factor, C, is called the capacitance of the capacitor, which for the plate capacitor has been estimated to be

$$C = \frac{\epsilon A}{d}. \tag{5.12}$$

The capacitance is the measure of the amount of charge on the capacitor, its capacity, for a given voltage between its terminals, or vice versa, its voltage drop for a given amount of charge. The capacitance is seen to be determined by the plate geometry and the dielectric constant of the material in between the plates (if there is vacuum between the plates, the dielectric constant of vacuum, $\epsilon = \epsilon_0$, appears). The unit of capacitance is called the *farad* (F), and equals one coulomb per volt, i.e. a capacitor of one farad at a potential of one volt carries a charge of one coulomb.

The linear relationship between voltage and charge of a capacitor (or conductors in general) is an exact relation: for the electric field to double, thereby doubling the potential difference, the surface charge density has to double. Recall that, in terms of the electrostatic potential, Gauss's law turns into Poisson's equation,

[2] The polarizable dielectric diminishes the electric field between the plates for the same amount of charge on the capacitor, diminishing the capacitor's voltage, and therefore raises the threshold for discharge by an internal spark.

$$\nabla \cdot \mathbf{E} = \frac{\rho}{\epsilon_0}, \qquad -\Delta_{\mathbf{r}}\varphi(\mathbf{r}) = \frac{\rho(\mathbf{r})}{\epsilon_0}. \tag{5.13}$$

For a point charge of unit charge located at $\mathbf{r}'$, corresponding to the charge distribution $\rho(\mathbf{r}) = \delta(\mathbf{r} - \mathbf{r}')$, the solution to Poisson's equation is the Coulomb potential, $G(\mathbf{r}, \mathbf{r}') \equiv \varphi_{\mathbf{r}'}(\mathbf{r})$,

$$-\Delta_{\mathbf{r}}G(\mathbf{r}, \mathbf{r}') = \frac{1}{\epsilon_0}\delta(\mathbf{r} - \mathbf{r}'), \qquad G(\mathbf{r}, \mathbf{r}') = \frac{1}{4\pi\epsilon_0|\mathbf{r} - \mathbf{r}'|} \tag{5.14}$$

the Green's function for Poisson's equation (see Appendix B). For a charge distribution consisting of point charges of different strengths and locations[3]

$$\rho(\mathbf{r}) = \sum_n q_n\delta(\mathbf{r} - \mathbf{r}_n). \tag{5.15}$$

The superposition principle for the electric field, that each point charge contributes additively with its strength according to Coulomb's law, dictates that the electrostatic potential is

$$\begin{aligned} \varphi(\mathbf{r}) &= \sum_n \frac{q_n}{4\pi\epsilon_0|\mathbf{r} - \mathbf{r}_n|} \\ &= \int d\mathbf{r}' \, \frac{\rho(\mathbf{r}')}{4\pi\epsilon_0|\mathbf{r} - \mathbf{r}'|}. \end{aligned} \tag{5.16}$$

The linearity between charge distribution and electrostatic potential embodied in the linearity of Poisson's equation is thus mathematically expressed via its Green's function,

$$\varphi(\mathbf{r}) = \int d\mathbf{r}' \, G(\mathbf{r}, \mathbf{r}')\,\rho(\mathbf{r}'). \tag{5.17}$$

According to Eq. (5.17), the linearity between the voltage and charge of a capacitor is exact, i.e. the capacitor's capacitance is independent of its charge. If the charge of the capacitor is changed, $q_n \to \lambda q_n$, the potential, and thereby the voltage, is, according to Eq. (5.17), similarly changed, $V \to \lambda V$.[4]

The expression for the capacitance in Eq. (5.12), the capacitance of an ideal plate capacitor, is only an approximation for a real finite plate capacitor. Owing to the strong Coulomb force, charges of equal sign spread out over the whole surface of a conductor, although the main charge for the plate capacitor with very close plates is on the inside, since plus

[3] The continuous charge density appearing in Maxwell's equations describing macroscopic objects is defined by adding up the point charges in a volume Ω (or surface)

$$\int_\Omega d\mathbf{r}\, \rho(\mathbf{r}) \equiv \sum_{\mathbf{r}_n \in \Omega} q_n.$$

[4] The conclusion also follows from the uniqueness of a solution of Laplace's equation satisfying a particular set of boundary conditions, so that, as a consequence: the potentials on N charged conductors are related to their charges according to a linear relation

$$\phi_i = \sum_{i=1}^{N} p_{ij}\, Q_j,$$

where the so-called coefficients of potential, p_{ij}, depend only on the geometry.

and minus charges attract each other. The neglect of the spill-out effect of charge means that the capacitance expression, Eq. (5.12), over-estimates the strength of the electric field and thereby the voltage; the estimate of the capacitance comes out a little too small, and introducing a larger *effective* plate area magnitude would give the correct value for the capacitance.

When the electric field changes in a capacitor, it induces a magnetic field. However, its magnitude is negligible except for very large frequencies. The electric field is therefore, with sufficient accuracy, curl-free, and the line integral of the electric field around a closed path vanishes

$$0 = \oint d\mathbf{l} \cdot \mathbf{E} = \int_{\substack{\text{along} \\ \text{wires}}} d\mathbf{l} \cdot \mathbf{E} + \int_{\substack{\text{between} \\ \text{plates}}} d\mathbf{l} \cdot \mathbf{E} + \int_{2}^{1} d\mathbf{l} \cdot \mathbf{E}, \tag{5.18}$$

where the last line integral is along a path outside the capacitor connecting terminals 2 to 1. The direction of the line integral has thus been chosen as the direction from the positively charged plate to the negatively charged plate and then further through the wires and outside the capacitor to close the loop. The wires are assumed perfect conductors, so the electric field vanishes inside it and we have

$$0 = \frac{Q}{C} - (\phi(1) - \phi(2)) \equiv \frac{Q}{C} - V, \tag{5.19}$$

where the voltage over the capacitor has been introduced,

$$V \equiv -\int_{2}^{1} d\mathbf{l} \cdot \mathbf{E} = \phi(1) - \phi(2) \tag{5.20}$$

viz. the electric potential difference between the *in* and *out* terminals (which calculated outside the capacitor equals *minus* the line integral between the *in* and *out* terminals calculated between the capacitor plates). The contribution to the line integral of the electric field picked up between the terminals is due to the bulging out of the field due to the spilled-out charge distribution to the outside surfaces of the capacitor plates and arms.

The voltage–charge relation for a capacitor is thus

$$V = \frac{Q}{C}, \tag{5.21}$$

where the exact value for the capacitance might not be obtainable analytically, but if needed can be obtained with sufficient accuracy by numerical computation.

The circuit theory symbol for an ideal capacitor, neglecting internal resistance and magnetic field properties, is depicted in Figure 5.7.

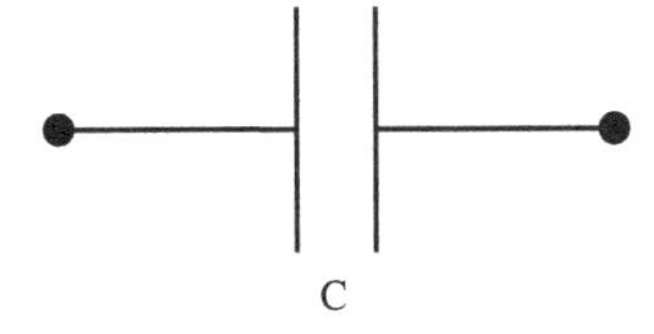

Figure 5.7 Circuit symbol for a capacitor.

Exercise 5.2 The Coulomb force is extremely strong compared to gravitation, which therefore can be neglected in electromagnetism. To get an appreciation, consider a conducting sphere with radius 1 cm charged by electrons to the electrostatic potential of –1 V. Calculate the density of excess electrons on the conductor, say, the number of electrons per square ångström (the permittivity of vacuum, its dielectric constant, has, to two-digit precision and in SI units, the magnitude $\epsilon_0 \simeq 8.8 \times 10^{-12}\,\mathrm{C}^2/\mathrm{N\,m}^2$).

A similar lesson that follows for the relevant voltages is that the charge per unit length on a wire, say, the capacitor arm, is very small when the radius of the wire is small. The electric field associated with the charge on a wire is only non-vanishing close to the wire.

A capacitor stores charge, which in turn means it stores electrical energy, the energy stored in the electric field in the capacitor. Let us calculate the amount of work, i.e. energy, needed to charge a capacitor. Assume the final charge state has been achieved by transferring small portions of the charge distribution, say electrons, in steps from one plate to the other. If the capacitor has the intermediate charge Q_i corresponding to voltage V_i, the transfer of the next small amount of charge, dQ, is done against an opposing electric field. The amount of work for each small surface charge being transported along its field line to its destination on the other plate is the magnitude of its charge times the line integral of the electric field. All these line integrals give the same value per unit charge, *viz.* the equipotential difference (the voltage V_i), and the work done is $V_i\,dQ$. The increase in electrostatic energy, dE_C, of the capacitor is thus

$$dE_C = V_i\,dQ. \tag{5.22}$$

Employing the voltage–charge relation for a capacitor, $V_i = Q_i/C$, gives, for the change in electrostatic energy due to a change in charge,

$$dE_C = \frac{Q_i}{C}\,dQ_i. \tag{5.23}$$

Integrating to the final charge value

$$\int_0^{E_C} dE_C = \int_0^{Q} dQ_i\,\frac{Q_i}{C} \tag{5.24}$$

gives, for the electrical energy stored in the capacitor with charge Q,

$$E_C = \frac{1}{2}\frac{Q^2}{C}. \tag{5.25}$$

According to Eq. (5.12), the capacitance is huge when the plates are very close. In that case, the amount of work required to charge up the capacitor is very small.

Exercise 5.3 Show that the electrical energy stored in the plate capacitor can be expressed by the electric field according to

$$E_C = \frac{\epsilon_0}{2} \int d\mathbf{x}\, \mathbf{E}^2. \tag{5.26}$$

Exercise 5.4 Assume capacitor plate distance 1 mm. Determine the plate area if the capacitance is one picofarad, $C = 1\,\text{pF} = 10^{-12}\,\text{F}$.

5.6 Capacitor Discharge

Consider the electric circuit obtained by connecting the terminals of a capacitor to those of a resistor. The circuit diagram representation of the physical circuit is depicted in Figure 5.8.

Before the switch is closed, the electrostatic potential is constant in each of the two disconnected parts of the circuit, but each has a different value, their difference being equal to the voltage of the capacitor, Q_0/C. The electrons in the negatively charged capacitor arm have the additional constant potential energy eQ_0/C relative to the electrons in the other arm. Everywhere in the disconnected circuit electrons are moving around as discussed in Section 4.8. Before the switch is closed, there is no electric field in the resistor, only around the capacitor and between the switch terminals (the line integral of the electric field around the open circuit vanishes for the electrostatic situation). When the switch is closed, a net amount of electrons rushes in from the high potential energy region, converting their potential energy into kinetic energy through experiencing the potential drop, thereby occupying wave packet states in the wire that were previously unoccupied.[5] A net current of electrons

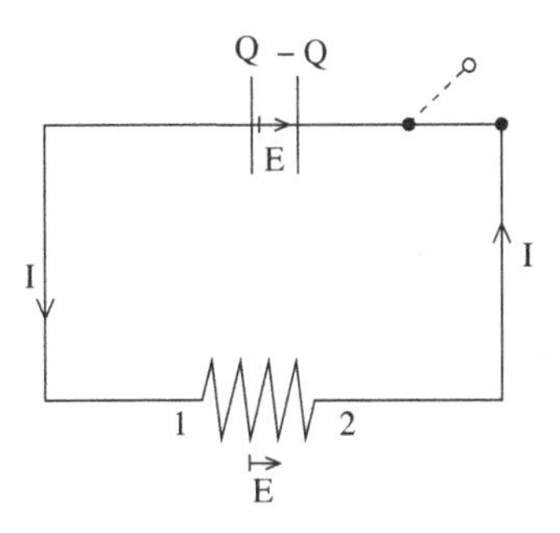

Figure 5.8 Discharge of capacitor through a resistor.

[5] To avoid a spark, vacuum tunneling, occurring just before closing impact, the switch should be constructed to be physically non-pointed, the opposite geometry to that of a scanning tunneling microscope tip, where a tunnel current is desired.

therefore flows from the negative capacitor electrode into the resistor terminal wire, the excess electron charge on its way producing the same potential in the tiny wire as in the negative capacitor arm. Owing to the strong Coulomb repulsion, a jolt of electron current appears in the whole circuit. Electrons entering the resistor are partly reflected and opposite charge distributions are established at the ends of the resistor, creating the necessary electric field to drive a current through the resistor, i.e. establishing the current flow implied by Ohm's law (the jump in electric field from zero in the wire to a finite value in the resistor is created by the charge located in the jump region, an excess or deficit of electrons at the respective ends). A steady current flows almost instantaneously in the entire circuit; the current is the same in the wires and resistor, the flow of the charged electrons being like an incompressible fluid. The time scale on which this is established can be estimated from the charge continuity equation

$$0 = \frac{\partial \rho(t)}{\partial t} + \nabla \cdot \mathbf{j} = \frac{\partial \rho(t)}{\partial t} + \sigma \nabla \cdot \mathbf{E}, \tag{5.27}$$

where the last equality follows from $\mathbf{j} = \sigma \mathbf{E}$, Ohm's law. Using Gauss's law $\nabla \cdot \mathbf{E} = \rho/\epsilon$, charge relaxation is seen to be governed by the equation

$$\frac{\partial \rho(t)}{\partial t} + \frac{\sigma}{\epsilon}\rho(t) = 0, \tag{5.28}$$

implying a charge equilibration time scale

$$\tau_{\text{eq}} = \frac{\epsilon}{\sigma} = \epsilon \rho. \tag{5.29}$$

Since typical values for the dielectric constant are at most up to 10 times that of vacuum, $\epsilon \sim 10\epsilon_0 \sim 10^{-10}\,\text{C}^2/\text{N}\,\text{m}^2$, even a material with resistivity as high as $10^8\,\Omega\,\text{m}$ gives electrostatic equilibration in a fraction of a second. In circuit theory, we are not interested in the initial non-equilibrium microscopic processes happening on this small time scale. Instead, we say that, at the time of switching, $t = 0$, the capacitor voltage, $V_0 = Q_0/C$, appears over the resistor through which the current according to Ohm's law starts to flow $RI(t = 0) = Q_0/C$. We shall say that the current instantaneously jumps from the value zero to this initial value, the current being the same throughout the circuit (except in between the capacitor plates where no current flows). Negative charge, electrons, leaves the negatively charged capacitor arm, and an equal amount of positive charge leaves the other plate, or rather the same amount of electrons are neutralizing the excess positive charge due to the initial deficit of electrons – the capacitor is discharging.

Consider at some instant, t, the line integral around the closed loop located inside the wires and the resistor and in between the capacitor plates. Since the magnetic field associated with the current is negligible, the electric field is *curl*-free and the line integral of the electric field through the circuit is zero at any moment of time,

$$0 = \frac{Q(t)}{C} + \int_2^1 d\mathbf{l} \cdot \mathbf{E}(\mathbf{r}, t), \tag{5.30}$$

where the first term on the right-hand side equals the line integral of the electric field between the capacitor plates at the moment its charge is $Q(t)$, and the integral is along any path inside the resistor starting at terminal 2 and ending at terminal 1. Let us assume that the resistor is a dirty metal where only electrons conduct. Electrons in the resistor are pushed by the electric field to move in the opposite direction of the field lines. The current density of the electrons is opposite to the electric field, whereas electric current density is parallel to the electric field due to the negative sign of the electron charge. The line integral of the electric field from terminal 2 to terminal 1 (from *out* to *in* terminal according to the wire direction chosen in Figure 5.8) is, according to Eq. (5.7), minus the voltage drop over the resistor, and Eq. (5.30) becomes

$$0 = \frac{Q(t)}{C} - RI(t). \tag{5.31}$$

With the chosen wire direction as depicted in Figure 5.8, the current is equal to minus the rate of change of charge leaving the positive capacitor plate,

$$I(t) = -\frac{dQ(t)}{dt}. \tag{5.32}$$

This equation is simply the charge continuity equation applied to the positively charged arm of the capacitor, and involves a chosen positive direction for the current as indicated by the wire arrows in Figure 5.8. A positive current flows in the direction indicated by the arrow.

The equation governing the capacitor charge dynamics is then

$$0 = \frac{Q(t)}{C} + R\frac{dQ(t)}{dt}, \tag{5.33}$$

which has the solution

$$Q(t) = Q_0\, e^{-t/RC}. \tag{5.34}$$

The capacitor is thus discharged through the resistor on the time scale $\tau_d = RC$.

Multiplying Eq. (5.31) with the current and using Eq. (5.32) gives the energy conservation equation

$$\frac{d}{dt}\left(\frac{Q^2(t)}{2C}\right) + RI^2(t) = 0. \tag{5.35}$$

The released electric field energy of the capacitor is dissipated in the resistor. The capacitor provides the resistor with the electric power (use Eq. (5.35))

$$V_C(t)I(t) = I(t)Q(t)/C = RI^2(t). \tag{5.36}$$

The initial electrostatic energy stored in the capacitor, $Q_0^2/2C$, is eventually dissipated in the resistor, producing the total amount of heat

$$W = \int_0^\infty dt'\, RI^2(t') = -\int_0^\infty dt'\, \frac{d}{dt'}\left(\frac{Q^2(t')}{2C}\right) = \frac{Q_0^2}{2C}. \tag{5.37}$$

For the simple circuit depicted in Eq. (5.8), we are able to describe its functioning, to a good approximation, in terms of what goes on inside the components, the charge dynamics of the components. However, we can also describe the circuit only in terms of what goes on outside the components, in terms of the current and the potential difference, or voltage drop, between terminals. This opposite way of thinking is the spirit of circuit theory: by idealizing the components, the need to deal with charge intricacies in devices is avoided.

For a capacitor, the standard convention is to have the wire arrow point toward the capacitor plate labeled by Q. With this convention, the amount of charge on a capacitor, i.e. the amount of charge on the capacitor part labeled by Q, is the result of its charging history,

$$Q(t) = \int^{t} dt' \, I(t'). \tag{5.38}$$

In this convention, the resistor–capacitor (RC) circuit will have wire arrows as depicted in Figure 5.9. In order to obtain the current dynamics, a time derivative is taken of circuit equations containing capacitors. Often the relevant lower time limit in Eq. (5.38) is irrelevant or unknown, and a time derivative thereby removes this initial information, the capacitor's charge history. The discharging of the capacitor is a counter-example where the initial information about the charge state is used. Nevertheless, let us analyze the RC circuit using the standard wire direction convention and circuit theory discourse.

Consider the closed curve depicted in Figure 5.9, lying inside the perfect wires but connecting terminals by going outside the components. Assuming the magnetic field to be negligible, the line integral of the electric field around this closed curve vanishes,

$$0 = \oint d\mathbf{l} \cdot \mathbf{E} = V_R + V_C, \tag{5.39}$$

where the voltage over the resistor according to Eq. (5.6) is

$$V_R = \int_{\text{in}}^{\text{out}} d\mathbf{l} \cdot \mathbf{E} = RI \tag{5.40}$$

and the voltage over the capacitor according to Eq. (5.20) is

$$V_C = \int_{\text{in}}^{\text{out}} d\mathbf{l} \cdot \mathbf{E} = \frac{Q}{C}. \tag{5.41}$$

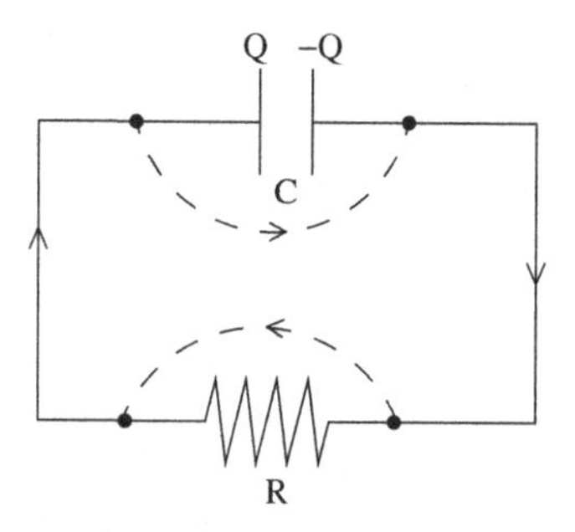

Figure 5.9 Capacitor–resistor loop.

In both cases the line integral of the electric field is between the *in* and *out* terminals taken outside the components.

The charge on the capacitor is specified by its charging history according to Eq. (5.38), giving

$$0 = RI(t) + \frac{\int^t dt' \, I(t')}{C}. \tag{5.42}$$

Taking the time derivative on both sides of Eq. (5.42) gives the equation for the current dynamics,

$$\frac{I(t)}{C} + R\,\frac{dI(t)}{dt} = 0, \tag{5.43}$$

which is immediately solved to give

$$I(t) = I_0\, e^{-t/RC}. \tag{5.44}$$

With the chosen wire direction, the initial voltage over the resistor is, according to Eq. (5.42),

$$RI(t=0) = -\frac{Q(t=0)}{C} = -\frac{Q_0}{C}, \tag{5.45}$$

giving for the current

$$I(t) = -\frac{Q_0}{RC}\, e^{-t/RC}, \tag{5.46}$$

the negative sign of the current signaling in this wire direction convention that the capacitor is discharging.

A capacitor has many applications in electric circuits. It can provide charge to another device, or provide a definite voltage in response to a change in charge. A touch-screen can, for example, function by the change of capacitance by a pushed plate of a capacitor, a voltage thereby fed to all the relevant input terminals of a processor. A capacitor evens out voltage changes and can thus be used to protect other components against too big voltage fluctuations.

5.7 Battery

Consider the circuit of the previous section, a charged capacitor in series with a resistor. In the limit of the capacitance and its charge becoming infinitely large in such a fashion that their ratio $Q/C \equiv V_C$ is constant, the circuit depicted in Figure 5.8 would have a constant current $I = V_C/R$. The *infinite* capacitor is a voltage source, always having the constant voltage, V_C, between its terminals. The device acts as a battery with voltage Q/C, providing the electric power IV_C to the load (for the circuit depicted in Figure 5.8 just the resistor). A real capacitor could be turned into a battery if some internal agents would ensure that, whenever an electron leaves the negatively charged plate, an electron arrives transported by an agent from the positively charged plate (or an external agent, say a friction-based device turning mechanical energy into electrical energy, such as a Van de Graaff generator). Thereby the charge state of the capacitor is kept the same, and the electrons in the negatively charged terminal are held at the additional potential energy eQ/C relative to the electrons in the positively charged terminal. Per unit charge transported, the agents would have to produce work equal to the voltage, $V = Q/C$, against the electric field inside the capacitor, turning *agent* energy into an equal amount of electrical energy of the capacitor. Batteries are devices that can turn one type of energy, say, chemical or solar, into electrical energy.

A d.c. battery is a device where, for example, 1.5 V is stamped on its body and two metallic contacts stick out. What that means is that one of the electrodes sticking out, denoted the *plus* terminal, has a higher electric potential of 1.5 V relative to the other electrode, the negative electrode or terminal. In the negative electrode, the electrons have the additional constant potential energy of 1.5 eV relative to the potential energy of the electrons in the positive electrode. Consider the line integral of the electric field along a closed curve imagined inside the battery and its terminals, and completed by the curve emanating from the positive terminal and reaching the negative terminal in the external region. This line integral will have the value of 1.5 V. In the open-circuit situation, the line integral only picks up a contribution from the path outside the battery, the electric field created by small amounts of surface charge on the terminals.

Consider now a battery connected to a load, say, a resistor. An ideal battery keeps the electric potential difference between its terminals at a constant value even when a load draws a current (as in the case of the *infinite* capacitor). A battery is said to be a source of electromotive force (e.m.f.). The line integral of the force per unit charge carrier, $\mathbf{F}_u$, around the just mentioned closed loop through the battery and connecting the terminals through an outer circuit, say, a resistor, is equal to the e.m.f. of the battery,

$$\mathcal{E} \equiv \oint d\mathbf{l} \cdot \mathbf{F}_u = 1.5\,\mathrm{V}, \tag{5.47}$$

but now the contribution to the line integral comes from a non-electric source. In the capacitor agent set-up *battery* with a resistor as load, the line integral contribution from the electric field, being *curl*-free, integrates to zero. The e.m.f. is the work per unit delivered charge that the *agents* perform, which is equal to the voltage of the capacitor. When there is an e.m.f., there is a net force on the charge carriers in a circuit and therefore a push of charges: the e.m.f. gives rise to a current flowing in the direction for which the line integral in Eq. (5.47) is positive.

To illustrate a real battery, consider the electrochemical cell consisting of two metal plates immersed in an acid solution, one plate made of zinc and the other of the non-reactive metal, platinum. The zinc plate is slowly dissolving zinc ions, Zn^{2+}, into the solution, leaving the zinc plate charged with electrons, the negative electrode. The hydrogen ions, H^+, of the acid (having divorced from a molecule, leaving a negative ion X^- in the solution) migrate and at the platinum plate combine with electrons forming hydrogen gas, H_2, bubbling up through the solution. Having had electrons extracted leaves the platinum plate positively charged. When the plates are not connected externally, the electric potential difference between the plates reaches a maximum value, the zinc plate becoming so negative

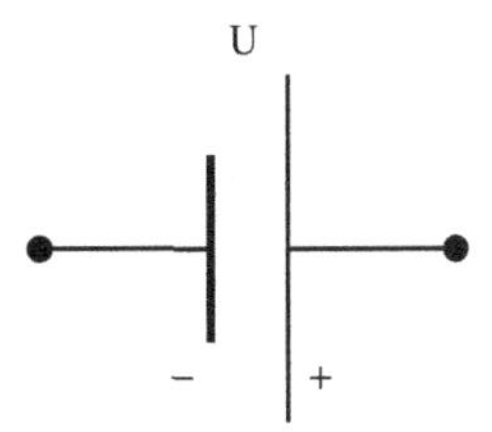

Figure 5.10 Circuit symbol for an ideal battery.

that zinc ions cannot leave the plate and the platinum plate being so positive that the protons, H^+, cannot capture electrons. The strong Coulomb force between the ions in the acid solution will make it overall charge neutral, and, according to Gauss's law, the electric field vanishes in the solution. When the plates are connected by a resistive wire, a current of electrons flows through it from the negative to the positive electrode and a corresponding equal amount of positive ionic current flows inside the battery. The cell keeps a constant voltage difference at its terminals, i.e. operates as a battery. The device stops operating as a battery when the zinc plate is dissolved.

The potential difference between the terminals of a real battery decreases when the current drawn increases; because the internal ionic transport process experiences resistance, the diffusing ions cannot completely keep up the short-circuit battery voltage. For small currents, this voltage drop is proportional to the current, and a real battery is thus represented as an ideal battery in series with a resistor R_i (if needed, or absorbed in the resistance of the external system). All the power of the battery is thus not available for electrical consumption between its terminals. A part is consumed by the internal charge transport processes and there transformed into heat; the battery heats up when a current is drawn.

The circuit symbol for an ideal battery is depicted in Figure 5.10, its e.m.f. or battery voltage, U, indicated as well as the positive and negative terminals, the latter by the shorter line.

The standard arrow directions for the battery terminals is pointing in to the negative terminal and out from the positive terminal. For a battery connected to, for example, a resistor, the I accompanying the arrow is then of positive value. The line integral of the electric field taken outside the battery from the negative to the positive terminal is $-U$, minus the electromotive force.

Consider a battery in a circuit loop. When a current I is drawn from the ideal battery, the battery delivers the electric power $P = UI$. The amount, I, of charge is, per unit time, transported around the loop having an electric potential difference U over the load, and the amount of work per unit time performed on the charges in the load is UI; the internal (say, chemical) battery energy is converted into electrical energy. The power delivered to the external system by a real battery is $P = (U - R_i I)I$.

Exercise 5.5 Consider connecting, at time $t = 0$, a battery of voltage U to an uncharged capacitor in series with a resistor. Show that the circuit equation is

$$0 = -U + \frac{Q}{C} + RI. \tag{5.48}$$

Show that the current is given by

$$I(t) = \frac{U}{R} e^{-t/\tau_c}, \qquad \tau_c = RC. \tag{5.49}$$

When $t \gg \tau_c$, the capacitor has been charged to the value $Q \simeq CU$ and the current vanishes. The capacitor is an open-circuit device, and for the present circuit eventually charged to the maximum value the battery can provide

$$Q_{\max} = \int_0^\infty dt\, I(t) = C\,U \tag{5.50}$$

counteracting the battery voltage, and there is no current flowing.

Note that the power provided by the battery, $UI(t)$, equals the power dissipated in the resistor plus the rate of change in the capacitor energy,

$$UI(t) = RI^2(t) + \frac{d}{dt}\left(\frac{Q^2(t)}{2C}\right). \tag{5.51}$$

5.8 Inductor

A current in a wire generates a magnetic field, its direction specified by the right-hand rule (the Biot–Savart law). If your right thumb points in the direction of the current, the fingers specify the direction of the magnetic field lines circling the wire. To get an appreciable magnetic field, the cumulative effect of many wires is needed. An inductor is the device obtained by winding many turns of wire into a coil and bringing out terminals away from the coil, as depicted schematically in Figure 5.11. In the spatial region inside the coil, a strong magnetic field is generated by the current in the many close windings of the coil.

We assume that no magnetic field is present in the region near the terminals. This can be achieved by circling the coil or solenoid up into the shape of a torus and twisting the

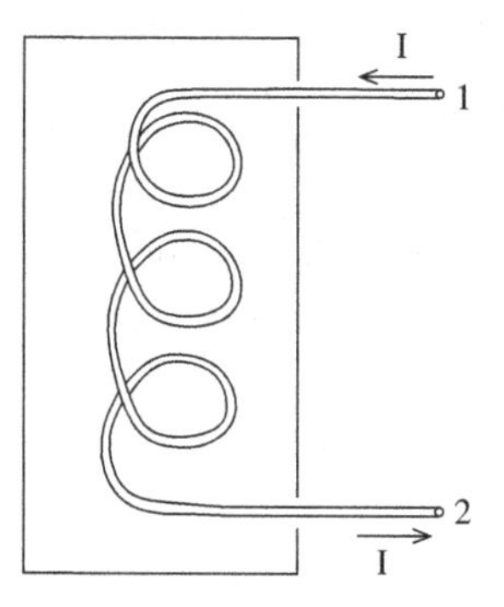

Figure 5.11 An inductor.

terminal wires together, or by encapsulating the coil inside a metal box, as depicted in Figure 5.11.

When the current changes, the magnetic field inside the coil changes and an electric field circling the magnetic field lines is created according to the law of induction,

$$\nabla \times \mathbf{E}(\mathbf{x}, t) = -\frac{\partial \mathbf{B}(\mathbf{x}, t)}{\partial t}. \tag{5.52}$$

Consider a closed curve winding inside the coil wire and externally connecting terminal 2 to terminal 1, thus specifying the direction of the curve. Assuming that the coil wire is a perfect conductor, the electric field vanishes inside. A surface charge distribution is therefore generated on the perfect coil wire in order to cancel out the electric field created by the time-varying magnetic field. A charge distribution is set up, which must be plus and minus charges on neighboring coil wires skewed in location relative to each other. Since these opposite charges are very close, due to the neighboring coil wires being very close, we can assume that these charges do not produce any electric field in the external region between the terminals. Thus by idealizing the coil wire to be a perfect conductor, we do not have to deal with its intricate charge distribution (at least, not for low-frequency currents as discussed in Section 5.11). The line integral along the described closed curve becomes, by applying Stokes' theorem,

$$\oint d\mathbf{l} \cdot \mathbf{E} = \int_S d\mathbf{s} \cdot \nabla \times \mathbf{E}(\mathbf{x}, t) = -\int_S d\mathbf{s} \cdot \frac{\partial \mathbf{B}(\mathbf{x}, t)}{\partial t} = -\frac{d}{dt} \int_S d\mathbf{s} \cdot \mathbf{B}(\mathbf{x}, t), \tag{5.53}$$

where the surface S is the surface of the circling coil. Its actual physical shape could be visualized by a soap foam clinging to the loops of the coil. The line integral is thus

$$\oint d\mathbf{l} \cdot \mathbf{E} = -\frac{d\Phi(t)}{dt}, \tag{5.54}$$

where Φ is the total magnetic flux through the coil loops. We note that for the chosen curve the vector $d\mathbf{s}$ according to the right-hand rule has the same direction as the magnetic field, and the change in magnetic flux in the coil thus induces an electromotive force determined by the rate of change of the flux, Faraday's law,

$$\mathcal{E}(t) = \oint d\mathbf{l} \cdot \mathbf{E} = \int_2^1 d\mathbf{l} \cdot \mathbf{E} = -\frac{d\Phi(t)}{dt}. \tag{5.55}$$

The second equality follows from the coil wire being a perfect conductor, so that only a line integral of the electric field between terminals 2 and 1 is picked up. The wire direction has been chosen such that a positive current gives a positive magnetic flux. If the coil is tightly wound so that the total area is, with sufficient accuracy, given by the area of a single winding times the number of windings, the total magnetic flux, Φ, is the sum of the flux through each of the coil loops. The magnetic flux in the coil is due to its current, $\Phi(t) = \Phi(I(t))$, and linearly related to the current,

$$\mathcal{E}(t) = -L\frac{dI(t)}{dt}, \qquad L = \frac{d\Phi}{dI}, \tag{5.56}$$

where the proportionality factor, L, is called the (self-)inductance of the inductor.

If the positive (negative) current through the coil is increased, the rate of change of the current is positive, and the e.m.f. of the coil, i.e. minus the flux change, is negative. The power delivered by the coil to the rest of the circuit is in that case negative, $\mathcal{E}I < 0$; the coil is absorbing energy from the electrical energy of the circuit, storing it in the increased magnetic field in the coil. If the current through the coil decreases, the induced e.m.f. is positive, $\mathcal{E} > 0$, and the coil delivers the power $\mathcal{E}I > 0$ to the circuit as the energy stored in the magnetic field in the coil decreases.

When a coil is present in a circuit, the current in that section of the circuit is a continuous function of time, as an infinite e.m.f. precludes a discontinuity in the current. An ideal coil allows a constant current to flow without any voltage difference between its terminals. However, in the case of a current *change*, an electromotive force is induced in the opposite direction to the current direction, i.e. the e.m.f. induces a current opposing the change. The strength of opposition is specified by the self-inductance, L, of the coil. Similarly, if the current decreases, the e.m.f. is positive, i.e. in the direction of the current and therefore again opposing the change in current. An inductor thus endows the current dynamics with *inertia* just as mass provides the inertia of the motion of a particle as dictated by Newton's equation: a change in the velocity of a particle requires a force, and the mass of the particle ensures the continuity of the particle trajectory as a function of time (as well as continuity of the velocity unless the pathological case of a delta-function kick is considered). Mathematically, the equations of circuit theory are thus formally identical to those of elementary mechanics, and knowledge of the solutions of the latter translates into solutions of circuit equations.

The behavior of the inductor can also be described in terms of the electric potential difference between its terminals, its voltage. The voltage induced at the terminals of the coil, the line integral of the electric field outside the coil from the *in* to the *out* terminal, is

$$V(t) \equiv \int_1^2 d\mathbf{l}\cdot\mathbf{E} = \varphi(1) - \varphi(2) = -\mathcal{E}(t) = L\frac{dI(t)}{dt}. \tag{5.57}$$

The circuit provides the power

$$V(t)\,I(t) = L\frac{dI(t)}{dt}\,I(t) = \frac{d}{dt}\left(\frac{LI^2(t)}{2}\right) \tag{5.58}$$

to the inductor. Thus in order to build up a current $I(t)$ in the inductor, the work

$$E_L = \int_0^t dt'\,\frac{d}{dt'}\left(\frac{LI^2(t')}{2}\right) \tag{5.59}$$

has to be done, i.e. the inductor has the magnetic energy

$$E_L = \tfrac{1}{2}LI^2(t). \tag{5.60}$$

The circuit diagram symbol for an ideal inductor is depicted in Figure 5.12.

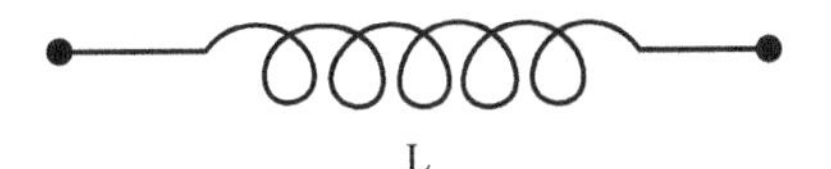

Figure 5.12 Circuit symbol for an inductor.

Exercise 5.6 Consider connecting, at time $t = 0$, a d.c. battery of voltage U to an inductor in series with a resistor. Show that the circuit equation is

$$0 = -U + L\frac{dI}{dt} + RI. \tag{5.61}$$

Show that the current is given by

$$I(t) = \frac{U}{R}(1 - e^{-t/\tau}), \qquad \tau = \frac{L}{R}. \tag{5.62}$$

When $t \gg \tau$, the maximum current is reached, $I_{\max} \simeq U/R$, and the voltage over the inductor vanishes.

Note that the power provided by the battery $UI(t)$ equals the power dissipated in the resistor plus the rate of change in the inductor energy,

$$UI(t) = RI^2(t) + \frac{d}{dt}\left(\frac{LI^2(t)}{2}\right). \tag{5.63}$$

5.9 Voltage Oscillator

Consider a capacitor connected to an inductor, as depicted in the circuit diagram of Figure 5.13. Assume initially that the capacitor is charged, $Q(t = 0) = Q_0$. When the switch is closed, at time $t = 0$, a current starts flowing, building up a magnetic field in the inductor.

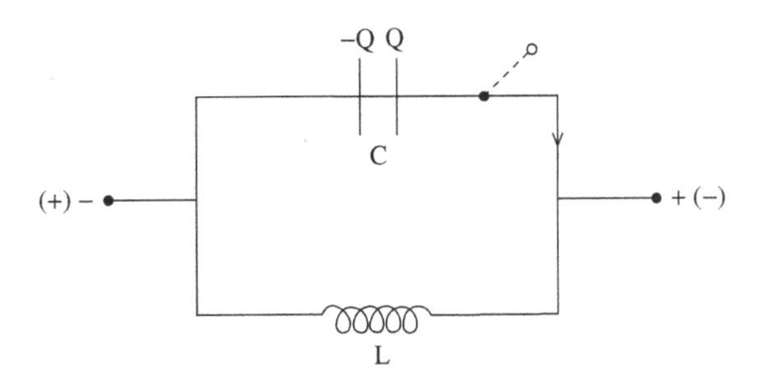

Figure 5.13 Charged capacitor connected to an inductor.

The potential difference of the inductor terminals, *in* minus *out*, is according to Eq. (5.57)

$$V(t) = \varphi_1 - \varphi_2 = L\frac{dI(t)}{dt}. \tag{5.64}$$

The potential difference $\varphi_1 - \varphi_2$, however, is also equal to the electric potential difference between the positive and negative capacitor plates, which, according to Eq. (5.21), is equal to $Q(t)/C$, giving the equation determining the dynamics of the circuit,

$$L\frac{dI(t)}{dt} = \frac{Q(t)}{C}. \tag{5.65}$$

We are using *discharge* current convention, as the wire direction in Figure 5.13 indicates, and

$$I(t) = -\frac{dQ(t)}{dt}, \tag{5.66}$$

and differentiating Eq. (5.65) gives

$$L\frac{d^2I(t)}{dt^2} + \frac{I(t)}{C} = 0. \tag{5.67}$$

Owing to the presence of the inductor, the current is a continuous function of time, giving the initial condition at the closing of the switch, $I(t = 0) = 0$, and the solution to Eq. (5.67) is

$$I(t) = I_0 \sin(\omega t), \tag{5.68}$$

where the current oscillation frequency is $\omega = 1/\sqrt{LC}$. The current amplitude at the time of switching is determined by Eq. (5.65),

$$L\frac{dI(t = 0)}{dt} = \frac{Q_0}{C}, \tag{5.69}$$

where Q_0 is the initial charge of the capacitor, giving the current amplitude, $I_0 = Q_0/\sqrt{LC}$, in terms of the circuit parameters and the initial charge on the capacitor. The current expressed in terms of the parameters of the problem is thus

$$I(t) = \frac{Q_0}{\sqrt{LC}} \sin(\omega t), \tag{5.70}$$

and the time dependence of the charge on the capacitor is then, according to Eq. (5.65),

$$Q(t) = Q_0 \cos(\omega t). \tag{5.71}$$

Multiplying Eq. (5.65) by the current gives the circuit's energy conservation equation,

$$0 = I(t)L\frac{dI(t)}{dt} - I(t)\frac{Q(t)}{C} = \frac{d}{dt}\left(\frac{L}{2}I^2(t)\right) + \frac{d}{dt}\left(\frac{Q^2(t)}{2C}\right), \tag{5.72}$$

which describes the oscillation of electromagnetic energy between the capacitor and inductor, the electric field energy in the capacitor and the magnetic field energy in the inductor. At time $t = 0$ the circuit's energy is completely in the capacitor's electrical energy; then it diminishes as it discharges and magnetic energy builds up in the inductor. At time $t = \pi/2\omega$, the circuit's energy is completely in the magnetic field of the inductor; and as the current changes sign, the capacitor starts to be oppositely charged until all the magnetic energy in the inductor is again converted to electric field energy in the capacitor at cycle time $T = 2\pi/\omega$.

The considered problem is mathematically identical to the oscillations of a harmonic oscillator, $m\ddot{x} = -kx$, its repeated oscillations of kinetic and potential energy. The inertia of the current due to the inductance plays the role of the mass inertia of the particle, $L \leftrightarrow m$; and the inverse of the capacitance plays the role of the force constant, $1/C \leftrightarrow k$. The circuit dynamics can similarly be viewed as the repeated oscillations between *kinetic* and *potential* energies, as the sum of the magnetic energy in the inductor $LI(t)^2/2$ and the electrical energy in the capacitor $Q(t)^2/2C$ is conserved and therefore, according to Eq. (5.72), equal to the initial electrostatic energy stored in the capacitor,

$$\frac{L}{2} I(t)^2 + \frac{1}{2C} Q(t)^2 = \frac{1}{2C} Q_0^2. \tag{5.73}$$

The considered ideal inductor–capacitor (LC) circuit has the interesting property that, as depicted in Figure 5.13, the voltage on the two terminals are opposite in sign and oscillating. Such a device could swap input values storing bit information. However, due to a real system exhibiting resistance, the oscillations will in practice quickly die out, as discussed in Section 5.10.

Exercise 5.7 Discuss the circuit in Figure 5.13 in standard circuit theory convention, i.e. assign the opposite current direction compared to that in Figure 5.13. Outside the inductor, the magnetic field is assumed to vanish, and the line integral of the electric field around the closed loop inside the wire and outside components connecting their terminals thus vanishes,

$$0 = \oint d\mathbf{l} \cdot \mathbf{E} = \int_{1,C}^{2,C} d\mathbf{l} \cdot \mathbf{E} + \int_{1,L}^{2,L} d\mathbf{l} \cdot \mathbf{E}. \tag{5.74}$$

Show that, according to Eqs. (5.20) and (5.57), this becomes the equation

$$L\frac{dI(t)}{dt} + \frac{Q(t)}{C} = 0, \tag{5.75}$$

with charge current convention given by Eq. (5.42). Show that the current is specified by

$$I(t) = -\frac{Q_0}{\sqrt{LC}} \sin(\omega t). \tag{5.76}$$

The circuit equation (5.65) can also be rewritten $-\mathcal{E}(t) + V_C(t) = 0$ and its content rephrased as *the sum of the voltages over the components add to zero*, and an e.m.f. is counted with a minus sign. This is a statement about a *curl*-free electric field, or energy conservation in the instantaneous electric field around a loop, i.e. Kirchhoff's second law: the line integral of the electric field around a closed loop taken between the terminals outside the components must vanish.

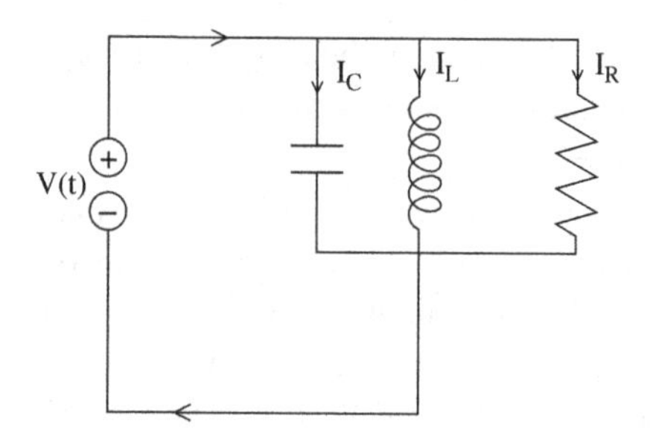

Figure 5.14 Externally excited LC oscillator.

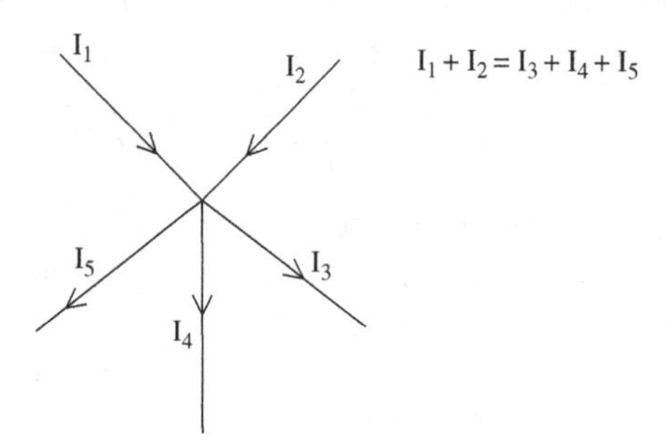

Figure 5.15 The algebraic sum of incoming current equals that of the outgoing.

An oscillator can also be excited externally. Consider, for example, the circuit in Figure 5.14. Assume a spark occurs at the $\pm$ terminals inducing currents in the incompressible electron liquid in the components as indicated in Figure 5.14. The voltage, $V(t)$, is the same over the three components,

$$V(t) = L \frac{dI_{\mathrm{L}}(t)}{dt} = \frac{Q}{C} = RI_{\mathrm{R}}, \tag{5.77}$$

giving for the currents

$$I_{\mathrm{R}} = \frac{V(t)}{R}, \qquad I_{\mathrm{C}} = C \frac{dV(t)}{dt}, \qquad I_{\mathrm{L}} = \frac{1}{L} \int^{t} dt' \, V(t'). \tag{5.78}$$

In the circuit in Figure 5.14, we observe the feature that a wire can split. A wire can in general split into any number of wires, fan out, and we thus encounter the termination of wires at a point, at a so-called circuit node as depicted in Figure 5.15. At a node, part of a perfect conductor, there can be no charge pile-up, and the algebraic sum of the currents in wires with their direction toward the node equals the algebraic sum of the currents in wires that have their direction pointing away from the node. The requirement of *no charge pile-up* at a node is referred to as Kirchhoff's first law and its content is encapsulated in the equation stated alongside the node depicted in Figure 5.15.

In the considered open-circuit situation in Figure 5.14, the currents through the components adds to zero

$$I_{\mathrm{C}} + I_{\mathrm{L}} + I_{\mathrm{R}} = 0. \tag{5.79}$$

Inserting into the current conservation equation, Eq. (5.79), the expressions from Eq. (5.78), and performing a time differentiation, the equation determining the voltage becomes

$$C\frac{d^2V}{dt^2} + \frac{1}{R}\frac{dV}{dt} + \frac{V(t)}{L} = 0. \tag{5.80}$$

Only for a lossless coil does the circuit exhibit persistent voltage oscillations, as in reality it has resistance and the solution is (a reminder of how to solve a linear second-order differential equation is given in the next section in connection with Eq. (5.85))

$$V(t) = V_0\, e^{-t/2RC}\cos(\omega t), \tag{5.81}$$

representing damped oscillations where the oscillation frequency is

$$\omega = \sqrt{\frac{1}{LC} + \frac{1}{(2RC)^2}}. \tag{5.82}$$

To have sustained voltage oscillation, one could provide a current pulse for each oscillation cycle. The oscillating charge on the terminals, forming a dipole, emits radiation, and the circuit can work as a transmitter; in fact, the set-up was used in early wireless transmitters. The circuit also works in the reverse mode, where incoming electromagnetic fields generate a voltage, for example, in electrical tuning where one wants optimal response at a particular frequency. Radios and cellular telephones and other wireless gadgets work on the same principle, except that electronic excitation of circuits is used.

5.10 Damped Oscillator

Even if an active element such as a d.c. battery is introduced into a circuit, resistance leads to damping. Consider, for example, the circuit depicted in Figure 5.16, and assume the switch is closed at time $t = 0$.

Now follow the standard procedure of circuit theory: form the vanishing line integral of the electric field around the closed loop between the terminals outside of the components. Thereby the circuit equation follows as

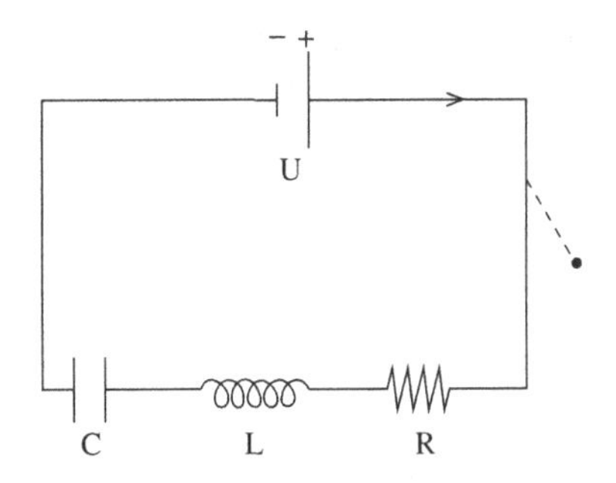

Figure 5.16 Capacitor, inductor and resistance in series connected to a battery.

$$0 = -U + L\frac{dI(t)}{dt} + RI(t) + \frac{Q(t)}{C} \tag{5.83}$$

or rewriting

$$U = L\frac{dI(t)}{dt} + RI(t) + \frac{Q(t)}{C}. \tag{5.84}$$

Thus Kirchhoff's second law can be rephrased: the battery voltage or electromotive force, U, is dropped over the components.

Taking the time derivative, the circuit equation becomes[6]

$$L\frac{d^2I(t)}{dt^2} + R\frac{dI(t)}{dt} + \frac{I(t)}{C} = 0. \tag{5.85}$$

Since upon differentiation with respect to $I(t)$ the equation demands the same function back, the solution must be the exponential function, and seeking the solution in the form $I(t) \propto \exp(\lambda t)$ turns Eq. (5.85) into the algebraic equation for λ,

$$L\lambda^2 + R\lambda + \frac{1}{C} = 0, \tag{5.86}$$

which has in general the two solutions

$$\lambda_\pm = -\frac{R}{2L} \pm \frac{R}{2L}\sqrt{1 - \frac{4L}{CR^2}}. \tag{5.87}$$

The general solution to the linear second-order differential equation (5.85) is thus specified by the two roots and

$$I(t) = I_1\, e^{\lambda_+ t} + I_2\, e^{\lambda_- t}. \tag{5.88}$$

Owing to the presence of the inductor, the sought solution should satisfy the initial condition $I(t = 0) = 0$, and thereby $I_2 = -I_1$, giving

$$I(t) = I_1 \exp\left(-\frac{R}{2L}t\right)\left[\exp\left(t\frac{R}{2L}\sqrt{1 - \frac{4L}{CR^2}}\right) - \exp\left(-t\frac{R}{2L}\sqrt{1 - \frac{4L}{CR^2}}\right)\right]. \tag{5.89}$$

The circuit equation thus allows for three different types of dynamics depending on parameter values. In the over-damped regime, $R^2 > 4L/C$, and for critical damping, $R^2 = 4L/C$, both roots are negative, and the current therefore vanishes exponentially (in the latter case the faster). When the resistance is small, $R^2 < 4L/C$, the under-damped regime is encountered, where the roots have imaginary parts,

$$\lambda_\pm = -\frac{R}{2L} \pm i\omega, \qquad \omega = \sqrt{\frac{1}{LC} - \frac{R^2}{4L^2}}, \tag{5.90}$$

and the current

$$I(t) = I_0\, e^{-(R/4L)t} \sin(\omega t) \tag{5.91}$$

[6] This equation has the well-known form from classical mechanics, *viz.* that of a harmonic oscillator with friction, the resistance R corresponding to the friction constant.

oscillates with frequency ω before eventually the exponential envelope kills the oscillation on the time scale $\tau \sim L/R$. Assuming initially that the capacitor is uncharged, $Q(t=0)=0$, Eq. (5.83) reduces at time $t=0$ to

$$L\frac{dI(t=0)}{dt} = U, \tag{5.92}$$

determining the current amplitude in terms of the circuit parameters to be

$$I_0 = \frac{U}{\sqrt{L/C - R^2/4}}. \tag{5.93}$$

5.11 Real Components

To circumvent the hopeless task of obtaining, from Maxwell's equations, the exact electromagnetic properties of components, with their intricate charge and current dynamics, approximations were introduced. The analysis of resistors, capacitors and inductors was thereby made simple, and the approximations made for real physical systems lead to the diagrammatic circuit theory of ideal components. However, any real component shows aspects of all three ideal elements. For example, in a real resistor, the magnetic field created by its current means that a real resistor must be ascribed an inductance. Also, the charges at the ends of the resistor form a capacitor, and the resistor must also be ascribed a capacitance. A real resistor should thus be represented by an ideal resistor in series with an ideal inductor, both in parallel with an ideal capacitor. Similarly, a change in charge of a capacitor changes its electric field, which in turn creates a magnetic field, so that a real capacitor also has inductance.

A real inductor is made of a real conductor, which is not perfect and thus exhibits resistance. Furthermore, opposite charges on opposing wire turns act as capacitors, so a real inductor should thus be modeled as an ideal inductor in series with an ideal resistor and both in parallel with an ideal capacitor. The question of when such realities become of practical importance should thus be addressed, and since their presence is related to the changes in current, it is a matter of frequency.

The obtained ideal I–V characteristics are also valid for a.c. currents, as long as the electric field is described with sufficient accuracy by an instantaneous electrostatic field, i.e. it is *curl*-free. In view of the linearity of a component's I–V characteristics and ensuing linearity of the circuit equations, it is convenient to introduce complex *currents* and *voltages* oscillating with the frequency in question (instead of the clumsier trigonometric functions),

$$I(t) = I_0\, e^{i\omega t}, \qquad V(t) = V_0\, e^{i\omega t}, \tag{5.94}$$

all physical currents and voltages then being represented by, say, the real part of the complex quantities. The complex current–voltage characteristics are then represented by so-called impedances of the ideal components,

$$Z_{\mathrm{R}} = R, \qquad Z_{\mathrm{L}} = i\omega L, \qquad Z_{\mathrm{C}} = \frac{1}{i\omega C}, \tag{5.95}$$

expressing the linear relationship between the complex *currents* and *voltages*,

$$V(t) = Z(\omega) I(t). \tag{5.96}$$

A real component is a black box, but by linearity voltage and current are related by Eq. (5.96), and its impedance $Z(\omega)$ can be represented in terms of the basic impedance expressions of Eq. (5.95).

Currents flow mainly through components with small impedance given the option. The importance of the various ideal component parts describing a real device thus depends on the frequency in question. Consider, for example, a real inductor for which its ideal *inductor* impedance magnitude varies as ωL. Then, at very low frequencies, the impedance of the real inductor is given by the coil's resistance, the change in current being so small that the induced e.m.f. is negligible. For higher frequencies, where ωL is bigger than the resistance, the coil is ideal. The opposite charge distributions on opposing coil turns form capacitors, adding up to an effective capacitance of the coil, and at very high frequencies it is the main component determining the real inductor impedance due to the smallness of the capacitor impedance at high frequencies, $C/i\omega$. The capacitor in parallel with the ideal inductor and resistor draws the current. Wire capacitors with very small impedance will therefore short-circuit the in-parallel inductor and resistor – the alternating current, instead of taking turns in the coil, passes through the capacitors.

Acceleration of charge produces electromagnetic radiation and an a.c. circuit thus loses electromagnetic energy to its surrounding. An electromagnetic wave of frequency ω has the wave length $\lambda = 2\pi c/\omega$, where c is the speed of light (in vacuum, $c = 1/\sqrt{\epsilon_0 \mu_0} = 299\,792\,458\,\text{m/s}$). Thus the criterion for neglecting a circuit's radiation loss (as done in circuit theory) is that its size, L, is much smaller than the wave length, $L \ll 2\pi c/\omega$, since in that case, for every element of the circuit with a certain current flow direction, there is another element close by where the current is in the opposite direction, and far from the circuit the radiation field from these two sources cancel each other, thus confining the electromagnetic field to the neighborhood of the circuit. To get appreciable electromagnetic radiation requires special design, the branch of electromagnetism called antenna technology.

5.12 Summary

The physics of so-called passive or linear electric circuit elements, resistor, capacitor and inductor, have been discussed. The word *passive* refers to the fact that the device simply responds to external conditions, whereas an active element such as a battery keeps a voltage at its terminals. Passive components have a common feature: two protruding terminals where at one end an electric current enters and at the other terminal the same amount leaves. Between the terminals there is an electric potential difference, the voltage over the component, and current and voltage have a linear relationship specified by a component constant. Their basic performance in electronic circuits in conjunction with the active element of a battery, a d.c. voltage source, were described in terms of the diagrammatic circuit language.

Two rules govern the performance of electric circuits, Kirchhoff's laws, expressing the relevant content of Maxwell's equations. Firstly, at a node of a circuit where wires meet, the amount of current flowing in to the node equals the amount flowing out: charge cannot pile up in conductors due to the multitude of free electrons. Secondly, the sum of the instantaneous voltage drops in a closed electric circuit loop passing through the wires and connecting the terminals of the components stretching outside the components add up to zero. This is the statement of energy conservation in the conservative electric force field. Equivalently stated: the voltage of a battery is dropped over the components in accordance with their I–V characteristics.

Any passive real component has aspects of all three ideal components. Usually it is relatively easy to judge when these additional aspects have to be taken into account. But one should always keep in mind that: Real life is complicated. Thank God, engineers are in demand!

6 Quantum Wells

In atoms, electrons are confined by the attractive Coulomb force of the atomic nucleus to mostly roam a region on the order of 10^{-8} cm. Atoms, consisting of opposite charges, generally attract each other, say, through electric dipole interaction, and when atoms are close together, electrons can tunnel from one to the other. The corresponding negative electron charge probability density in between the atoms counteracts the repulsive force between the positively charged nuclei, thereby binding atoms into molecules. The same mechanism, chemical bonding, is at play when a gigantic number of atoms form semiconductor crystal structures. A qualitative understanding of why matter lumps together is sufficient for our purpose, and this can be provided by solving simple one-dimensional models where the potentials are stepwise constant. Nowadays, these potentials are not only of academic interest, since by modern technology, for example, molecular beam epitaxy, such potentials can be created for electrons in so-called heterostructures, as discussed in Chapter 12.

6.1 Symmetric Well

Consider a particle of mass m in a one-dimensional symmetric well of extension a, i.e. the potential

$$V(x) = \begin{cases} V, & x < -\frac{1}{2}a, \\ 0, & -\frac{1}{2}a < x < \frac{1}{2}a, \\ V, & x > \frac{1}{2}a, \end{cases} \tag{6.1}$$

where $V > 0$, i.e. the potential depicted in Figure 6.1. This potential could be a crude model of an electron bound in an atom, or a quite realistic model of a quantum well created inside a heterostructure, as discussed in Chapter 12.

The solutions to the time-independent Schrödinger equation corresponding to the continuous part of the energy spectrum, $E > V$, are plane waves, corresponding to different constant potential energy inside and outside the well, however, and thereby different wave numbers, due to the jump in potential at the two points, $x = \pm a/2$, defining the well.[1] The

[1] The potential describes a delta function force at the two points of the well both with directions pointing into the well (see Eq. (B.22)). The solutions of Newton's equation thus allow the particle classically to have any energy value $E \geq 0$, the lowest energy, $E = 0$, corresponding to the particle at rest somewhere inside the well, and for energies $0 < E < V$, the particle is bouncing back and forth. For energies $E > V$, a particle is moving from one region outside of the well to the other: incoming particle with a constant velocity, then

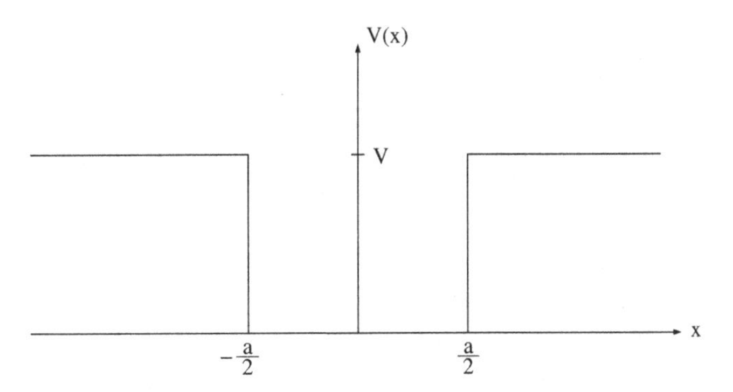

Figure 6.1 Square well potential.

extended plane waves describe the reflection and transmission properties of the well as discussed in Chapter 2, the non-normalizable solutions of the time-independent Schrödinger equation being referred to as scattering states.

Here interest will be in possible solutions with energies in the interval $0 < E < V$. In either of the regions of space outside the well, $|x| > a/2$, the time-independent Schrödinger equation takes the form

$$-\frac{\hbar^2}{2m}\frac{d^2\psi(x)}{dx^2} + V\psi(x) = E\psi(x) \tag{6.2}$$

or

$$\frac{d^2\psi(x)}{dx^2} = \kappa^2\psi(x), \qquad \kappa \equiv \kappa(E) = \sqrt{\frac{2m(V-E)}{\hbar}}. \tag{6.3}$$

The equation determining the energy eigenfunctions outside the well, Eq. (6.3), asks for a function that, when differentiated twice, produces the same function up to an overall multiplying positive constant. This is the defining characteristic of the exponential function, and the two independent solutions are $\exp(\pm\kappa(E)x)$. The general solution outside the well, $|x| > a/2$, thus has, for an energy lower than the confining potential, $E < V$, the form

$$\psi_E(x) = A\, e^{\kappa x} + D\, e^{-\kappa x}. \tag{6.4}$$

The regions outside the well are unbounded, and one solution is decaying and the other diverging. Owing to the physical requirement of a non-divergent wave function, as discussed in general in Appendix E, only the decaying normalizable part of the solution in either region is physically relevant, as displayed in Eq. (6.9).

Inside the well, $-a/2 < x < a/2$, a possible energy eigenfunction satisfies the equation

$$-\frac{\hbar^2}{2m}\frac{d^2\psi(x)}{dx^2} = E\psi(x) \tag{6.5}$$

constant but increased velocity in the well region, due to the push of the delta spike force, and finally the same constant velocity as the initial velocity but now moving in the region on the other side of the well. Potential energy outside the well region is turned into kinetic energy inside and converted again. The classical motion for $E > V$ is thus also non-localized just as in the quantum case. In quantum mechanics there appears the novel phenomenon of possible reflection of a particle with energies above the maximum value of the potential as discussed in Exercise 2.9 on page 36.

or

$$\frac{d^2\psi(x)}{dx^2} = -k^2\psi(x), \qquad k \equiv k(E) = \sqrt{2mE}/\hbar. \tag{6.6}$$

A general solution of the time-independent Schrödinger equation inside the box with eigenvalue E is thus a superposition of the two linearly independent plane wave solutions,

$$\psi_E(x) = B_1\, e^{ikx} + B_2\, e^{-ikx}. \tag{6.7}$$

Since the Hamiltonian is real, $\hat{H}^* = \hat{H}$, the energy eigenfunctions can, according to Eq. (E.31), be chosen as real functions, i.e. choose $B_2 = B_1^*$ in Eq. (6.7), and B_1 real or purely imaginary gives the cosine and sine functions, respectively. The solution inside the well can thus be sought real, i.e. in the form

$$\psi_E(x) = B\cos kx + C\sin kx. \tag{6.8}$$

Since the origin has been chosen such that the potential is inversion-symmetric, $V(-x) = V(x)$, the energy eigenfunctions can according to Eq. (E.35) be chosen so that they are either even or odd functions. We can thus seek the solutions assuming either a cosine or sine function inside the well corresponding to the solutions being either even or odd functions. A solution to the considered time-independent Schrödinger equation for $E < V$ can therefore be sought in the form

$$\psi(x) = \begin{cases} A\, e^{\kappa x}, & x < -\frac{1}{2}a, \\ B\cos kx,\; C\sin kx, & -\frac{1}{2}a < x < \frac{1}{2}a, \\ D e^{-\kappa x}, & x > \frac{1}{2}a, \end{cases} \tag{6.9}$$

where the cos and sin choice correspond to the possible even and odd eigenfunctions, respectively. The wave function *leaks* into the classically forbidden regions, and the extent of the leakage is a measure of the energy of the state in question according to Eq. (6.3), or, more precisely, the binding energy, $V - E$, of the state.[2]

From the orthogonality of energy eigenfunctions, a general feature of their form can be deduced. In order for the wave function overlap integral, Eq. (1.56), to vanish, the wave function of the first excited state (the state of next lowest energy) must have regions where its sign is opposite to that of the ground-state, lowest energy, wave function. In particular, it must have a node, i.e. at some point it must vanish. Wave functions of ever higher energy must therefore wiggle ever more in order to be orthogonal to the states of lower energy. Clearly, the more a wave function oscillates, the higher its corresponding kinetic energy will be, as also exemplified by the energy eigenfunctions of a free particle, for which $E \propto 1/\lambda^2$.

As discussed in Appendix E, the wave function and its first derivative must be continuous functions everywhere, leading to matching conditions at the points $x = \pm a/2$. The matching conditions at $x = -a/2$ give

$$\psi\left(-\frac{a}{2} - 0\right) = \psi\left(-\frac{a}{2} + 0\right), \qquad \psi'\left(-\frac{a}{2} - 0\right) = \psi'\left(-\frac{a}{2} + 0\right). \tag{6.10}$$

[2] If the well potential is chosen to vanish outside the well, corresponding to the change in Hamiltonian $\hat{H} \to \hat{H} - V$, a similar shift in energy values would occur and the sought energies would be negative. The binding energy of a state is, of course, unchanged.

A possible cosine solution can lead to an even solution without a node, whereas a sine solution has at least one node at $x = 0$. In view of the general feature of energy eigenfunctions, the more wiggly, the higher the energy; the ground-state wave function is therefore expected to be of the former kind. Considering first possible even solutions, i.e. the wave function in the well region is $B\cos kx$, the matching conditions (6.10) then become

$$A\,e^{-\kappa a/2} = B\cos(ka/2), \qquad \kappa A\,e^{-\kappa a/2} = kB\sin(ka/2). \tag{6.11}$$

Matching at $x = a/2$ gives the relations

$$B\cos(ka/2) = D\,e^{-\kappa a/2}, \qquad kB\sin(ka/2) = \kappa D\,e^{-\kappa a/2}. \tag{6.12}$$

One sees immediately from the first equations in Eqs. (6.11) and (6.12) that the continuity of the wave function gives $A = D$, not unexpectedly since we are searching for an even solution, for which $\psi(-x) = \psi(x)$.[3]

Since the exponential function is non-vanishing, the matching equations in Eq. (6.11) (or Eq. (6.12), as the two sets of equations contain the same information) can be divided, giving the requirement for the existence of an *even* energy state with energy E as

$$\tan\left(\frac{k(E)a}{2}\right) = \frac{\kappa(E)}{k(E)}. \tag{6.13}$$

Only for wave lengths for which the smoothness relationship, Eq. (6.13), is fulfilled does an even solution exist. A possible energy eigenfunction oscillates according to Eq. (6.9) inside the well, as characterized by the wave number $k(E)$, or equivalently the wave length $\lambda(E) = 2\pi/k(E)$, and joins up smoothly with the exponentially decaying functions outside the well. The common value of the energy eigenfunction at the two end points of the well is determined by normalization of the wave function, information, however, that is not needed to determine the allowed energies in view of the additional requirement of continuity of the derivative.

The possible energy values, as expected, do not depend on the location of the well. Had its left and right positions been arbitrary, x_l and $x_r = x_l + a$, the algebra would have been messier because of not taking advantage of the symmetry of the problem, but the equation determining the energy values would come out the same, determined according to Eq. (6.13) solely by $\hbar$ and the parameters defining the Hamiltonian, m, a and V.

The matching condition, say, at $x = a/2$, gives for possible *odd* energy eigenfunctions the matching relations

$$C\sin(ka/2) = D\,e^{-\kappa a/2}, \qquad kC\cos(ka/2) = -\kappa D\,e^{-\kappa a/2}, \tag{6.14}$$

[3] Only matching at one of the points is needed since the inversion symmetry of the solutions is already built in and guarantees proper smoothness at the other matching point. Had we started with the general complex solution in the well region, Eq. (6.7), all four matching conditions are needed, and the arithmetic of the problem will force out the symmetry properties of the functions: two types of solutions appear, one where $B_1 = B_2$ and $A = D$, and one where $B_1 = -B_2$ and $A = -D$, i.e. the more complex arithmetic would grind out the consequences of parity and time reversal symmetry inherent in the time-independent Schrödinger equation.

and the smoothness requirement on wave functions requires possible odd energies to be determined by

$$-\cot\left(\frac{k(E)a}{2}\right) = \frac{\kappa(E)}{k(E)}. \tag{6.15}$$

A graphical solution of the transcendental equations (6.13) and (6.15) gives a qualitative understanding of the energy spectrum of bound states. A dimensionless energy variable is introduced for convenience, dictated by the argument of the trigonometric functions,

$$\xi \equiv \frac{ka}{2} = \frac{1}{\hbar}\sqrt{\frac{ma^2E}{2}} = \sqrt{\frac{E}{E_*}}, \qquad E_* \equiv \frac{2\hbar^2}{ma^2}. \tag{6.16}$$

The trigonometric functions are then sketched as depicted in Figure 6.2.

Since

$$\frac{\kappa(E)}{k(E)} = \sqrt{\frac{V-E}{E}} = \sqrt{\frac{V/E_* - \xi^2}{\xi^2}}, \tag{6.17}$$

the equation determining the energies of possible even states, Eq. (6.13), then becomes

$$\tan\xi = \sqrt{\frac{\xi_*^2 - \xi^2}{\xi^2}}, \tag{6.18}$$

where

$$\xi_*^2 = \frac{V}{E_*} = \frac{ma^2V}{2\hbar^2}, \qquad \xi = \xi_*\sqrt{\frac{E}{V}}. \tag{6.19}$$

The equation determining the energies of possible odd energy states, Eq. (6.15), similarly becomes

$$-\cot\xi = \sqrt{\frac{\xi_*^2 - \xi^2}{\xi^2}}. \tag{6.20}$$

Now introduce

$$g(\xi) = \sqrt{\frac{\xi_*^2 - \xi^2}{\xi^2}} \tag{6.21}$$

for the function on the right-hand side of Eqs. (6.18) and (6.20). For values of ξ less than ξ_*, $0 \le \xi < \xi_*$, $g(\xi)$ is positive before vanishing at ξ_*, $g(\xi_*) = 0$. For values $\xi > \xi_*$, $g(\xi)$ is purely imaginary, and for $\tan(\xi_*\sqrt{E/V})$ to be purely imaginary would require E to be purely imaginary ($\tan(ir) = i\tanh(r)$), which is impossible as energy values are real. Thus ξ_* sets the upper limit for searching for possible energies of bound states. Further noticing the properties of the function g at the limiting points,

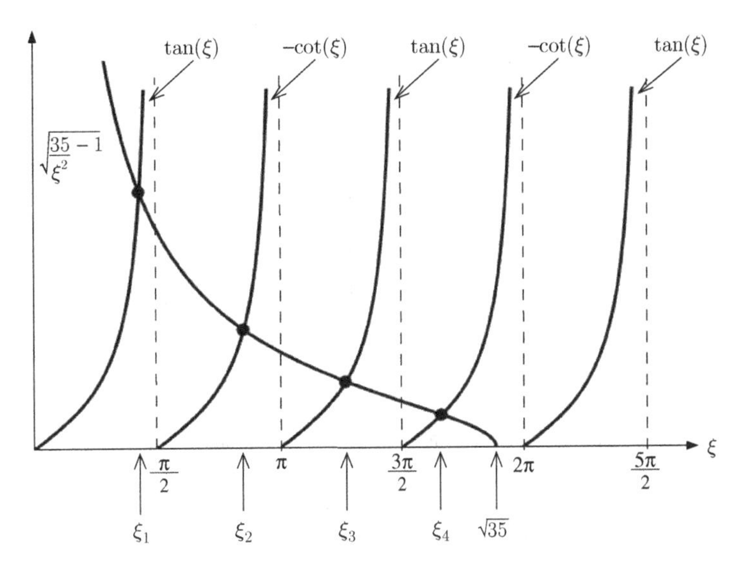

Figure 6.2 Graphical solution for the bound-state energies in a quantum well for the parameter value $\xi_* = \sqrt{35}$.

$$g(\xi) \simeq \begin{cases} \dfrac{\xi_*}{\xi}, & \xi \searrow 0, \\ \sqrt{\dfrac{2}{\xi_*}}\,\sqrt{\xi_* - \xi}, & \xi \nearrow \xi_*, \end{cases} \tag{6.22}$$

a drawing of the smooth function, g, can immediately be done as depicted in Figure 6.2.

A graphical solution of Eq. (6.18) is achieved by noting that the possible energy values for an even eigenstate occur whenever the function on the right-hand side crosses one of the arms of the tangent function, since then the right and left sides of Eq. (6.18) take on the same values. Since g and the tangent function always cross each other no matter how small the value of ξ_*, there is always a ground state for the one-dimensional quantum well no matter what the parameter values of a and V, i.e. no matter how shallow or narrow the well is (a feature not carried over to higher dimensions).

A graphical solution of Eq. (6.20) for the possible energy values for odd eigenstates occurs whenever the function on the right-hand side crosses one of the arms of minus the cotangent function. If the dimensionless parameter ξ_* is less than $\pi/2$, $0 < \xi_* < \pi/2$, then, as *minus* the cotangent function is negative in that region, there is no crossing of the two functions $g(\xi)$ and $-\cot(\xi)$, and no odd energy eigenstate exists. Only if $\xi_* > \pi/2$, corresponding to $\hbar^2/2ma^2 < V/\pi^2$, do odd solutions exist. The value of ξ_* determines where the function $g(\xi)$ vanishes, and thus determines the number of crossings, and thereby the number of bound states. For the case $\xi_* = \sqrt{35}$, there exist exactly two even and two odd energy eigenfunctions, as seen from Figure 6.2. In general, if ξ_* lies in the interval $(n-1)\pi/2 < \xi_* < n\pi/2$, there will be n solutions. If n is odd, there will be one less odd solution compared to the number of even solutions; and if n is even, there is an equal number $n/2$ of even and odd bound states.

The wave functions for the ground state and first excited state for a particle in a symmetric quantum well are depicted in Figure 6.3.

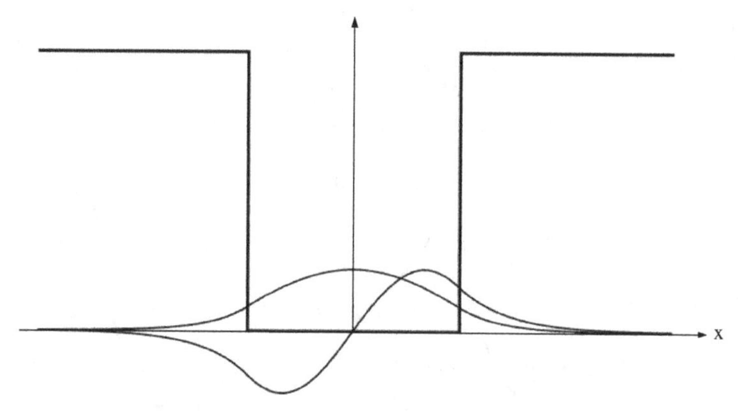

Figure 6.3 Ground-state and first excited-state (one node) wave functions.

Exercise 6.1 Obtain from the graphical solution the criterion (expressed in the parameters of the problem) for the existence of exactly three even and two odd bound states. Use this information to sketch the corresponding wave functions. Note the number of nodes corresponding to an energy eigenfunction.

We note that the energy spectrum of bound states corresponds to discrete values, $E_n = E_* \xi_n^2$, i.e. energy is quantized. For bound states, the walls of the well confine the wave function, and discrete values of the allowed modes is a general feature of confined waves, as discussed in Appendix E.[4] The shape of a bound-state wave function is determined by its value at location $x = a/2$, where the argument of the trigonometric function is $k_n a/2 = \xi_n$, ξ_n being the dimensionless energy value. Since the value of k_n determines the number of oscillations of the corresponding trigonometric function in the well, the number of its nodes, it determines the shape of the wave function. Since $(n-1)\pi/2 < \xi_n < n\pi/2$, the corresponding energy eigenfunction has $n - 1$ nodes.

Exercise 6.2 Obtain from the graphical solution of the finite quantum well the analytical expressions for the possible energy eigenvalues in the limit where $V \to \infty$, i.e. for a particle in a box. For an alternative way of obtaining the result, consult Appendix H.

6.2 Asymmetric Well

Consider a particle of mass m in an asymmetric one-dimensional quantum well, in fact, the most asymmetric one with one infinite potential wall, i.e. the potential

[4] This should be contrasted with the classical motion of the particle for $0 < E < V$, which can execute oscillations with any frequency $\nu = \sqrt{2E/m}/a$.

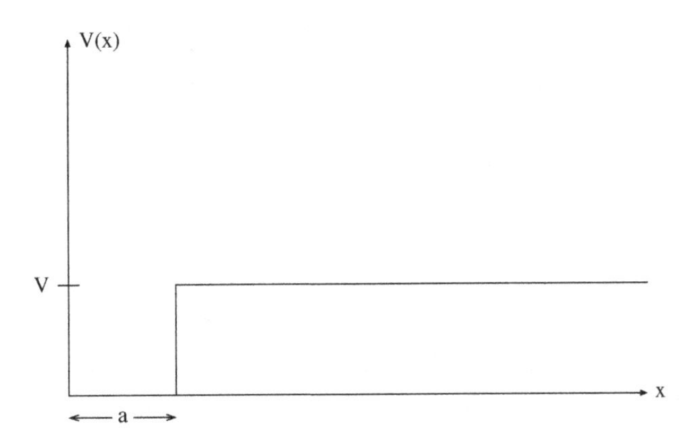

Figure 6.4 Asymmetric well.

$$V(x) = \begin{cases} \infty, & x \leq 0, \\ 0, & 0 < x < a, \\ V, & x > a, \end{cases} \tag{6.23}$$

where $V > 0$, the potential depicted in Figure 6.4. As discussed in Appendix E, the infinite wall makes the wave function vanish at the boundary and beyond, giving the boundary condition $\psi(x = 0) = 0$.

The energies and eigenfunctions for the problem defined by the asymmetric potential, Eq. (6.23), can now easily be assessed in terms of the symmetric well studied in the previous section. The symmetric well had even and odd solutions, but the solutions of the previous problem that do not vanish at $x = 0$, i.e. the even solutions described by the cosine functions inside the well, are no longer solutions to the asymmetric well case. That leaves only the odd solutions, i.e. those described by the sine functions inside the well, as they satisfy the boundary condition $\psi(x = 0) = 0$ (but now, of course, only being non-vanishing in the region $x > 0$). Since the size of the asymmetric well is denoted a, the comparison with the symmetric well should be done for well size $2a$, and the possible bound-state solutions for the asymmetric well are thus determined by the equation

$$-\cot(k(E)a) = \frac{\kappa(E)}{k(E)}. \tag{6.24}$$

A bound-state solution thus only exists for the asymmetric well if the parameter values satisfy $\hbar^2/2ma^2 < 4V/\pi^2$.

Exercise 6.3 Obtain the criterion for the existence of exactly two bound states in the asymmetric potential, Eq. (6.23), and use this to sketch the two energy eigenfunctions. Note their number of nodes.

6.3 Double Well

Consider a particle of mass m in the one-dimensional symmetric double well, the potential

$$V(x) = \begin{cases} V, & |x| < \frac{1}{2}l - \frac{1}{2}a, \\ 0, & \frac{1}{2}l - \frac{1}{2}a < |x| < \frac{1}{2}l + \frac{1}{2}a, \\ \infty, & |x| > \frac{1}{2}l + \frac{1}{2}a, \end{cases} \tag{6.25}$$

where $V > 0$, the potential depicted in Figure 6.5. Since quantum tunneling of the particle between the two wells is to be studied, the confining potential walls have for simplicity been chosen infinite.

The double-well potential can be viewed as being built by two of the asymmetric potential wells of size a of Figure 6.4 facing each other and whose centers are separated by the distance l. If the two wells are widely separated, the ground state of the particle would be the ground state of either of the separated wells, assuming of course that the asymmetric well has at least one bound state, as done in the following, i.e. the parameters are assumed to satisfy $\hbar^2/2ma^2 < 4V/\pi^2$. The particle in its ground state would thus be in either of the degenerate ground-state levels of the separated asymmetric wells as depicted in Figure 6.6.

The double well can be viewed as a simplified model for the ionic hydrogen molecule, H_2^+, where an electron experiences the Coulomb potential from two protons as depicted in Figure 6.7.[5] The electron in a low-energy state for the potential in Figure 6.7 thus experiences two huge potential walls and a tunneling barrier in between.

In the limit of large separation or a high barrier, the ground state of the double well is the ground state of either of the two separated wells. If the distance, l, between the two wells

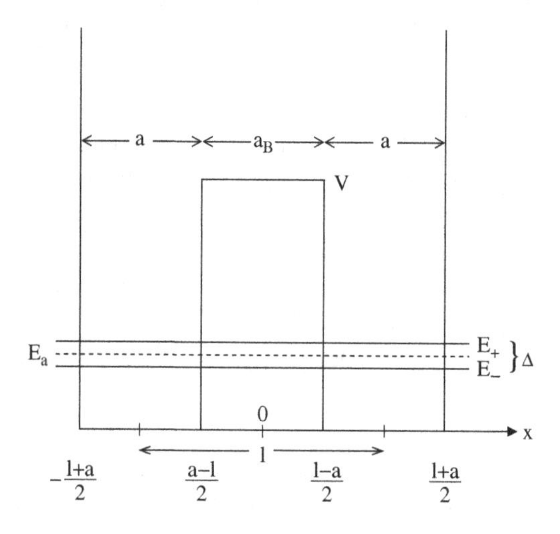

Figure 6.5 Double-well potential.

[5] Even for this simplest of molecules, the three-body Hamiltonian with its constituent Coulomb interactions cannot be solved exactly. Approximation schemes are necessary.

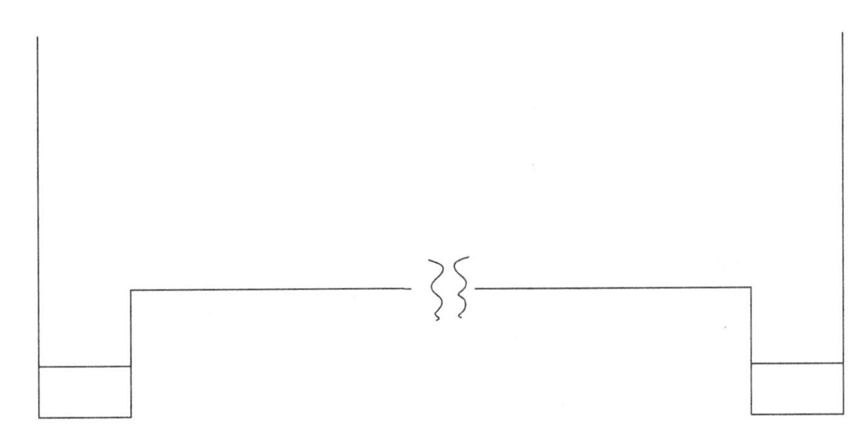

Figure 6.6 Widely separated wells.

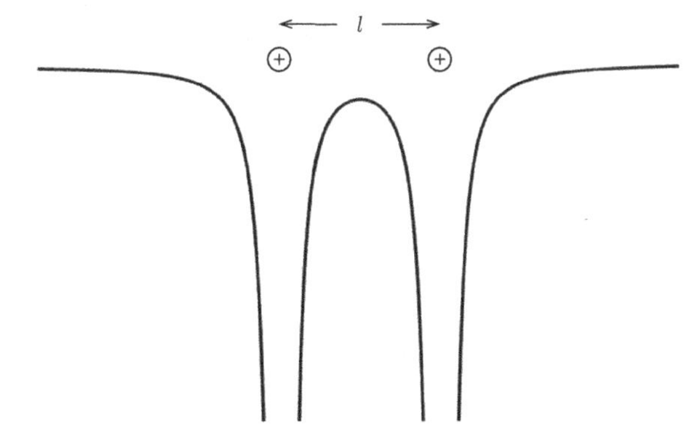

Figure 6.7 Potential produced by two protons.

becomes smaller, the tunneling barrier width, $a_B = l - a$, also gets smaller, and the mechanism of tunneling starts to become operative. When the particle by tunneling is roaming both wells, it is less confined and the ground-state energy of the double well should, in view of Heisenberg's uncertainty relation, be smaller than the ground-state energy of the asymmetric wells it is built from. The assumption that the single well has a ground state thus ensures the ground state of the double well.[6]

An energy eigenfunction in the double well, for an energy in the interval $0 < E < V$, can immediately be written down in terms of exponential functions in the three regions. However, respecting the infinite wall boundary conditions, $\psi(x = \pm(l + a)/2) = 0$, turns the function in the well regions conveniently into sine functions,

[6] The influence of the tunneling mechanism on the degenerate level is equivalent to adding a coupling term for a two-level system, $\epsilon\sigma_0 \to \epsilon\sigma_0 + v\sigma_x$, σ_0 denoting the unit 2×2 matrix, the degenerate level thus being split due to the off-diagonal elements of the Pauli matrix σ_x. The degenerate ground state of the isolated wells is thus expected to split into a lower double-well ground-state level and a first excited state, the general phenomenon of level repulsion.

$$\psi(x) = \begin{cases} A \sin\left(k\left(x + \frac{a+l}{2}\right)\right), & -\frac{l+a}{2} < x < \frac{a-l}{2}, \\ B\, e^{\kappa x} + C\, e^{-\kappa x}, & |x| < \frac{l-a}{2}, \\ D \sin\left(k\left(x - \frac{a+l}{2}\right)\right), & \frac{l-a}{2} < x < \frac{l+a}{2}, \end{cases} \tag{6.26}$$

where $k \equiv k(E) \equiv \sqrt{2mE}/\hbar$ and $\kappa = \kappa(E) = \sqrt{2m(V-E)}/\hbar$. In each of the quantum wells, the wave function starts out as a sine function, vanishes at the infinite wall, goes smoothly through the barrier region and connects with the sine function of the other well as demanded by the matching conditions.

The matching conditions for the wave function at the barrier, $x = \pm a_B/2 = \pm(l-a)/2$, lead to four equations determining the four unknown coefficients in Eq. (6.26). Solving these four equations would reveal that the solutions are either even or odd, i.e. for fulfilment the equations would demand that either $A = D$ and $B = C$ or $A = -D$ and $B = -C$. We spare ourselves this exercise, as its conclusion is a consequence of the double-well potential having been chosen to be inversion-symmetric, $V(-x) = V(x)$. The origin has been chosen such that the potential is inversion-symmetric, and taking immediate advantage of this symmetry, the solutions to the time-independent Schrödinger equation can be chosen as even or odd functions (as shown in Appendix E). This determines the two forms of the wave function in the barrier region, and the even or odd relationship demand of the coefficients in the two well regions. The solutions can thus be sought in the even and odd forms

$$\psi_\pm(x) = \begin{cases} \mp A_\pm \sin\left(k\left(x + \frac{a+l}{2}\right)\right), & -\frac{l+a}{2} < x < \frac{a-l}{2}, \\ B_\pm(e^{\kappa x} \mp e^{-\kappa x}), & |x| < \frac{l-a}{2}, \\ A_\pm \sin\left(k\left(x - \frac{a+l}{2}\right)\right), & \frac{l-a}{2} < x < \frac{l+a}{2}. \end{cases} \tag{6.27}$$

We have thus chosen notation such that ψ_- is an even function and ψ_+ is an odd function with at least a node at the origin. We expect the even function to correspond to the lowest energy. Our interest will be in the splitting of the ground-state level of the isolated wells, but the discussion is seen to be quite general for any level existing in the isolated wells as long as the level results in weak tunneling, the basis for the following result.

Having invoked the inversion symmetry of the potential, matching is only needed at one of the barrier points, say, $x = a_B/2 = (l-a)/2$, giving the matching equations

$$B_\pm(e^{\kappa a_B} \mp e^{-\kappa a_B}) = \pm A_\pm \sin(-ka) \tag{6.28}$$

and

$$\kappa B_\pm(e^{\kappa(l-a)/2} \pm \kappa\, e^{-\kappa(l-a)/2}) = \mp k A_\pm \cos(-ka) \tag{6.29}$$

for even and odd solutions, respectively. Dividing these continuity condition equations for the wave function and its derivative, the following equations determine the possible energy eigenvalues in the double well with values $E < V$, for even and odd wave functions, respectively,

$$\cot ka = -\frac{\kappa}{k} \times \begin{cases} \tanh(\kappa a_B/2), & \text{even solution } \psi_-, \\ \coth(\kappa a_B/2), & \text{odd solution } \psi_+. \end{cases} \tag{6.30}$$

If the distance between the two wells, l, or the barrier height, V, become very large, the two equations in Eq. (6.30) merge into one equation, as both the hyperbolic tangent, $\tanh x$, and cotangent, $\coth x$, are functions equal to one for large argument, $x = \kappa a_B/2 \gg 1$. Equation (6.24) determining the possible energy levels in the two separate asymmetric wells, discussed in the previous section, is then recovered.

Introducing the dimensionless energy variables $\xi = ka = \sqrt{2ma^2E}/\hbar$ and $\xi_*^2 \equiv 2ma^2 V/\hbar^2$, the equations (6.30) determining the energy eigenvalues corresponding to the even and odd energy eigenfunctions take (recall Eq. (6.17)) the form

$$\cot \xi = -\sqrt{\frac{\xi_*^2 - \xi^2}{\xi^2}} \times \begin{cases} \tanh(\kappa(\xi)a_B/2), & \text{even}, \\ \coth(\kappa(\xi)a_B/2), & \text{odd}, \end{cases} \tag{6.31}$$

and κ in terms of ξ is given by $\kappa = \kappa(\xi) = \sqrt{\xi_*^2 - \xi^2}/a$.

In the graphical representation of the left- and right-hand sides of Eq. (6.31), the tanh factor, being smaller than unity, will shift the curve representing the even solution on the right-hand side of Eq. (6.31) downward, leading to an energy eigenvalue, $E_- = E_a - \Delta_-$, that is lower than the value for the level in the isolated well E_a; and vice versa, the coth factor leads to an energy value, $E_+ = E_a + \Delta_+$, that is larger than the value for the isolated *atom*. This is displayed in Figure 6.8, where the indicated values of course are ξ_a and $\xi_\pm$, but correspond in turn to the energies as indicated in Figure 6.8.

Introducing the notation

$$f(E) \equiv \frac{k(E)}{\kappa(E)} \cot(k(E)a), \tag{6.32}$$

Eq. (6.24), determining the degenerate energy level, E_a, in the two isolated wells,

$$\cot(k(E_a)a) = -\frac{\kappa(E_a)}{k(E_a)}, \tag{6.33}$$

is thus specified by the equation

$$f(E_a) = -1. \tag{6.34}$$

The equation determining the possible energy levels, $0 < E < V$, in the double well corresponding to even wave functions then becomes, $E_\pm = E_a \pm \Delta_\pm$,

$$f(E_-) = -\tanh\left(\frac{a_B}{2}\kappa(E_-)\right) \tag{6.35}$$

according to Eq. (6.30), and the possible energy levels corresponding to odd wave functions is determined by the equation

$$f(E_+) = -\coth\left(\frac{a_B}{2}\kappa(E_+)\right). \tag{6.36}$$

Assuming that the barrier is wide or its height appreciable compared to the energy in question, the problem exhibits a small parameter, *viz.* $e^{-a_B\kappa(E_\pm)} \ll 1$, corresponding to the weak tunneling regime where the tunneling probability of the square barrier is given by

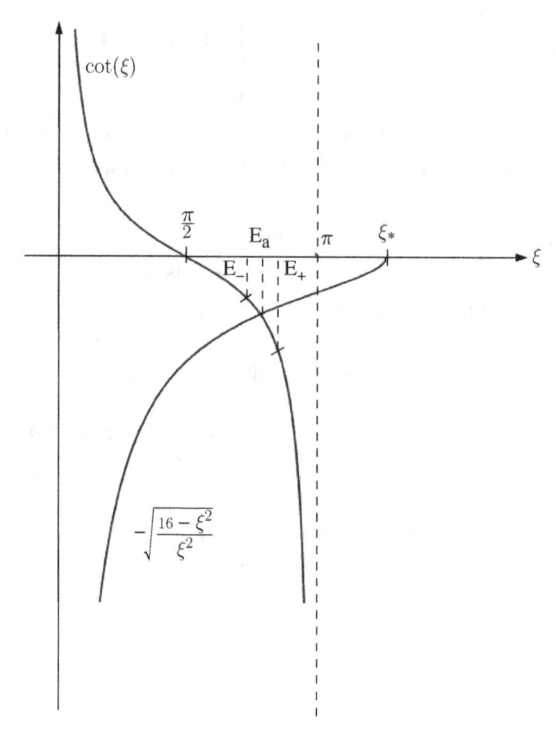

Figure 6.8 Graphical solution for the lowest energy levels in a double well for the parameter value $\xi_* = 4$.

Eq. (2.89). The two equations determining the even and odd state energy levels therefore with exponential accuracy become[7]

$$f(E_-) = -1 + 2\, e^{-a_B\kappa(E_-)} \tag{6.37}$$

and

$$f(E_+) = -1 - 2\, e^{-a_B\kappa(E_+)}. \tag{6.38}$$

According to Eq. (6.34) we can rewrite the equations as

$$f(E_a - \Delta_-) = f(E_a) + 2\, e^{-a_B\kappa(E_a - \Delta_-)} \tag{6.39}$$

and

$$f(E_a + \Delta_+) = f(E_a) - 2\, e^{-a_B\kappa(E_a + \Delta_+)}. \tag{6.40}$$

Owing to the correction terms on the right-hand side being exponentially small, the shifts in energies, $\Delta_\pm$, are small in comparison to E_a and $V - E_a$. Thus using the fact that $\Delta_\pm \ll E_a, (V - E_a)$, Taylor-expanding to lowest order is sufficient, giving

$$-f'(E_a)\, \Delta_- = 2\, e^{-a_B\kappa(E_a)}\, e^{a_B\kappa'(E_a)\Delta_-} \tag{6.41}$$

[7] For $x \gg 1$, $\tanh x = (e^x - e^{-x})/(e^x + e^{-x}) = (1 - e^{-2x})/(1 + e^{-2x}) \simeq (1 - e^{-2x})(1 - e^{-2x}) \simeq (1 - 2e^{-2x})$ and $\coth x = 1/\tanh x \simeq (1 + 2e^{-2x})$.

and

$$f'(E_{\rm a})\,\Delta_+ = -2\,e^{-a_B\kappa(E_{\rm a})}\,e^{-a_B\kappa'(E_{\rm a})\Delta_+}. \tag{6.42}$$

Inserting the derivative of $\kappa(E_{\rm a})$,

$$\kappa'(E_{\rm a}) = -\frac{\kappa(E_{\rm a})}{2(V-E_{\rm a})}, \tag{6.43}$$

the exponentials become

$$\exp[\pm a_B\kappa'(E_{\rm a})\Delta_\mp] = \exp\left(\mp\frac{a_B\kappa(E_{\rm a})\Delta_\mp}{2(V-E_{\rm a})}\right) \tag{6.44}$$

and thereby

$$\begin{aligned}\exp[-a_B\kappa(E_{\rm a})]\exp[\pm a_B\kappa'(E_{\rm a})\Delta_\mp] &= \exp\left[-a_B\kappa(E_{\rm a})\left(1\mp\frac{\Delta_\mp}{2(V-E_{\rm a})}\right)\right]\\ &\simeq \exp[-a_B\kappa(E_{\rm a})]\end{aligned} \tag{6.45}$$

and with exponential accuracy the two shifts are identical, $\Delta_+ = \Delta_-$, and determined by

$$f'(E_{\rm a})\,\Delta_+ = -2\,e^{-a_B\kappa(E_{\rm a})}. \tag{6.46}$$

The magnitude of the level splitting, $\Delta = \Delta_+ + \Delta_- = E_+ - E_-$, thus becomes

$$\Delta = -\frac{4\,e^{-a_B\kappa(E_{\rm a})}}{f'(E_{\rm a})}. \tag{6.47}$$

Evaluating the derivative ($k \equiv k(E_{\rm a})$ and $\kappa \equiv \kappa(E_{\rm a})$),

$$\begin{aligned}f'(E_{\rm a}) &= \left(\frac{k(E_{\rm a})}{\kappa(E_{\rm a})}\right)'\cot(k(E_{\rm a})a) + \frac{k(E_{\rm a})}{\kappa(E_{\rm a})}\frac{-1}{\sin^2(k(E_{\rm a})a)}k'(E_{\rm a})a\\ &= -\frac{\kappa}{k}\frac{(\kappa k' - k\kappa')}{\kappa^2} - \frac{akk'}{\kappa}(1+\cot^2(ka)),\end{aligned} \tag{6.48}$$

the second equality follows by rewriting the first term using Eq. (6.34), and using it again on the trivially rewritten second term, $1/\sin^2 x = (\cos^2 x + \sin^2 x)/\sin^2 x = 1 + \cot^2 x$, gives (as $1+\cot^2 ka = 1 + \kappa^2/k^2$),

$$\begin{aligned}f'(E_{\rm a}) &= -\frac{k'}{k} + \frac{\kappa'}{\kappa} - \frac{akk'}{\kappa} - \frac{a\kappa k'}{k}\\ &= -\frac{1}{2E_{\rm a}} - \frac{1}{2(V-E_{\rm a})} - \frac{ak^2(E_{\rm a})}{2E_{\rm a}\kappa(E_{\rm a})} - \frac{a\kappa(E_{\rm a})}{2E_{\rm a}}\\ &= -\frac{V}{2E_{\rm a}(V-E_{\rm a})} - \frac{a\kappa(E_{\rm a})}{2E_{\rm a}}\frac{V}{2(V-E_{\rm a})}\end{aligned} \tag{6.49}$$

or

$$f'(E_{\rm a}) = -\frac{V(1+a\kappa(E_{\rm a}))}{2E_{\rm a}(V-E_{\rm a})}. \tag{6.50}$$

The level split thus becomes

$$\Delta = 8\,\frac{E_{\rm a}(V-E_{\rm a})}{V(1+a\kappa(E_{\rm a}))}\,e^{-a_B\kappa(E_{\rm a})}. \tag{6.51}$$

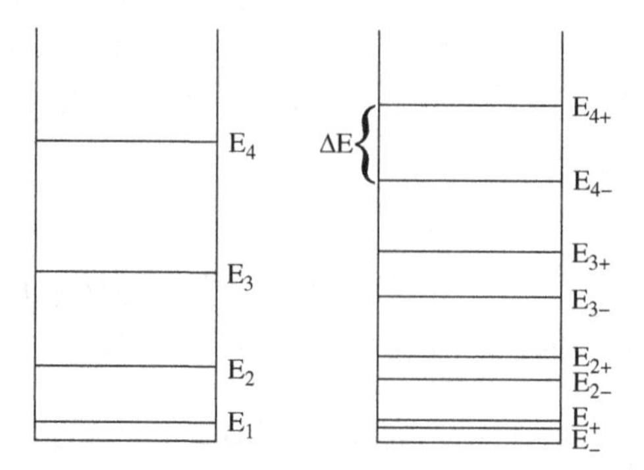

Figure 6.9 Level splitting in a double well assuming the existence of four bound states in the corresponding asymmetric wells.

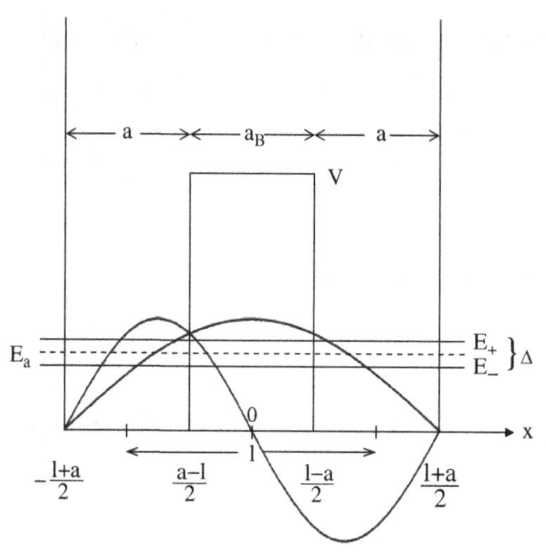

Figure 6.10 Lowest-energy wave functions in the double well.

Indeed, the level split is small, $\Delta \ll E_a, (V - E_a)$, controlled by the exponentially small parameter. The double-well energy levels are then given by

$$E_\pm - E_a = \pm\frac{\Delta}{2} = \pm 4\,\frac{E_a(V - E_a)}{V(1 + a\kappa(E_a))}\, e^{-a_B \kappa(E_a)}. \tag{6.52}$$

The ground state thus has a lower energy, E_-, than the ground-state energy of the isolated *atom* state, E_a.

The level or tunneling splitting can, according to Eq. (2.89), be expressed in terms of the tunneling probability of the square barrier at energy E_a,

$$\Delta = \frac{2V\, e^{a_B \kappa(E_a)}}{1 + a\kappa(E_a)}\, T(E_a). \tag{6.53}$$

In Figure 6.9, the energy levels depicted to the left are those of a particle trapped in either of the separated asymmetric potential wells, and to the right is shown the level splitting in the double well.

The shape of the energy eigenfunctions, their number of nodes, can be determined by the value of the sine function at the left of the barrier ($x = -a_B/2 = (a - l)/2$), for which the sine function becomes $\sin(ak(E_\pm))$, for the even and odd functions, respectively. According to Figure 6.8, for the ground and first excited states we have $\pi/2 < \xi_\mp^{(0)} = ak(E_\mp^{(0)}) < \pi$, i.e. the sine functions $\sin(ak(E_\pm^{(0)}))$ are positive, and the even ground-state function has no nodes, whereas the first excited state has its node at $x = 0$. If the next two level-split states exist, $3\pi/2 < \xi_\mp^{(1)} = ak(E_\mp^{(1)}) < 2\pi$, the sine functions have a node in each well, giving two nodes for the even function, and the odd third excited state has its additional node at $x = 0$, thus having three nodes. This feature clearly repeats according to Figure 6.2: for the next higher bound state, the wave function has an additional node; in fact, this is a general feature of bound states in a potential. The wave functions for the ground and first excited states are depicted in Figure 6.10.

6.4 Chemical Bond

The analysis of the double-well potential allows a qualitative understanding of an all-pervasive feature of the world around us, *viz.* that atoms lump together. As noted in Section 6.3, the double well can be thought of as a crude model of the hydrogen ion, H_2^+, the Coulomb bound state of two protons and one electron. The ground state of the double well has a lower energy, E_-, than the ground-state energy of the isolated *ion–atom* state, E_a; it takes energy to break up the double-well *molecular ion*. The electron in the double well is less localized, thereby having lower kinetic energy.

The tunneling mechanism binds atoms into molecules by sharing electrons: so-called monovalent bonding in the case of H_2^+, since only one electron is involved; and covalent bonding in, say, the case of the hydrogen molecule H_2, if two electrons are involved. The average inter-proton distances for H_2^+ and H_2 are 1.06 Å and 0.74 Å, respectively. When the two protons share two electrons, the electrons are given larger space in which to roam, thus, according to the uncertainty principle, lowering their kinetic energy, and this generally outweighs the change in potential energy, to make covalent bonding energetically favorable. The two protons in the hydrogen molecule repel each other, and the probability for electrons to be in between the protons must be large so that their negative charge density provides the *glue* to keep the molecule together.[8] In order for the two electrons to be simultaneously in between the nuclei, their spin states should be the antisymmetric singlet state

$$\chi(s_z^{(1)}, s_z^{(2)}) = \frac{1}{\sqrt{2}}[\chi_\uparrow(s_z^{(1)})\,\chi_\downarrow(s_z^{(2)}) - \chi_\downarrow(s_z^{(1)})\,\chi_\uparrow(s_z^{(2)})], \tag{6.54}$$

allowing for the symmetric spatial wave function $\psi(\mathbf{x}_1, \mathbf{x}_2) \simeq \psi_1(\mathbf{x}_1)\psi_2(\mathbf{x}_2) + \psi_2(\mathbf{x}_1)\psi_1(\mathbf{x}_2)$, where to a first approximation we can take $\psi_i(\mathbf{x})$ as hydrogen wave functions for the two atoms. The spatially symmetric wave function gives a large electron probability in between the ions, providing electrostatic attraction of the protons. The covalent bond is

[8] This chemist's argument seems to assume that the wave function represents something real.

stronger than the monovalent one, and the binding or bond energies of H_2^+ and H_2 are 256 and 433 kJ mol^{-1}, respectively. Covalent bonds are said to be saturated: the exclusion principle does not allow the existence of the three-electron molecule of hydrogen and helium (HHe).

If you open a book on chemistry, you can find simple pictures like

$$\mathrm{H{=}H} \quad \text{and} \quad \mathrm{H{-}H}$$

representing the covalent and monovalent bonding of the hydrogen molecule H_2 and hydrogen ion H_2^+, the pictorial representation language of chemistry. This is a very useful mental picture of the chemical bonding of atoms: the first states that a hydrogen molecule, H_2, can be formed by a covalent bond, and the second that the hydrogen ion, H_2^+, can be formed by a monovalent bond. This is the way a chemist keeps track of how very complicated molecules, built out of hundreds of atoms, are held together. Chemical formulas can take on incredible complexity.[9]

Understanding chemistry is important for understanding life around us. For example, ozone, the molecule O_3, consisting of three oxygen atoms, is a nuisance in urban atmospheres, but is essential to life on Earth. Ozone, kept by gravity 25 km up in the atmosphere, is the molecule that filters out most of the biologically lethal ultraviolet radiation from our otherwise life-giver, the Sun.

Another kind of strong chemical bond exists, the ionic bond, which is purely electrostatic in nature. For example, if a sodium atom (having one valence electron outside a closed shell) meets a chlorine atom (missing one electron to complete a closed shell), a chemical reaction occurs in which the sodium valence electron transfers to the chlorine atom. The formed ions, Na^+ and Cl^-, attract each other by Coulomb interaction and form the NaCl molecule, the state being energetically favorable compared to that of the two separate atoms.

Sodium and chlorine can also coexist in solid form. Take a drop of salt water and allow it to be heated up by sunlight. Soon the water evaporates and a white grain is left, the crystal structure of sodium and chlorine formed by ionic bonding, the mineral rock salt that we consume daily.

6.5 Tunneling Frequency

To understand double-well tunneling dynamics, consider a particle launched in a superposition of the ground state and the first excited state (an overall normalization factor $1/\sqrt{2}$ is left implicit for simplicity):

$$\psi(x) = \psi_-(x) \pm \psi_+(x). \tag{6.55}$$

Since the weights of the wave functions under the barrier are small and the two wave functions are even and odd, respectively (recall Figure 6.10), the superposition corresponds

[9] The chemist's tool kit consists of more tools, reflecting hybridization of bonds, triple bonds (involving six electrons), etc. However, for our purposes, only a basic understanding of the chemical bond is needed.

to a state with almost only non-vanishing wave function in one of the wells (either the right or left well depending on the choice of $\pm$ in the superposition). The particle is initially with almost certainty in one of the two wells. The time evolution of an energy eigenfunction is provided by its energy phase factor. At subsequent times, the wave function, Eq. (6.55), will thus, according to the Schrödinger equation, be the superposition (choosing as initial state the superposition with the plus sign)

$$\begin{aligned} \psi(x,t) &= \psi_-(x)\, e^{-(i/\hbar)E_- xt} + \psi_+(x)\, e^{-(i/\hbar)E_+ t} \\ &= e^{-(i/\hbar)E_+ t}[\psi_+(x) + e^{-(i/\hbar)\Delta t}\psi_-(x)]. \end{aligned} \tag{6.56}$$

After a time $t = \pi\hbar/\Delta$, the relative sign in the superposition is reversed and the particle is almost with certainty in the other well; thus a particle launched in one initial well state, Eq. (6.55), oscillates, tunneling back and forth between the two wells with a frequency determined by the level splitting, $\nu = \Delta/\hbar$. The tunneling frequency is thus a direct measure of the binding energy (and proportional to the tunneling probability).

If the two protons in a hydrogen molecule, H_2, are 5 nm apart, the electrons switch wells only after 10^{12} years. At a separation of 0.2 nm, the transfer frequency reaches $10^{14}\,s^{-1}$; and at the internuclear distance of the hydrogen molecule H_2, the two electrons have equal probability to be in either of the proton wells, their exchange forming a covalent bond.

6.6 From Atoms to Solids

Matter is found not only in the gas phase, such as our atmosphere of mainly nitrogen and oxygen, where the lumping of atoms is at most into small molecules, O_2 and N_2, but also as dense assemblies of an enormous number of atoms: liquids, glasses and crystalline solid-state objects, the latter being the ones of our interest. The physical properties of solid-state objects vary tremendously, and have various thermodynamic stable and metastable states depending on temperature and pressure, varying in electronic, photonic, magnetic, etc. properties. Our interest will be solid-state objects at ambient pressure and temperature, in particular, the crystalline solids of importance for electronics: metals, insulators and semiconductors.

Of particular interest will be covalently bonded solids. When molecules can be formed by bonding, so can solids, as the trick can be repeated, and many elements of importance for technology solidify by covalent bonding. Consider, for example, the element silicon. This element is abundant around us, for example, on beaches, and constitutes almost 30% by mass of the elements in the Earth's crust. The element silicon is an atom with 14 protons in its nucleus and a corresponding number of electrically neutralizing electrons. It has 10 electrons in closed shells and four loosely bound electrons, the so-called valence electrons, outside its closed shell. These four valence electrons are energetically eager to participate with the valence electrons in other silicon atoms to form covalent bonds, in this way hooking silicon atoms together. So if we have a gathering of silicon atoms and the temperature is low enough that chemical bonds will not be broken by thermal agitation, covalent bonding of the valence electrons will take place as each silicon atom surrounds itself with four

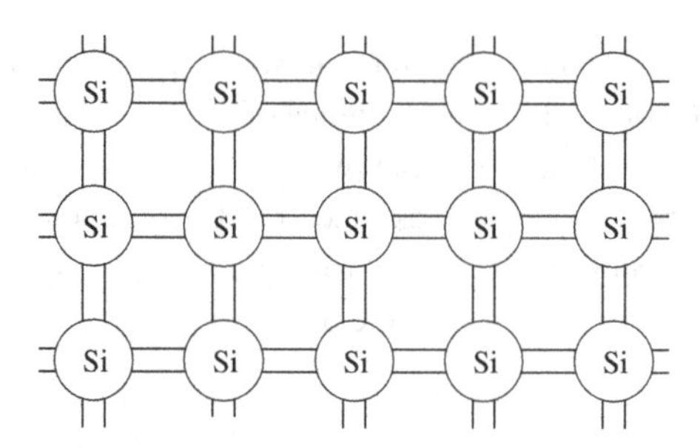

Figure 6.11 A two-dimensional rendition of a covalently bonded solid.

nearest-neighbor silicon atoms and forms a crystal structure, as shown schematically in a two-dimensional equivalent in Figure 6.11. Neighboring silicon atoms contribute one electron each to their covalent bond. The electronic density in a covalently bonded crystal is thus non-uniform, valence electrons mainly being located in the covalent bonds.

The energetically favorable crystal structure of a covalently bonded crystal is not a cubic lattice, as alluded to in Figure 6.11, but in fact the diamond structure (the crystal structure of the insulator diamond built by carbon atoms), and to stress this we draw a two-dimensional rendition of silicon as in Figure 10.1 later. The same covalent crystal bonding happens for other elements with four valence electrons, for example, germanium or III–V or II–VI semiconductor compounds, as discussed in Chapter 12.

Another important binding mechanism leading to crystal structures is that of metallic bonding. Typical examples are sodium or copper, having one valence electron, whose whereabouts is more or less evenly spread out over the ionic crystal lattice, consisting of regularly spaced nuclei with tightly bound core electrons (almost in the shell states of the atom). The spread-out electronic charge provides the electrostatic force, the glue, to keep the repulsion between the positively charged ions in check. The generic situation for valence electrons in metals is therefore that of motion in a repetitive structure, a periodic structure.

6.7 Summary

In a covalently bonded crystal, a finite energy is needed to break a bond in order to get a net motion of electrons through the crystal, and if this energy is large, we expect no current at small force, i.e. insulating behavior. If the bonding energy is sufficiently low, so that the lattice vibrations at room temperature can knock electrons to sites from where they can move, interstitial sites, we expect a behavior somewhere in between a conductor and an insulator, a so-called semiconductor.

For a double well, the degenerate level of separated wells splits into two close energy levels. For a triple well, three levels occur; and with ever more wells, a whole band of close energies occurs, their number equaling the number of wells. For an infinite number of identical wells and barriers, a periodic potential emerges. To gain a quantitative understanding of the motion of electrons in periodic structures, we leave the chemist's local point of view of chemical bonding, and in the next chapter we take the global crystal view of electrons experiencing motion in a periodic potential. Incorporating the essential feature of electrons being indistinguishable fermions will then give the quantitative distinction between the electrical properties of crystal structures: metals, semiconductors and insulators.

7 Particle in a Periodic Potential

Particles in periodic potentials occur in many contexts, for example, ultra-cold atoms in optical lattices created by interfering laser light, or electrons in crystalline substances such as metals and semiconductors. Before discussing three-dimensional crystals in Chapter 9, a particle in a one-dimensional periodic potential is considered, since it illustrates in a simple fashion the quantum physics of periodicity. Although it sounds artificial, the study of one-dimensional crystals is not only of academic interest, as it is relevant for studying organic conductors, quasi-one-dimensional structures. Furthermore, a general analysis of the energy spectrum can be carried through.[1]

7.1 Periodic Potential

Consider the time-independent Schrödinger equation

$$\hat{H}\psi(x) = \epsilon\,\psi(x), \qquad \hat{H} = -\frac{\hbar^2}{2m}\frac{d^2}{dx^2} + V(x) \tag{7.1}$$

for a particle in a periodic potential, i.e. the potential is characterized by its periodicity, say a,

$$V(x+a) = V(x) \tag{7.2}$$

for an arbitrary value of x. A periodic potential thus stretches from minus to plus infinity, the real line.[2] As an example, the periodic potential consisting of square barriers is displayed in Figure 7.1.

The stepwise constant potential depicted in Figure 7.1 could represent approximately the potential felt by an electron in a layered heterostructure, a so-called superlattice, as discussed in Chapter 12, or be a crude model of a one-dimensional crystal where the wells

[1] The reader preferring a discussion of the three-dimensional case immediately can jump directly to Chapter 9. However, the following chapters can be understood from the basic feature of the existence of energy gaps in the spectrum of electrons in periodic potentials, and the reader interested only in the essential features can avoid the additional indices brought about by two or three dimensions. Alternatively, this chapter can be considered as a warm-up for the three-dimensional case, since the techniques introduced here can be simply generalized to three dimensions.

[2] A particle in a finite *periodic* potential with imposed periodic boundary condition is considered in Section 7.12, a case simply solved by Fourier analysis, and the reader interested only in this limited case can proceed directly there.

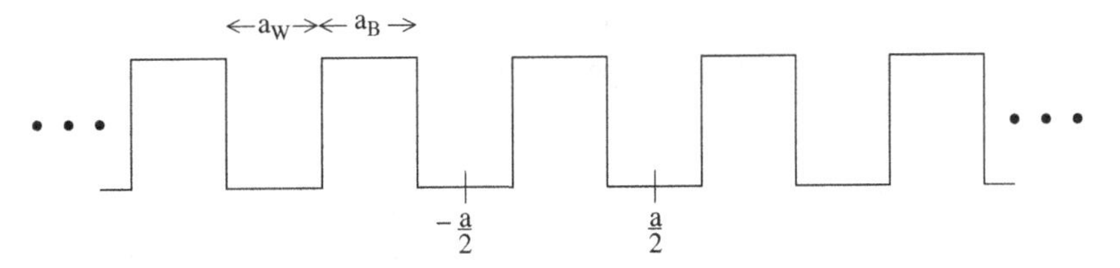

Figure 7.1 Example of a periodic potential.

represent the attraction of an electron by the ions of a one-dimensional crystal, the Kronig–Penney model.

Specifying a potential in a region of length a, and then translating this unit cell of potential profile any multiple in both directions, any periodic potential of periodicity a is created. Conversely, given a periodic potential of periodicity a, any choice of interval of length a provides, in view of Eq. (7.2), a unit cell. In Figure 7.1 is displayed one choice of unit cell for the square barrier periodic potential, the one for which the potential is inversion-symmetric around the center of the chosen unit cell.

In the following, an arbitrary periodic potential is considered, and our aim thus is not *solving* the time-independent Schrödinger equation, but seeking the characteristic features of the energy spectrum and eigenfunctions. This is an example of exploring the properties of solutions to differential equations without knowing the solutions themselves but only their symmetry properties. Such a qualitative analysis is often enough for an understanding of the basic physical properties described by a differential equation.

7.2 Crystal Translations

Consider the crystal[3] translation operator, $\hat{T}_a$, translating the periodicity of the periodic potential, i.e. the operator operating on an arbitrary function, f, according to

$$\hat{T}_a f(x) = f(x+a). \tag{7.3}$$

The value of the function $\hat{T}_a f$ at position x is given by the value of f at $x+a$.

Operating with the crystal translation operator on the left-hand side of the time-independent Schrödinger equation (7.1) gives

$$\hat{T}_a \hat{H} \psi(x) = -\frac{\hbar^2}{2m}\psi''(x+a) + V(x+a)\,\psi(x+a). \tag{7.4}$$

The opposite order of operation gives

$$\hat{H}\hat{T}_a \psi(x) = \hat{H}\psi(x+a) = -\frac{\hbar^2}{2m}\psi''(x+a) + V(x)\,\psi(x+a). \tag{7.5}$$

[3] *Crystal* is in this chapter used synonymously for *periodic potential*.

Owing to the periodicity of the potential, the right-hand sides of Eqs. (7.4) and (7.5) are identical. The translation operator thus commutes with the Hamiltonian, i.e. for an arbitrary function ψ,

$$\hat{H}\hat{T}_a \psi = \hat{T}_a \hat{H} \psi, \tag{7.6}$$

or, stated as an operator identity, $[\hat{H}, \hat{T}_a] = 0$, the commutator of the two operators vanishes. Consequently, if ψ is an energy eigenfunction corresponding to the energy value ϵ, so is the translated function, as $\hat{H}(\hat{T}_a\psi) = \hat{T}_a(\hat{H}\psi) = \hat{T}_a(\epsilon\psi) = \epsilon(\hat{T}_a\psi)$.

Since differentiation and translation are commuting operations, i.e. interchangeable without consequence, commutation of the crystal translation operator and the periodic potential Hamiltonian is simply the result of the commutation with the potential (multiplication operator), $[\hat{T}_a, V] = 0$.

7.3 Crystal Translation Eigenvalues

The time-independent Schrödinger equation for a particle in a one-dimensional potential is a linear differential equation of *second* order. It has therefore, for each possible eigenvalue, ϵ, in general at most two independent solutions, say, ψ_1 and ψ_2, i.e. eigenfunctions of the Hamiltonian, $\hat{H}\psi_i = \epsilon\psi_i$. Their independence means that they are not proportional, $\psi_1 \neq c\psi_2$, and thus describe two different physical states with the same energy. Since the crystal translation operator commutes with the Hamiltonian, the solutions can, according to Appendix P, be chosen also as eigenfunctions of the crystal translation operator, i.e.

$$\hat{T}_a\psi_1(x) = \lambda_1\psi_1(x), \qquad \hat{T}_a\psi_2(x) = \lambda_2\psi_2(x). \tag{7.7}$$

The equations state that the translated energy eigenfunctions are proportional to the untranslated ones, i.e. for all x,

$$\psi_1(x+a) = \lambda_1\psi_1(x), \qquad \psi_2(x+a) = \lambda_2\psi_2(x). \tag{7.8}$$

The eigenvalues are different from zero, $\lambda_i \neq 0,\ i = 1, 2$, as otherwise $\psi_{1,2} = 0$.

A translation can be repeated, giving

$$\psi_i(x+na) = (\hat{T}_a)^n\psi_i(x) = (\lambda_i)^n\psi_i(x), \qquad i = 1, 2. \tag{7.9}$$

The inverse of the translation operator, $\hat{T}_a^{-1}$, $\hat{T}_a^{-1}\hat{T}_a = \hat{I}$, has, as

$$\psi_i = \hat{T}_a^{-1}\hat{T}_a\psi_i = \hat{T}_a^{-1}\lambda_i\psi_i = \lambda_i\hat{T}_a^{-1}\psi_i, \qquad i = 1, 2, \tag{7.10}$$

the inverse eigenvalues

$$\hat{T}_a^{-1}\psi_i = \lambda_i^{-1}\psi_i, \qquad i = 1, 2. \tag{7.11}$$

The inverse of the translation operator, $\hat{T}_a^{-1}$, equals $\hat{T}_{-a}$, as the translation operator $\hat{T}_{-a}$ undoes the translation of $\hat{T}_a$, $\hat{T}_a^{-1} = \hat{T}_{-a}$. Repeated translations in the negative direction therefore give

$$\psi_i(x-na) = (\hat{T}_{-a})^n\psi_i(x) = (\hat{T}_a^{-1})^n\psi_i(x) = (\lambda_i)^{-n}\psi_i(x), \qquad i = 1, 2. \tag{7.12}$$

If $|\lambda_i| \neq 1$, the energy eigenfunctions according to Eqs. (7.9) and (7.12) diverge in either of the limits $x \to \pm\infty$. Such divergent functions are discarded as unphysical solutions to the problem of a particle in a periodic potential (see the general discussion of wave function properties in Appendix E).[4] Thus $|\lambda_i| = 1$, i.e.

$$\lambda_1 = e^{i\phi_1}, \qquad \lambda_2 = e^{i\phi_2}, \tag{7.13}$$

where ϕ_1 and ϕ_2 are real. As expected, the probabilities for particle location at equivalent points, from where the whole periodic potential looks the same, are identical, $|\psi_{1,2}(x+a)|^2 = |\psi_{1,2}(x)|^2$.

As shown in Appendix E, the Wronskian of the time-independent Schrödinger equation,

$$W(x) \equiv \psi_1(x)\psi_2'(x) - \psi_2(x)\psi_1'(x), \tag{7.14}$$

is a *constant*; in particular, therefore, $W(x) = W(x+a) = \hat{T}_a W(x)$. Using Eq. (7.8) it follows that

$$\begin{aligned} W(x) = \hat{T}_a W(x) &= \psi_1(x+a)\psi_2'(x+a) - \psi_2(x+a)\psi_1'(x+a) \\ &= \lambda_1\psi_1(x)\lambda_2\psi_2'(x) - \lambda_2\psi_2(x)\lambda_1\psi_1'(x) = \lambda_1\lambda_2 W(x), \end{aligned} \tag{7.15}$$

and since the Wronskian for two independent solutions is a constant *different* from zero, it follows that their translation operator eigenvalues are reciprocals,

$$\lambda_1\lambda_2 = 1. \tag{7.16}$$

For the physical solutions for a particle in a periodic potential, it follows, according to Eqs. (7.16) and (7.13), that

$$\lambda_1 = e^{i\phi}, \qquad \lambda_2 = e^{-i\phi}. \tag{7.17}$$

Energy eigenfunctions have been tagged by the eigenvalues of the crystal translation operator, values located on the unit circle and each other's complex conjugate. In case for certain energies the solutions are dependent, $\psi_2 = c\psi_1$, it follows from Eq. (7.7) that the crystal translation eigenvalues are equal, i.e. corresponding to either of the two possibilities $\lambda_2 = \pm 1 = \lambda_1$.

7.4 Conjugation Property

Complex conjugation and translation commute as

$$(\hat{T}_a\psi(x))^* = \psi^*(x+a) = \hat{T}_a\psi^*(x). \tag{7.18}$$

[4] This does not mean that these mathematical solutions are not of use in other situations. If a potential is periodic only in the region $[0,\infty]$ and for example constant in the other half region, i.e. vacuum in one half-space and crystal in the other half-space, the non-divergent solution in the region $[0,\infty]$ is a physically relevant one, being the evanescent real wave function of a reflected electron impinging on the crystal, as discussed further in Section 8.3. Another situation where these solutions are of importance is for the finite insulator region in a tunnel junction. The following analysis shows that the states with real reciprocal translation eigenvalues correspond to *imaginary* Bloch numbers and (according to Eq. (7.46) for $k \to -ik$) are (divergent) solutions corresponding to all energies in the energy gap region, so-called gap states, whose wave functions can be chosen real. As an exercise, verify these statements.

Complex conjugation of Eq. (7.7) therefore gives

$$\hat{T}_a \psi_1^*(x) = \lambda_1^* \psi_1^*(x). \tag{7.19}$$

Since the energy eigenvalue, ϵ, is real, the function $\psi_1^*(x)$ is also an energy eigenfunction corresponding to the energy in question. Expressing ψ_1^* in terms of the considered two independent solutions,

$$\psi_1^* = a_1 \psi_1 + a_2 \psi_2, \tag{7.20}$$

demands $(a_1, a_2) \neq (0, 0)$, as ψ_1 is not the zero function. According to Eqs. (7.19) and (7.20), and using Eq. (7.7),

$$\begin{aligned} \lambda_1^* \psi_1^* = \hat{T}_a \psi_1^* &= \hat{T}_a(a_1 \psi_1 + a_2 \psi_2) = a_1 \lambda_1 \psi_1 + a_2 \lambda_2 \psi_2 \\ &= a_1 \lambda_1 \psi_1 + a_2 \lambda_1^* \psi_2. \end{aligned} \tag{7.21}$$

Substituting Eq. (7.20) also in the left-hand side of Eq. (7.21) gives

$$\lambda_1^*(a_1 \psi_1 + a_2 \psi_2) = a_1 \lambda_1 \psi_1 + a_2 \lambda_1^* \psi_2 \tag{7.22}$$

and thereby

$$a_1(\lambda_1^* - \lambda_1)\psi_1 = 0. \tag{7.23}$$

Assume $\lambda_1 \neq \lambda_1^*$ $(= \lambda_2)$, equivalent to $\lambda_1 \neq \pm 1$ in view of $|\lambda_1| = 1$; then if $a_1 \neq 0$ this leads to $\psi_1 = 0$, contradicting it being a wave function for the particle. Thus $a_1 = 0$ and thereby $a_2 \neq 0$ and Eq. (7.20) becomes

$$\psi_1^* = a_2 \psi_2. \tag{7.24}$$

An eigenvalue equation only defines the eigenfunction up to an overall constant, and in Eq. (7.7) we can equally well make the choice $\psi_2 \to a_2^{-1} \psi_2$. Thus, according to Eq. (7.24), two independent energy eigenfunctions for a particle in a periodic potential can be chosen as each other's complex conjugate

$$\psi_2(x) = \psi_1^*(x). \tag{7.25}$$

When the translation eigenvalues are different, $\lambda_2 \neq \lambda_1 = \lambda_2^*$, the eigenfunctions are each other's complex conjugate just like the crystal translation eigenvalues.[5] The energy eigenfunctions are, by being chosen as also eigenfunctions of the crystal translation operator, thus the ones connected by time reversal symmetry (like the plane wave energy eigenfunctions of a free particle). The case of $\lambda_1 = \lambda_2$, i.e. $\lambda_1 = \pm 1 = \lambda_2$, corresponds to non-degenerate energy eigenfunctions, as discussed in Section 7.10.

[5] Implying that the corresponding eigenfunctions are not real and thus giving a non-vanishing current, as studied in Chapter 8.

7.5 Bloch's Theorem

The crystal translation eigenvalue dependence on the periodicity, a, has so far been suppressed, $\lambda = \lambda(a) = \exp(i\phi(a))$, a dependence implied by the eigenvalue equation

$$\hat{T}_a \psi(x) = \lambda(a)\psi(x) \tag{7.26}$$

due to the appearance of the periodicity on the left-hand side of the equation. Translating twice, the periodicity is assigned the eigenvalue of twice the periodicity

$$\hat{T}_{a+a}\psi(x) = \lambda(a+a)\psi(x). \tag{7.27}$$

Equation (7.27) can be interpreted as translating twice the periodicity, which is equivalent to two consecutive translations,

$$\hat{T}_{2a}\psi(x) = \psi(x+2a) = \hat{T}_a \hat{T}_a \psi(x) = \lambda(a)\lambda(a)\psi(x), \tag{7.28}$$

and according to Eq. (7.27),

$$\lambda(2a) = \lambda(a)\lambda(a), \qquad e^{i\phi(2a)} = e^{i\phi(a)}\, e^{i\phi(a)} = e^{i2\phi(a)}. \tag{7.29}$$

Since in Eq. (7.27) any integer could be used instead of 2, and a can be any real number, $\phi(a)$ must be a linear function of the periodicity, $\phi(a) = ka$, where k is a real number. The dependence of the eigenvalues on the periodicity is thus

$$\lambda_1(a) = e^{ika}, \qquad \lambda_2(a) = e^{-ika}, \tag{7.30}$$

i.e. for the two independent solutions corresponding to the same energy,

$$\psi_1(x+a) = e^{ika}\psi_1(x), \qquad \psi_2(x+a) = e^{-ika}\psi_2(x). \tag{7.31}$$

An energy eigenfunction can therefore be labeled by a real quantum number k, $\psi_k(x)$ (corresponding to the crystal translation eigenvalue $\lambda_k = \exp(ika)$), thereby also labeling the energy, $\epsilon = \epsilon(k)$. The other independent solution corresponding to the energy in question is, according to Eq. (7.31), labeled by the number $-k$ and, recalling Eq. (7.25),

$$\psi_{-k}(x) = \psi_k^*(x), \tag{7.32}$$

thus time reversal symmetry is the reason for the spectrum property $\epsilon(-k) = \epsilon(k)$.

We have arrived at Bloch's theorem: an energy eigenfunction for a particle in a periodic potential can be labeled by the real Bloch number, k, describing its phase change between equivalent locations,

$$\psi_k(x+a) = e^{ika}\psi_k(x). \tag{7.33}$$

Functions satisfying this Bloch condition are called Bloch functions, crystal translation and energy eigenfunctions for a particle in a periodic potential. Periodicity of the potential thus leads to a phase constraint on energy eigenfunctions.

7.6 Bloch Waves

Consider the scattering properties of a unit cell of an arbitrary periodic potential, choosing for simplicity the unit cell to be the interval $[0, a]$. Since the energy spectrum is bounded from below, a constant can always be added to the periodic potential Hamiltonian so that its eigenvalues are positive. Outside the interval $[0, a]$, the free regions, the potential is chosen to vanish. The general solution to the time-independent Schrödinger equation for the specified scattering situation is thus, in the free regions, the superposition of two independent plane waves,

$$\psi_\epsilon(x) = \begin{cases} A_r^L\, e^{iqx} + A_l^L\, e^{-iqx}\,, & x < 0, \\ A_r^R\, e^{iqx} + A_l^R\, e^{-iqx}\,, & x > a, \end{cases} \tag{7.34}$$

where the real number q denotes the free particle wave number for the energy in question, $q = \sqrt{2m\epsilon}/\hbar$. The amplitudes are connected by the transfer matrix (recall Chapter 2 and see the general discussion in Appendix F),

$$\mathbf{A}^R = \mathcal{T}(a, 0)\, \mathbf{A}^L, \tag{7.35}$$

and a solution to the time-independent scattering scenario Schrödinger equation in question exists for any energy $\epsilon \geq 0$, since both types of independent solutions, Bloch type or *diverging/decaying* functions in the unit cell region, are at our disposal for matching of solutions for the scattering scenario.

Next we require that the scattering solution should be produced by using only one of the Bloch-type solutions in the unit cell, i.e. only one Bloch function is allowed for matching: the wave function in region $[0, a]$ is demanded to be proportional to a Bloch function, i.e. $B\psi_k$. The function ψ_k translated from the unit cell to the real axis according to the Bloch condition is then the solution to the periodic potential Hamiltonian corresponding to energy $\epsilon(k) = \epsilon = \hbar^2 q^2/2m$. Then the matching conditions at $x = 0$ can, according to Eq. (2.59), be expressed as

$$\begin{pmatrix} 1 & 1 \\ iq & -iq \end{pmatrix} \begin{pmatrix} A_r^L \\ A_l^L \end{pmatrix} = \begin{pmatrix} \psi_k(0) & 0 \\ \psi_k'(0) & 0 \end{pmatrix} \begin{pmatrix} B \\ 0 \end{pmatrix} \tag{7.36}$$

and at $x = a$ as

$$\begin{pmatrix} e^{iqa} & e^{-iqa} \\ iq\, e^{iqa} & -iq\, e^{-iqa} \end{pmatrix} \begin{pmatrix} A_r^R \\ A_l^R \end{pmatrix} = \begin{pmatrix} \psi_k(a) & 0 \\ \psi_k'(a) & 0 \end{pmatrix} \begin{pmatrix} B \\ 0 \end{pmatrix}$$
$$= e^{ika} \begin{pmatrix} \psi_k(0) & 0 \\ \psi_k'(0) & 0 \end{pmatrix} \begin{pmatrix} B \\ 0 \end{pmatrix}, \tag{7.37}$$

where the last equality follows from the Bloch condition and its consequence, $\psi_k'(a) = e^{ika}\psi_k'(0)$. Combining Eqs. (7.36) and (7.37) gives the amplitude relationship for such a matching demand,

$$\mathbf{A}^R = e^{ika} \mathcal{R}(a) \mathbf{A}^L, \tag{7.38}$$

where

$$\mathcal{R}(a) = \begin{pmatrix} e^{-iqa} & 0 \\ 0 & e^{iqa} \end{pmatrix}. \tag{7.39}$$

Combining the transfer matrix relationship, Eq. (7.35), and the amplitude relationship that follows when only the Bloch function ψ_k is available for matching and only its Bloch condition has been utilized, Eq. (7.37), gives the requirement

$$[\mathcal{T}(a,0) - e^{ika}\mathcal{R}(a)]\mathbf{A}^L = \begin{pmatrix} 0 \\ 0 \end{pmatrix} \tag{7.40}$$

for the existence of a Bloch function corresponding to energy $\epsilon(k) = \epsilon = \hbar^2 q^2/2m$.

To have a non-vanishing wave function demands that the matrix appearing in Eq. (7.40) is not invertible, which is equivalent to its determinant vanishing,

$$\det(\mathcal{T} - e^{ika}\mathcal{R}) = 0. \tag{7.41}$$

Writing out the components in Eq. (7.41), the determinant is immediately calculated as

$$\begin{vmatrix} T_{11} - e^{i(k-q)a} & T_{12} \\ T_{21} & T_{22} - e^{i(k+q)a} \end{vmatrix} = \det \mathcal{T} - T_{11}\, e^{i(k+q)a} - T_{22}\, e^{i(k-q)a} + e^{2ika}. \tag{7.42}$$

Multiplying Eq. (7.41) by e^{-ika} and (see Appendix F) using the fact that the determinant of the transfer matrix is one, $\det \mathcal{T} = 1$, Eq. (7.41) leads to

$$0 = e^{ika} + e^{-ika} - T_{11}\, e^{iqa} - T_{22}\, e^{-iqa} \tag{7.43}$$

or equivalently (according to Appendix F, the relationship $T_{22} = T_{11}^*$ is generally valid as a consequence of time reversal symmetry)

$$\cos(ka) = \frac{T_{11}\, e^{iqa} + (T_{11}\, e^{iqa})^*}{2} = \Re(T_{11}\, e^{iqa}), \tag{7.44}$$

the appearance of the cosine function being a direct consequence of the Bloch condition. Equation (7.44) can be expressed in terms of the crystal translation eigenvalue in question,

$$\Re\lambda_k = \Re(T_{11}\, e^{iqa}). \tag{7.45}$$

Introducing $T_{11}(q) \equiv |T_{11}(q)|\, e^{i\chi(q)}$, the existence of a Bloch state with energy $\epsilon(k) = \hbar^2 q^2/2m$ requires the equation

$$\cos(ka) = |T_{11}(q)| \cos(qa + \chi(q)), \tag{7.46}$$

the Bloch energy condition, to be satisfied.

The symmetry constraint on allowed energy eigenfunctions, the Bloch condition, has thus been turned into an equation relating Bloch numbers and crystal structure (in one dimension trivially represented by the periodicity a) to the scattering properties of the unit cell potential: only if a real k can satisfy Eq. (7.46) does a Bloch state exist with the corresponding energy $\epsilon(k) = \hbar^2 q^2/2m$. The consequence of the Bloch function phase rigidity on the possible allowed energy values can now be investigated.

Exercise 7.1 Insert Eq. (2.69) into Eq. (7.44) and obtain for the Kronig–Penney model the condition for the existence of a Bloch state of energy $\epsilon(k) = \epsilon = \hbar^2 q^2/2m$, $0 < \epsilon < V$, i.e. show that Eq. (7.46) becomes

$$\cos(ka) = \cosh(\kappa a_B)\cos(q(a - a_B)) - \frac{q^2 - \kappa^2}{2q\kappa}\sinh(\kappa a_B)\sin(q(a - a_B)), \tag{7.47}$$

where $q = \sqrt{2m\epsilon}/\hbar$ and $\kappa = \sqrt{2m(V - \epsilon)}/\hbar$.

Show that, for $\epsilon > V$ (i.e. $\kappa \to -ik_V$, where $k_V = \sqrt{2m(\epsilon - V)}/\hbar$), the Bloch energy condition becomes (recalling Figure 7.1, $a - a_B$ is the size of the well of the potential, $a_W = a - a_B$)

$$\cos(ka) = \cos(k_V a_B)\cos(q a_W) - \frac{q^2 + k_V^2}{2qk_V}\sin(k_V a_B)\sin(q a_W). \tag{7.48}$$

Show that the right-hand sides of Eqs. (7.47) and (7.48) are equal for $\epsilon = V$.

Exercise 7.2 Use the result of Exercise 2.7 on page 31 to show that the 11-component of the transfer matrix for the delta potential $V(x) = w\delta(x)$ is

$$T_{11}(q) = \sqrt{1 + \frac{m^2 w^2}{\hbar^4 q^2}}\, e^{i\chi(q)}, \tag{7.49}$$

where the phase is

$$\chi(q) = \arctan\left(-\frac{mw}{\hbar^2 q}\right). \tag{7.50}$$

For the case $w > 0$, show that $\chi(q \to 0) \to 3\pi/2$, and note that in the limit $q \to 0$ (for $x \ll -1$, $\arctan(x) \simeq 3\pi/2 - 1/x$) the right-hand side of Eq. (7.46) becomes

$$\lim_{q \to 0} |T_{11}(q)| \cos(qa + \chi(q)) = 1 + \frac{maw}{\hbar^2}. \tag{7.51}$$

As a check of correctness, show that this is the same limit as the result obtained by taking the delta function limit of the Kronig–Penney potential on the right-hand side of Eq. (7.47).

Show that, for $w > 0$, the phase $\chi(q)$ varies smoothly from $3\pi/2$ to 2π when q varies from 0 to ∞. Show that, for $w < 0$, the phase $\chi(q)$ varies smoothly from $\pi/2$ to π when q varies from 0 to ∞.

Exercise 7.3 Show that, for a periodic array of delta function spikes,

$$V(x) = \sum_{n=-\infty}^{\infty} w\delta(x - na), \qquad w > 0, \tag{7.52}$$

the Bloch energy condition, Eq. (7.44), takes, upon using Eq. (2.75), the form

$$\cos(ka) = \cos(qa) + \frac{mw}{\hbar^2 q}\sin(qa). \tag{7.53}$$

Show that the result also follows from Eq. (7.47) when taking the delta function limit of the Kronig–Penney model.

Exercise 7.4 For the Kronig–Penney model, in the limit of large periodicity, lattice constant $a \to \infty$ and well size a_W kept constant, show that Eq. (7.47) turns into the condition for an energy level in the isolated well. The energy bands shrink into single degenerate energy levels, the quantized energy levels of the quantum well.

7.7 Band Structure

Analyzing Eq. (7.46) graphically gives a qualitative understanding of the energy spectrum of a particle in an arbitrary periodic potential. Owing to the phase shift, the function $|T_{11}(q)| \cos(qa + \chi(q))$ is a stretched cosine function as a function of q, with an amplitude given by the square root of the inverse of the tunneling probability or transmission coefficient, $T(\epsilon = \hbar^2 q^2/2m)$, as, according to Eq. (2.78), $|T_{11}(q)|^2 = 1/T(\epsilon)$. The amplitude of the cosine function on the right-hand side of Eq. (7.46) is therefore greater than one, $|T_{11}(q)| \geq 1$, as the transmission probability satisfies $0 \leq T(\epsilon) \leq 1$. Thus $|T_{11}(q)|$ diverges to plus infinity for small q values, since the tunneling probability vanishes at zero energy (e.g. recall for the square barrier that it diverges as $1/q$), and approaches one for large q values, as the transmission probability for very large energies approaches one. The function on the right-hand side of Eq. (7.46) is therefore, at small q values, larger than one, and has the shape sketched in Figure 7.2. The value of the right-hand side function, $|T_{11}(q)| \cos(qa + \chi(q))$, for the value $q = 0$ is model-dependent. For the Kronig–Penney model, according to Eq. (7.47), the value is equal to $\cosh(\kappa_0 a_B) + \frac{1}{2}\kappa_0(a - a_B)\sinh(\kappa_0 a_B)$, with $\kappa_0 \equiv \sqrt{2mV}/\hbar$; and for the delta model, for $q = 0$, according to Eq. (2.75) or (7.53), $\Re(T_{11}(q)\, e^{iqa})$ attains the value $1 + maw/\hbar^2$. In both cases this is a value larger than one, as to be expected in view of the diverging amplitude.

The left-hand side of Eq. (7.46), the cosine function specified by the real Bloch number, can only take on values between ± 1. Displaying the constant functions, ± 1, in Figure 7.2 as well, the graphical solution thus establishes the qualitative features of the band structure, as there can only be equality to the left-hand side in Eq. (7.46) if $|T_{11}(q)| \cos(qa + \chi(q))$ lies within the range ± 1.

At a certain smallest q value, q_1, the right-hand side of Eq. (7.46) equals one, $|T_{11}(q_1)| \cos(q_1 a + \chi(q_1)) = 1$, due to the cosine function attaining the inverse value $1/|T_{11}(q_1)|$ less than one. The curve representing the right-hand side of Eq. (7.46) then hits the $+1$ line for the first time corresponding, according to Eq. (7.45), to a Bloch state with crystal translation eigenvalue equal to one, $\lambda = 1$. We can choose the Bloch number $k = 0$ to correspond to this solution, as $\lambda_0 = 1$ or equivalently $\cos(0 \times a) = 1$. The corresponding lowest Bloch state energy is $\epsilon(k = 0) = \hbar^2 q_1^2/2m$. As q increases, the right-hand side of Eq. (7.46) is smaller than one in absolute magnitude, and positive Bloch numbers and energies can satisfy Eq. (7.46) until the right-hand side of Eq. (7.46) crosses the value minus one, say at q value q_2, corresponding to Bloch number $k = \pi/a$ and energy $\epsilon(k = \pi/a) = \hbar^2 q_2^2/2m$. In the whole interval from $k = 0$ to $k = \pi/a$ there are thus

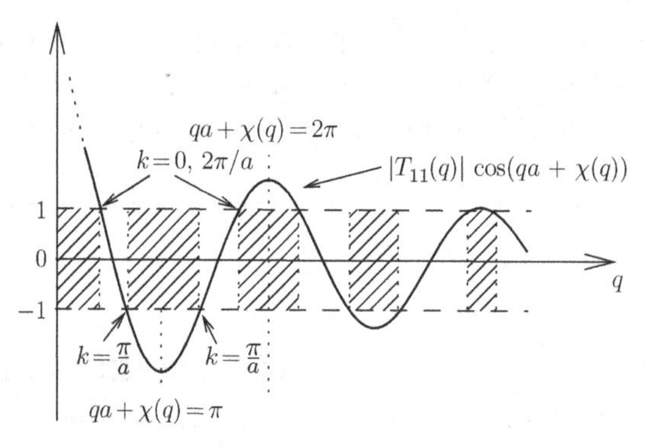

Figure 7.2 Graphical solution of Eq. (7.46).

solutions to Eq. (7.46), i.e. corresponding real k and q values for which the expressions on the left- and right-hand sides of Eq. (7.46) are equal. As q increases, the energy spectrum, $\epsilon(k(q)) = \hbar^2 q^2/2m$, is a monotonic increasing function of the Bloch number k as depicted in Figure 7.3. When q increases beyond the value corresponding to the right-hand side of Eq. (7.46) crossing the -1 line, there are no acceptable solutions to Eq. (7.46) and there will be a corresponding region of forbidden energy values.[6] Not until the curve representing the right-hand side of Eq. (7.46) again attains the value minus one, crossing again the -1 line, does a solution exist. Thus $k = \pi/a$ (or the equally proper choice $k = -\pi/a$ as discussed in Section 7.9) is again a solution; however, now it corresponds to a larger energy value, since q has increased, i.e. an energy gap appears. As q increases, there are solutions until the right-hand side function of Eq. (7.46) hits the $+1$ line, i.e. allowed energies $\epsilon(k)$ for the whole k interval from $k = \pi/a$ to $k = 2\pi/a$ as depicted in Figure 7.3. This scenario of regions of forbidden energies, which appears each time the function on the right-hand side of Eq. (7.46) overshoots the ± 1 lines, now repeats itself, giving the next range of energies in the k interval from $k = 2\pi/a$ to $k = 3\pi/a$, etc., as depicted in Figure 7.3. The continuous parabolic energy spectrum of a free particle is, by the presence of the periodic potential, broken up into piecewise, monotonically increasing, continuous parts with an energy gap appearing at Bloch numbers $\pi/a,\ 2\pi/a,\ 3\pi/a, \ldots$, the so-called Bragg points.[7]

Since cosine is an even function, for a k value solution to Eq. (7.46), the value $-k$ will also satisfy the equation, the $\pm k$ solutions corresponding to the same q value, i.e. the $\pm k$ Bloch states, $\psi_{\pm k}$, have the same energy, $\epsilon(-k) = \epsilon(k) = \hbar^2 q^2/2m$. This symmetry was already picked out by the Bloch functions also being eigenfunctions of the crystal translation operator, as $\psi_{-k} = \psi_k^*$. Accompanying the lowest range of energies corresponding to Bloch numbers from $k = 0$ to $k = \pi/a$ there is thus a mirror-symmetric range of energies corresponding to Bloch numbers from $k = 0$ to $k = -\pi/a$, and there

[6] There are mathematical solutions corresponding to energies in the gap region, but they are here physically unacceptable as they violate the Bloch condition. The solutions to equation (7.46) in the energy gap region have imaginary values of k (recall $\cos(\pi + iu) = -\coth u$, u real) thus giving the divergent/decaying solutions discussed in Section 7.3.

[7] When the periodic potential vanishes, the Bloch energy constraint, Eq. (7.46), becomes $\cos(ka) = \cos(qa)$, i.e. disappears as $k = q$ and the free particle energy parabola emerges.

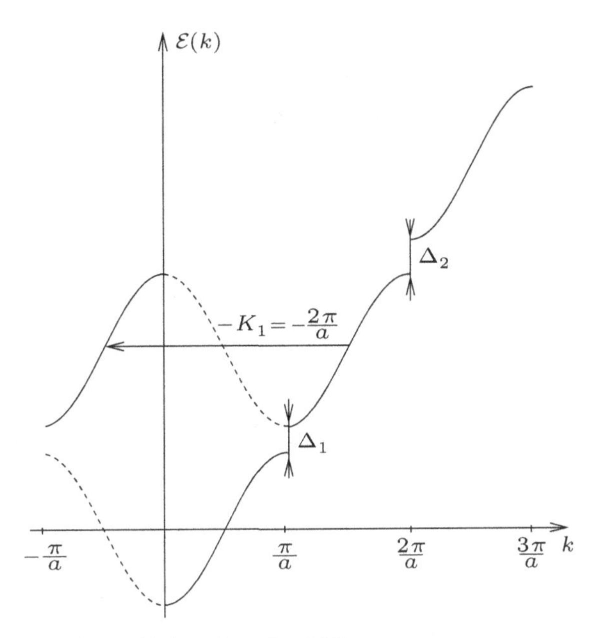

Figure 7.3 Band structure obtained from the graphical solution of Eq. (7.46).

is thus a whole band of lowest energies as the Bloch number changes from $k = -\pi/a$ to $k = \pi/a$, as depicted in Figure 7.3, the lowest energy band. The energy spectrum for all negative Bloch numbers is similarly obtained by mirror-reflecting the spectrum for the positive values. Representing the energy spectrum in this way, i.e. Bloch numbers extending along the real axis, is referred to as the extended zone scheme.

The Bloch condition thus leads to a qualitative understanding of the energy spectrum for an arbitrary periodic potential. The characteristic feature of a quantum particle in a periodic potential is the existence of ranges of energies that are not attainable by the particle, so-called energy gaps. No quantum state of the particle in the periodic potential exists corresponding to an energy value in a gap region.[8]

At larger energies, the band gaps become smaller, as the transmission probability generally gets larger with increasing energy. Consider the Kronig–Penney model, Eq. (7.48), $\epsilon > V$. For very large energies, $\epsilon \gg V$,

$$\frac{q^2 + k_V^2}{2qk_V} = \frac{q^2 + q^2(1 - V/\epsilon)}{2q^2\sqrt{1 - V/\epsilon}} = 1 + \frac{5}{8}\left(\frac{V}{\epsilon}\right)^2 + \mathcal{O}\left(\left(\frac{V}{\epsilon}\right)^3\right) \tag{7.54}$$

[8] This is in contrast to the classical scenario. The classical dynamics of a particle in a periodic potential has nothing this spectacular about it. In a sinusoidal potential, say, the state of lowest energy is the particle at rest at the bottom of one of the wells. States of continuously higher energy correspond to oscillations around the bottom of the well, and for energies higher than the maximum of the potential the particle will move in either of the two directions.

and Eq. (7.48) becomes

$$\cos(ka) = \cos(qa_W)\cos(k_V a_B) - \left(1 - \frac{5}{8}\left(\frac{V}{\epsilon}\right)^2\right)\sin(qa_W)\sin(k_V a_B). \tag{7.55}$$

Dropping the small $(V/\epsilon)^2$ term, corresponding to approximating k_V by q, gives

$$\cos(ka) \simeq \cos(q(a_W + a_B)) = \cos(qa), \tag{7.56}$$

showing that in this limit the band gap narrows to a point. Thus, at high energies, band gaps are generally progressively smaller.

7.8 Bloch Function Properties

Introducing

$$u_k(x) \equiv e^{-ikx}\psi_k(x), \tag{7.57}$$

the Bloch function is expressed as a modulated plane wave,

$$\psi_k(x) = e^{ikx}u_k(x). \tag{7.58}$$

From the Bloch condition, Eq. (7.33), it follows that

$$u_k(x+a) = u_k(x), \tag{7.59}$$

i.e. the function has the periodicity of the potential (i.e. just like the potential, an eigenfunction of the crystal translation operator with eigenvalue one). The function u_k therefore has the Fourier expansion (see Appendix C)

$$u_k(x) = \sum_K u_k(K)\, e^{iKx}, \tag{7.60}$$

where ($K = K_l$ for short)

$$K_l = \frac{2\pi}{a}l, \qquad l = 0, \pm1, \pm2, \pm3, \ldots, \tag{7.61}$$

are the so-called reciprocal lattice numbers, satisfying for lattice equivalent distances

$$e^{iKR} = 1, \qquad R = ma, \qquad m = 0, \pm1, \pm2, \pm3, \ldots. \tag{7.62}$$

By inverse Fourier transformation,

$$u_k(K) = \frac{1}{a}\int_{-a/2}^{a/2} dx\, e^{-ixK}u_k(x). \tag{7.63}$$

The Fourier expansion of the Bloch function,

$$\psi_k(x) = \sum_K e^{i(k+K)x}u_k(K), \tag{7.64}$$

thus involves only plane waves with wave numbers k and k *plus* any reciprocal lattice number.

Expressing Bloch's theorem in the form Eq. (7.58) splits the Bloch function into a plane wave and a part determined by the periodic potential. Namely, observing the identity (obtained by explicitly differentiating the exponential function)

$$-\frac{\partial^2}{\partial x^2}\,e^{ikx}u_k(x) = -i\frac{\partial}{\partial x}\,e^{ikx}\left(-i\frac{\partial}{\partial x}+k\right)u_k(x) = e^{ikx}\left(-i\frac{\partial}{\partial x}+k\right)^2 u_k(x), \tag{7.65}$$

the Hamiltonian operates on a Bloch function according to

$$\left(-\frac{\hbar^2}{2m}\frac{\partial^2}{\partial x^2}+V(x)\right)e^{ikx}u_k(x) = e^{ikx}\left(\frac{\hbar^2}{2m}\left(-i\frac{\partial}{\partial x}+k\right)^2+V(x)\right)u_k(x), \tag{7.66}$$

and the exponential function in Eq. (7.1) can be factorized and divided out. The periodic part of the Bloch function thus satisfies

$$\left\{-\frac{\hbar^2}{2m}\left(\frac{\partial}{\partial x}+ik\right)^2+V(x)\right\}u_k(x) = \epsilon(k)\,u_k(x). \tag{7.67}$$

Complex conjugating Eq. (7.67), it follows in view of $\epsilon(-k)=\epsilon(k)$ that

$$u_{-k}(x) = u_k^*(x), \tag{7.68}$$

a property also directly seen from Eq. (7.32).

Since k is a continuous parameter and $\epsilon(k)$ is continuous near $k=0$, it follows from Eq. (7.67) that u_k and therefore ψ_k are continuous functions of the Bloch number. Then $\psi_{+0}=\psi_{-0}=\psi_0^*$, i.e. the lowest energy level, $\epsilon(0)$, is non-degenerate, as the Bloch function is real (just as in the case of a free particle).

In view of the boundary condition, Eq. (7.59), the problem in Eq. (7.67) is analogous to the *guitar string* problem: a function satisfying a wave equation with amplitude specified at the boundaries. We therefore expect, as for any type of confined waves, that for each k an infinite hierarchy of solutions exists (higher harmonics) with energies $\epsilon_n(k)$, $n=1,2,3,\ldots$, like the resonances of a vibrating string. This suggests representing the energy spectrum in a different way than the extended zone scheme.[9]

Exercise 7.5 Show that

$$\bar{T}_k \equiv \frac{1}{2m}\int_0^a dx\;\psi_k^*(x)\left(\frac{\hbar}{i}\frac{\partial}{\partial x}\right)^2\psi_k(x) = \frac{1}{2m}\int_0^a dx\;u_{nk}^*(x)\left(\frac{\hbar}{i}\frac{\partial}{\partial x}+\hbar k\right)^2 u_{nk}(x)$$

$$= a\sum_K \frac{\hbar^2(k+K)^2}{2m}\,|u_k(K)|^2, \tag{7.69}$$

[9] Since k is a continuous parameter in Eq. (7.67), for each n a continuous band of energies is then obtained, $\epsilon_n(k)$. Equation (7.67) and the boundary condition (7.59) do not say whether these energy bands overlap, suggesting there will be regions of values for which there are no corresponding energy states. This is precisely the scenario established explicitly in Section 7.7 by studying the implication of the Bloch condition on the properties of matter waves in a periodic potential.

and thereby express the Bloch energy as the sum of the average kinetic and the average potential energy parts

$$\epsilon(k) = \bar{T}_k + \bar{V}_k. \tag{7.70}$$

Hint: See Section 8.2.

7.9 Reduced and Repeated Zone Schemes

The k values can be restricted to any interval of length $2\pi/a$, say, from $-\pi/a$ to π/a corresponding to our choice of the lowest energy state being represented by Bloch number $k = 0$.[10] Namely, the range of energies from $k = \pi/a$ to $k = 2\pi/a$ in the extended zone scheme could equally well have been depicted as stretching from $k = -\pi/a$ to $k = 0$, i.e. translated the distance $2\pi/a$ to the left, as depicted in Figure 7.3 (or chosen like that to begin with). The energy values are the same and therefore also the energy eigenfunctions, and the Bloch condition is unchanged with this new Bloch labeling, $k' = k - 2\pi/a$, as $\hat{T}_a\psi_{k-2\pi/a} = \exp(ika)\psi_{k-2\pi/a}$, or $\lambda_{k-2\pi/a} = \lambda_k$. However, these energies and energy eigenfunctions must, of course, be distinguished from those in the lowest band, i.e. $\epsilon_2(k') \equiv \epsilon(k)$ and $\psi_{2,k'} \equiv \psi_k$. The energies and energy eigenfunctions for the lowest band of energies are now denoted $\epsilon_1(k)$ and ψ_{1k}. The next band of energies and energy eigenfunctions again corresponding to the range from $-\pi/a$ to π/a are denoted $\epsilon_2(k)$ and ψ_{2k}, the so far missing range from $k = 0$ to $k = \pi/a$ obtained by translating the range from $-2\pi/a$ to $-\pi/a$ the distance $2\pi/a$ to the right as dictated by the crystal translation eigenvalues (equivalent to using the mirror symmetry $\epsilon_2(-k) = \epsilon_2(k)$). When the Bloch number traverses from $-\pi/a$ to π/a, the crystal translation eigenvalues traverse the unit circle from -1 counterclockwise to -1. In both of the bands, the Bloch states labeled by k have crystal translation eigenvalue $\lambda_k = \exp(ika)$. The third band of energies is obtained by translating the range of energies from $k = 2\pi/a$ to $k = 3\pi/a$ the distance $2\pi/a$ to the left and the range of energies from $k = -3\pi/a$ to $k = -2\pi/a$ the distance $2\pi/a$ to the right, as dictated by the crystal translation eigenvalues. In this way, the energy spectrum in the extended zone scheme is translated into the so-called first Brillouin zone, Bz_1, the infinity of energies for the same Bloch number now distinguished by a band index, $\epsilon_n(k)$, $n = 1, 2, 3, \ldots$.

In Figure 7.4 are depicted the two lowest bands in this so-called reduced zone scheme. The states labeled by the zone center value, $k = 0$, correspond to crystal translation eigenvalue $\lambda = 1$, and those at the zone boundaries, $k = \pm\pi/a$, to translation eigenvalue $\lambda = -1$. The energy value at the zone center is alternately a minimum and maximum by construction, starting with a minimum in the lowest band. The energy values in the second

[10] The so-called first Brillouin zone, the choice seen to respect the time reversal symmetry property of the energy spectrum, $\epsilon(-k) = \epsilon(k)$.

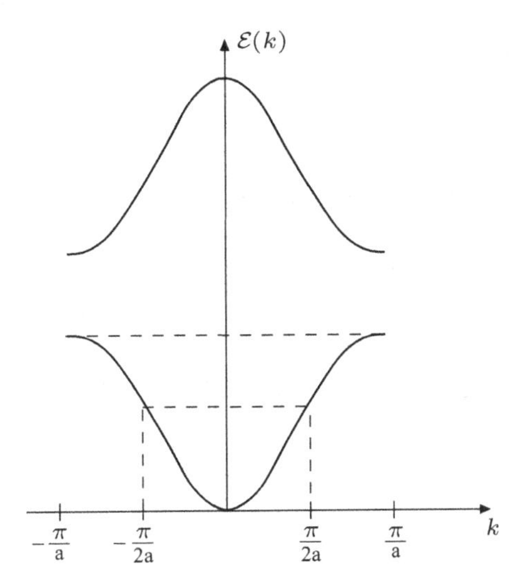

Figure 7.4 The two lowest bands separated by the energy gap.

and first bands, $\epsilon_2(k)$ and $\epsilon_1(k)$, are closest at the Bragg points, $k = \pm\pi/a$, whereas the third and second bands, $\epsilon_3(k)$ and $\epsilon_2(k)$, are closest at the zone center, $k = 0$, etc.

By construction the energies and eigenfunctions in the reduced and extended zone schemes are related according to (with $k \in [-\pi/a, \pi/a]$)

$$\epsilon_n(k) = \epsilon\left(k + \frac{(n-1)\pi}{a}\,\mathrm{sign}(k)\right), \qquad \psi_{nk}(x) = \psi_{k+\mathrm{sign}(k)(n-1)\pi/a}(x), \tag{7.71}$$

for n odd, and

$$\epsilon_n(k) = \epsilon\left(k - \frac{n\pi}{a}\,\mathrm{sign}(k)\right), \qquad \psi_{nk}(x) = \psi_{k-\mathrm{sign}(k)n\pi/a}(x), \tag{7.72}$$

for n even. According to Eqs. (7.71), (7.72) and (7.32),

$$\psi_{n-k}(x) = \psi^*_{nk}(x). \tag{7.73}$$

Introducing

$$u_{nk}(x) \equiv e^{-ikx}\psi_{nk}(x), \qquad u_{n-k}(x) = u^*_{nk}(x), \tag{7.74}$$

the Bloch functions in the reduced zone scheme, labeled by Bloch and band numbers,[11]

$$\psi_{nk}(x) = u_{nk}(x)\,e^{ikx}, \qquad k \in \left]-\frac{\pi}{a}, \frac{\pi}{a}\right], \qquad n = 1, 2, 3, \ldots, \tag{7.75}$$

satisfy the Bloch condition

$$\psi_{nk}(x+a) = e^{ika}\psi_{nk}(x), \qquad u_{nk}(x+a) = u_{nk}(x). \tag{7.76}$$

According to Eqs. (7.71), (7.72) and (7.57),

$$u_{nk}(x) = u_{k+\mathrm{sign}(k)(n-1)\pi/a}(x)\,e^{ix\,\mathrm{sign}(k)(n-1)\pi/a} \tag{7.77}$$

[11] Anticipating that in each band only one of the states corresponding to $k = \pm\pi/a$ should be included in the Brillouin zone, as they correspond to the same state, $\psi_{n\pi/a}(x) = \psi_{n-\pi/a}(x)$, as shown in Section 7.10.

for n odd, and

$$u_{nk}(x) = u_{k-\mathrm{sign}(k)n\pi/a}(x)\, e^{ix\,\mathrm{sign}(k)n\pi/a} \tag{7.78}$$

for n even.

The plane wave part of the Bloch function has a wave length at least on the order $\lambda \geq 2a$, since $|k| \leq \pi/a$, whereas the periodic part, $u_{nk}(x)$, has short wave length oscillations, being built by plane waves with wave vectors, $K = 2\pi m/a$.

As for example in the next section, it can be useful to employ a highly redundant exposition of the energy spectrum and eigenfunctions, expanding each band to all k values according to the redundancy prescription, the so-called repeated zone scheme,

$$\epsilon_n(k+K) \equiv \epsilon_n(k), \qquad \psi_{nk+K}(x) \equiv \psi_{nk}(x), \tag{7.79}$$

which is possible because

$$\epsilon_n(\pi/a) = \epsilon_n(-\pi/a + 2\pi/a), \qquad \psi_{n\pi/a}(x) = \psi_{n-\pi/a}(x). \tag{7.80}$$

From the Bloch condition it follows that

$$u_{nk+K}(x) = e^{-iKx} u_{nk}(x), \tag{7.81}$$

the phase factor compensating the explicit k term in the equation determining u_{nk},

$$\left\{ -\frac{\hbar^2}{2m}\left(\frac{\partial}{\partial x} + ik\right)^2 + V(x) \right\} u_{nk}(x) = \epsilon_n(k)\, u_{nk}(x). \tag{7.82}$$

7.10 Energy Band Properties

Even general quantitative features of the energy spectrum follow from the Bloch condition. The derivatives of $\epsilon_n(k)$ are, in view of $\epsilon_n(k + 2\pi/a) = \epsilon_n(k)$, equal at $k = -\pi/a$ and $k = \pi/a$,

$$\frac{\partial \epsilon_n(\pi/a)}{\partial k} = \frac{\partial \epsilon_n(-\pi/a)}{\partial k}. \tag{7.83}$$

Using the property $\epsilon(-k) = \epsilon(k)$,

$$\begin{aligned} \frac{\partial \epsilon_n(-\pi/a)}{\partial k} &= \frac{\epsilon_n(-\pi/a + \Delta k) - \epsilon_n(-\pi/a)}{\Delta k} \\ &= -\frac{\epsilon_n(\pi/a) - \epsilon_n(\pi/a - \Delta k)}{\Delta k} = -\frac{\partial \epsilon_n(\pi/a)}{\partial k}, \end{aligned} \tag{7.84}$$

the identities

$$\frac{\partial \epsilon_n(\pm\pi/a)}{\partial k} = -\frac{\partial \epsilon_n(\pm\pi/a)}{\partial k} \tag{7.85}$$

are obtained, and the derivatives at Bragg points thus vanish,

$$\frac{\partial \epsilon_n(\pi/a)}{\partial k} = 0 = \frac{\partial \epsilon_n(-\pi/a)}{\partial k}. \tag{7.86}$$

The energy therefore has extrema at the zone edges, as also depicted in Figure 7.4.

Denote by $F(\epsilon)$ the function on the right-hand side of Eq. (7.46) as a function of energy, and Eq. (7.46) can be rewritten, with $\epsilon(k) = \epsilon = \hbar^2 q^2/2m$, as

$$\cos(ka) = F(\epsilon). \tag{7.87}$$

The energy value for which F first equals one is $\epsilon = \epsilon_1(k = 0) = \hbar^2 q_1^2/2m$, i.e. $F(\epsilon_1(k = 0)) = 1$. Taylor-expanding for small k on the right-hand side of Eq. (7.87) gives

$$\cos(ka) = F\left(\epsilon_1(0) + \frac{\partial \epsilon_1(0)}{\partial k} k + \mathcal{O}(k^2)\right) = 1 + F'(\epsilon_1(0)) \frac{\partial \epsilon_1(0)}{\partial k} k + \mathcal{O}(k^2). \tag{7.88}$$

Since the energies for which $F'(\epsilon) = 0$, according to Figure 7.2 (as $F'(\epsilon) = \tilde{F}'(q)m/\hbar^2 q$), only lie in the gap regions, we conclude that $F'(\epsilon(0)) \neq 0$. The derivative of the energy must therefore vanish at the zone center, since the cosine function around $k = 0$ starts with a quadratic term. The analysis is identical for each band, as $\cos(ka) = 1$ at zone centers, $k = 0$, and the energy therefore has extrema at the zone center,

$$\frac{\partial \epsilon_n(k = 0)}{\partial k} = 0. \tag{7.89}$$

The band energies thus have extremum values at the bottom and top of each band, as also depicted in Figure 7.4. The bands have by construction alternately minimum and maximum at the zone center starting out with a minimum for the lowest band according to the analysis resulting from Figure 7.2.

As shown in the next chapter, the derivative of the Bloch energy gives the Bloch state probability current, and a vanishing derivative thus means that the corresponding Bloch function can be chosen to be real and therefore, according to Eq. (7.73),

$$\psi_{n-\pi/a}(x) = \psi^*_{n\pi/a}(x) = \psi_{n\pi/a}(x). \tag{7.90}$$

As the Bloch condition suggests, $\lambda_{-\pi/a} = -1 = \lambda_{\pi/a}$ (anti-periodic boundary condition $\psi_{n\pm\pi/a}(x + a) = -\psi_{n\pm\pi/a}(x)$), the Bloch functions corresponding to the Bragg points in a band are thus identical, and the state corresponding to energy $\epsilon = \epsilon_n(\pm\pi/a)$ is non-degenerate. That the energy derivative vanishes at the zone center follows in this light from the already observed continuity property $\psi^*_{n0} = \psi_{n0}$.

Introduce a band mass to describe the curvature of the quadratic dispersion at an energy extremum. For the case of a minimum at the zone center (the discussion is analogous for the case of a minimum at the Bragg points, $k = \pm\pi/a$)

$$\frac{1}{m_*} \equiv \frac{1}{\hbar^2} \frac{\partial^2 \epsilon_n(k = 0)}{\partial k^2}, \qquad \epsilon_n(k) = \epsilon_n(0) + \frac{\hbar^2 k^2}{2m_*} + \cdots. \tag{7.91}$$

Near a band minimum, a Bloch electron has the same dispersion as a free electron except that its mass is changed by the periodic potential (and a constant energy shift occurs). Evaluating the coefficients of the k^2 terms on the left- and right-hand sides of Eq. (7.88) gives the expression for the zone center band mass as

$$m_* = \frac{\hbar^2}{a^2} F'(\epsilon_n(0)). \tag{7.92}$$

If the energy at the zone center has a minimum, at the Bragg points there is then a maximum (and vice versa), and to describe the negative curvature of the quadratic dispersion at Bragg point π/a (or equivalently $-\pi/a$) a positive band mass is defined by

$$\frac{1}{m^*} \equiv -\frac{1}{\hbar^2}\frac{\partial^2 \epsilon_n(k = \pi/a)}{\partial k^2}, \qquad \epsilon_n(k) = \epsilon_n(\pi/a) - \frac{\hbar^2 k^2}{2m^*} + \cdots . \tag{7.93}$$

Evaluating the coefficients of the k^2 terms on the left- and right-hand sides of Eq. (7.88) close to $k = \pi/a$ gives the expression for the band mass,

$$m^* = -\frac{\hbar^2}{a^2} F'(\epsilon_n(\pi/a)). \tag{7.94}$$

The negative curvature of the energy spectrum, absent for a free particle, is of importance for the motion of a particle in a periodic potential.

Denote by ϵ_0 an energy value for which $F(\epsilon_0) = 0$. Taylor-expanding Eq. (7.87) for energies, ϵ, close to ϵ_0 gives

$$\cos(ka) \simeq F'(\epsilon_0)(\epsilon - \epsilon_0) \tag{7.95}$$

and thereby the cosine dispersion for energies close to ϵ_0,

$$\epsilon(k) = \epsilon = \epsilon_0 + \frac{1}{F'(\epsilon_0)} \cos(ka). \tag{7.96}$$

This is the generic dispersion for bands of small width, the width seen to be equal to $2/|F'(\epsilon_0)|$. The bands have alternately minimum or maximum at the zone center, corresponding to the alternating sign of $F'(\epsilon_0)$, starting out with a minimum for the lowest band, as seen from Figure 7.2, i.e. $F'(\epsilon_0)$ starting out negative.

Exercise 7.6 Show that the energy for Bloch number $k = \pi/2a$, midway between the zone center and the Bragg point, has the following properties:

$$\frac{\partial \epsilon_n(\pi/2a)}{\partial k} = (-1)^n \frac{a}{F'(\epsilon_n(\pi/2a))} \neq 0 \tag{7.97}$$

and

$$\frac{\partial^2 \epsilon_n(\pi/2a)}{\partial k^2} = 0. \tag{7.98}$$

At $k = \pi/2a$, the energy thus has a reflection point, the curvature of $\epsilon_n(k)$ changing sign. In the region close to $k = \pi/2a$, the energy spectrum is therefore linear,

$$\epsilon_n(k) = \epsilon_n\left(\frac{\pi}{2a}\right) + (-1)^n \frac{a}{F'(\epsilon_n(\pi/2a))} k. \tag{7.99}$$

7.11 Completeness of Bloch Functions

The periodic part of the Bloch function satisfies the equation

$$\hat{H}(k)\, u_{nk}(x) = \epsilon_n(k)\, u_{nk}(x), \tag{7.100}$$

where the *Hamiltonian* (recall Eq. (7.67))

$$\hat{H}(k) = \frac{1}{2m}\left(\frac{\hbar}{i}\frac{\partial}{\partial x} + \hbar k\right)^2 + V(x) \tag{7.101}$$

depends on the Bloch number. Since k enters as a continuous parameter, the dependence of u_{nk} on the Bloch number is smooth in each band.

The operator $\hat{H}(k)$ is seen to be a hermitian operator for functions which, like $u_{nk}(x)$, satisfy the unit cell periodic boundary condition, $f(x+a) = f(x)$. The eigenfunction u_{nk} is thus orthogonal to $u_{n'k}$ when $n \neq n'$, since the band energies for different bands are different (in fact, to any $u_{n'k'}$). For a given k, the set of functions $\{u_{nk}\}_{n=1,2,3,\dots}$ is thus orthogonal. According to Eqs. (7.60) and (7.76),

$$u_{nk}(x) = \sum_K u_{nk}(K)\, e^{iKx}, \tag{7.102}$$

where the coefficient $u_{nk}(K)$ according to Eq. (7.63) equals the projection of u_{nk} onto the plane wave specified by wave number K, $u_{nk}(K) = \langle K \mid u_{nk}\rangle$ (and the projection of the plane waves on states u_nk are $\langle u_{nk} \mid K\rangle = \langle K \mid u_{nk}\rangle^* = u^*_{nk}(K)$). The functions $\{u_{nk}\}_{n=1,2,3,\dots}$ are a complete set for functions on the unit cell satisfying the periodic boundary condition, a rotation of the complete basis of plane waves. The expansion coefficients of function $f(x)$ on the two basis sets are related according to

$$f(K) = \sum_{n=1}^{\infty} f_n u_{nk}(K), \qquad f_n = \sum_K u^*_{nk}(K) f(K). \tag{7.103}$$

The completeness relation on the real axis is

$$\sum_{n=1}^{\infty} u^*_{nk}(x)\, u_{nk}(x') = \sum_{m=-\infty}^{\infty} \delta(x - x' + ma), \tag{7.104}$$

the sum reflecting the periodicity of the u functions.

According to Eq. (7.104),

$$\begin{aligned}\sum_{n=1}^{\infty} \int_{-\pi/a}^{\pi/a} dk\, \psi^*_{nk}(x)\, \psi_{nk}(x') &= \int_{-\pi/a}^{\pi/a} dk\, e^{-ik(x-x')} \sum_n u^*_{nk}(x)\, u_{nk}(x') \\ &= \int_{-\pi/a}^{\pi/a} dk\, e^{ikma} \sum_{m=-\infty}^{\infty} \delta(x - x' + ma) \\ &= \sum_{m=-\infty}^{\infty} \frac{e^{im\pi} - e^{-im\pi}}{ima}\, \delta(x - x' + ma),\end{aligned} \tag{7.105}$$

and, as only the $m = 0$ term gives a non-vanishing contribution,

$$\sum_{n=1}^{\infty} \int_{-\pi/a}^{\pi/a} dk\ \psi_{nk}^*(x)\, \psi_{nk}(x') = \frac{2\pi}{a}\, \delta(x - x'). \tag{7.106}$$

The Bloch functions therefore constitute a complete orthogonal set and any state of a particle in a periodic potential is a superposition of Bloch states,

$$\psi(x) = \sum_{n=1}^{\infty} \int_{-\pi/a}^{\pi/a} dk\ a_{nk}\, \psi_{nk}(x) = \sum_{n=1}^{\infty} \int_{-\pi/a}^{\pi/a} dk\ a_{nk}\, e^{ikx} u_{nk}(x). \tag{7.107}$$

Bloch function overlaps can be evaluated by summing the contributions from each unit cell by the substitution $x' \equiv x - ma$,

$$\begin{aligned} \int_{-\infty}^{\infty} dx\ \psi_{nk}^*(x)\, \psi_{n'k'}(x) &= \int_{-\infty}^{\infty} dx\ e^{ix(k'-k)} u_{nk}^*(x)\, u_{n'k'}(x) \\ &= \sum_{m=-\infty}^{\infty} e^{ima(k'-k)} \int_{-a/2}^{a/2} dx'\ e^{ix'(k'-k)} u_{nk}^*(x')\, u_{n'k'}(x'). \end{aligned} \tag{7.108}$$

The m summation gives $\delta(k - k')$ multiplied by $2\pi/a$ (see Appendix C) and thereby the normalization condition

$$\int_{-\infty}^{\infty} dx\ \psi_{nk}^*(x)\, \psi_{n'k'}(x) = \frac{2\pi}{a}\, \delta(k - k')\, \delta_{nn'} \int_{-a/2}^{a/2} dx\ |u_{nk}(x)|^2, \tag{7.109}$$

and thus the Bloch functions, being infinitely extended, are *delta*-normalized (analogous to the plane waves).

We note that u_{nk} is only determined up to a phase factor, which can depend on the Bloch number and the band index, as the transformation $u_{nk} \to e^{i\varphi_n(k)} u_{nk}$ produces a solution to Eq. (7.100) with the same energy. The Bloch function is thereby transformed as $\psi_{nk} \to e^{i\varphi_n(k)} \psi_{nk}$. That ψ_{nk} can thereby lose its repeated zone scheme periodicity property, $\psi_{nk+K} \neq \psi_{nk}$, need not be of concern for simplifying problem solving, and this freedom of phase choice is employed in Section 8.6.

7.12 Finite Crystal

A real crystalline solid is finite, thus giving rise to deviation from periodicity. The electrons in a finite crystal are confined to be inside the crystal except for an exponentially small leakage, similarly as in the Sommerfeld model discussed in Section 3.2. Since our interest lies in electron transport, periodic boundary conditions providing current-carrying energy eigenstates are more convenient, as also discussed in Appendix H. The bulk of the crystal, where allowed wave functions are Bloch functions, determines its electronic transport properties. The finite crystal is assumed to consist of N unit cells, its total length being $L = x_r - x_l = Na$, i.e. surface reconstruction effects are neglected, and the macroscopic periodic boundary condition respecting the periodicity of the potential is conveniently employed, the Born–von Karman boundary condition[12]

$$\psi(x_l + L) = \psi(x_l). \tag{7.110}$$

The continuous wave function can therefore be represented by its Fourier expansion (see Appendix C)[13]

$$\psi(x) = \sum_k \psi(k)\, e^{ikx}, \tag{7.111}$$

where the wave numbers in the plane wave expansion are given by

$$k = \frac{2\pi}{L} m = \frac{2\pi}{Na} m, \qquad m = 0, \pm 1, \pm 2, \ldots, \tag{7.112}$$

the set of wave numbers in the so-called Born–von Karman (BvK) set.

The microscopic periodicity of the potential, $V(x + a) = V(x)$, gives the Fourier series expansion

$$V(x) = \sum_K V(K)\, e^{iKx}, \tag{7.113}$$

with Fourier expansion coefficients only for reciprocal lattice numbers, Eq. (7.61).

As usual, in this context, we stick to the notation that a lower case k denotes a number in the BvK set, Eq. (7.112), whereas a capital K denotes a reciprocal lattice number. The latter set are equidistantly spaced points on a line, a distance $2\pi/a$ apart, with the BvK numbers k spaced in between. In the bulk limit, where the number of *atoms* in the crystal is astronomically large, $N \sim 10^{22}$, the BvK numbers lie extremely densely together, as depicted in Figure 7.5.

The rather trivial observation that the plane wave expansion of a periodic potential, Eq. (7.113), only has components for wave lengths that respect the periodicity of the

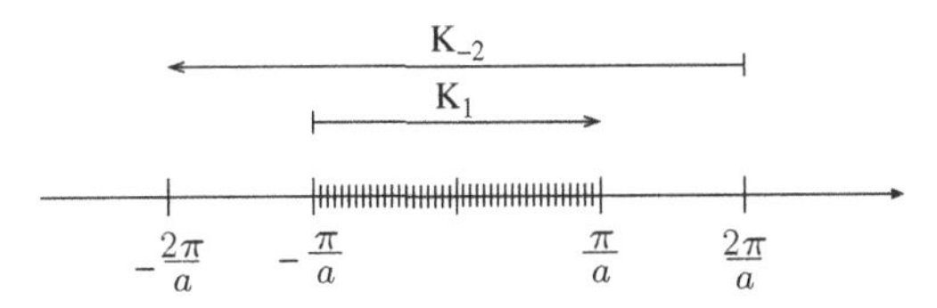

Figure 7.5 Reciprocal lattice numbers and Born–von Karman numbers displayed in the first Brillouin zone.

[12] In one dimension, the periodic boundary condition (7.110) can be made physical, as x can be imagined as the position variable on a circle of perimeter L, i.e. in fact an angle variable. The boundary condition (7.110) then expresses the single-valuedness of a wave function, as x_l and $x_l + L$ label the same point on the circle and the wave function therefore has the same value.

[13] We are starting from scratch to solve the problem, ignoring our knowledge of Bloch's theorem.

potential has, combined with the macroscopic periodic boundary condition, far-reaching consequences for the allowed energy eigenfunctions.

In terms of their Fourier expansions, the potential term in the time-independent Schrödinger equation becomes

$$V(x)\,\psi(x) = \sum_{K,k} V(K)\,\psi(k)\,e^{ix(k+K)} = \sum_{K,k'} e^{ik'x} V(K)\,\psi(k'-K), \tag{7.114}$$

where the last equality follows from the substitution of k' for k according to $k' = k + K$, which can be done, since, if k is of the form $2\pi m/Na$, so clearly is k', i.e. if k is a BvK number, so is $k' = k + K$.

The kinetic energy term takes the Fourier expansion form

$$-\frac{\hbar^2}{2m}\frac{d^2\psi(x)}{dx^2} = \sum_k \frac{\hbar^2 k^2}{2m}\,\psi(k)\,e^{ikx}, \tag{7.115}$$

and the time-independent Schrödinger equation for a particle in a finite periodic potential thus takes the form, in terms of the Fourier expansions,

$$\sum_k \left(\frac{\hbar^2 k^2}{2m}\,\psi(k) + \sum_K V(K)\,\psi(k-K)\right) e^{ikx} = \epsilon \sum_k \psi(k)\,e^{ikx}. \tag{7.116}$$

From the uniqueness of a Fourier expansion, or taking the scalar product with the plane wave state labeled by k on both sides of Eq. (7.116), the Fourier coefficients of an energy eigenstate of a particle in a periodic potential satisfy the equation

$$\left(\frac{\hbar^2 k^2}{2m} - \epsilon\right)\psi(k) + \sum_K V(K)\,\psi(k-K) = 0. \tag{7.117}$$

Owing to the periodicity of the potential, Fourier coefficients for all BvK k numbers are not involved in the expansion of an energy eigenfunction, Eq. (7.111), only those that differ, relative to a given k number, by reciprocal lattice numbers. Each solution of a periodic potential time-independent Schrödinger equation satisfying the BvK boundary condition is thus a superposition of plane waves containing components with k and k *plus* possibly any K number. If for a given energy value ϵ a set of amplitudes, $\{\psi(k-K)\}_K$, that satisfies Eq. (7.117) can be found, a solution to the time-independent Schrödinger equation is according to Eq. (7.111) given by the Fourier expansion

$$\psi(x) = \sum_K \psi(k-K)\,e^{i(k-K)x}. \tag{7.118}$$

This energy eigenfunction is thus specified by k, the Bloch number,

$$\psi_k(x) = e^{ikx} \sum_K \psi(k-K)\,e^{-iKx}, \tag{7.119}$$

and the corresponding energy eigenvalue is therefore specified in terms of this quantum number, $\epsilon = \epsilon(k)$.

We have thus arrived at Bloch's theorem: the wave function corresponding to a given BvK k value is, according to Eq. (7.119) and the property of a reciprocal lattice number, $\exp(-iKa) = \exp(-i2\pi l) = 1$, seen to satisfy the Bloch condition

$$\psi_k(x+a) = e^{ika}\psi_k(x). \tag{7.120}$$

The analysis in Section 7.6 of the energy spectrum can then be taken over, the only change being that Bloch numbers and energies now are discrete, and the energy spectrum in the extended zone scheme is obtained.

Introducing the function

$$u_k(x) \equiv \sum_K \psi(k-K)\, e^{-iKx}, \tag{7.121}$$

Bloch's theorem can, according to Eq. (7.119), alternatively be stated as: the wave function of a particle in a periodic potential has the form

$$\psi_k(x) = e^{ikx}u_k(x), \tag{7.122}$$

where the function u_k, according to $e^{-iKa} = 1$, has the periodicity of the potential

$$u_k(x+a) = u_k(x). \tag{7.123}$$

The function u_k satisfies Eq. (7.67) and for a given k value there is an infinite set of confined wave solutions, u_{nk}, $n = 1, 2, \ldots$, and Bloch functions

$$\psi_{nk}(x) = e^{ikx}u_{nk}(x), \qquad k \in \mathrm{Bz}_1, \qquad n = 1, 2, 3, \ldots, \tag{7.124}$$

and corresponding energies $\epsilon_n(k)$ labeled by a band index. The reduced zone scheme representation is obtained for the discrete k numbers.

Changing the Bloch number k of a given Bloch function by a reciprocal lattice number, $k \to k+K$, the corresponding Bloch function (recall Eq. (7.119)) transforms according to

$$\psi_{nk+K}(x) = \sum_{K'} \psi_n(k+K-K')\, e^{i(k+K-K')x} = e^{ikx}\sum_{K''}\psi_n(k-K'')\, e^{-iK''x}, \tag{7.125}$$

where the last equality follows from the substitution $K'' = K' - K$, simply a relabeling of the reciprocal lattice numbers. The function ψ_{nk+K} is thus the same function as $\psi_{nk}(x)$, $\psi_{nk+K}(x) = \psi_{nk}(x)$, and the reduced zone scheme can be extended to the repeated zone scheme where $\epsilon_n(k+K) = \epsilon_n(k)$.

The BvK restriction on the finite crystal Bloch functions also follows immediately from the analysis of the infinite crystal. The macroscopic periodic boundary condition, $\psi(x) = \psi(x+L)$, becomes, in view of the Bloch condition,

$$\psi_k(x) = \psi_k(x+Na) = e^{ikNa}\psi_k(x), \tag{7.126}$$

the constraint allowing only the discrete BvK numbers out of the continuum of Bloch numbers for the infinite crystal. The constructed Bloch function, Eq. (7.124), equals the corresponding Bloch number function of the infinite system on the considered interval, and is periodically repeated for positions chosen outside the interval.

7.13 Orthonormality of Bloch Functions

For a finite crystal, the scalar product between Bloch states is

$$\begin{aligned}\langle nk \mid n'k'\rangle &\equiv \int_0^L dx\, \psi^*_{nk}(x)\, \psi_{n'k'}(x)\\ &= \int_0^L dx\, e^{i(k'-k)x} u^*_{nk}(x)\, u_{n'k'}(x)\\ &= \delta_{n,n'} \int_0^L dx\, e^{i(k'-k)x} u^*_{nk}(x)\, u_{nk'}(x),\end{aligned} \tag{7.127}$$

where the Kronecker function in the band indices is due to the fact that energy bands in one spatial dimension do not overlap, and energy functions corresponding to different energies are orthogonal (recall Section 1.5.1). The integral (7.127) can be evaluated by splitting it up into integrals over each of the N unit cells of the crystal, $x = X_m + x'$, $X_m = ma$, as depicted in Figures 7.6 and 9.7.

Employing the periodicity property of the u_{nk} functions, Eq. (7.127) becomes

$$\langle nk \mid n'k'\rangle = \delta_{n,n'} \sum_{m=1}^{N} e^{i(k'-k)am} \int_0^a dx\, e^{i(k'-k)x} u^*_{nk}(x)\, u_{nk'}(x), \tag{7.128}$$

N being the number of unit cells constituting the crystal, $L = Na$. The summation over the cell index, m, produces the Kronecker function $\delta_{k,k'}$ multiplied by N, assuming $k' - k < 2\pi/a$ (see Appendix C), and thereby

$$\langle nk \mid n'k'\rangle = N\delta_{n,n'}\delta_{k,k'} \int_0^a dx\, u^*_{nk}(x)\, u_{nk}(x). \tag{7.129}$$

The degenerate states ψ_{nk} and ψ_{n-k} are thus seen to be orthogonal (except for $k = \pm\pi/a$),

$$\langle nk \mid n-k\rangle \propto \delta_{k,-k} = 0. \tag{7.130}$$

The finite crystal Bloch functions can be chosen to be orthonormal,

$$\langle nk \mid n'k'\rangle = \delta_{n,n'}\delta_{k,k'}, \tag{7.131}$$

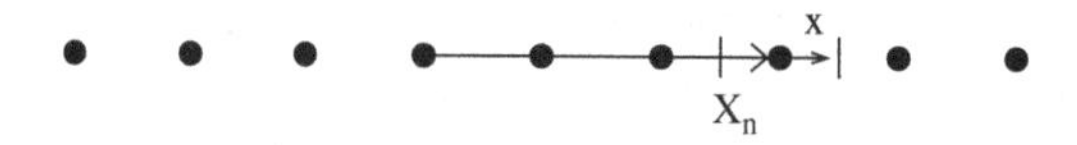

Figure 7.6 Splitting the length of the crystal into its unit cells.

if the periodic part of the Bloch function according to Eq. (7.129) is normalized,

$$N \int_0^a dx\, |u_{nk}(x)|^2 = \int_0^L dx\, |u_{nk}(x)|^2 = 1. \tag{7.132}$$

As discussed in Section 7.11, for a given k, the set $\{u_{nk}\}_{n=1,2,3,...}$ is a complete set for functions on the unit cell satisfying the unit cell periodic boundary condition. The Bloch functions with discrete Bloch numbers constitute a complete orthonormal set for the functions satisfying the Born–von Karman boundary condition, and a state of a particle is a superposition of the corresponding Bloch states,

$$\psi(x) = \sum_{nk \in \mathrm{Bz}_1} a_{nk} \psi_{nk}(x) = \sum_{nk \in \mathrm{Bz}_1} a_{nk}\, e^{ikx} u_{nk}(x). \tag{7.133}$$

7.14 Metals and Insulators

The conduction properties of electrons in crystalline solids are in the main approximation well modeled as independent electrons in a periodic potential. The presence of the resulting energy gaps leads to other properties than just metallic. The transport properties of one-dimensional crystals are now briefly discussed before being investigated in detail in the next chapter.

The ground state of electrons in a periodic potential is obtained by filling up the states of higher energy in accordance with the exclusion principle. The filling is determined by the allowed number of states per unit length and energy, just as in the Sommerfeld model, except for the states now being Bloch states. The counting of Bloch states is facilitated by the periodic boundary condition. The allowed Bloch numbers are then discrete, $k \in \mathrm{BvK}$, and the distance between two adjacent allowed k values is, according to Eq. (7.112), $2\pi/L$. There is thus one allowed k state for each interval $\Delta k = 2\pi/L$, and the number of allowed k states in the Brillouin zone, $]-\pi/a, \pi/a]$, is $(2\pi/a)/\Delta k = L/a = N$, i.e. equal to the number N of unit cells of the crystal, the number of *atoms* making up the crystal. This gives a rigid relationship between the number of electrons in the crystal and the filling fraction of the bands.

If our one-dimensional crystal is imagined to be built up by hydrogen atoms, giving one electron per unit cell, the ground state for the electrons in the crystal corresponds to filling up half of the Bloch states in the Brillouin zone due to the spin degeneracy, i.e. the region $k \in\,]-\pi/2a, \pi/2a]$, as depicted in Figure 7.7.[14] If the crystal is imagined to be built up by helium atoms, having two electrons, the lowest band can exactly accommodate all the electrons, as depicted in Figure 7.7. The scheme repeats. If the one-dimensional

[14] As discussed in Appendix M, the number of Bloch states per unit length and energy for an infinite system is the same as the one obtained by boundary condition state counting. So in the infinite periodic potential case, the same band occupation occurs; say, for *hydrogen*, the lowest band will be the half-filled continuum of Bloch states from $-\pi/a$ to π/a.

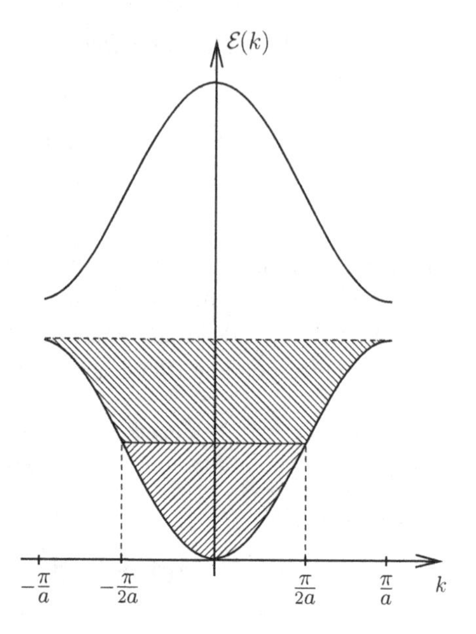

Figure 7.7 Filled and half-filled bands, respectively.

crystal is imagined to be built up by lithium atoms, having three electrons, the first band is completely filled and the second band half-filled.

For an even number of electrons, the unoccupied states will thus be separated from the occupied states in the ground state by an energy gap. In the presence of an electric field, the unoccupied states can only be reached by the mechanism of Zener tunneling, as discussed in Section 8.13. For a band gap on the order of 10 eV, the electric current for the relevant magnitudes of electric field is there shown to be vanishing – the system is an insulator.

If the crystal atoms have an odd number of electrons, the unoccupied states are not separated by an energy gap from the occupied states in the ground state, and a small electric field leads, just as in the Sommerfeld model, to a current-carrying state – the crystal is a metal. As discussed in Chapter 8, the transport properties of electrons in periodic structures are in that case analogous to those of the Sommerfeld model, except that the energy dispersion is changed from a quadratic dependence to a nonlinear dependence on the Bloch number.

If the band gap is small enough that at room temperature electrons thermally excited across the gap can give rise to a measurable current, the material is called a semiconductor. These materials can be electronically engineered to become good conductors by introducing special foreign atoms into the semiconductor crystal, and even made to behave as if positive charges are the charge carriers. Semiconductors are discussed in Chapters 8 and 10.

In a one-dimensional world, at zero temperature, crystals would have only filled bands or one band half-filled, depending on whether the atoms building the crystal have an even or odd number of electrons. In the three-dimensional world, crystal structure allows for less rigid behavior, as discussed in Chapter 9.

The antisymmetric basis functions are Slater determinants of Bloch (and spin) functions. For an N-electron system, say,

$$\psi_{n_1,k_1,s_1^z;\ldots;n_N,k_N,s_N^z}(x_1,s_1^{z\prime};\ldots;x_N,s_N^{z\prime})$$
$$= \frac{1}{\sqrt{N!}} \begin{vmatrix} \psi_{n_1,k_1,s_1^z}(x_1,s_1^{z\prime}) & \psi_{n_1,k_1,s_1^z}(x_2,s_2^{z\prime}) & \ldots & \psi_{n_1,k_1,s_1^z}(x_N,s_N^{z\prime}) \\ \psi_{n_2,k_2,s_2^z}(x_1,s_1^{z\prime}) & \psi_{n_2,k_2,s_2^z}(x_2,s_2^{z\prime}) & \ldots & \psi_{n_2,k_2,s_2^z}(x_N,s_N^{z\prime}) \\ \vdots & \vdots & \ddots & \vdots \\ \psi_{n_N,k_N,s_N^z}(x_1,s_1^{z\prime}) & \psi_{n_N,k_N,s_N^z}(x_2,s_2^{z\prime}) & \ldots & \psi_{n_N,k_N,s_N^z}(x_N,s_N^{z\prime}) \end{vmatrix}, \tag{7.134}$$

where

$$\psi_{n,k,s^z}(x,s^{z\prime}) = \psi_{n,k}(x)\,\chi_{s^z}(s^{z\prime}) \tag{7.135}$$

is the product of a Bloch function and the spin function.

The average particle density in the state given by Eq. (7.134) is (recall Exercises 3.1 and 3.2 on page 44)

$$n(x) = \sum_{i=1}^{N} |\psi_{n_i,k_i}(x)|^2 = \sum_{i=1}^{N} |u_{n_i,k_i}(x)|^2 \tag{7.136}$$

and the probability current is

$$S(x) = \frac{\hbar}{im}\sum_{i=1}^{N} \psi^*_{n_i,k_i}(x)\,\overset{\leftrightarrow}{\nabla}_x\,\psi_{n_i,k_i}(x), \qquad \overset{\leftrightarrow}{\nabla}_x \equiv \frac{1}{2}\left(\frac{\overrightarrow{\partial}}{\partial x} - \frac{\overleftarrow{\partial}}{\partial x}\right). \tag{7.137}$$

Integrating the density gives the particle number. Consider, for example, the ground state for the half-filled band case,

$$\int_0^L dx\, n(x) = 2\sum_{|k|<\pi/2a}\int_0^L dx\, |u_{n_i,k_i}(x)|^2 = 2\sum_{|k|<\pi/2a} 1, \tag{7.138}$$

where the factor of 2 accounts for spin degeneracy, and the second equality arises from the normalization, Eq. (7.132), of the Bloch function. Recalling the interspacing of the Bloch numbers gives

$$\int_0^L dx\, n(x) = 2\,\frac{2\times(\pi/2a)}{\Delta k} = 4\,\frac{\pi/2a}{2\pi/L} = \frac{L}{a} = N, \tag{7.139}$$

or equivalently one electron per unit cell, as assumed.

7.15 Summary

Quantum states in a periodic potential are described by Bloch functions, a periodic function modulated by a plane wave. The presence of the periodic part of the Bloch function, $u_{nk}(x)$, signifies that translation invariance is broken by the periodic potential. The energy eigenfunctions are not simply plane waves, as for a free particle, which are eigenfunctions with respect to any translation operator

$$\hat{T}_{x_0}\, e^{ikx} = e^{ikx_0}\, e^{ikx}. \tag{7.140}$$

Only for a smaller group of translations, those that respect the periodicity and the inverse and repetitions, is a Bloch function an eigenfunction, as specified by the Bloch condition

$$\hat{T}_a \psi_{nk}(x) = e^{ika} \psi_{nk}(x). \tag{7.141}$$

This rigidity imposed on the energy eigenfunctions has the profound consequence that there are regions of forbidden energies, called energy gaps. In one dimension, a complete qualitative analysis of the energy spectrum was possible for an arbitrary periodic potential in terms of the scattering properties of the unit cell potential by the use of the transfer matrix introduced in Chapter 2 to study tunneling and scattering. Whether unoccupied states in a one-dimensional periodic potential are separated from the ground state by an energy gap or not depends on the oddness/evenness of the number of electrons in the unit cell.

It is amazing that the wave nature of electron dynamics in a periodic potential has as far-reaching consequences as being the foundation of our information technology based on room temperature semiconductors, the technology realizing age-old ideas of calculating mechanical machines, the all-pervasive computer. The pushing around of electric currents will be explored in the following chapters, and in the next chapter we study the motion of Bloch electrons and the peculiarities of their contribution to the electric current in a crystal.

8 Bloch Currents

Energy band structure gives rise to characteristic particle dynamics in periodic potentials. The response of Bloch electrons to applied electric fields determines the relevance of crystalline solids for electronics. The energy gaps determine whether a crystalline solid behaves electrically as a metal or an insulator, or in between, a semiconductor.

8.1 Bloch State Current

The stationary Bloch state

$$\psi_{nk}(x,t) = \psi_{nk}(x)\, e^{-(i/\hbar)\epsilon_n(k)t} = u_{nk}(x)\, e^{ikx}\, e^{-(i/\hbar)\epsilon_n(k)t} \tag{8.1}$$

has associated with it the probability current

$$S_{nk}(x,t) = \frac{\hbar}{2im}\left(\psi^*_{nk}(x,t)\,\frac{\partial\psi_{nk}(x,t)}{\partial x} - \psi_{nk}(x,t)\,\frac{\partial\psi^*_{nk}(x,t)}{\partial x}\right), \tag{8.2}$$

which, for a stationary state, of course, is independent of time,

$$S_{nk}(x) = \frac{\hbar}{2im}\left(\psi^*_{nk}(x)\,\frac{d\psi_{nk}(x)}{dx} - \psi_{nk}(x)\,\frac{d\psi^*_{nk}(x)}{dx}\right), \tag{8.3}$$

and according to the continuity equation, Eq. (1.38),

$$\frac{dS_{nk}(x)}{dx} = 0, \tag{8.4}$$

also independent of position, $S_{nk}(x) = S_{nk}$. In one dimension, a particle in a stationary state, here a Bloch state, has at each point the same probability current.

The Bloch function is infinitely extended and its absolute square has only a relative probability interpretation, just as for the free particle plane wave as discussed in Section 1.3. However, for a Bloch state a reference volume presents itself, the unit cell. The expression in Eq. (8.3) becomes, in terms of the u function,

$$\begin{aligned} S_{nk} &= \frac{1}{2m}\left(u^*_{nk}(x)\left(\frac{\hbar}{i}\frac{\partial}{\partial x} + \hbar k\right)u_{nk}(x) - u_{nk}(x)\left(\frac{\hbar}{i}\frac{\partial}{\partial x} - \hbar k\right)u^*_{nk}(x)\right) \\ &= \frac{\hbar}{2im}\left(2ik\,|u_{nk}(x)|^2 + u^*_{nk}(x)\frac{\partial u_{nk}(x)}{\partial x} - u_{nk}(x)\frac{\partial u^*_{nk}(x)}{\partial x}\right), \end{aligned} \tag{8.5}$$

which explicitly only expresses invariance with respect to the periodicity. The expression in Eq. (8.5) is therefore integrated over a unit cell,

$$S_{nk} = \frac{1}{a}\int_0^a dx\, S_{nk}. \tag{8.6}$$

Inserting for the integrand the expression in Eq. (8.5) and performing a partial integration gives (the boundary terms vanish due to the periodicity of the u functions)

$$S_{nk} = \frac{1}{ma}\int_0^a dx\, u^*_{nk}(x)\left(\frac{\hbar}{i}\frac{\partial}{\partial x} + \hbar k\right) u_{nk}(x). \tag{8.7}$$

The probability current can be related to the energy. Recall the energy eigenvalue equation for the u function, Eq. (7.100). Since the *Hamiltonians* for two Bloch numbers close in value are related according to

$$\begin{aligned}\hat{H}(k+\Delta k) &= \hat{H}(k) + \frac{\hbar}{m}\Delta k\left(\frac{\hbar}{i}\frac{\partial}{\partial x} + \hbar k\right) + \mathcal{O}((\Delta k)^2)\\ &\equiv \hat{H}(k) + \hat{H}'(k)\Delta k + \mathcal{O}((\Delta k)^2),\end{aligned} \tag{8.8}$$

the energy eigenvalue of $\hat{H}(k+\Delta k)$, $\epsilon_n(k+\Delta k)$, is in lowest-order perturbation theory equal to the energy eigenvalue of the unperturbed Hamiltonian, $\hat{H}(k)$, i.e. $\epsilon_n(k)$, plus the first-order correction,

$$\epsilon_n(k+\Delta k) = \epsilon_n(k) + \langle u_{nk}|\hat{H}'(k)|u_{nk}\rangle\,\Delta k + \mathcal{O}((\Delta k)^2), \tag{8.9}$$

which is the matrix element of the perturbation evaluated in the normalized unperturbed eigenstates of $\hat{H}(k)$, i.e. u_{nk} in Eq. (8.11) should be normalized in the unit cell as required by perturbation theory,

$$\int_0^a dx\, |u_{nk}(x)|^2 = 1. \tag{8.10}$$

The shift in energy eigenvalue is thus

$$\epsilon_n(k+\Delta k) = \epsilon_n(k) + \frac{\hbar}{m}\Delta k\int_0^a dx\, u^*_{nk}(x)\left(\frac{\hbar}{i}\frac{\partial}{\partial x} + \hbar k\right) u_{nk}(x) + \mathcal{O}((\Delta k)^2), \tag{8.11}$$

thereby giving the formula

$$\frac{\hbar}{m}\int_0^a dx\, u^*_{nk}(x)\left(\frac{\hbar}{i}\frac{\partial}{\partial x} + \hbar k\right) u_{nk}(x) = \frac{\partial \epsilon_n(k)}{\partial k} \tag{8.12}$$

or according to Eqs. (8.7) and (8.3)

$$\frac{\hbar}{2im}\left(\psi^*_{nk}(x)\frac{d\psi_{nk}(x)}{dx} - \psi_{nk}(x)\frac{d\psi^*_{nk}(x)}{dx}\right) = \frac{1}{a\hbar}\frac{\partial \epsilon_n(k)}{\partial k}. \tag{8.13}$$

The probability current in a Bloch state is thus specified by the derivative of its energy,

$$aS_{nk} = \frac{1}{\hbar}\frac{\partial\epsilon_n(k)}{\partial k}. \tag{8.14}$$

In comparison to the free electron case, the effect of the periodic potential is to change the linear dependence on the wave number of the probability current to the nonlinear relation specified by the Bloch energy.

Bloch states corresponding to the Bragg points, $k = -\pi/a$ and $k = \pi/a$, have vanishing probability current, as the derivative of the energy, according to Eq. (7.86), vanishes. The corresponding Bloch functions can therefore be chosen to be real, $\psi^*_{n\pi/a} = \psi_{n\pi/a}$ and $\psi^*_{n-\pi/a} = \psi_{n-\pi/a}$. As $\psi_{n-k} = \psi^*_{nk}$, the Bloch functions $\psi_{n-\pi/a}$ and $\psi_{n\pi/a}$ are thus identical, $\psi_{n-\pi/a} = \psi_{n\pi/a}$, as already anticipated in Eq. (7.90); the energy in question is non-degenerate. For zone center states, we observed in Section 7.10 that their wave functions, by continuity in the Bloch number, are real, $\psi^*_{n0} = \psi_{n0}$. Thus, in view of Eq. (8.14), this provides a different reason (other than the one in Section 7.10 based on $F'(\epsilon(0)) \neq 0$) for the vanishing derivative of the energy at zone centers, relating it instead to a wave function property, $u^*_{n0} = u_{n0}$.

Bloch states are, according to Eq. (8.14), in general current-carrying. A Bloch electron thus propagates unimpeded in the periodic potential. A perfect crystal is a perfect conductor, a hallmark of the quantum theory of solids. The resistance of a crystalline solid is thus due to its imperfections, such as missing atoms, different atoms (impurities) and thermal vibration of the ions.

8.2 Bloch Velocity

A particle in a Bloch state does not have a definite velocity, but is in a superposition of states with definite velocities, as the Bloch function is a superposition of plane waves (recall Eq. (7.121))

$$\psi_{nk}(x) = \sum_K u_{nk}(K)\, e^{i(k+K)x}, \qquad K = \frac{2\pi}{a}\, l, \qquad l = 0, \pm 1, \pm 2, \ldots, \tag{8.15}$$

where (recall Eq. (7.63))

$$u_{nk}(K) = \frac{1}{a}\int_0^a dx\, e^{-ixK}\, u_{nk}(x). \tag{8.16}$$

The normalization condition on the coefficients $\{u_{nk}(K)\}_K$ is obtained by any normalization of $u_{nk}(x)$; for example, the normalization condition (8.10) gives ($u_{nk}(l) \equiv u_{nk}(K)$, $K = 2\pi l/a$)

$$1 = \int_0^a dx\, |u_{nk}(x)|^2 = \int_0^a dx \sum_{l=-\infty}^{\infty} e^{-i(2\pi l/a)x} u^*_{nk}(l) \sum_{l'=-\infty}^{\infty} e^{i(2\pi l'/a)x} u_{nk}(l'). \tag{8.17}$$

The integral gives $a\delta_{l,l'}$, and thereby the normalization condition for the velocity distribution probabilities is

$$1 = a \sum_{l=-\infty}^{\infty} |u_{nk}(l)|^2 = a \sum_K |u_{nk}(K)|^2. \tag{8.18}$$

The probability that a particle in Bloch state ψ_k has velocity value $\hbar(k+K)/m$ is thus $a|u_{nk}(K)|^2$. The average velocity for a particle in a Bloch state is accordingly

$$\bar{v}_{nk} = a \sum_K \frac{\hbar(k+K)}{m} |u_{nk}(K)|^2. \tag{8.19}$$

Inserting Eq. (8.16) gives upon interchange of integration and summation

$$\begin{aligned} \bar{v}_{nk} &= \frac{\hbar}{ma} \int_0^a dx \int_0^a dx' \; u^*_{nk}(x)\, u_{nk}(x') \sum_K e^{ixK} (k+K)\, e^{-ix'K} \\ &= \frac{\hbar}{ma} \int_0^a dx \int_0^a dx' \; u^*_{nk}(x)\, u_{nk}(x') \sum_K e^{ixK} \left(k + i\frac{\partial}{\partial x'}\right) e^{-ix'K}. \end{aligned} \tag{8.20}$$

Performing a partial integration gives (noting that the boundary terms cancel each other)

$$\bar{v}_{nk} = \frac{\hbar}{ma} \int_0^a dx \int_0^a dx' \; u^*_{nk}(x) \sum_K e^{iK(x-x')} \left(k - i\frac{\partial}{\partial x'}\right) u_{nk}(x'). \tag{8.21}$$

The sum over the reciprocal lattice numbers gives a delta spike of strength a in the unit cell, $a\delta(x-x')$ (see Appendix C), and the x' integration is then trivial, giving

$$\bar{v}_{nk} = \frac{1}{m} \int_0^a dx\; u^*_{nk}(x) \left(\frac{\hbar}{i}\frac{\partial}{\partial x} + k\right) u_{nk}(x), \tag{8.22}$$

and the average value of the velocity in a Bloch state is thus, according to Eq. (8.12), specified by its energy

$$\bar{v}_{nk} = \frac{1}{\hbar} \frac{\partial \epsilon_n(k)}{\partial k}. \tag{8.23}$$

The relationship between the average velocity and the derivative of the energy, Eq. (8.23), is identical to the formula for a free particle (recall Eq. (1.50)), but now the spectrum is not the free energy parabola. Furthermore, the relation between (average) velocity and energy for a Bloch electron is analogous to that of a classical particle where the so-called *crystal* momentum, $p \equiv \hbar k$, plays the role of momentum.

Not surprisingly, the average value of the velocity in a Bloch state becomes, according to Eq. (8.22), in terms of the Bloch function,

$$\bar{v}_{nk} = \frac{1}{m} \int_0^a dx\; \psi^*_{nk}(x) \frac{\hbar}{i}\frac{\partial}{\partial x} \psi_{nk}(x) = \langle \hat{v} \rangle_{nk}, \tag{8.24}$$

the general formula for the average velocity in terms of the expectation value of the velocity operator, where normalization is such that the Bloch function, according to Eq. (8.17), is normalized in a unit cell.

The probability current, Eq. (8.14), and average velocity are proportional,

$$S_{nk} = \frac{1}{a}\,\bar{v}_{nk}. \tag{8.25}$$

A particle in a Bloch state normalized in the unit cell and corresponding to average velocity $\bar{v}_{nk}$ has the probability current $\bar{v}_{nk}/a$, the flux entering and leaving the unit cell at this rate, the inverse of the time it takes the electron to traverse the unit cell.

Exercise 8.1 Show that a superposition of Bloch functions within a given band n with Bloch vectors close to k is a wave packet moving with velocity

$$v_n(k) = \frac{1}{\hbar}\frac{\partial \epsilon_n(k)}{\partial k}. \tag{8.26}$$

The average velocity in a Bloch state thus equals the group velocity of its corresponding Bloch wave packet.

8.3 Crystal Diffraction

The vanishing of the probability current for states corresponding to Bloch numbers at the zone boundaries, $k = \pm\pi/a$, can also be related to the occurrence of the energy gap at a zone boundary and thereby to a physical phenomenon: total reflection by a crystal. Consider an electron (or another charged particle) with an energy corresponding to a value in a gap region impinging from vacuum on a *periodic* potential filling only the half-space $[\,0,\infty\,[$. The solution for an energy in the gap corresponds in the periodic potential region to an imaginary Bloch number and the wave function in the periodic potential region is a spatially decaying function (as discussed in Section 7.3). The current in such an evanescent wave in the region $[\,0,\infty\,[$ thus vanishes at infinity, and since the current for an energy eigenstate is spatially constant, it vanishes everywhere in the crystal region. The incoming particle in a plane wave state corresponding to an energy value in a gap region is thus totally reflected, so-called Bragg-reflected, in analogy with X-ray diffraction.

As discussed in Section 7.10, at very high energies, the energy gaps narrow almost to points, and the Bloch energy condition becomes $\cos(ka) \simeq \cos(qa)$ quite generally (recall the Kronig–Penney model case, Eq. (7.56) or the fact that at very high energies different periodic potentials have the same scattering properties). The band-gap energy at which the electron is totally reflected thus corresponds to the wave number, q, of the incoming wave being equal to the band edge Bloch number,

$$qa \simeq k_{\mathrm{be}}a. \tag{8.27}$$

A diffraction experiment, here observing when the electron is reflected, thus measures the wave vector of the incoming electron.

The band gaps are (recall Section 7.7) located at Bloch numbers (here the extended zone scheme is the natural choice, as q can take on any real value)

$$k = \frac{\pi}{a}\, m, \qquad m = 1, 2, 3, \ldots, \tag{8.28}$$

and introducing the wave length of the incoming particle $\lambda = 2\pi/q$, the condition for total reflection, Eq. (8.27), becomes

$$m\lambda = 2a, \tag{8.29}$$

the condition for Bragg scattering in the one-dimensional case (only normal incidence). A high-energy electron impinging from vacuum on a periodic potential with a wave length equal to twice the lattice periodicity, or fractions thereof, is thus totally reflected. Measuring the energy for which the electron is reflected thus provides a measurement of the crystal structure, in the present one-dimensional case simply a measurement of the lattice constant.[1]

Exercise 8.2 Calculate the wave length of an electron of energy 1 keV.

8.4 Forced Bloch Particle

Consider a Bloch electron in an electric field, i.e. in a tilted periodic potential, a washboard potential, where, in addition to the periodic potential, a constant force is applied. For the case of a Bloch electron, a constant electric field, E, is present, corresponding to a potential energy $V_{\rm app}(x) = eEx$. The Hamiltonian for an electron in a periodic potential and constant force, $F = -eE$, thus takes the form[2]

$$\hat{H}_{\rm F} = -\frac{\hbar^2}{2m}\frac{d^2}{dx^2} + V(x) - Fx = \hat{H}_0 - Fx. \tag{8.30}$$

A particle initially in the state specified by the Bloch function $\psi_{n_0k_0}(x)$, say, at time $t = 0$, when the applied field is turned on, evolves by the evolution operator to the wave function at time t,

$$\psi(x,t) = e^{-(i/\hbar)t\hat{H}_{\rm F}}\,\psi_{n_0k_0}(x) = e^{-(i/\hbar)t(\hat{H}_0 - Fx)}\,\psi_{n_0k_0}(x). \tag{8.31}$$

[1] Historically, the serendipitous diffraction of high-energy electrons by a nickel crystal by Davisson and Germer in 1927 established the wave nature of a massive particle (as envisioned by de Broglie in 1924), a manifestation of Schrödinger's wave mechanics from 1926, and was seminal in giving credence to Schrödinger's equation.

[2] The problem, involving a singular perturbation, should be approached with care. The topic has, for the field of physics, a long and controversial history and a vast literature [2].

Consider the evolution operator sandwiched by the crystal translation operator and its inverse,

$$\hat{T}_a\, e^{-(i/\hbar)t\hat{H}_\mathrm{F}}\,\hat{T}_{-a} = \hat{T}_a\, e^{-(i/\hbar)t(\hat{H}_0 - Fx)}\,\hat{T}_{-a}$$

$$= \sum_{n=0}^{\infty} \frac{(-it/\hbar)^n}{n!}\, \hat{T}_a (\hat{H}_0 - Fx)^n\, \hat{T}_{-a}. \tag{8.32}$$

Inserting $1 = \hat{T}_{-a}\hat{T}_a$ in between the Hamiltonian factors, we need to evaluate

$$\hat{T}_a\hat{H}_\mathrm{F}\hat{T}_{-a} = \hat{T}_a(\hat{H}_0 - Fx)\hat{T}_{-a} = \hat{T}_a\hat{H}_0\hat{T}_{-a} - \hat{T}_a Fx\,\hat{T}_{-a} = \hat{H}_0 - \hat{T}_a Fx\,\hat{T}_{-a}, \tag{8.33}$$

the last equality being valid since the periodic potential Hamiltonian commutes with the crystal translation operator. Operating on an arbitrary function, one sees that

$$\hat{T}_a Fx\,\hat{T}_{-a} = (x+a)F, \tag{8.34}$$

i.e.

$$\hat{T}_a(\hat{H}_0 - Fx)\,\hat{T}_{-a} = \hat{H}_0 - F(x+a) \tag{8.35}$$

or

$$\hat{T}_a\hat{H}_\mathrm{F}\hat{T}_{-a} = \hat{H}_\mathrm{F} - Fa, \tag{8.36}$$

expressing the fact that the presence of the constant electric field spoils the commutation of the Hamiltonian and the crystal translation operator. The expression in Eq. (8.32) can now be re-exponentiated, giving

$$\hat{T}_a\, e^{-(i/\hbar)t\hat{H}_\mathrm{F}}\,\hat{T}_{-a} = \sum_{n=0}^{\infty} \frac{(-it/\hbar)^n}{n!}\,(\hat{H}_\mathrm{F} - Fa)^n = e^{-(i/\hbar)t(\hat{H}_\mathrm{F} - Fa)}. \tag{8.37}$$

Since

$$\hat{T}_a\,\psi(x,t) = \hat{T}_a\, e^{-(i/\hbar)t\hat{H}_\mathrm{F}}\,\hat{T}_{-a}\hat{T}_a\,\psi_{n_0k_0}(x) = e^{-(i/\hbar)t(\hat{H}_\mathrm{F} - Fa)}e^{ik_0a}\psi_{n_0k_0}(x)$$

$$= e^{i[k_0 + (tF/\hbar)]a}e^{-(i/\hbar)t\hat{H}_\mathrm{F}}\,\psi_{n_0k_0}(x) = e^{i[k_0 + (tF/\hbar)]a}\,\psi(x,t), \tag{8.38}$$

the evolved function $\psi(x,t)$ satisfies the Bloch condition corresponding to Bloch number $k_0 + tF/\hbar$. In the reduced zone scheme, the function $k(t) = k_0 + tF/\hbar$ is a sawtooth-shaped function, increasing linearly in time until $k(t) = \pi/a$ ($F > 0$), at which time it starts all over traversing again the first Brillouin zone. The general function with the property in Eq. (8.38) is the superposition of Bloch functions corresponding to the Bloch number $k_0 + Ft/\hbar$, and the function $\psi(x,t)$ has the form, choosing now the arbitrary initial time at $t = t_i$ ($k_i \equiv k(t_i)$),

$$\psi(x,t) = \sum_{n=1}^{\infty} c_n(t)\,\psi_{n,(k_i + F(t-t_i)/\hbar)}(x), \qquad c_n(t = t_i) = \delta_{nn_i}. \tag{8.39}$$

In the presence of a force, interband transitions thus occur. The question is, at what rate?

8.5 Interband Transitions

The wave function in Eq. (8.39) by construction satisfies Eq. (8.31), the Schrödinger equation

$$i\hbar\frac{\partial\psi(x,t)}{\partial t} = \hat{H}_{\mathrm{F}}\,\psi(x,t), \tag{8.40}$$

the solution at the initial time reducing to the Bloch state labeled by $n_i k_i$, $\psi(x,t_i) = \psi_{n_i k_i}(x)$.

To expediently reveal the interband coupling, phase factors are taken out explicitly, rewriting the solution, Eq. (8.39), in the form (i.e. redefining $c_n(t)$)

$$\psi(x,t) = \sum_{n=1}^{\infty} a_n(t)\, e^{ik(t)x} u_{nk(t)}(x)\, e^{-(F/\hbar)\int_{t_i}^{t} dt'\, \gamma_{nn}(k(t'))}\, e^{-(i/\hbar)\int_{t_i}^{t} dt'\, \epsilon_n(k(t'))}, \tag{8.41}$$

where the notation $k(t) = k_i + F(t-t_i)/\hbar$ has been introduced ($a_n(t_i) = \delta_{nn_i}$), and

$$\gamma_{nn'}(k(t)) \equiv \left\langle u_{nk} \left| \frac{\partial u_{n'k}}{\partial k} \right.\right\rangle\Bigg|_{k=k(t)} = \int_0^a dx\, u_{nk}^*(x)\frac{\partial}{\partial k}u_{n'k}(x)\Bigg|_{k=k(t)}. \tag{8.42}$$

We note that $\gamma_{nn'}(k) = \gamma_{n'n}^*(k)$, and that from the normalization of u_{nk} it follows that $\gamma_{nn}(k)$ is purely imaginary.

Inserting Eq. (8.41), the right-hand side of Eq. (8.40) gives (letting $n \to n'$)

$$\begin{aligned}(\hat{H}_0 - Fx)\,\psi(x,t) = \sum_{n'=1}^{\infty} & a_{n'}(t)(\epsilon_{n'}(k(t)) - Fx)\, e^{ik(t)x} u_{n'k(t)}(x) \\ & \times e^{-(F/\hbar)\int_{t_i}^{t} dt'\, \gamma_{n'n'}(k(t'))}\, e^{-(i/\hbar)\int_{t_i}^{t} dt'\, \epsilon_{n'}(k(t'))},\end{aligned} \tag{8.43}$$

and the left-hand side of Eq. (8.40) becomes, $\dot{k}(t) = F/\hbar$,

$$\begin{aligned}i\hbar\frac{\partial\psi(x,t)}{\partial t} &= i\hbar\sum_{n'=1}^{\infty}\left(\dot{a}_{n'}(t) + a_{n'}(t)\left(i\dot{k}(t)x - \frac{F}{\hbar}\gamma_{n'n'}(k(t)) - \frac{i}{\hbar}\epsilon_{n'}(k(t))\right)\right) \\ &\quad \times e^{ik(t)x} u_{n'k(t)}(x)\, e^{-(F/\hbar)\int_{t_i}^{t} dt'\, \gamma_{n'n'}(k(t'))}\, e^{-(i/\hbar)\int_{t_i}^{t} dt'\, \epsilon_{n'}(k(t'))} \\ &\quad + i\hbar\sum_{n'=1}^{\infty} a_{n'}(t)\, e^{ik(t)x}\, e^{-(F/\hbar)\int_{t_i}^{t} dt'\, \gamma_{n'n'}(k(t'))} \\ &\quad \times e^{-(i/\hbar)\int_{t_i}^{t} dt'\, \epsilon_{n'}(k(t'))}\, \dot{k}(t)\frac{\partial}{\partial k}u_{n'k}(x)\Bigg|_{k=k(t)}.\end{aligned} \tag{8.44}$$

The $\epsilon_{n'}(k(t))$ and Fx terms on the two sides of Eq. (8.40) cancel, i.e. the right-hand side is canceled. Taking the scalar product with $u_{nk(t)}$ and using the orthonormality in the band index of the u_n functions gives

$$0 = i\hbar\left(\dot{a}_n(t) - \frac{F}{\hbar}\gamma_{nn}(k(t))\,a_n(t)\right)e^{ik(t)x}\,e^{-(F/\hbar)\int_{t_i}^{t}dt'\,\gamma_{nn}(k(t'))}\,e^{-(i/\hbar)\int_{t_i}^{t}dt'\,\epsilon_n(k(t'))}$$
$$+ i\hbar\sum_{n'=1}^{\infty} a_{n'}(t)\,e^{ik(t)x}\,e^{-(F/\hbar)\int_{t_i}^{t}dt'\,\gamma_{n'n'}(k(t'))}\,e^{-(i/\hbar)\int_{t_i}^{t}dt'\,\epsilon_{n'}(k(t'))}\frac{F}{\hbar}\gamma_{nn'}(k(t)). \tag{8.45}$$

Canceling the overall factor $\exp(ik(t)x)$ and noting that the $n' = n$ term in the sum is canceled, the equation determining the wave function evolution becomes

$$\dot{a}_n(t) = -\sum_{n'=1\,(n'\neq n)}^{\infty} a_{n'}(t)\frac{F}{\hbar}\gamma_{nn'}(k(t))$$
$$\times\, e^{(F/\hbar)\int_{t_i}^{t}dt'\,(\gamma_{nn}(k(t'))-\gamma_{n'n'}(k(t')))}\,e^{(i/\hbar)\int_{t_i}^{t}dt'\,(\epsilon_n(k(t'))-\epsilon_{n'}(k(t')))}. \tag{8.46}$$

8.6 Transition Rate

The exact solution, Eq. (8.41), is a superposition of Bloch functions of different band indices. If energies are far apart, the energy factor in Eq. (8.46) is wildly oscillating, leading to a tiny contribution from far-away bands. In one dimension, we shall therefore consider the case where only two bands are relevant, the two-band model. Then the amplitudes are specified by (labeling the bands by 1 and 2)

$$\dot{a}_1(t) = -a_2(t)\frac{F}{\hbar}\gamma_{12}(k(t))\,e^{(F/\hbar)\int_{t_i}^{t}dt'\,(\gamma_{11}(k(t'))-\gamma_{22}(k(t')))}\,e^{(i/\hbar)\int_{t_i}^{t}dt'\,(\epsilon_1(k(t'))-\epsilon_2(k(t')))} \tag{8.47}$$

$$\dot{a}_2(t) = -a_1(t)\frac{F}{\hbar}\gamma_{21}(k(t))\,e^{(F/\hbar)\int_{t_i}^{t}dt'\,(\gamma_{22}(k(t'))-\gamma_{11}(k(t')))}\,e^{(i/\hbar)\int_{t_i}^{t}dt'\,(\epsilon_2(k(t'))-\epsilon_1(k(t')))}. \tag{8.48}$$

Consider the particle initially in the lowest band, $a_n(t_i) = \delta_{n1}$. Lowest-order perturbation theory is the approximation $a_1(t) \simeq 1$, whereby

$$\dot{a}_2(t) = -\frac{F}{\hbar}\gamma_{21}(k(t))\,e^{(F/\hbar)\int_{t_i}^{t}dt'\,(\gamma_{22}(k(t'))-\gamma_{11}(k(t')))}\,e^{(i/\hbar)\int_{t_i}^{t}dt'\,(\epsilon_2(k(t'))-\epsilon_1(k(t')))} \tag{8.49}$$

and the equation can be immediately integrated. Since $dk = F\,dt/\hbar$, the progression of time can be measured in units of k and Eq. (8.49) can be rewritten as

$$\frac{da_2(k)}{dk} = -\gamma_{21}(k)\,e^{\int_{k_i}^{k}dk'\,(\gamma_{22}(k')-\gamma_{11}(k'))}\,e^{(i/F)\int_{k_i}^{k}dk'(\epsilon_2(k')-\epsilon_1(k'))}. \tag{8.50}$$

As noted at the end of Section 7.11, the function u_{nk} can always be subjected to the phase transformation

$$u_{nk} \to e^{i\varphi_n(k)}u_{nk}. \tag{8.51}$$

In that case,

$$\gamma_{nn}(k) = \int_0^a dx\; u_{nk}^*(x) \frac{\partial}{\partial k}\, u_{nk}(x) \to \gamma_{nn}(k) + i \frac{\partial \varphi_n(k)}{\partial k} \tag{8.52}$$

whereas

$$\gamma_{21}(k) = \int_0^a dx\; u_{2k}^*(x) \frac{\partial}{\partial k}\, u_{1k}(x) \to \gamma_{21}(k)\, e^{i(\varphi_1(k) - \varphi_2(k))} \tag{8.53}$$

due to the orthogonality of the u functions. The factor of interest thus transforms according to

$$\gamma_{21}(k)\, e^{\int_{k_i}^k dk'\,(\gamma_{11}(k') - \gamma_{22}(k'))}$$
$$\to e^{i(\varphi_1(k)-\varphi_2(k))} \gamma_{21}(k)\, e^{\int_{k_i}^k dk'\,[\gamma_{11}(k') + i\,\partial\varphi_1(k')/\partial k' - \gamma_{22}(k') - i\,\partial\varphi_2(k')/\partial k']} \tag{8.54}$$

and by choosing

$$\varphi_n(k) = \frac{i}{2} \int_{k_i}^k dk'\; \gamma_{nn}(k'), \tag{8.55}$$

the phase factor is transformed away,

$$\gamma_{21}(k)\, e^{\int_{k_i}^k dk'\,(\gamma_{22}(k') - \gamma_{11}(k'))} \to \gamma_{21}(k). \tag{8.56}$$

The transformation (8.51) thus transforms (8.50) into

$$\frac{da_2(k)}{dk} = -\gamma_{21}(k)\, e^{(i/F) \int_{k_i}^k dk'\,(\epsilon_2(k') - \epsilon_1(k'))}, \tag{8.57}$$

which is immediately integrated to

$$a_2(k) = \int_{k_i}^k dk''\; \gamma_{21}(k'')\, e^{(i/F) \int_{k_i}^{k''} dk'\,(\epsilon_2(k') - \epsilon_1(k'))}, \tag{8.58}$$

as the initial condition dictates $a_2(k_i) = a_2(t_i) = 0$.

The upper band amplitude after $k(t)$ traverse of the Brillouin zone, say, choosing $k_i = -\pi/a$ and then correspondingly $k_f = \pi/a$, is

$$a_2(k_f = \pi/a) = \int_{-\pi/a}^{\pi/a} dk''\; \gamma_{21}(k'')\, e^{(i/F) \int_{-\pi/a}^{k''} dk'\,(\epsilon_2(k') - \epsilon_1(k'))}. \tag{8.59}$$

Introducing

$$\chi(k) \equiv \frac{1}{F} \int_0^k dk'\; (\epsilon_2(k') - \epsilon_1(k')) \tag{8.60}$$

gives

$$a_2(\pi/a) = e^{i\varphi} \int_{-\pi/a}^{\pi/a} dk\ \gamma_{21}(k)\, e^{i\chi(k)}, \tag{8.61}$$

where

$$e^{i\varphi} = e^{(i/F)\int_{-\pi/a}^{0} dk\,(\epsilon_2(k)-\epsilon_1(k))} = e^{(i/F)\int_{0}^{\pi/a} dk\,(\epsilon_2(k)-\epsilon_1(k))} = e^{i\chi(\pi/a)} \tag{8.62}$$

is an overall phase factor.

Since the electric field strengths typically obtainable in crystals are small compared to the energy gap, the phase factor $e^{i\chi(k)}$ is due to the smallness of F wildly oscillating except close to possible stationary points where $d\chi(k)/dk = 0$, or equivalently $\epsilon_2(k) - \epsilon_1(k) = 0$. Since the energies in different bands are different (in one dimension), stationary points are only possible in the complex k plane, i.e. only a complex q can solve the analytically continued equation $\epsilon_2(q) = \epsilon_1(q)$ (as shown in Appendix Q, there is one stationary point, q, in the upper half-plane and one, q^*, in the lower half-plane). The scenario thus suggests the stationary phase approximation, where the integration contour in Eq. (8.61) is shifted into the complex plane passing through a stationary point. However, as shown in Appendix Q, $\gamma_{21}(k)$ is singular at the stationary point, as

$$\gamma_{21}(k) = \int_0^a dx\ u^*_{2k}(x) \frac{\partial u_{1k}(x)}{\partial k} = \frac{1}{\epsilon_2(k) - \epsilon_1(k)} \frac{i\hbar}{m} \int_0^a dx\ u_{2-k}(x) \frac{\partial u_{1k}(x)}{\partial x}, \tag{8.63}$$

where in the last equation the non-analytic operation of complex conjugation is avoided, anticipating the analytical continuation of $\gamma_{21}(k)$ into the complex plane. In Appendix Q, it is shown that the interband coupling term has a simple pole in the upper half-plane at the stationary point,

$$\gamma_{21}(k) \simeq \frac{-i/4}{k-q}, \qquad k \simeq q = i\frac{\sqrt{m_0 \Delta}}{\hbar}, \tag{8.64}$$

where Δ is the energy gap and m_0 the reduced band mass. The singularity of $\gamma_{21}(k)$ at the stationary point thus precludes the standard steepest descent calculation. Since q is a branch point for the energy spectra, $\epsilon_n(k)$, the contour should take a semicircular detour just below the stationary point. The integration along the real line is thus shifted to the parallel contour in the complex plane determined by the stationary point at q but avoiding this singularity by taking a small semicircular detour below it. The integral in Eq. (8.61) is estimated by the contribution from the semicircle,

$$a_2(\pi/a) \simeq e^{i\varphi}\, e^{i\chi(q)} \int_{\cup} dk\ \gamma_{21}(k) = -\frac{i}{4}\, e^{i\varphi}\, e^{i\chi(q)} \int_{\cup} dk\ \frac{1}{k-q}, \tag{8.65}$$

and the semicircular integral picks up the factor $i\pi$, giving

$$a_2(\pi/a) \simeq \tfrac{1}{4}\pi\, e^{i\varphi}\, e^{i\chi(q)}. \tag{8.66}$$

In Appendix Q, the band energies are obtained using the $\mathbf{k}\cdot\mathbf{p}$ method, and the integral becomes

$$\chi(q) = \frac{1}{F}\int_0^q dk\,(\epsilon_2(k) - \epsilon_1(k)) = \frac{1}{F}\int_0^{im_0\Delta/\hbar^2} dk\,\sqrt{\Delta^2 + \frac{\hbar^2\Delta}{m_0}k^2}. \tag{8.67}$$

Substituting $u = -ik$, the elementary integral

$$\chi(q) = \frac{1}{F}\int_0^{\sqrt{m_0\Delta/\hbar^2}} du\,\sqrt{\frac{m_0\Delta}{\hbar^2} - u^2} \tag{8.68}$$

gives $\chi(q) = i\pi\sqrt{m_0}\Delta^{3/2}/2\hbar F$, and

$$e^{i\chi(q)} = e^{-(\pi/2\hbar F)\sqrt{m_0}\Delta^{3/2}}. \tag{8.69}$$

The Bloch number progresses with constant speed, $\dot{k}(t) = F/\hbar$, and Brillouin zone traversal therefore takes the time $T = 2\pi\hbar/aF$. Introducing the rate of transition per unit time of traversal of the Brillouin zone gives

$$\Gamma \equiv \frac{|a_2(\pi/a)|^2}{T} = \frac{\pi a F}{32\hbar}\,e^{-(\pi/\hbar F)\sqrt{m_0}\Delta^{3/2}}. \tag{8.70}$$

For wide band gap and low field strengths, interband transitions are completely negligible. Only if huge electric field strengths are achievable can the threshold of the exponential be overcome.

8.7 Bloch Oscillation

For low electric field strengths, where interband transitions can be neglected, the wave function, Eq. (8.41) or (8.39), becomes the Houston function

$$\begin{aligned}\psi(x,t) \simeq \psi_n^{\text{Hou}}(x,t) &\equiv \psi_{nk(t)}(x)\,e^{-(i/\hbar)\int_{t_i}^t dt'\,\epsilon_n(k(t'))}\\ &= e^{ik(t)x}u_{nk(t)}(x)\,e^{-(i/\hbar)\int_{t_i}^t dt'\,\epsilon_n(k(t'))}.\end{aligned} \tag{8.71}$$

The Houston function is a solution of the equation

$$\begin{aligned}i\hbar\frac{\partial\psi_n^{\text{Hou}}(x,t)}{\partial t} &= \hat{H}_{\text{F}}\,\psi_n^{\text{Hou}}(x,t)\\ &\quad - i\sum_{n=1}^{\infty}\left(F\frac{\partial}{\partial k}u_{n'k}(x)\bigg|_{k=k(t)}\right)e^{ik(t)x}u_{nk(t)}(x)\,e^{-(i/\hbar)\int_{t_i}^t dt'\,\epsilon_{n'}(k(t'))},\end{aligned} \tag{8.72}$$

i.e. except for the last terms (the interband coupling terms), solving the Schrödinger equation for a Bloch particle driven by a constant force. If interband transitions can be neglected, Bloch dynamics is thus specified by the time dependence of the Bloch number, $k(t)$, as described by the Houston function. The equation of motion for the Bloch number, $\hbar\dot{k}(t) = F$, is analogous to that of the wave number of a particle subjected to a constant force but otherwise free, in the latter case describing the evolution of the wave number of the plane wave (in the free case this evolution is exact, as discussed in Appendix N). However, driven Bloch dynamics is very different from driven *free* particle motion because of Bragg reflection due to the periodic potential. If interband transitions are negligible, the band index is a constant of motion, and a Bloch particle traverses the band it started out in, i.e. the Bloch number progresses with constant speed through the Brillouin zone, and is Bragg-reflected when $k(t)$ reaches the Bragg point $k = \pi/a$ $(F > 0)$, i.e. the particle traverses the band all over again.[3] The particle performs Bloch oscillation with frequency $\nu = 1/T = aF/2\pi\hbar$. The probability of particle location oscillates in space according to

$$|\psi_{n_0k(t)}(x)|^2 = |u_{n_0k(t)}(x)|^2. \tag{8.73}$$

If there is only a single Bloch electron in a band, the corresponding charge oscillation radiates electromagnetic waves, being a Bloch oscillator antenna able to emit electromagnetic radiation in the gigahertz regime. It is difficult to observe Bloch oscillations in crystals due to frequent intraband scattering by impurities preventing completion of even a single oscillation period. In heterostructure superlattices, the Bloch period can be tuned, making Bloch oscillations easier. In a pure system, such as cold atoms in optical lattices, Bloch oscillations are observed.

Exercise 8.3 Assume a lattice constant of one Ångström, $a = 1$Å, and an electric field of ten thousand volts per centimeter, $E = 10^4$ V/cm. Calculate the corresponding Bloch oscillation frequency.

8.8 Inert Bands

In a one-dimensional crystal, the ground state corresponds to filled or half-filled bands corresponding to whether the number of electrons per unit cell is even or odd. Consider first the even case where the ground state has completely filled bands. If interband transitions can be neglected, the Bloch states in the presence of a constant electric field evolve simply according to the rate of change of the Bloch number, $\dot{k}(t) = -eE/\hbar$, as described by the Houston function, Eq. (8.71), and depicted in Figure 8.1.

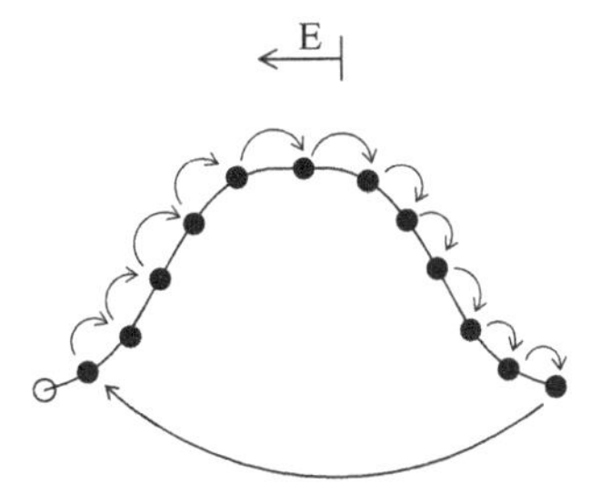

Figure 8.1 Filled band inert to an electric field.

[3] The classical dynamics is not that spectacular: for low energies the particle is trapped in a local minimum executing oscillation, but above a threshold energy the particle will rush down the tilted periodic potential.

A Bloch state contributes its probability current with associated average electric current,

$$I_{nk} = -eS_{nk} = \frac{-e}{a\hbar}\frac{\partial\epsilon_n(k)}{\partial k}. \tag{8.74}$$

The total contribution to the current from a filled band vanishes,

$$I = \int_{-\pi/a}^{\pi/a} dk\, I_{nk} = \frac{-e}{a\hbar}\int_{-\pi/a}^{\pi/a} dk\, \frac{\partial\epsilon_n(k)}{\partial k} = 0, \tag{8.75}$$

since $\epsilon_n(\pi/a) = \epsilon_n(-\pi/a)$. Or, more to the point: the current in Bloch state k cancels the current in Bloch state $-k$, the velocities $v_n(k)$ on the positive branch, $k > 0$, being exactly compensated by the velocities on the negative branch $k < 0$, $v_n(-k) = -v_n(k)$ by time reversal symmetry (or equivalently $S_{n-k} = -S_{nk}$).

Even though the Bloch states change Bloch vector, as indicated in Figure 8.1, Bragg reflection keeps the band filled at all times. The filled band case in the absence of interband transitions thus corresponds to an insulator: no current response to an applied electric field. An insulator thus only allows a current to be pushed through it if interband transitions can provide a measurable current, but for field strengths below threshold the current is negligible.[4]

In the half-filled case, just as in the Sommerfeld model, scattering dominates the transport properties of a Bloch metal. To discuss transport in general, the extended Bloch states are not suitable. Just as the plane waves of the Sommerfeld model were abandoned for the Gaussian wave packets, what is needed is localized Bloch wave packets. Once the Bloch wave packet description of electrons in a periodic potential is established, Drude transport results of relevance are directly taken over, the only difference being the difference in energy dispersion of the electrons whose manifestation in metals leads only to a Fermi surface average. In crystals with moderate band gaps, so-called semiconductors, charge carriers are provided by doping, as discussed in Chapter 10, and the generic case is Bloch electron motion in spatially varying potentials, for which the local transport description is needed.

8.9 Bloch Wave Packets

Consider superposing Bloch waves of Bloch numbers close to k_0 in band n,

$$\psi^{(n)}_{x_0k_0}(x) = \int_{\mathrm{Bz}_1} dk\; g(k-k_0)\,\psi_{nk}(x) = \int_{-\pi/a}^{\pi/a} dk\; g(k-k_0)\,u_{nk}(x)\,e^{ikx}$$

[4] Interband transition was originally suggested by Zener as an explanation of dielectric breakdown, i.e. currents occurring at very high electric fields in insulators (or semiconductors), and the mechanism is also referred to as Zener tunneling. The effect is also referred to as internal field emission, the analogous effect taking place when an applied electric field leads to emission of electrons from a metal due to tunneling.

$$\simeq u_{nk_0}(x) \int_{-\pi/a}^{\pi/a} dk\, g(k-k_0)\, e^{ikx}. \tag{8.76}$$

The approximation is justified by choosing the width, δk, of the function g around k_0 very small, as u_{nk} is a smooth function of k. Employing (in order to have definite expressions)

$$g_{x_0k_0}(k) = e^{-(k-k_0)^2/(4\delta k^2)-ikx_0} \tag{8.77}$$

and taking the width δk very small compared to the size of the Brillouin zone, $\delta k \ll 2\pi/a$, the integration limits can be taken to infinity and the Gaussian integral performed (the result follows directly from the identical integration in Appendix M, Eqs. (M.10) and (M.1))

$$\psi^{(n)}_{x_0k_0}(x) = \mathcal{N}\, u_{nk_0}(x)\, e^{ik_0x} e^{-(x-x_0)^2/(4\delta x^2)} = \mathcal{N}\, \psi_{nk_0}(x)\, e^{-(x-x_0)^2/(4\delta x^2)}, \tag{8.78}$$

where $\delta x^2 = 1/4\delta k^2$ and $\mathcal{N}$ has been inserted for normalization. Instead of having the plane wave in front of the Gaussian envelope function as in the free particle case, Eq. (4.2), there is instead the Bloch function in question. The Bloch Gaussian wave packet is labeled by its characteristics: band and Bloch number and wave packet location.

For convenience, consider a finite stretch of *periodic* potential. The normalization constant is then specified by

$$1 = \int_{-L/2}^{L/2} dx\, |\psi^{(n)}_{x_0k_0}(x)|^2 = |\mathcal{N}|^2 \int_{-L/2}^{L/2} dx\, |u_{nk_0}(x)|^2\, e^{-(x-x_0)^2/(2\delta x^2)}. \tag{8.79}$$

Since δk is much smaller than the width of the Brillouin zone, $\delta k \ll 2\pi/a$, the width of the Gaussian function, δx, is much larger than the unit cell, $\delta x = 1/2\delta k \gg a$. The integrand is thus a product of a periodic function, $|u_{nk_0}(x+a)|^2 = |u_{nk_0}(x)|^2$, and a slowly varying function on the scale of a. The integral doubling formula of Appendix T can therefore be applied to obtain

$$\int_{-L/2}^{L/2} dx\, |\psi^{(n)}_{x_0k_0}(x)|^2 \simeq |\mathcal{N}|^2 \int_{-L/2}^{L/2} dx\, e^{-(x-x_0)^2/(2\delta x^2)} \frac{1}{L} \int_{-L/2}^{L/2} dx\, |u_{nk_0}(x)|^2. \tag{8.80}$$

The Gaussian integral is evaluated with sufficient accuracy by extending the integration limits to infinity since L is assumed large compared to δx, and the other integral gives, according to the unit cell normalization choice, Eq. (8.10), the value N, the number of unit cells. Using that $L = Na$ gives

$$\int_{-L/2}^{L/2} dx\, |\psi^{(n)}_{x_0k_0}(x)|^2 \simeq |\mathcal{N}|^2 \sqrt{2\pi\delta x^2}\, \frac{1}{a}, \tag{8.81}$$

determining the dimensionless normalization constant to be $\mathcal{N} = (2\pi\delta x^2)^{-1/4} a^{1/2}$. Just as for the free case, the Bloch Gaussian state carries no reference to the finite, but large, size of the system.

Consider the expression

$$\sum_n \int_{-L/2}^{L/2} dx_0 \int_{-\pi/a}^{\pi/a} \frac{dk_0}{2\pi}\, \psi^{(n)*}_{x_0k_0}(x)\, \psi^{(n)}_{x_0k_0}(x')$$

$$= \mathcal{N}^2 \sum_n \int_{-L/2}^{L/2} \int_{-\pi/a}^{\pi/a} \frac{dx_0\, dk_0}{2\pi}\, u^*_{nk_0}(x)\, u_{nk_0}(x') e^{-ik_0(x-x')}\, e^{-(x-x_0)^2/(4\delta x^2)} e^{-(x'-x_0)^2/(4\delta x^2)}$$

$$= \mathcal{N}^2 \int_{-L/2}^{L/2} \int_{-\pi/a}^{\pi/a} \frac{dx_0\, dk_0}{2\pi} \sum_{m=-N/2}^{N/2} \delta(x - x' + ma) e^{ik_0am}\, e^{-(x-x_0)^2/(4\delta x^2)} e^{-(x'-x_0)^2/(4\delta x^2)}, \tag{8.82}$$

the last equality following from the completeness relation for the u functions, Eq. (7.133). The k_0 integration is now trivially performed, giving

$$\mathcal{N}^2 \sum_n \int_{-L/2}^{L/2} \int_{-\pi/a}^{\pi/a} \frac{dx_0 dk_0}{2\pi}\, \psi^{(n)*}_{x_0k_0}(x)\, \psi^{(n)}_{x_0k_0}(x')$$

$$= \mathcal{N}^2 \sum_{m=-N/2}^{N/2} \delta(x - x' + ma)\, \frac{e^{i\pi m} - e^{-i\pi m}}{iam} \int_{-L/2}^{L/2} \frac{dx_0}{2\pi}\, e^{-(x-x_0)^2/(4\delta x^2)}\, e^{-(x'-x_0)^2/(4\delta x^2)} \tag{8.83}$$

and only the $m = 0$ term is non-vanishing, so that

$$\sum_n \int_{-L/2}^{L/2} \int_{-\pi/a}^{\pi/a} \frac{dx_0\, dk_0}{2\pi}\, \psi^{(n)*}_{x_0k_0}(x)\, \psi^{(n)}_{x_0k_0}(x') = \frac{1}{a}\, \delta(x - x') \int_{-L/2}^{L/2} \frac{dx_0}{2\pi}\, e^{-(x-x_0)^2/(2\delta x^2)}. \tag{8.84}$$

The x_0 integration can again be approximately evaluated, thereby making the overall factor equal to one, i.e.

$$\sum_n \int_{-L/2}^{L/2} \int_{-\pi/a}^{\pi/a} \frac{dx_0\, dk_0}{2\pi}\, \psi^{(n)*}_{x_0k_0}(x)\, \psi^{(n)}_{x_0k_0}(x') = \delta(x - x'). \tag{8.85}$$

The set of Bloch wave packets, $\{\psi^{(n)}_{x_0k_0}\}_{x_0k_0}$, is thus complete.

The overlap integral between Bloch wave packets can be evaluated using the integral doubling formula

$$\langle nx_0k_0 \mid n'x'_0k'_0\rangle \equiv \int_{-L/2}^{L/2} dx\, \psi^{(n)*}_{x_0k_0}(x)\, \psi^{(n')}_{x'_0k'_0}(x)$$

$$= \mathcal{N}^2 \int_{-L/2}^{L/2} dx\, u^*_{nk_0}(x)\, u_{n'k'_0}(x)\, e^{-ix(k_0-k'_0)} e^{-(x-x_0)^2/(4\delta x^2)} e^{-(x-x'_0)^2/(4\delta x^2)}$$

$$\simeq \mathcal{N}^2 \frac{1}{L} \int_{-L/2}^{L/2} dx\, u^*_{nk_0}(x)\, u_{n'k'_0}(x) \int_{-L/2}^{L/2} dx\, e^{-ix(k_0-k'_0)} e^{-(x-x_0)^2/(4\delta x^2)} e^{-(x-x'_0)^2/(4\delta x^2)}$$

$$= \delta_{nn'} \frac{\mathcal{N}^2}{L} \int_{-L/2}^{L/2} dx\, u^*_{nk_0}(x)\, u_{nk'_0}(x) \int_{-L/2}^{L/2} dx\, e^{-ix(k_0-k'_0)} e^{-(x-x_0)^2/(4\delta x^2)} e^{-(x-x'_0)^2/(4\delta x^2)}, \quad (8.86)$$

where in the last equality we have used that u functions corresponding to different bands are orthogonal. The Gaussian integral is immediately evaluated with sufficient accuracy, giving

$$\langle n x_0 k_0 \mid n' x'_0 k'_0 \rangle \simeq \delta_{nn'} |\mathcal{N}|^2 \sqrt{2\pi \delta x^2}\, e^{-(x_0-x'_0)^2/(8\delta x^2)} e^{-(k_0-k'_0)^2/(8\delta k^2)}$$
$$\times e^{-(i/2)(k_0-k'_0)(x_0+x'_0)} \frac{1}{L} \int_{-L/2}^{L/2} dx\, u^*_{nk_0}(x)\, u_{nk'_0}(x). \quad (8.87)$$

Since the Gaussian function is only non-vanishing for $k'_0 - k_0 \sim \delta k \ll 2\pi/a$ and the u function depends smoothly on k'_0, we can substitute $k'_0 \to k_0$ in $u_{nk'_0}$, giving

$$\langle n x_0 k_0 \mid n' x'_0 k'_0 \rangle \simeq \delta_{nn'}\, a\, e^{-(x_0-x'_0)^2/(8\delta x^2)} e^{-(k_0-k'_0)^2/(8\delta k^2)} e^{-(i/2)(k_0-k'_0)(x_0+x'_0)}$$
$$\times \frac{1}{L} \int_{-L/2}^{L/2} dx\, u^*_{nk_0}(x)\, u_{nk_0}(x)$$
$$= \delta_{nn'}\, e^{-(x_0-x'_0)^2/(8\delta x^2)} e^{-(k_0-k'_0)^2/(8\delta k^2)} e^{-(i/2)(k_0-k'_0)(x_0+x'_0)}, \quad (8.88)$$

the last equality following from the unit cell normalization choice, Eq. (8.10). The Bloch wave packet states are thus only orthogonal if they refer to different bands. Within a band, the Bloch wave packets are, in view of Eq. (8.88), an over-complete set, which just like the Gaussian wave packets can be pruned to completeness. As shown in Appendix M, to each (x, k) cell of size $\Delta x_c\, \Delta k_c = 2\pi$ is associated a Bloch wave packet $\psi^{(n)}_{x_0 k_0}$, Eq. (8.78), corresponding to the lattice point (x_0, k_0) (and band) in question, a lattice Bloch Gaussian state.

Exercise 8.4 Show that

$$-\frac{\hbar^2}{2m} \frac{\partial^2 \psi^{(n)}_{x_i k_i}(x)}{\partial x^2} = -\frac{\hbar^2}{2m} \mathcal{N} e^{-(x-x_i)^2/(4\delta x^2)} \frac{\partial^2 \psi_{nk_i}(x)}{\partial x^2} + \frac{\hbar^2}{4m\delta x^2} \psi^{(n)}_{x_i k_i}(x)$$
$$+ \frac{\hbar^2}{2m} \frac{x - x_i}{\delta x^2} \mathcal{N} e^{-(x-x_i)^2/(4\delta x^2)} \frac{\partial \psi_{nk_i}(x)}{\partial x}$$
$$- \frac{\hbar^2}{2m} \frac{(x - x_i)^2}{4\delta x^4} \psi^{(n)}_{x_i k_i}(x) \quad (8.89)$$

and thereby that (with $V_p(x)$ being the periodic potential in question)

$$\left(-\frac{\hbar^2}{2m}\frac{\partial^2}{\partial x^2}+V_p(x)\right)\psi^{(n)}_{x_ik_i}(x)=\epsilon_n(k_i)\,\psi^{(n)}_{x_ik_i}(x)+\frac{\hbar^2}{4m\delta x^2}\,\psi^{(n)}_{x_ik_i}(x)$$
$$+\frac{\hbar^2}{2m}\frac{x-x_i}{\delta x^2}\,\mathcal{N}e^{-(x-x_i)^2/(4\delta x^2)}\frac{\partial\psi_{nk_i}(x)}{\partial x}$$
$$-\frac{\hbar^2}{2m}\frac{(x-x_i)^2}{4\delta x^4}\,\psi^{(n)}_{x_ik_i}(x). \tag{8.90}$$

8.10 Local Fermi Sea

The Slater determinants built by lattice Bloch Gaussian wave packets are a complete set of many-body basis states for electrons in a periodic potential, and their properties can be assessed by the same arguments used in the free particle case in Chapter 4, in fact, taken literally over.

The average energy in a Bloch wave packet state is, according to Eq. (8.90),

$$\int_{-L/2}^{L/2} dx\,\psi^{(n)*}_{x_0k_0}(x)\,\hat{H}\,\psi^{(n)}_{x_0k_0}(x)$$
$$=\epsilon_n(k_0)+\frac{\hbar^2}{4m\delta x^2}+\frac{\hbar^2}{2m}|\mathcal{N}|^2\int_{-L/2}^{L/2}dx\,\frac{x-x_0}{\delta x^2}\,e^{-(x-x_0')^2/(2\delta x^2)}\psi^*_{x_0k_0}(x)\,\frac{\partial\psi_{x_0k_0}(x)}{\partial x}$$
$$-\frac{\hbar^2}{2m}|\mathcal{N}|^2\int_{-L/2}^{L/2}dx\,\left(\frac{x-x_0}{2\delta x^2}\right)^2 e^{-(x-x_0')^2/(2\delta x^2)}|\psi_{x_0k_0}(x)|^2, \tag{8.91}$$

which upon using the integral doubling formula gives (the third term on the right-hand side then vanishes due to the oddness of the Gaussian term)

$$\int_{-L/2}^{L/2} dx\,\psi^{(n)*}_{x_0k_0}(x)\,\hat{H}\,\psi^{(n)}_{x_0k_0}(x)$$
$$\simeq\epsilon_n(k_0)+\frac{\hbar^2}{4m\delta x^2}-\frac{\hbar^2}{2m}|\mathcal{N}|^2\int_{-L/2}^{L/2}dx\left(\frac{x-x_0}{2\delta x^2}\right)^2 e^{-(x-x_0)^2/(2\delta x^2)}\frac{1}{L}\int_{-L/2}^{L/2}dx\,|u_{nk_0}(x)|^2$$
$$=\epsilon_n(k_0)+\frac{\hbar^2}{4m\delta x^2}-\frac{\hbar^2}{8m\delta x^2}$$

$$= \epsilon_n(k_0) + \frac{\hbar^2}{8m\delta x^2}$$
$$= \epsilon_n(k_0) + \frac{\hbar^2 \delta k^2}{2m}. \tag{8.92}$$

Shifting the Hamiltonian by a constant term, the average energy in the Bloch wave packet state is thus equal to the corresponding Bloch energy $\epsilon_n(k_0)$, and independent of the envelope location x_0.

The local Fermi sea of Bloch wave packets can now be established analogously to the Sommerfeld free particle case. The periodic potential is of length L, so the phase or (x, k) space is the square of positions from $-L/2$ to $L/2$ and Bloch numbers in the first Brillouin zone, i.e. k values from $-\pi/a$ to π/a. The state labels reside on a lattice (x_0, k_0), each cell of which being of size $\Delta x_c \, \Delta k_c = 2\pi$. The state of lowest average energy is obtained by filling up Bloch wave packet states of increasing energy, which means states of Bloch label numbers of increasing magnitude. Denote by k_0 the Bloch number of the populated Bloch wave packet that is maximum in energy. The number of boxes in the k direction is then $2k_0/\Delta k_c$, and in the x direction the number of boxes is $L/\Delta x_c$. Consider first the case of one electron per unit cell, i.e. there are $N = L/a$ electrons to accommodate. Then, due to spin degeneracy,

$$\frac{N}{2} = \frac{L}{\Delta x_c} \frac{2k_0}{\Delta k_c} = \frac{L}{\pi} k_0 \tag{8.93}$$

or

$$k_0 = \frac{N}{2} \frac{\pi}{L} = \frac{\pi}{2a} = \frac{\pi}{2} n, \tag{8.94}$$

which equals the Fermi wave vector, as the ground state of the Bloch energy eigenstates corresponds to the band being half-filled, $k_0 = k_{\mathrm{F}} = \pi/2a$. The Fermi energy is determined by the Fermi wave vector, $\epsilon_{\mathrm{F}} = \epsilon_1(\pi/2a)$.

The number of states, $N_{\Delta x_c}$, in volume Δx_c is obtained by counting the number of Δk_c boxes,

$$N_{\Delta x_c} = 2\frac{2k_0}{\Delta k_c}, \tag{8.95}$$

the prefactor accounting for spin degeneracy, and the density of electron states in the cell volume is

$$n_{\Delta x_c} = \frac{N_{\Delta x_c}}{\Delta x_c} = 2\frac{2k_0}{\Delta k_c \, \Delta x_c} = \frac{1}{a} = \frac{N}{L}. \tag{8.96}$$

The local Fermi sea (or column) of electron states thus has the same density as N electrons in the interval L, $n_{\Delta x_c} = N/L = n$.

Since

$$\sum_i \theta(\epsilon_{\mathrm{F}} - \epsilon_1(k_i)) = \#\text{ of } k\text{-space cells in the local Fermi sea}$$
$$= \frac{2k_{\mathrm{F}}}{\Delta k_i}, \tag{8.97}$$

the sum over all occupied phase space cells gives ($\Delta k_c \equiv \Delta k_i$)

$$\sum_c \frac{\Delta x_c \, \Delta k_c}{2\pi} \theta(\epsilon_F - \epsilon_1(k_c)) = L \frac{\Delta k_i}{2\pi} \sum_i \theta(\epsilon_F - \epsilon_1(k_i)), \tag{8.98}$$

as each spatial cell gives the same contribution, and the space cell intervals, Δx_c, add up to the system length L. According to Eq. (8.97) the expression equals

$$\sum_c \frac{\Delta x_c \, \Delta k_c}{2\pi} \theta(\epsilon_F - \epsilon_1(k_c)) = L \frac{\Delta k_i}{2\pi} \frac{2k_F}{\Delta k_i} = \frac{N}{2}. \tag{8.99}$$

The counting of occupied states in the Bloch wave packet *ground* state can therefore be rephrased in terms of the density of electrons, n, as the density is equal to the sum over occupied phase space cells according to

$$\begin{aligned} n &= \frac{2}{L} \sum_c \frac{\Delta x_c \, \Delta k_c}{2\pi} \theta(\epsilon_F - \epsilon_1(k_c)) \\ &= 2 \sum_i \frac{\Delta k_i}{2\pi} \theta(\epsilon_F - \epsilon_1(k_i)). \end{aligned} \tag{8.100}$$

The above consideration was for one electron per unit cell, the Bloch metal, where the lowest band is half-filled. When, say, the lowest band is completely filled and the second band is half-filled, the Fermi energy is $\epsilon_F = \epsilon_2(\pi/2a)$. The filled band is inert and the density of Bloch wave packet electrons in the second band is the conduction electron density.

8.11 Bloch Wave Packet Density of States

Next we want to obtain, for given spin, the density of Bloch wave packet states (BWPS) with average energy in a small energy interval $\Delta\epsilon$. Introducing the number, $N(\epsilon, \epsilon + \Delta\epsilon)$, of Bloch wave packet states with average energy between ϵ and $\epsilon + \Delta\epsilon$,

$$\begin{aligned} N(\epsilon, \epsilon + \Delta\epsilon) &\equiv \text{\# of BWPS with average energy between } \epsilon \text{ and } \epsilon + \Delta\epsilon \\ &= \text{\# of BWPS with momentum labels between } k \text{ and } k + \Delta k \\ &\quad + \text{\# of BWPS with momentum labels between } -k \text{ and } -k - \Delta k, \end{aligned} \tag{8.101}$$

where

$$\epsilon = \epsilon_n(k), \qquad \epsilon + \Delta\epsilon \simeq \epsilon_n(k) + \frac{\partial \epsilon_n(k)}{\partial k} \Delta k = \epsilon + \frac{\partial \epsilon_n(k)}{\partial k} \Delta k. \tag{8.102}$$

Since in the one-dimensional case the bands do not overlap, the magnitude of the energy by itself locates the band in question. Equation (8.101) can be rewritten as

$$N(\epsilon, \epsilon + \Delta\epsilon) = 2 \times (\text{\# of BWPS with Bloch numbers between } k \text{ and } k + \Delta k), \tag{8.103}$$

where the factor of 2 counts the two Bloch lattice numbers $\pm k$ corresponding to the same energy value, $\epsilon_n(\pm k) = \epsilon(k) = \epsilon$. For each spatial cell Δx_c, there are $\Delta k/\Delta k_c$ Bloch wave packet states with Bloch numbers between k and $k + \Delta k$, and therefore $2\,\Delta k/\Delta k_c$ Bloch wave packet states with energies between ϵ and $\epsilon + \Delta\epsilon$. Since there are $L/\Delta x_c$ spatial cells, Eq. (8.103) becomes

$$N(\epsilon, \epsilon + \Delta\epsilon) = \frac{2\,\Delta k}{\Delta k_c}\frac{L}{\Delta x_c}. \tag{8.104}$$

The density of Bloch wave packet states per unit length of the system and per unit energy is thus

$$N(\epsilon) = \frac{N(\epsilon, \epsilon + \Delta\epsilon)}{\Delta\epsilon\, L} = \frac{2\,\Delta k}{\Delta\epsilon}\frac{1}{\Delta x_c\,\Delta k_c} = \frac{2\,\Delta k}{\Delta\epsilon}\frac{1}{2\pi}. \tag{8.105}$$

Since, according to Eq. (8.102),

$$\frac{\Delta k}{\Delta\epsilon} = \frac{1}{|\partial\epsilon_n(k)/\partial k|}, \tag{8.106}$$

the Bloch wave packet density of states in one dimension is

$$N(\epsilon) = \frac{1}{\pi\hbar v_n(\epsilon)}, \qquad v_n(\epsilon) \equiv \frac{1}{\hbar}\left|\frac{\partial\epsilon_n(k)}{\partial k}\right|\Bigg|_{\epsilon_n(k)=\epsilon}, \tag{8.107}$$

with $v_n(\epsilon)$ the magnitude of the average velocity in Bloch state nk.

The density of states for the Bloch energy eigenstates can be obtained using the Born–von Karman boundary condition. The number of Bloch energy eigenstates in energy interval $\Delta\epsilon$ is then

$$N(\epsilon, \epsilon + \Delta\epsilon) = \frac{2\,\Delta k}{\Delta k_L}, \qquad \Delta k_L = \frac{2\pi}{L}, \tag{8.108}$$

which according to Eq. (8.105) then gives the same density of states as for Bloch wave packet states.

Despite the different way of counting states, the density of Bloch energy states (BES) equals the density of Bloch wave packet states (BWPS) for the following reason. Consider the energy interval $\Delta\epsilon$ with its corresponding associated doublet Bloch number interval Δk. Bloch number values for Bloch energy states are macroscopically close, their separation being $\Delta k_L = 2\pi/L$, and

$$\text{\# of BES with energy in interval } \Delta\epsilon = \frac{2\,\Delta k}{\Delta k_L} = \frac{L}{\pi}\,\Delta k = \frac{L}{\pi}\frac{\Delta\epsilon}{\hbar v(\epsilon)}. \tag{8.109}$$

The Bloch number labels of the Bloch wave packet states are less dense, their separation being the distance between neighboring Bloch number cells, $\Delta k_c = 2\pi/\Delta x_c$, but the Bloch wave packet states are replicated $L/\Delta x_c$ times, corresponding to the different spatial cells, so that

$$\text{\# of BWPS with average energy in interval } \Delta\epsilon = \frac{2\,\Delta k}{\Delta k_c}\,\frac{L}{\Delta x_c} = \frac{L}{\pi}\,\frac{\Delta\epsilon}{\hbar v(\epsilon)}. \tag{8.110}$$

The macroscopic degeneracy in location of Bloch wave packet states thus compensates to give exactly the same density of states as for Bloch energy eigenstates.

8.12 Equilibrium State

The significance of the Bloch wave packet *ground* state is not at this point clear, but it will be shown to serve as reference state. We would like to obtain a description of thermal equilibrium in terms of the Gaussian Bloch states. However, from the statistical argument in Appendix K for obtaining the Fermi–Dirac distribution, energy eigenstates are called for, which the Gaussian Bloch states are not. To the rescue comes the degeneracy of the Bloch metal, as only states near the Fermi energy, as in the Sommerfeld case, contribute to physical quantities.

Consider a Bloch Gaussian state labeled by (n, x_i, k_i). Such a state is approximately an energy eigenstate if the spatially dependent terms are negligible compared to the $\epsilon_n(k_i)\,\psi^{(n)}_{x_i k_i}(x)$ term in Eq. (8.90). Since the spatial extension of the Gaussian function is on the order of its variance, this requires $\hbar^2/4m\delta x^2 \ll \epsilon_n(k_i)$, or equivalently $\hbar^2\delta k^2/m \ll \epsilon_n(k_i)$. Since δk is much smaller than the size of the Brillouin zone, $\delta k \ll \pi/a$, and the Fermi wave number is $k_{\rm F} = \pi/2a$, estimation using $\epsilon_n(k_i) \sim \hbar^2 k_i^2/m$ gives that the criterion is well satisfied for Bloch numbers near the Fermi wave number. Bloch Gaussian states with Bloch numbers on the order of (or larger than) the Fermi wave number, $|k_i| \sim k_{\rm F}$, can thus be considered approximate eigenstates of the periodic potential Hamiltonian

$$\left(-\frac{\hbar^2}{2m}\frac{\partial^2}{\partial x^2} + V_p(x)\right)\psi^{(n)}_{x_i,k_i}(x) \simeq \epsilon_n(k_i)\,\psi^{(n)}_{x_i,k_i}(x). \tag{8.111}$$

The energy weight counting argument in Appendix K leading to the thermal equilibrium distribution, the Fermi–Dirac distribution, involves the energy eigenvalues of the Hamiltonian. However, for energy intervals near the Fermi energy, the counting is indifferent to using the energies associated with the exact Bloch energy states or the Bloch Gaussian approximate energy eigenstates, as their energy values and density of states are the same. When using the Bloch Gaussian states, the counting is done for the states in each cell interval Δx_c, for each local Fermi sea. The counting argument in Appendix K thus associates the Bloch Gaussian state labeled by (n, x, k), where $|k| \sim p_{\rm F}$, with the probability of being occupied as specified by the Fermi function,

$$f_n(x,k) = \frac{1}{e^{(\epsilon_n(k)-\epsilon_{\rm F})/kT} + 1} = f(\epsilon_n(k)), \tag{8.112}$$

the Fermi distribution function for Bloch Gaussian states.

Summing the distribution function in Eq. (8.112) over all phase space cells gives

$$\frac{2}{L}\sum_c f_n(x_c,k_c) = \frac{2}{L}\sum_c \frac{\Delta x_c\,\Delta k_c}{2\pi}\,\frac{1}{e^{(\epsilon_n(k_c)-\epsilon_{\rm F})/kT}+1} = 2\int_0^\infty d\epsilon\; N_n(\epsilon) f(\epsilon) = n, \tag{8.113}$$

as the phase space cell summation according to Section 8.11 can be turned into energy integration, with $N_n(\epsilon)$ being the density of states, and therefore gives the density for the band in question. Each spatial cell, Δx_c, gives the same contribution, and since the sum of space cell intervals, Δx_c, adds up to the system length L, Eq. (8.113) can be rewritten as

$$\frac{2}{L}\sum_c f_n(x_c,k_c) = 2\sum_{k_i} \frac{\Delta k_c}{2\pi}\,\frac{1}{e^{(\epsilon_n(k_i)-\epsilon_{\rm F})/kT}+1}. \tag{8.114}$$

In the following, we assume the nth band half-filled at zero temperature, this band thus providing the conduction electron density of the Bloch metal. We would like to turn the summation into an integral, thus rewriting the formula for the density provided by the nth band, Eq. (8.113), as

$$n = 2\int \frac{dk}{2\pi}\,\frac{1}{e^{(\epsilon_n(k)-\epsilon_{\rm F})/kT}+1}, \tag{8.115}$$

even when thinking in terms of occupied Bloch Gaussian states. However, such a conversion is valid only for Bloch numbers large compared to the size of Δk_c, i.e. $|k| \gg \Delta k_c$, such that Δk_c can be considered infinitesimal. However, we can safely write (recall Eq. (8.100))

$$0 = 2\int \frac{dk}{2\pi}\left(\frac{1}{e^{(\epsilon_n(k)-\epsilon_{\rm F})/kT}+1} - \theta(\epsilon_{\rm F}-\epsilon_n(k))\right) \tag{8.116}$$

and interpret the distribution functions as denoting occupation of Gaussian Bloch states in a given spatial cell since the integrand is only non-zero in the thin thermal layer around the Fermi surface. The formula (8.115) is kept but with the proviso that a subtraction procedure must be employed in order that the interpretation is only applied to Bloch Gaussian states that are approximate energy eigenstates, i.e. to states near the Fermi surface.

If the electrons experience an additional applied potential, $V(x)$, which varies slowly on the scale of the position variance of the Bloch Gaussian state, the Bloch Gaussian wave packet satisfies with high accuracy

$$V(x)\,\psi^{(n)}_{x_i,k_i}(x) \simeq V(x_i)\,\psi^{(n)}_{x_i,k_i}(x) \tag{8.117}$$

and the Bloch Gaussian state for $k_i \sim k_{\rm F}$ is an approximate eigenstate for the Bloch electron in the applied potential

$$\left(-\frac{\hbar^2}{2m}\frac{\partial^2}{\partial x^2} + V_p(x) + V(x)\right)\psi^{(n)}_{x_i,k_i}(x) \simeq (\epsilon_n(k_i)+V(x_i))\,\psi^{(n)}_{x_i,k_i}(x) \tag{8.118}$$

corresponding to energy eigenvalue $E_n(x_i, k_i) = \epsilon_n(k_i) + V(x_i)$. The counting argument in Appendix K then gives that the probability that the Bloch Gaussian state $\psi^{(n)}_{x,k}$ is occupied is

$$f_n(x,k) = \frac{1}{e^{(\epsilon_n(k)+V(x)-\epsilon_F)/kT} + 1}, \tag{8.119}$$

and the occupation of each local Fermi sea is given by the local Fermi function.

The argument for expressing the density in terms of populated Bloch Gaussian states can now be taken over from Section 4.5, and the thermal equilibrium conduction electron density becomes, in the presence of an applied potential,

$$n(x) = 2\int \frac{dk}{2\pi} \frac{1}{e^{(\epsilon_n(k)+V(x)-\epsilon_F)/kT} + 1}, \tag{8.120}$$

of course again with the proviso that the formula is only used for the part of the Bloch energy spectrum for which the Bloch Gaussian states are approximate energy eigenstates.

The discussion of impurity screening in the Sommerfeld metal in Section 4.6 can now be carried over verbatim for the Bloch metal, the role of momentum labeling of energies being taken by the Bloch number or so-called crystal momentum.

8.13 Bloch Wave Packet Dynamics

Consider the Bloch wave packet superposition

$$\psi^{(n)}_{x_0k_0}(x,t) = \int dk\; g(k)\, \psi_{nk}(x)\, e^{-(i/\hbar)\epsilon_n(k)t} = \int dk\; g(k)\, u_{nk}(x)\, e^{ikx}\, e^{-(i/\hbar)\epsilon_n(k)t}, \tag{8.121}$$

which clearly is a solution to the Schrödinger equation for the particle in the periodic potential. The weight function is taken very peaked at k_0, so that

$$\psi^{(n)}_{x_0k_0}(x,t) \simeq u_{nk_0}(x) \int dk\; g(k)\, e^{ikx}\, e^{-(it/\hbar)[\epsilon_n(k_0)+(k-k_0)\partial\epsilon_n(k_0)/\partial k]}$$

$$= u_{nk_0}(x)\, e^{ik_0x}\, e^{-(i/\hbar)\epsilon_n(k_0)t} \int dk\; g(k)\, e^{ix(k-k_0)}\, e^{-it(k-k_0)v_n(k_0)}. \tag{8.122}$$

Employing the Gaussian weight

$$g_{x_0k_0}(k) = \mathcal{N}\, e^{-(k-k_0)^2/(4\delta k^2) - i(k-k_0)x_0}, \tag{8.123}$$

the Gaussian integral is immediately approximately performed, giving

$$\psi^{(n)}_{x_0k_0}(x,t) = \mathcal{N}\, \psi_{nk_0}(x)\, e^{-(i/\hbar)\epsilon_n(k_0)t}\, e^{-(x-x_0-v_n(k_0)t)^2/(4\delta x^2)}, \tag{8.124}$$

where $\mathcal{N}$ is the normalization factor, $\mathcal{N} = (2\pi\delta x^2)^{-1/4} a^{1/2}$, and $\delta x^2 = 1/4\delta k^2$. The Schrödinger dynamics thus makes the Bloch wave packet envelope move with the Bloch velocity. The constructed solution, at time $t = 0$, is the Bloch Gaussian state labeled by (nx_0k_0).

The current associated with the Bloch wave packet $\psi^{(n)}_{x_0k_0}(x,t)$ is

$$S^{(n)}_{x_0k_0}(x,t) = \frac{\hbar}{2im}\,\psi^{(n)*}_{x_0k_0}(x,t)\,\frac{\partial}{\partial x}\,\psi^{(n)}_{x_0k_0}(x,t) + \text{c.c.} \tag{8.125}$$

The term coming from differentiating the Gaussian part is purely imaginary and therefore canceled by the complex conjugation term, giving

$$\begin{aligned} S^{(n)}_{x_0k_0}(x,t) & \\ &= \mathcal{N}^2\, e^{-(x-x_0-v_n(k_0)t)^2/(2\delta x^2)}\,\frac{\hbar}{2im}\left(\psi^*_{nk_0}(x)\,\frac{\partial}{\partial x}\,\psi_{nk_0}(x) - \psi_{nk_0}(x)\,\frac{\partial}{\partial x}\,\psi^*_{nk_0}(x)\right) \\ &= \mathcal{N}^2\, e^{-(x-x_0-v_n(k_0)t)^2/(2\delta x^2)}\,\frac{1}{a\hbar}\frac{\partial \epsilon_n(k_0)}{\partial k}, \end{aligned} \tag{8.126}$$

where the last equality follows from Eq. (8.13), and thereby

$$S^{(n)}_{x_0k_0}(x,t) = v_n(k_0)\,\frac{1}{\sqrt{2\pi\delta x^2}}\,e^{-(x-x_0-v_n(k_0)t)^2/(2\delta x^2)}. \tag{8.127}$$

The probability current in a Bloch Gaussian state is thus determined by the velocity of the corresponding Bloch state (analogous to the free particle case, Eq. (4.106)).

The average current for a Bloch metal can now be expressed in terms of the Bloch Gaussian states in a similar fashion as for the Sommerfeld case of Section 4.8.3, and a simple picture of the Bloch electron gas appears: the electrons can be thought of as occupying different Bloch Gaussian states as dictated by the physical situation in question, each such adding its contribution to the average current. Each cell volume in real space, Δx_c, contains a local Fermi sea of independent electrons occupying Bloch Gaussian states with different Bloch lattice numbers describing electrons moving in either of the two directions with different velocities.

The above discussion referred explicitly to the one-dimensional case, but is straightforwardly generalized to two or three dimensions, and the quasi-classical picture of Bloch electron dynamics emerges: each cell volume in real space of a Bloch electron gas, $\Delta \mathbf{x}_c$, contains independent electrons occupying Bloch Gaussian states with different Bloch lattice vectors describing Bloch electrons moving in all directions, as depicted in Figure 4.2. The evolved quasi-classical picture of Bloch Gaussian states describing the motion of Bloch electrons in a metal is thus the same as that provided by classical physics, except for the feature that, instead of classical particles moving on straight lines in between scattering, quantum particles in Gaussian Bloch wave packets have their centers move as dictated by classical mechanics.

Exercise 8.5 Calculate, as in Section 4.8.3, the Drude conductivity for a Bloch metal.

8.14 Summary

The energy eigenstates of a particle in a periodic potential are in general current-carrying, similar to free particles but with non-quadratic dispersion. The current response of a Bloch electron to an applied electric field is for very strong fields determined by interband transitions, whereas for typical field strengths it is determined by Bragg reflection. In the latter case, the total current produced by the electrons in a crystal depends on whether its energy bands are only partially filled since filled bands are not current-carrying, and this leads to understanding the electric classification of crystalline solids.

The exact energy eigenstates of Bloch electrons in the presence of external fields, which determine the current of a system in a non-equilibrium steady state, are not available. A different approach is therefore called for to cope with such situations, especially relevant later for understanding the electric current properties of semiconductors. The complete set of Gaussian Bloch states were therefore constructed and used to determine density and current expressions.

9 Crystalline Solids

A crystalline solid, such as a metal or a semiconductor, is an assembly of a huge number of nuclei and electrons. From a first principles point of view, the dynamics of such a system constitutes an unsolvable many-body problem, as the number of particles involved is astronomical, on the order of 10^{23}. We shall therefore be interested in an approximate description of the system relevant for the particular type of phenomenon of interest.

X-ray and electron diffraction experiments reveal the periodic arrangement of matter in, for example, a metal, i.e. its crystal structure. The electric charge of the atoms in the metal is separated spatially into two: a regular array of ions; and conduction electrons. The former are represented by the nuclei and the tightly bound electrons (core electrons concentrated in spatially well-localized regions almost as in the isolated atom), and the latter by the atomic valence electrons, which in a metal have their probability density spread almost evenly throughout the solid, thereby providing the *glue* that prevents the positively charged ions from being pushed apart. The description of crystalline solids takes as its starting point the idealization of a perfect crystal. This is a point of view well taken, as the characteristic feature of a perfect crystal, its periodic structure, sets profound constraints on the properties of the electrons in real-world non-ideal crystalline solids.

9.1 Mean Field Potential

The Hamiltonian for the conduction electrons of a crystal, relevant for our purposes, is

$$\hat{H} = \sum_{i_e} -\frac{\hbar^2 \Delta_{i_e}}{2m} + \frac{1}{2} \sum_{i_e \neq j_e} \frac{e^2}{4\pi\epsilon_0 |\mathbf{r}_{i_e} - \mathbf{r}_{j_e}|} + \sum_{n,i_e} V_{i-e}(\mathbf{r}_n - \mathbf{r}_{i_e}). \tag{9.1}$$

The ions are assumed structureless charged point particles fixed in space at regular positions $\mathbf{r}_n$.[1] The first and second terms in Eq. (9.1) are the kinetic energy term of the conduction electrons and their mutual Coulomb repulsion, and the last term is the screened conduction electron–ion interaction.[2] The Coulomb interaction between the electrons of course makes an exact analytical treatment impossible.

[1] Taking into account their kinetic energy leads to quantized collective motion of the lattice, which at finite temperature leads to scattering of electrons.

[2] We should of course treat all the electrons identically, i.e. regularly spaced nuclei and *all* the electrons. However, since the core electrons are localized, this will not be necessary, as our interest is electronics, which is manifested by electrons of highest energy, and not, for example, core spectroscopy. The core electrons, in the periodic potential description, give rise to low-lying flat tight-binding bands. The localized core electrons have

The defining property of a perfect infinite crystal is its periodicity: when viewed from a set of certain points, the positions of the ions in the crystal appear located at the same relative positions, i.e. from these *equivalent* points in space the crystal looks exactly the same (a simple example is depicted in Figure 9.1). The ionic potential experienced by an electron at two equivalent points, separated by a vector **R**, is thus identical,

$$\sum_n V_{i-e}(\mathbf{r}_n - (\mathbf{r}_{i_e} + \mathbf{R})) = \sum_n V_{i-e}(\mathbf{r}_n - \mathbf{r}_{i_e}). \tag{9.2}$$

If not for the electron–electron interaction, the electrons would experience a periodic potential.

The Hamiltonian describing the electrons in the crystal, Eq. (9.1), is invariant with respect to a translation of all the electrons by any vector that respects the symmetry of the crystal, i.e. the Hamiltonian commutes with this operation. Translating only one of the electrons is expected to be an almost commuting operation in view of the fact that only $N-1$ of the Coulomb terms violate the symmetry out of the total number of $N(N-1)$ terms, N being the number of electrons, which for a crystal is large (infinite for an infinite crystal). In view of this, it is assumed that the effect of the Coulomb potential from the myriad of other electrons can be described by an average potential. Since the mutual electronic Coulomb interaction is translation-invariant, this average potential should be endowed with the symmetry property of the ionic lattice: at sites from which the crystal looks the same, the average potential takes on the same value. A conduction electron is thus assumed to experience an average potential, including now also the potential from the ions, with the periodicity property

$$V(\mathbf{r}+\mathbf{R}) = V(\mathbf{r}), \tag{9.3}$$

where **R** denotes any vector that connects equivalent points in the crystal. A two-dimensional rectangular lattice is depicted in Figure 9.1, in which case **R** can be any multiple and sum of vectors formed by the vectors $\mathbf{a}_1$ and $\mathbf{a}_2$. The considered approximation is referred to as the mean field description, but in all honesty we are now studying a different problem, the hope being that the perturbation of the real intractable many-body problem, $V_{\mathrm{mb}} - V$, is qualitatively irrelevant and even quantitatively negligible.[3]

The intractable N-body problem has thus been traded for N one-body problems of a particle moving in a mean field potential V,

$$\hat{H}_{\mathrm{MF}} = -\frac{\hbar^2}{2m}\Delta_{\mathbf{r}} + V(\mathbf{r}). \tag{9.4}$$

To be studied is the time-independent Schrödinger equation

$$\left(-\frac{\hbar^2}{2m}\Delta_{\mathbf{r}} + V(\mathbf{r})\right)\psi(\mathbf{r}) = \epsilon\,\psi(\mathbf{r}) \tag{9.5}$$

little contribution to the screening of the Coulomb interaction of the conduction electrons. In elements with heavy nuclei, the relativistic spin–orbit interaction needs, for certain properties, to be included.

[3] A more sophisticated handling of the electron–electron interaction is the Landau–Silin version of the phenomenological Fermi-liquid theory. However, as to be anticipated by the screening of the long-range Coulomb interaction, the effective electron–electron interaction becomes weak and leads to insignificant corrections to the transport phenomena of interest.

for a particle in a periodic potential, i.e. the potential has the property (9.3) for an arbitrary crystal structure. But before that, the mathematical formalism describing three-dimensional crystal structures is introduced.[4]

9.2 Direct and Reciprocal Lattices

A crystal consists of identical repetition of the same atomic arrangement. The periodicity of a perfect infinite crystal is specified by its Bravais lattice, defined as an array of points having the following property: viewed from any of the points in the Bravais lattice, $\mathcal{L}$, the crystal looks the same,

$$\mathcal{L} = \{\mathbf{R} \mid \text{for which the crystal looks the same from points } \mathbf{r} \text{ and } \mathbf{r} + \mathbf{R}\,\}. \tag{9.6}$$

The Bravais lattice specifies how a basic atomic arrangement generates a periodic structure by Bravais lattice translations. Crystal structure complexity ranges from the simplest case of single identical ions located at the Bravais lattice points, to complicated arrangement of atoms, ions and molecules, a lattice with a so-called basis, being periodically repeated as determined by the Bravais lattice.

Imagine a two-dimensional crystal where identical ion locations are on a rectangular lattice as depicted in Figure 9.1. The infinite crystal looks the same from points that can be connected by integer steps of the vectors $\mathbf{a}_1$ and $\mathbf{a}_2$ connecting nearest ions.

In general, a Bravais lattice can be specified by a set of so-called primitive vectors, $\{\,\mathbf{a}_i\}_i$, and integers. A three-dimensional Bravais lattice is specified by the points

$$\mathcal{L} = \{\mathbf{R_n} \mid \mathbf{R_n} = n_1\mathbf{a}_1 + n_2\mathbf{a}_2 + n_3\mathbf{a}_3;\ \mathbf{n} \equiv (n_1, n_2, n_3),\ n_1, n_2, n_3 \in \mathcal{Z}\}. \tag{9.7}$$

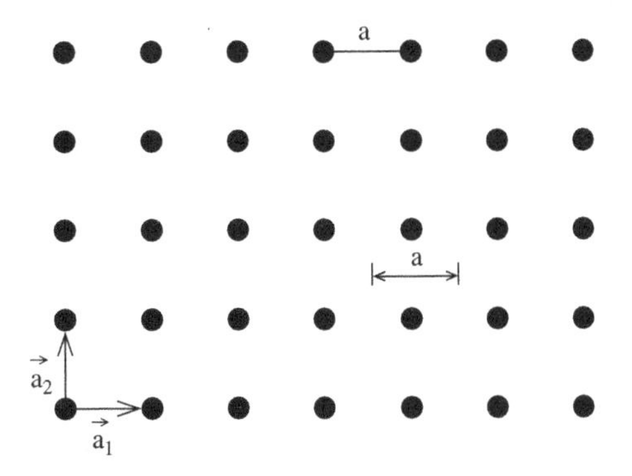

Figure 9.1 Portion of a two-dimensional rectangular Bravais lattice.

[4] The reader only interested in the essential feature of energy gaps in the spectrum of a particle in a periodic potential, has enough knowledge from the one-dimensional discussion to understand the basic properties of electron transport in crystals. Three dimensions leads in general to a diversity of possible band structures, a topic of solid-state physics.

It might not be as easy as for the quadratic lattice to see that the two different definitions of a Bravais lattice, Eqs. (9.6) and (9.7), are identical, but for the few lattices of our interest, sketching a little portion of the lattice based on the second definition, Eq. (9.7), reveals that it fulfills definition one, Eq. (9.6). There exist 14 different kinds of Bravais lattices, differing by their symmetry properties, as concluded by Bravais.

The volume spanned by the primitive vectors, $\mathbf{a}_1 \cdot (\mathbf{a}_2 \times \mathbf{a}_3)$, constitutes a choice of unit cell, i.e. a volume which, translated through all lattice vectors $\mathbf{R_n}$, covers without overlap all of space. The choice of basis vectors is not unique, but their spanned volume is. Any unit cell has the same volume as, translated by the same translations, it exactly fills the volume of the finite crystal.

Associated with any Bravais lattice is its so-called reciprocal lattice $\tilde{\mathcal{L}}$, the points or vectors, $\mathbf{K}$, satisfying the property

$$\forall\, \mathbf{R_n} \in \mathcal{L} : \mathbf{K} \cdot \mathbf{R_n} = 2\pi p_{\mathbf{n}}, \qquad p_{\mathbf{n}} \text{ integer.} \tag{9.8}$$

The sum of two reciprocal lattice vectors again satisfies Eq. (9.8), and the reciprocal lattice is therefore a vector space spanned by a basis, the set of vectors

$$\tilde{\mathcal{L}} = \{\mathbf{K_n} \mid \mathbf{K_n} = n_1\mathbf{b}_1 + n_2\mathbf{b}_2 + n_3\mathbf{b}_3,\ \mathbf{n} \equiv (n_1, n_2, n_3),\ n_1, n_2, n_3 = 0, \pm 1, \pm 2, \ldots\}. \tag{9.9}$$

A possible choice of basis vectors specifying the reciprocal lattice is

$$\begin{pmatrix} \mathbf{b}_1 \\ \mathbf{b}_2 \\ \mathbf{b}_3 \end{pmatrix} \equiv \frac{2\pi}{\mathbf{a}_1 \cdot (\mathbf{a}_2 \times \mathbf{a}_3)} \begin{pmatrix} \mathbf{a}_2 \times \mathbf{a}_3 \\ \mathbf{a}_3 \times \mathbf{a}_1 \\ \mathbf{a}_1 \times \mathbf{a}_2 \end{pmatrix}, \tag{9.10}$$

as clearly

$$\mathbf{a}_i \cdot \mathbf{b}_j = 2\pi\, \delta_{ij} \tag{9.11}$$

and the vectors in Eq. (9.9) are then precisely the vectors having property (9.8).

Exercise 9.1 Consider the simple case where the basis vectors $\mathbf{a}_1$, $\mathbf{a}_2$ and $\mathbf{a}_3$ are orthogonal, forming a rectangular parallelepiped with edges of any length. Show that the reciprocal lattice basis vectors are

$$\begin{pmatrix} \mathbf{b}_1 \\ \mathbf{b}_2 \\ \mathbf{b}_3 \end{pmatrix} \equiv 2\pi \begin{pmatrix} \mathbf{a}_1/a_1^2 \\ \mathbf{a}_2/a_2^2 \\ \mathbf{a}_3/a_3^2 \end{pmatrix}, \tag{9.12}$$

Note that this leaves the reciprocal lattice of the same type with also orthogonal basis vectors.

9.3 Bloch Functions

A real crystalline solid is finite, giving rise to deviation from periodicity.[5] Inside the crystal, the allowed wave functions are determined by the periodic potential; and outside, the wave

[5] In fact, the following discussion of the three-dimensional case is the simple generalization of the one-dimensional case presented in detail in Section 7.12.

functions of the electrons rapidly vanish. The presence of a surface of a finite crystal can for bulk properties thus be neglected, and the periodic boundary condition respecting the periodicity of the potential is conveniently employed, the Born–von Karman boundary condition,

$$\psi(\mathbf{r} + N_i\,\mathbf{a}_i) = \psi(\mathbf{r}), \qquad i = 1, 2, 3, \tag{9.13}$$

where $L_i = N_i\, a_i$ is the length of the crystal in the directions of the respective basis vectors. The finite crystal has the volume $\Omega = |N_1\mathbf{a}_1 \cdot (N_2\mathbf{a}_2 \times N_3\mathbf{a}_3)| = N|\mathbf{a}_1 \cdot (\mathbf{a}_2 \times \mathbf{a}_3)|$, the crystal volume being equal to N times the volume of the unit cell, $N = N_1N_2N_3$. A wave function satisfying the Born–von Karman boundary condition has a Fourier expansion (as discussed in Appendix B)

$$\psi(\mathbf{r}) = \sum_{\mathbf{k}\in \mathrm{BvK}} \psi_{\mathbf{k}}\, e^{i\mathbf{k}\cdot\mathbf{r}}, \tag{9.14}$$

where the $\mathbf{k}$ sum runs over the discrete set of vectors in the Born–von Karman (BvK) lattice, determined by the crystal structure,

$$\mathrm{BvK} \equiv \left\{ \mathbf{k} \,\middle|\, \mathbf{k} = \frac{n_1}{N_1}\mathbf{b}_1 + \frac{n_2}{N_2}\mathbf{b}_2 + \frac{n_3}{N_3}\mathbf{b}_3;\; n_i = 0, \pm 1, \pm 2, \ldots \right\}, \tag{9.15}$$

since the Born–von Karman boundary condition, Eq. (9.13), leads to the relation for the Fourier coefficients

$$\psi_{\mathbf{k}} = \psi_{\mathbf{k}}\, e^{iN_i\,\mathbf{k}\cdot\mathbf{a}_i}, \qquad i = 1, 2, 3, \tag{9.16}$$

and therefore only non-zero expansion coefficients, $\psi_{\mathbf{k}}$, for $\mathbf{k}$ vectors satisfying

$$e^{iN_i\,\mathbf{k}\cdot\mathbf{a}_i} = 1, \qquad i = 1, 2, 3, \tag{9.17}$$

i.e. the vector space defined by Eq. (9.15). The Born–von Karman lattice includes as a subset the reciprocal lattice, but is much denser and becomes a continuum in the limit of an infinite crystal.

The mean field crystal potential is characterized by its periodic property

$$V(\mathbf{r} + \mathbf{R_n}) = V(\mathbf{r}) \tag{9.18}$$

valid for all the Bravais lattice vectors, $\mathbf{R_n}$, since they specify the equivalent points of the crystal structure in question. Any periodic function has a Fourier series expansion

$$V(\mathbf{r}) = \sum_{\mathbf{K}} V_{\mathbf{K}}\, e^{i\mathbf{K}\cdot\mathbf{r}} \tag{9.19}$$

and the periodicity condition, Eq. (9.18), and the uniqueness of the Fourier expansion (the completeness of the plane waves) lead to the requirement on allowed expansion $\mathbf{K}$ vectors (recall the one-dimensional case discussed in Section 7.12)

$$V_{\mathbf{K}} = e^{i\mathbf{K}\cdot\mathbf{R_n}}\, V_{\mathbf{K}} \tag{9.20}$$

for all the Bravais lattice vectors $\mathbf{R_n}$. The periodic potential therefore only has non-zero Fourier expansion coefficients for $\mathbf{K}$ vectors satisfying Eq. (9.8), i.e. allowed $\mathbf{K}$ vectors are the reciprocal lattice vectors.

The rather trivial observation that the plane wave expansion of a periodic potential,

$$V(\mathbf{r}) = \sum_{\mathbf{K}\in\mathcal{L}} V_{\mathbf{K}}\, e^{i\mathbf{K}\cdot\mathbf{r}}, \tag{9.21}$$

only has components for vectors that respect the symmetry of the potential, the reciprocal lattice vectors, has, combined with the macroscopic BvK boundary condition, far-reaching consequences for the allowed energy eigenfunctions and energies for an electron in such a potential.

In terms of their Fourier expansions, the potential term in the time-independent Schrödinger equation becomes

$$V(\mathbf{r})\,\psi(\mathbf{r}) = \sum_{\mathbf{K}\in\tilde{\mathcal{L}},\,\mathbf{k}\in\text{BvK}} V_{\mathbf{K}}\,\psi_{\mathbf{k}}\,e^{i\mathbf{r}\cdot(\mathbf{k}+\mathbf{K})} = \sum_{\mathbf{K}\in\tilde{\mathcal{L}},\,\mathbf{k}'\in\text{BvK}} e^{i\mathbf{k}'\cdot\mathbf{r}}\,V_{\mathbf{K}}\,\psi_{\mathbf{k}'-\mathbf{K}}, \tag{9.22}$$

the last equality following from the substitution of $\mathbf{k}'$ for $\mathbf{k}$ according to $\mathbf{k}' = \mathbf{k} + \mathbf{K}$, which can be done since adding a reciprocal lattice vector to a BvK vector gives a BvK vector. In the following it is tacitly assumed that capital $\mathbf{K}$ are reciprocal lattice vectors and lower-case $\mathbf{k}$ belong to the BvK lattice.

The kinetic energy term takes the Fourier expansion form

$$-\frac{\hbar^2}{2m}\,\Delta_{\mathbf{r}}\,\psi(\mathbf{r}) = \sum_{\mathbf{k}\in\text{BvK}} \frac{\hbar^2\mathbf{k}^2}{2m}\,\psi_{\mathbf{k}}\,e^{i\mathbf{k}\cdot\mathbf{r}} \tag{9.23}$$

and the time-independent Schrödinger equation for a particle in a periodic potential, Eq. (9.5), thus takes the form in terms of the Fourier expansions,

$$\sum_{\mathbf{k}\in\text{BvK}} \left(\frac{\hbar^2\mathbf{k}^2}{2m}\,\psi_{\mathbf{k}} + \sum_{\mathbf{K}\in\tilde{\mathcal{L}}} V_{\mathbf{K}}\,\psi_{\mathbf{k}-\mathbf{K}}\right) e^{i\mathbf{k}\cdot\mathbf{r}} = \epsilon \sum_{\mathbf{k}\in\text{BvK}} \psi_{\mathbf{k}}\,e^{i\mathbf{k}\cdot\mathbf{r}}. \tag{9.24}$$

From the uniqueness of the Fourier expansion, or taking the scalar product with the plane wave labeled by $\mathbf{k}$ on both sides of Eq. (9.24), the Fourier coefficients of an energy eigenstate of a particle in a periodic potential are seen to satisfy the equation

$$\left(\frac{\hbar^2\mathbf{k}^2}{2m} - \epsilon\right)\psi_{\mathbf{k}} + \sum_{\mathbf{K}\in\tilde{\mathcal{L}}} V_{\mathbf{K}}\,\psi_{\mathbf{k}-\mathbf{K}} = 0. \tag{9.25}$$

Owing to the periodicity of the potential, Fourier coefficients for all BvK vectors are not involved in the expansion of an energy eigenfunction, Eq. (9.14); only those that differ, relative to a given BvK vector $\mathbf{k}$, by reciprocal lattice vectors enter. A solution of the time-independent Schrödinger equation, Eq. (9.24), is thus a superposition of plane waves containing components with $\mathbf{k}$ and $\mathbf{k}$ *plus* possibly any reciprocal lattice vector. If for a given energy value ϵ a set of amplitudes $\{\psi_{\mathbf{k}-\mathbf{K}}\}_{\mathbf{K}}$ satisfies Eq. (9.25), a solution to the time-independent Schrödinger equation, Eq. (9.5), is found as

$$\psi_{\mathbf{k}}(\mathbf{r}) = \sum_{\mathbf{K}\in\tilde{\mathcal{L}}} \psi_{\mathbf{k}-\mathbf{K}}\,e^{i(\mathbf{k}-\mathbf{K})\cdot\mathbf{r}} = e^{i\mathbf{k}\cdot\mathbf{r}} \sum_{\mathbf{K}\in\tilde{\mathcal{L}}} \psi_{\mathbf{k}-\mathbf{K}}\,e^{-i\mathbf{K}\cdot\mathbf{r}}. \tag{9.26}$$

The energy eigenvalue in question is, as dictated by the wave function, a function of the $\mathbf{k}$ vector in question, $\epsilon = \epsilon(\mathbf{k})$. The BvK vectors are also referred to as the Bloch vectors.

The solution, Eq. (9.26), corresponding to Bloch vector $\mathbf{k}$ is called a Bloch function, and has according to Eqs. (9.26) and (9.8) the property

$$\psi_{\mathbf{k}}(\mathbf{r}+\mathbf{R_n}) = e^{i\mathbf{k}\cdot\mathbf{R_n}}\psi_{\mathbf{k}}(\mathbf{r}) \tag{9.27}$$

valid for all Bravais lattice vectors $\mathbf{R_n}$ of the crystal, i.e. the probability amplitudes at *equivalent* points differ by a phase factor determined by the Bloch vector and the lattice translation vector in question, Bloch's theorem.

Taking the complex conjugate of Eq. (9.25), and using that $V^*_{\mathbf{K}} = V_{-\mathbf{K}}$ since $V(\mathbf{r})$ is a real function, the set of amplitudes $\{\psi^*_{\mathbf{k}+\mathbf{K}}\}_{\mathbf{K}}$ is seen to satisfy Eq. (9.25) for the same energy value. The function

$$\psi^*_{\mathbf{k}}(\mathbf{r}) \equiv \sum_{\mathbf{K}\in\tilde{\mathcal{L}}} \psi^*_{\mathbf{k}+\mathbf{K}}\, e^{-i\mathbf{r}\cdot(\mathbf{k}+\mathbf{K})} = e^{-i\mathbf{r}\cdot\mathbf{k}} \sum_{\mathbf{K}\in\tilde{\mathcal{L}}} \psi^*_{\mathbf{k}+\mathbf{K}}\, e^{-i\mathbf{r}\cdot\mathbf{K}} \tag{9.28}$$

is thus a solution to Eq. (9.5) corresponding to energy eigenvalue $\epsilon(\mathbf{k})$ provided the function in Eq. (9.26) is also such a solution.

The Bloch functions constitute a highly over-complete set. Shifting the Bloch vector $\mathbf{k}$ in the Bloch function, Eq. (9.26), by a reciprocal lattice vector, $\mathbf{k} \to \mathbf{k}+\mathbf{K}'$, the corresponding Bloch function transforms according to

$$\begin{aligned}\psi_{\mathbf{k}+\mathbf{K}'}(\mathbf{r}) &= \sum_{\mathbf{K}\in\tilde{\mathcal{L}}} \psi_{\mathbf{k}+\mathbf{K}'-\mathbf{K}}\, e^{i(\mathbf{k}+\mathbf{K}'-\mathbf{K})\cdot\mathbf{r}} \\ &= \sum_{\mathbf{K}''\in\tilde{\mathcal{L}}} \psi_{\mathbf{k}-\mathbf{K}''}\, e^{i(\mathbf{k}-\mathbf{K}'')\cdot\mathbf{r}} = \psi_{\mathbf{k}}(\mathbf{r})\end{aligned} \tag{9.29}$$

where the second last equality follows from the substitution $\mathbf{K}'' = \mathbf{K}-\mathbf{K}'$, simply a relabeling of the reciprocal lattice vectors. Bloch functions, Eq. (9.26), differing by a reciprocal lattice vector are thus identical, and the energy eigenvalues are therefore also periodic,

$$\epsilon(\mathbf{k}+\mathbf{K}) = \epsilon(\mathbf{k}). \tag{9.30}$$

Describing the quantum states of a particle in a periodic potential in terms of the whole over-complete set of Born–von Karman vectors is called the repeated zone scheme.

Introducing the function

$$u_{\mathbf{k}}(\mathbf{r}) \equiv \sum_{\mathbf{K}\in\mathcal{L}} \psi_{\mathbf{k}-\mathbf{K}}\, e^{-i\mathbf{K}\cdot\mathbf{r}}, \tag{9.31}$$

Bloch's theorem, or equivalently Eq. (9.26), can alternatively be stated as: the wave function of a particle in a periodic potential has the form

$$\psi_{\mathbf{k}}(\mathbf{r}) = e^{i\mathbf{k}\cdot\mathbf{r}}u_{\mathbf{k}}(\mathbf{r}), \tag{9.32}$$

where the function $u_{\mathbf{k}}$ according to Eq. (9.26) has the periodicity property of the lattice,

$$u_{\mathbf{k}}(\mathbf{r}+\mathbf{R_n}) = u_{\mathbf{k}}(\mathbf{r}). \tag{9.33}$$

The periodic part of the Bloch function, $u_{\mathbf{k}}(\mathbf{r})$, in the energy eigenfunction signifies that translation invariance is broken by the periodic potential. The energy eigenfunctions are not simply plane waves as for a free particle. The plane wave is an eigenfunction for any translation, i.e. the relation (9.27) is satisfied for any translation, and there is equal probability for the particle to be at any point for such a state of definite momentum. In the presence of a periodic potential, only a smaller group of translations is respected by the energy eigenfunctions as specified by the Bloch condition, Eq. (9.27), the translations that respect the symmetry of the periodic potential.

Observing the identity (obtained by explicitly differentiating the exponential function)

$$\left(\frac{\hbar^2}{2m}(-i\nabla_{\mathbf{r}})^2 + V(\mathbf{r})\right) e^{i\mathbf{k}\cdot\mathbf{r}} u_{\mathbf{k}}(\mathbf{r}) = e^{i\mathbf{k}\cdot\mathbf{r}} \left(\frac{\hbar^2}{2m}(-i\nabla_{\mathbf{r}} + \mathbf{k})^2 + V(\mathbf{r})\right) u_{\mathbf{k}}(\mathbf{r}), \tag{9.34}$$

the periodic part of the Bloch function is, from the time-independent Schrödinger equation,

$$\left(\frac{\hbar^2}{2m}(-i\nabla_{\mathbf{r}})^2 + V(\mathbf{r})\right) \psi_{\mathbf{k}}(\mathbf{r}) = \epsilon(\mathbf{k})\, \psi_{\mathbf{k}}(\mathbf{r}) \tag{9.35}$$

seen to be determined by the equation

$$\left\{\frac{\hbar^2}{2m}(-i\nabla_{\mathbf{r}} + \mathbf{k})^2 + V(\mathbf{r})\right\} u_{\mathbf{k}}(\mathbf{r}) = \epsilon(\mathbf{k})\, u_{\mathbf{k}}(\mathbf{r}). \tag{9.36}$$

In view of the boundary condition, Eq. (9.33), and the Bloch condition, Eq. (9.27), Eq. (9.36) need only be solved for the spatial variable $\mathbf{r}$ restricted to a unit cell. The problem is thus analogous to the *drum head* or *guitar string* problem: a wave equation with the amplitude specified at the boundary. We therefore expect, as for any type of confined waves, that for each Bloch vector $\mathbf{k}$ an infinity of solutions exists (higher harmonics), $\epsilon_n(\mathbf{k})$, $n = 1, 2, 3, \ldots$, where the label n is called the band index, with corresponding solutions $u_{n\mathbf{k}}(\mathbf{r})$. A Bloch function thus depends on the Bloch vector and a band index,

$$\psi_{n\mathbf{k}}(\mathbf{r}) = e^{i\mathbf{k}\cdot\mathbf{r}} u_{n\mathbf{k}}(\mathbf{r}), \qquad u_{n\mathbf{k}}(\mathbf{r} + \mathbf{R}) = u_{n\mathbf{k}}(\mathbf{r}), \tag{9.37}$$

and exhibits redundancy in the repeated zone scheme,

$$\psi_{n\mathbf{k}+\mathbf{K}}(\mathbf{r}) = \psi_{n\mathbf{k}}(\mathbf{r}), \qquad \epsilon_n(\mathbf{k} + \mathbf{K}) = \epsilon_n(\mathbf{k}). \tag{9.38}$$

Since the Bloch vector $\mathbf{k}$ is a parameter in the equation determining the energy eigenvalues, Eq. (9.36), the allowed energies thus form bands, $\epsilon_n(\mathbf{k})$, for each value of the band index (in the limit of an infinite crystal, $\epsilon_n(\mathbf{k})$ is for a given band index a continuous function of $\mathbf{k}$). In general, there is no reason that different bands should overlap, and regions of forbidden energies are anticipated, i.e. there can be energy gaps between different bands. This fact will have far-reaching consequences, since it will give rise to quite different transport properties of crystalline solids depending on the existence of a gap between occupied and unoccupied Bloch states in the ground state.

The *Hamiltonian*

$$\hat{H}_{\mathbf{k}} = \frac{\hbar^2}{2m}(-i\nabla_{\mathbf{r}} + \mathbf{k})^2 + V(\mathbf{r}) \tag{9.39}$$

is a hermitian operator on functions on the unit cell of the crystal which satisfies the periodicity condition, Eq. (9.33). For any vector $\mathbf{k}$, the functions $\{u_{n\mathbf{k}}(\mathbf{r})\}_{n=1,2,3,\ldots}$ constitute a complete orthogonal basis set of functions for the linear vector space of functions satisfying the boundary condition in Eq. (9.33), since they are the eigenfunctions of a hermitian operator. Since the Bloch vector is a parameter in the *Hamiltonian*, Eq. (9.39), the dependence of $u_{\mathbf{k}}$ on the Bloch vector is therefore smooth.

9.4 Wigner–Seitz Unit Cell

The Bloch vectors can in view of Eq. (9.38) be restricted to a unit cell of the reciprocal lattice, the so-called reduced zone scheme. The unit cell spanned by the reciprocal lattice basis vectors $\{\mathbf{b}_{i=1,2,3}\}$ spans a unit cell that has the disadvantage of not in general displaying the full symmetry of the lattice. A convenient choice of unit cell in **k**-space is constructed in the following way. Choose the origin in **k**-space, $\mathbf{k} = \mathbf{0}$, at one of the reciprocal lattice points. Draw the connecting lines to all nearest, next nearest, etc. (usually only a few are needed) lattice points in the reciprocal lattice. A two-dimensional example for the case of a square lattice is displayed in Figure 9.2. Think about the planes cutting these lines orthogonally at the mid-point. The distance between these planes will be the length of the basis vectors, for the considered lattice $\mathbf{b}_1$, $\mathbf{b}_2$, $\mathbf{b}_3$, and sums of these. Since both $\pm\mathbf{b}_i$ are lattice points, the planes comes in pairs of opposing planes. The volume consisting of the points that lie closer to the chosen origin than to any other reciprocal lattice point will therefore be enclosed by sections of these orthogonal mid-point planes (so-called Bragg planes). By construction, on translations by reciprocal lattice vectors, this cell will exactly cover the whole **k**-space, and is thus a unit cell in the reciprocal lattice, the Wigner–Seitz unit cell. By construction, if a **k** value lies in the Wigner–Seitz cell, so does $-\mathbf{k}$. The Wigner–Seitz unit cell can of course be constructed for any lattice, and not just for a **k**-space lattice. Our choice of unit cell for lattices will be the Wigner–Seitz unit cell.

In reciprocal space, the Wigner–Seitz unit cell is called the first Brillouin zone, Bz_1. The nth Brillouin zone is defined as the set of points that can be reached from the origin by crossing exactly $n - 1$ Bragg planes. All Brillouin zones have the same volume, and

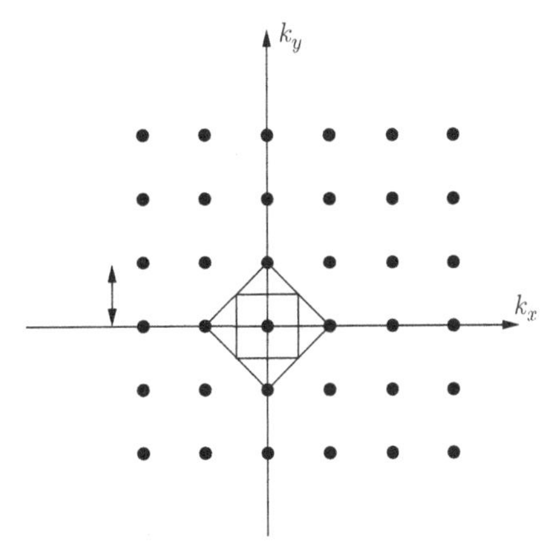

Figure 9.2 Construction of the first two Brillouin zones for the square lattice.

translating a Brillouin zone back into the first zone exactly covers it. The first two Brillouin zones for the case of a two-dimensional square lattice are displayed in Figure 9.2. The energy spectrum displayed as band structure in the first Brillouin zone can be translated by associated reciprocal lattice vectors into the next Brillouin zones, thus displaying the energy spectrum in the extended zone scheme (recall also the discussion in Section 7.7).

The Bloch vectors in the first Brillouin zone allowed by the Born–von Karman boundary condition are, assuming N_i even,

$$\mathrm{Bz}_1 = \left\{ \mathbf{k} \,\middle|\, \mathbf{k} = \frac{n_1}{N_1}\mathbf{b}_1 + \frac{n_2}{N_2}\mathbf{b}_2 + \frac{n_3}{N_3}\mathbf{b}_3;\ n_i = -\frac{N_i}{2} + 1, -\frac{N_i}{2} + 2, \ldots, \frac{N_i}{2} \right\}. \quad (9.40)$$

The Bloch states specified by $-N_i/2$ and $N_i/2$ are identical since they differ by a reciprocal lattice vector (recall Eq. (9.38)), and only one of them should be included. The number of n_i corresponding to Bloch vectors in the first Brillouin zone is N_i, and the first Brillouin zone thus contains $N = N_1N_2N_3$ Bloch vectors. The volume of the Wigner–Seitz unit cell, Bz_1, equals the volume spanned by the basis vectors,

$$\tilde{v} = \mathbf{b}_1 \cdot (\mathbf{b}_2 \times \mathbf{b}_3) = \left(\frac{2\pi}{v}\right)^2 \mathbf{b}_1 \cdot ((\mathbf{a}_3 \times \mathbf{a}_1) \times (\mathbf{a}_1 \times \mathbf{a}_2)) = \frac{(2\pi)^2}{v}\mathbf{b}_1 \cdot \mathbf{a}_1 = \frac{(2\pi)^3}{v}, \quad (9.41)$$

where use has been made of the identity $\mathbf{a} \times (\mathbf{b} \times \mathbf{c}) = (\mathbf{a} \cdot \mathbf{c})\mathbf{b} - (\mathbf{a} \cdot \mathbf{b})\mathbf{c}$, and $v = \mathbf{a}_1 \cdot (\mathbf{a}_2 \times \mathbf{a}_3)$ denotes the volume of the unit cell of the real space crystal, the Bravais lattice unit cell, $v = \Omega/N$, where N is the number of points (or equivalently unit cells) in the Bravais lattice of volume Ω. According to Eq. (9.15), there is one Born–von Karman vector in each **k**-space volume $\Delta\mathbf{k}$, where $\Delta\mathbf{k} = \mathbf{b}_1 \cdot (\mathbf{b}_2 \times \mathbf{b}_3)/N_1N_2N_3$, i.e. $N\Delta\mathbf{k} = \tilde{v}$, or $\Delta\mathbf{k} = (2\pi)^3/Nv = (2\pi)^3/\Omega$. The counting of Bloch states is thus identical to the counting of allowed free particle, plane wave, states in the Sommerfeld model as a consequence of the macroscopic, BvK, boundary condition being the same.

Exercise 9.2 Show that if N_i is odd, the allowed states in the first Brillouin zone are specified by $n_i = -\frac{1}{2}(N_i - 1), -\frac{1}{2}(N_i - 1) + 1, \ldots, -1, 0, 1, \ldots, \frac{1}{2}(N_i - 1)$.

By construction, the points on opposing faces of a Wigner–Seitz unit cell can be connected by a Bravais lattice vector. The function $\exp(i\mathbf{K} \cdot \mathbf{r})$ therefore, as a function of **r**, takes on the same value at opposing real space Wigner–Seitz unit cell faces for any reciprocal lattice vector **K**, $\exp\{i\mathbf{K} \cdot (\mathbf{r} + \mathbf{R})\} = \exp(i\mathbf{K} \cdot \mathbf{r})$. Integration of the function over the real space Wigner–Seitz crystal unit cell gives the Kronecker function

$$\int_v d\mathbf{r}\, e^{i\mathbf{K}\cdot\mathbf{r}} = v\, \delta_{\mathbf{K},\mathbf{0}}, \quad (9.42)$$

since the real and imaginary parts of the function $\exp(i\mathbf{K}\cdot\mathbf{r}) = \cos\mathbf{K}\cdot\mathbf{r} + i\sin\mathbf{K}\cdot\mathbf{r}$ oscillate a full number of wave lengths and thus are as equally positive as negative, except for the case $\mathbf{K} = \mathbf{0}$. Analogously, integrating over the Brillouin zone gives

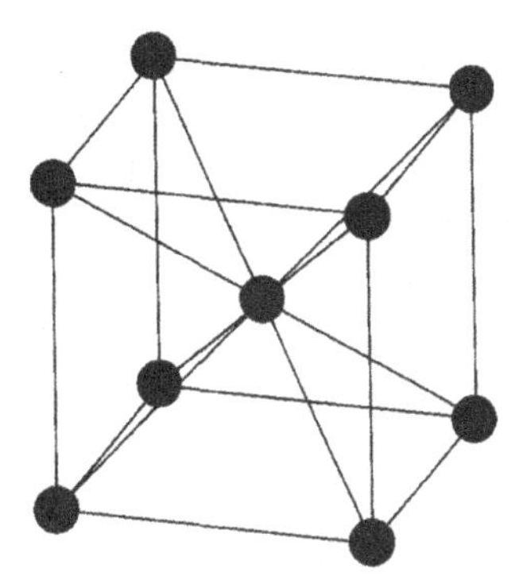

Figure 9.3 Conventional unit cell for the bcc lattice.

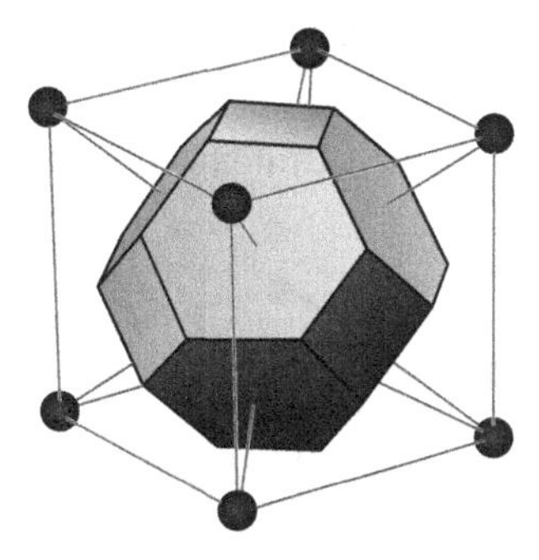

Figure 9.4 Wigner–Seitz unit cell for the bcc lattice.

$$\int_{\mathrm{Bz_I}} d\mathbf{k}\, e^{i\mathbf{k}\cdot\mathbf{R}} = \tilde{v}\, \delta_{\mathbf{R},\mathbf{0}}. \tag{9.43}$$

To give a flavor of crystal structures, a few common ones are depicted. In Figure 9.3 is depicted the *conventional* unit cell of the body-centered cubic (bcc) lattice, where, in addition to lattice points occupying the corners of a cube, there is an additional lattice point at its center. Repeating a conventional unit cell gives a quick way of constructing a finite portion of the lattice. Although at first glance one could be inclined to get the impression that center points are not equivalent to corner points, this lattice is indeed a Bravais lattice. This can be seen by drawing a portion of the lattice; a few repetitions of the conventional unit cell should convince one, as discussed in Appendix R.

The Wigner–Seitz unit cell for the bcc lattice is depicted in Figure 9.4. Its volume is half that of the conventional unit cell, and contains by construction exactly one lattice point, at $\mathbf{k} = \mathbf{0}$, the zone center.

Another common crystal structure is the face-centered cubic (fcc) lattice, which, instead of an additional point located at the center of the cube, has six points located at the centers of the cube's six faces. Its conventional unit cell is depicted in Figure 9.5.

In Figure 9.5(a), a primitive unit cell is depicted, six parallelogram faces. It encloses no lattice points, having all its corners at lattice points. The volume of the primitive unit cell is one-quarter of the cube's volume. It does not have the full symmetry of the fcc lattice. In Figure 9.5(b), parallel equidistantly spaced ionic planes are depicted for the fcc lattice, an example of planes, Bragg planes, identifiable by the crystal's X-ray diffraction.

When growing crystals, the directions perpendicular to such planes are the possible growth directions. Crystal structures can nowadays be built in machines by putting atoms together one by one, as discussed in Chapter 12. For some atoms, the process resembles that of stacking cannon balls in layers on top of each other, the planes in Figure 9.5(b) visualizing such a case in question.[6]

The Wigner–Seitz unit cell for the fcc lattice is depicted in Figure 9.6. One can show that the reciprocal lattice to the bcc lattice is an fcc lattice (and vice versa). The reader interested in a more detailed description of bcc and fcc lattices can consult Appendix R or the numerous books on solid-state physics.

The ground state of electrons in a periodic potential is obtained by filling up **k** states of higher energy. Consider, for example, a monovalent metal such as sodium (a bcc crystal structure). Its energy spectrum is the trivial generalization of the one-dimensional case; the energy increases as a function of the Bloch vector, only slightly differently in different directions. Owing to the spin degeneracy, all the N valence or conduction electrons can be accommodated in the first Brillouin zone, occupying half of the **k** states in the Brillouin

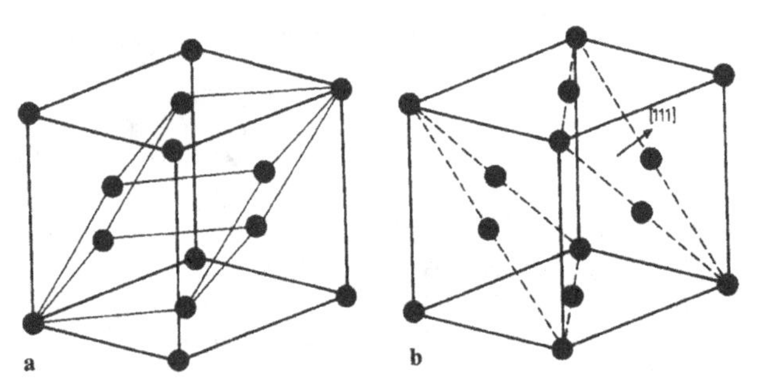

Figure 9.5 The fcc Bravais lattice.

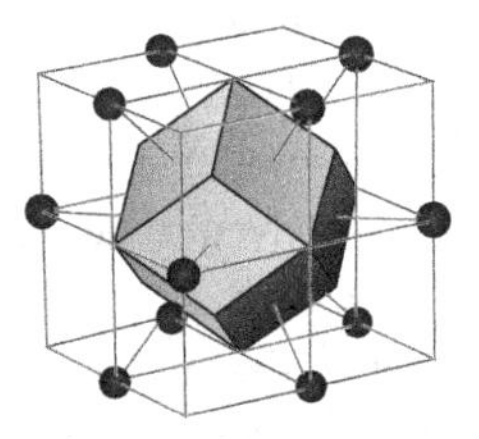

Figure 9.6 Wigner–Seitz unit cell for the fcc lattice.

[6] Make a drawing of a line-up of circles, and then four more rows in the same plane in such a way that the centers of the circles are as close as possible, close packing the circles. View the circles as representing balls, and then stack on new layers of balls in the close-packed fashion (or, better still, view an animation on the internet). The balls in the second layer occupy interstitial positions. When coming to the third layer, balls can be placed directly above the positions of the balls in the first layer, leaving the fourth layer directly above the balls in the second. Continuing this way gives the hexagonal close-packed lattice (hcp lattice), with layer structure form denoted {...ABABABAB...}. If the balls in the third layer instead are placed above the interstices in the first layer that were not covered in the second layer, and then balls in the fourth layer placed as the balls in the first layer, and balls in the fifth layer above those in the second layer, the fcc lattice results, the {...ABCABCABCABC...} close-packed structure. There are infinitely many close-packing ways as each successive layer has two choices of interstices. Only the fcc lattice is a Bravais lattice; the others are Bravais lattices with a basis.

zone due to the spin degeneracy, the $N/2$ $\mathbf{k}$ states of lowest energies $\epsilon_1(\mathbf{k})$ around the center $\mathbf{k} = \mathbf{0}$. A surface in $\mathbf{k}$-space separates occupied and unoccupied states, the Fermi surface, as specified by $\epsilon_1(\mathbf{k}) = \epsilon_F$. In the case of sodium, the Fermi surface is an only slightly distorted sphere, and sodium is described quite well by the Sommerfeld model. For other materials, Fermi surfaces can take on more complicated geometric structures. The more valence electrons there are to be accommodated, the higher in energy are the bands that become involved, and these will often only be partially filled, as determined by the Bloch dispersion. This leads to pockets of electrons occupying only parts of a Brillouin zone, and the corresponding enclosing Fermi surfaces can take on quite intricate shapes. However, for our purposes, only band structures of the simplest kind will be of interest.

9.5 Properties of Bloch Functions

For a finite crystal, the scalar product between Bloch functions is

$$\langle n\mathbf{k} \mid n'\mathbf{k}'\rangle = \int_\Omega d\mathbf{r}\; \psi^*_{n\mathbf{k}}(\mathbf{r})\, \psi_{n'\mathbf{k}'}(\mathbf{r}), \tag{9.44}$$

where Ω is the volume of the crystal. The Bloch functions will conveniently be chosen to have the form

$$\psi_{n\mathbf{k}}(\mathbf{r}) = \frac{1}{\sqrt{\Omega}}\, e^{i\mathbf{k}\cdot\mathbf{r}}\, u_{n\mathbf{k}}(\mathbf{r}), \qquad \mathbf{k} \in \text{BvK}, \qquad n = 1, 2, 3, \ldots. \tag{9.45}$$

The volume of the system is exhibited explicitly, stressing that the wave function has tiny magnitude everywhere but is extended in equal proportion throughout the unit cells of the large system (turning off the periodic potential gives normalized free particle plane waves).

The scalar product

$$\langle n\mathbf{k} \mid n'\mathbf{k}'\rangle = \frac{1}{\Omega} \int_\Omega d\mathbf{r}\; e^{i(\mathbf{k}'-\mathbf{k})\cdot\mathbf{r}}\, u^*_{n\mathbf{k}}(\mathbf{r})\, u_{n'\mathbf{k}'}(\mathbf{r}) \tag{9.46}$$

can be reduced by splitting the integral up into integrals over each of the N unit cells of the crystal. Any location in the crystal can be reached by first going to the nearest unit cell center and from there to the point in question, $\mathbf{r} = \mathbf{R}_m + \mathbf{r}'$, as depicted in Figure 9.7.

Employing the periodicity property of the $u_{n\mathbf{k}}$ functions, Eq. (9.33), the scalar product becomes

$$\langle n\mathbf{k} \mid n'\mathbf{k}'\rangle = \frac{1}{\Omega} \sum_{m=1}^{N} e^{i(\mathbf{k}'-\mathbf{k})\cdot\mathbf{R}_m} \int_v d\mathbf{r}'\; e^{i(\mathbf{k}'-\mathbf{k})\cdot\mathbf{r}'}\, u^*_{n\mathbf{k}}(\mathbf{r}')\, u_{n'\mathbf{k}'}(\mathbf{r}'), \tag{9.47}$$

where v denotes the volume of the unit cell, $Nv = \Omega$, N being the number of unit cells constituting the crystal. The summation over the cell index produces the Kronecker function multiplied by N (see Appendix C),

$$\langle n\mathbf{k} \mid n'\mathbf{k}'\rangle = \frac{N}{\Omega}\, \delta_{\mathbf{k},\mathbf{k}'} \int_v d\mathbf{r}'\; u^*_{n\mathbf{k}}(\mathbf{r}')\, u_{n'\mathbf{k}}(\mathbf{r}'). \tag{9.48}$$

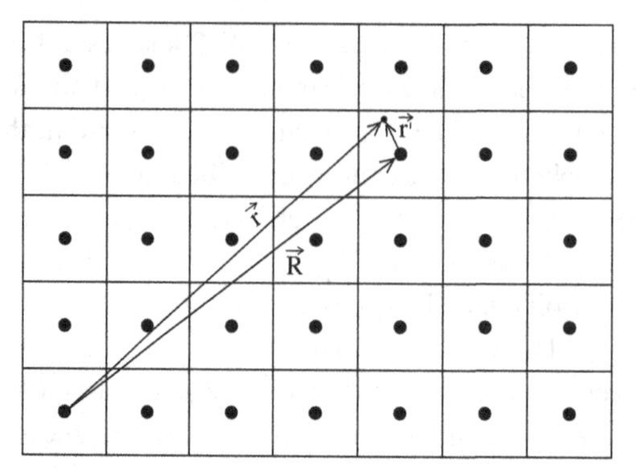

Figure 9.7 Splitting the volume of the crystal into its unit cells.

The degenerate states with the wave functions $\psi_{n\mathbf{k}}$ and $\psi_{n-\mathbf{k}}$, corresponding to the same energy, are thus, according to Eq. (9.48), orthogonal. Different bands can overlap, i.e. have the same energies but in different regions of **k**-space, as for example depicted in Figure 9.9. The degenerate energy eigenfunctions are, according to Eq. (9.48), orthogonal. When bands intersect, i.e. when $\epsilon_{n'\mathbf{k}}$ and $\epsilon_{n\mathbf{k}}$ are equal, orthogonality might have to be constructed, a peculiar case we shall not consider (recall that $u_{n\mathbf{k}}(\mathbf{r})$ and $u_{n'\mathbf{k}}(\mathbf{r})$ are eigenfunctions of the hermitian Hamiltonian $\hat{H}(k)$, and therefore typically correspond to different energies). The energy eigenstates can thus be chosen to be orthonormal,

$$\int_{\Omega} d\mathbf{r}\; \psi^*_{n\mathbf{k}}(\mathbf{r})\, \psi_{n'\mathbf{k}'}(\mathbf{r}) = \delta_{nn'}\, \delta_{\mathbf{k}\mathbf{k}'}. \tag{9.49}$$

With the chosen Bloch function normalization prefactor, the periodic part of a Bloch function satisfies, according to Eq. (9.48), the normalization condition

$$\frac{1}{v}\int_{v} d\mathbf{r}\; |u_{n,\mathbf{k}}(\mathbf{r})|^2 = 1 = \frac{1}{\Omega}\int_{\Omega} d\mathbf{r}\; |u_{n,\mathbf{k}}(\mathbf{r})|^2. \tag{9.50}$$

The periodic part of a Bloch function is then dimensionless, and the order of magnitude is, from the normalization condition, Eq. (9.50), seen to be typically of order one, $|u_{n\mathbf{k}}(\mathbf{r})| \sim 1$.

The Bloch states constitute a complete set, and any state of a particle in the periodic potential is a superposition of the corresponding Bloch states,

$$\psi(\mathbf{r}) = \sum_{n,\mathbf{k}\in \mathrm{Bz}_1} a_{n,\mathbf{k}}\, \psi_{n,\mathbf{k}}(\mathbf{x}), \tag{9.51}$$

where the n summation is the summation over the band indices.

9.6 Bloch Current

For a particle in the stationary Bloch state,

$$\psi_{n\mathbf{k}}(\mathbf{r},t) = \psi_{n\mathbf{k}}(\mathbf{r})\, e^{-(i/\hbar)\epsilon_{n\mathbf{k}}t}, \tag{9.52}$$

the probability current density is

$$\mathbf{S}_{n\mathbf{k}}(\mathbf{r}) = \frac{\hbar}{2im}\left(\psi^*_{n\mathbf{k}}(\mathbf{r},t)\frac{\partial \psi_{n\mathbf{k}}(\mathbf{r},t)}{\partial \mathbf{r}} - \psi_{n\mathbf{k}}(\mathbf{r},t)\frac{\partial \psi^*_{n\mathbf{k}}(\mathbf{r},t)}{\partial \mathbf{r}}\right)$$
$$= \frac{\hbar}{2im}\left(\psi^*_{n\mathbf{k}}(\mathbf{r})\frac{\partial \psi_{n\mathbf{k}}(\mathbf{r})}{\partial \mathbf{r}} - \psi_{n\mathbf{k}}(\mathbf{r})\frac{\partial \psi^*_{n\mathbf{k}}(\mathbf{r})}{\partial \mathbf{r}}\right), \tag{9.53}$$

which for a stationary state, of course, is independent of time, and therefore, according to the continuity equation, Eq. (1.38), satisfies

$$\nabla_{\mathbf{r}} \cdot \mathbf{S}_{n\mathbf{k}}(\mathbf{r}) = 0, \tag{9.54}$$

i.e. the current density is divergence-free.

The current density

$$S_{n\mathbf{k}}(\mathbf{r}) = \frac{1}{2m\Omega}\left(u^*_{n\mathbf{k}}(\mathbf{r})\left(\frac{\hbar}{i}\frac{\partial}{\partial \mathbf{r}} + \hbar\mathbf{k}\right)u_{n\mathbf{k}}(\mathbf{r}) - u_{n\mathbf{k}}(\mathbf{r})\left(\frac{\hbar}{i}\frac{\partial}{\partial \mathbf{r}} - \hbar\mathbf{k}\right)u^*_{n\mathbf{k}}(\mathbf{r})\right)$$
$$= \frac{\hbar}{2im\Omega}\left(2i\mathbf{k}|u_{n\mathbf{k}}(\mathbf{r})|^2 + u^*_{n\mathbf{k}}(\mathbf{r})\frac{\partial u_{n\mathbf{k}}(\mathbf{r})}{\partial \mathbf{r}} - u_{n\mathbf{k}}(\mathbf{r})\frac{\partial u^*_{n\mathbf{k}}(\mathbf{r})}{\partial \mathbf{r}}\right) \tag{9.55}$$

has the symmetry of the lattice, i.e. it repeats itself in each unit cell, as seen explicitly, $S_{n\mathbf{k}}(\mathbf{r}+\mathbf{R}) = S_{n\mathbf{k}}(\mathbf{r})$.

Integrating the probability current density over the crystal volume, and performing a partial integration, the boundary terms vanish due to the periodicity of the Bloch function, giving

$$\langle \mathbf{S}\rangle_{n\mathbf{k}} \equiv \frac{1}{\Omega}\int_{\Omega} d\mathbf{r}\, S_{n\mathbf{k}}(\mathbf{r}) = \frac{1}{\Omega m}\int_{\Omega} d\mathbf{r}\, \psi^*_{n\mathbf{k}}(\mathbf{r})\frac{\hbar}{i}\frac{\partial \psi_{n\mathbf{k}}(\mathbf{r})}{\partial \mathbf{r}} = \frac{1}{\Omega}\langle \hat{\mathbf{v}}\rangle_{n\mathbf{k}}, \tag{9.56}$$

i.e. $\langle \mathbf{S}\rangle_{n\mathbf{k}}$ is specified in terms of the average velocity, the expectation value of the velocity operator, in the Bloch state in question (recall Section 8.2). The average velocity can be rewritten as

$$\mathbf{v}_n(\mathbf{k}) \equiv \langle \hat{\mathbf{v}}\rangle_{n\mathbf{k}} = \frac{1}{m}\int_{\Omega} d\mathbf{r}\, \psi^*_{n\mathbf{k}}(\mathbf{r})(-i\hbar\nabla_{\mathbf{r}})\,\psi_{n\mathbf{k}}(\mathbf{r})$$
$$= \frac{1}{m\Omega}\int_{\Omega} d\mathbf{r}\, e^{-i\mathbf{k}\cdot\mathbf{r}}u^*_{n\mathbf{k}}(\mathbf{r})(-i\hbar\nabla_{\mathbf{r}})\,e^{i\mathbf{k}\cdot\mathbf{r}}u_{n\mathbf{k}}(\mathbf{r})$$
$$= \frac{1}{m\Omega}\int_{\Omega} d\mathbf{r}\, u^*_{n\mathbf{k}}(\mathbf{r})(\hbar\mathbf{k} - i\hbar\nabla_{\mathbf{r}})\,u_{n\mathbf{k}}(\mathbf{r}). \tag{9.57}$$

The average value of the velocity operator can be related to the band energy. Since the *Hamiltonians* of Eq. (9.39) for two Bloch vectors close in value are related according to

$$\hat{H}(\mathbf{k}+\Delta\mathbf{k}) = \hat{H}(\mathbf{k}) + \frac{\hbar^2}{m}\Delta\mathbf{k}\cdot(-i\nabla_{\mathbf{r}} + \mathbf{k}) + \mathcal{O}((\Delta\mathbf{k})^2), \tag{9.58}$$

the energy value $\epsilon_n(\mathbf{k}+\Delta\mathbf{k})$ can be calculated in lowest-order perturbation theory as eventually $\Delta\mathbf{k}$ will approach zero. The energy eigenvalue of $\hat{H}(\mathbf{k}+\Delta\mathbf{k})$ is then the energy eigenvalue of the unperturbed Hamiltonian, $\hat{H}(\mathbf{k})$, i.e. $\epsilon_n(\mathbf{k})$, plus the first-order correction, which is the matrix element of the perturbation evaluated in the normalized unperturbed eigenstates of $\hat{H}(\mathbf{k})$, i.e. $u_{n\mathbf{k}}/\sqrt{\Omega}$. The shift in energy eigenvalue is thus

$$\epsilon_n(\mathbf{k}+\Delta\mathbf{k}) = \epsilon_n(\mathbf{k}) + \frac{\hbar^2}{m}\Delta\mathbf{k}\cdot\int_\Omega \frac{d\mathbf{r}}{\Omega}\, u^*_{n\mathbf{k}}(\mathbf{r})(-i\nabla_{\mathbf{r}}+\mathbf{k})\, u_{n\mathbf{k}}(\mathbf{r}) + \mathcal{O}((\Delta\mathbf{k})^2). \tag{9.59}$$

The relationship between the band energy and the expectation value of the velocity operator in the corresponding Bloch state is then, in view of Eq. (9.57), established:

$$\mathbf{v}_n(\mathbf{k}) = \frac{1}{\hbar}\frac{\partial\epsilon_n(\mathbf{k})}{\partial\mathbf{k}}. \tag{9.60}$$

Bloch states are thus in general current-carrying. A Bloch electron propagates unimpeded through the crystal. The resistance of a real crystalline solid is thus due to its imperfections, such as missing atoms, different atoms (impurities) or vibrations of the ions at finite temperature.

The total particle current density from a filled band,

$$\begin{aligned}\mathbf{S}_n &= 2\sum_{\mathbf{k}\in\mathrm{Bz}_1}\langle\mathbf{S}\rangle_{n\mathbf{k}} = \frac{2}{\Omega}\sum_{\mathbf{k}\in\mathrm{Bz}_1}\mathbf{v}_n(\mathbf{k}) = 2\sum_{\mathbf{k}\in\mathrm{Bz}_1}\frac{\Delta\mathbf{k}}{(2\pi)^3}\mathbf{v}_n(\mathbf{k}) \\ &= \frac{2}{\hbar}\int_{\mathrm{Bz}_1}\frac{d\mathbf{k}}{(2\pi)^3}\frac{\partial\epsilon_n(\mathbf{k})}{\partial\mathbf{k}} = 0,\end{aligned} \tag{9.61}$$

vanishes since $\epsilon_n(\mathbf{k})$ is a periodic function, or since $\mathbf{v}_n(-\mathbf{k}) = -\mathbf{v}_n(\mathbf{k})$ these contributions cancel each other. Electrical current can thus only come from partially filled bands.

Exercise 9.3 Show that a superposition of Bloch functions within a given band n with Bloch vectors close to $\mathbf{k}$ is a wave packet moving with the group velocity

$$\mathbf{v}_n(\mathbf{k}) = \frac{1}{\hbar}\frac{\partial\epsilon_n(\mathbf{k})}{\partial\mathbf{k}}. \tag{9.62}$$

9.7 Band Structure Properties

The Hamiltonian for a particle in a potential is real, $\hat{H}^* = \hat{H}$, and the complex conjugate of a Bloch function, $\psi_{n,\mathbf{k}}(\mathbf{r})$, therefore has the same energy eigenvalue,

$$\hat{H}\,\psi^*_{n,\mathbf{k}}(\mathbf{r}) = \epsilon_n(\mathbf{k})\,\psi^*_{n,\mathbf{k}}(\mathbf{r}). \tag{9.63}$$

In view of the Bloch condition, Eq. (9.27), the Bloch function $\psi_{n,\mathbf{k}}(\mathbf{r})$ is a complex function, and the complex conjugated wave function $\psi^*_{n,\mathbf{k}}(\mathbf{r})$ does not represent the same state as the wave function $\psi_{n,\mathbf{k}}(\mathbf{r})$, i.e. they are independent solutions. From the Bloch condition,

Eq. (9.27), it follows that the complex conjugate of a Bloch function satisfies the following condition:

$$\psi^*_{n,\mathbf{k}}(\mathbf{r}+\mathbf{R}) = e^{-i\mathbf{k}\cdot\mathbf{R}}\psi^*_{n,\mathbf{k}}(\mathbf{r}). \tag{9.64}$$

The wave function $\psi^*_{n,\mathbf{k}}(\mathbf{r})$ thus satisfies the Bloch condition corresponding to the Bloch vector $-\mathbf{k}$, and is the state corresponding to Bloch vector $-\mathbf{k}$, $\psi^*_{n,\mathbf{k}}(\mathbf{r}) = \psi_{n,-\mathbf{k}}(\mathbf{r})$ (as also established directly according to Eq. (9.28)). As a consequence of time reversal symmetry, the energy spectrum satisfies $\epsilon_n(-\mathbf{k}) = \epsilon_n(\mathbf{k})$, valid irrespective of the crystal structure. The energy spectrum for a particle in a periodic potential thus respects inversion symmetry in **k**-space. By construction, the Wigner–Seitz unit cell respects this symmetry.

Typical crystal lattices often have symmetries that manifest themselves in additional properties of the band structure. For example, let us assume that the Brillouin zone has mirror symmetry with respect to the plane that is perpendicular to the k_x axis and passes through the point $k_x = 0$. A cross-section of such a mirror-symmetric Brillouin zone is shown in Figure 9.8, and two symmetric Bragg plane points, $\mathbf{k}_1$ and $\mathbf{k}_2$, are indicated.

The energy eigenvalues on two mirror-symmetric Bragg planes are equal, $\epsilon_n(\mathbf{k}_2) = \epsilon_n(\mathbf{k}_1)$, since the corresponding Bloch vectors differ by a reciprocal lattice vector, *viz.* the shortest reciprocal lattice vector along the k_x axis. In view of the general identity of energies differing by a reciprocal lattice vector, $\epsilon_n(\mathbf{k}+\mathbf{K}) = \epsilon_n(\mathbf{k})$, the derivatives along the k_x direction are therefore equal,

$$\frac{\partial \epsilon_n(\mathbf{k}_2)}{\partial k_x} = \frac{\partial \epsilon_n(\mathbf{k}_1)}{\partial k_x}. \tag{9.65}$$

By virtue of the inversion symmetry, the derivative at the mirror-symmetric points $\mathbf{k}_1$ and $\mathbf{k}_2$ also satisfies (recall Eq. (7.84))

$$\frac{\partial \epsilon_n(\mathbf{k}_2)}{\partial k_x} = -\frac{\partial \epsilon_n(\mathbf{k}_1)}{\partial k_x} \tag{9.66}$$

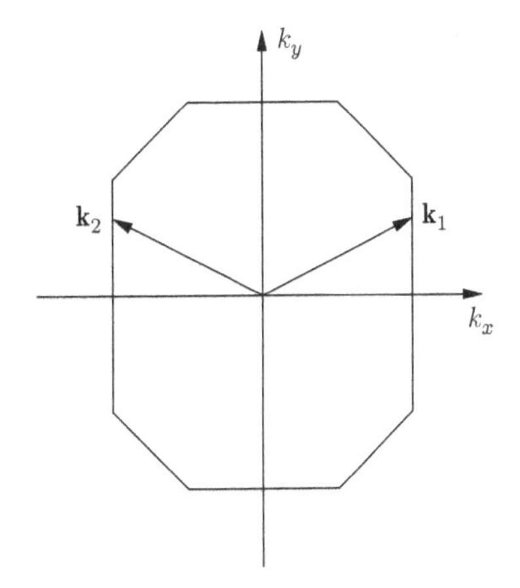

Figure 9.8 Brillouin zone with mirror symmetry.

and the derivative on the two mirror-symmetric Bragg planes thus vanishes,

$$\left.\frac{\partial \epsilon_n(\mathbf{k})}{\partial k_x}\right|_{\mathbf{k}=\mathbf{k}_{1,2}} = 0. \tag{9.67}$$

If a band edge is located at a Bragg plane, the band energy thus has an extremum there, and the energy dispersion near the band edge can be described in the effective mass approximation. Typically the band structure at a zone center is translated from a Bragg plane, leading to the same scenario.

The Bloch density of states (per spin) for a given band, $N_n(\epsilon)$, the number of Bloch energy levels per unit volume and unit energy in the nth band, is a quantity of interest when evaluating typically occurring integrals. Consider the constant energy surface in **k**-space, $S_n(\epsilon)$, determined by $\epsilon_n(\mathbf{k}) = \epsilon$, and the constant energy surface in close proximity determined by the energy $\epsilon + \Delta\epsilon$. Introducing the vector $\Delta\mathbf{k}$ by $\epsilon_n(\mathbf{k} + \Delta\mathbf{k}) = \epsilon + \Delta\epsilon$, then to first order

$$\Delta\epsilon = \nabla_{\mathbf{k}}\, \epsilon_n(\mathbf{k}) \cdot \Delta\mathbf{k} = |\nabla_{\mathbf{k}}\, \epsilon_n(\mathbf{k})|\, \Delta k, \tag{9.68}$$

where in the last equality $\Delta\mathbf{k}$ has been chosen perpendicular to the constant energy surface and we recall that the gradient, $\nabla_{\mathbf{k}}\, \epsilon_n(\mathbf{k})$, is perpendicular to the (*equipotential*) energy surface, and $\Delta k \equiv |\Delta\mathbf{k}|$. Since there is one **k** state per **k**-space volume of size $(2\pi)^3/\Omega$, the number of **k** states in the volume, $ds\, \Delta k$, determined by the surface element ds and Δk, is

$$\frac{\Delta k\, ds}{(2\pi)^3/\Omega} = \frac{\Omega}{(2\pi)^3} \frac{\Delta\epsilon\, ds}{|\nabla_{\mathbf{k}}\, \epsilon_n(\mathbf{k})|}, \tag{9.69}$$

where the last equality follows from Eq. (9.68). The Bloch density of states at energy ϵ is thus given by the surface integral along the constant energy surface

$$N_n(\epsilon) = \frac{1}{(2\pi)^3} \int_{S_n(\epsilon)} ds\, \frac{1}{|\nabla_{\mathbf{k}}\, \epsilon_n(\mathbf{k})|}. \tag{9.70}$$

The Bloch density of states can also be obtained from the counting formula

$$N_n(\epsilon) = \frac{1}{\Omega} \sum_{\mathbf{k} \in \mathrm{Bz}_1} \delta(\epsilon - \epsilon_n(\mathbf{k})) = \frac{1}{(2\pi)^3} \int_{\mathrm{Bz}_1} d\mathbf{k}\; \delta(\epsilon - \epsilon_n(\mathbf{k})), \tag{9.71}$$

since by Taylor-expanding

$$\epsilon_n(\mathbf{k}) = \epsilon_n(\mathbf{k}_0) + \nabla_{\mathbf{k}}\, \epsilon_n(\mathbf{k}_0) \cdot (\mathbf{k} - \mathbf{k}_0) \tag{9.72}$$

and setting $\epsilon_n(\mathbf{k}_0) = \epsilon$, the **k** integration becomes

$$\int_{\mathrm{Bz}_1} d\mathbf{k}\; \delta(\epsilon - \epsilon_n(\mathbf{k})) = \frac{1}{(2\pi)^3} \int_{\mathrm{Bz}_1} d\mathbf{k}\; \delta(\nabla_{\mathbf{k}}\, \epsilon_n(\mathbf{k}_0) \cdot (\mathbf{k} - \mathbf{k}_0)). \tag{9.73}$$

Performing the integration in the direction perpendicular to the constant energy surface then gives the result in Eq. (9.70).

Often one encounters functions that depend only on the Bloch vector through the energy

$$\frac{1}{\Omega}\sum_{\mathbf{k}\in Bz_1} F(\epsilon_n(\mathbf{k})) = \frac{1}{(2\pi)^3}\int_{\mathbf{k}\in Bz_1} d\mathbf{k}\, F(\epsilon_n(\mathbf{k})), \tag{9.74}$$

where the last equality follows in the large volume limit from the fact that there is one $\mathbf{k}$ state per $\mathbf{k}$-space volume of size $(2\pi)^3/\Omega$. The $\mathbf{k}$-space volumes are taken as those determined by the constant energy surfaces and the perpendicular direction, i.e. $d\mathbf{k} = ds\,\Delta k$, and the perpendicular length converted to energy according to Eq. (9.68), $ds\,\Delta k = ds\,\Delta\epsilon/|\nabla_{\mathbf{k}}\,\epsilon_n(\mathbf{k})|$. The $\mathbf{k}$ integration is thereby turned into the energy integration

$$\frac{1}{(2\pi)^3}\int_{\mathbf{k}\in Bz_1} d\mathbf{k}\, F(\epsilon_n(\mathbf{k})) = \int_0^\infty d\epsilon\, F(\epsilon)\frac{1}{(2\pi)^3}\int_{S_n(\epsilon)} ds\, \frac{1}{|\nabla_{\mathbf{k}}\epsilon_n(\mathbf{k})|} = \int_{\epsilon_{min}}^{\epsilon_{max}} d\epsilon\, N_n(\epsilon)\, F(\epsilon), \tag{9.75}$$

where the integration is from the lowest energy of $\epsilon_n(\mathbf{k})$ in the Brillouin zone to the highest.

9.8 Characteristic Band Edges

Band structure gives rise to different types of conduction properties depending on whether or not the excited states above the ground state are separated by a gap. Three-dimensional crystals can exhibit an enormous variety of fillings of bands in the ground state. In Figure 9.9, a situation with overlapping bands is displayed. A material with such a band edge configuration is a semi-metal. It has a smaller density of states at the Fermi energy compared to a half-filled band metal.

In a system with an energy gap, several bands are completely filled in the ground state. In the simplest case, there will be one band of highest energy, the valence band, which in the ground state is completely filled, and separated from a band of unoccupied states, the conduction band, by the energy gap. The energy gap can occur in different ways. The case

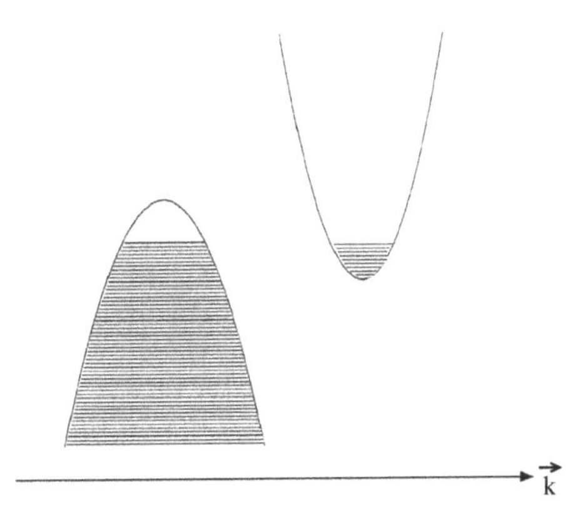

Figure 9.9 Band structure of a semi-metal.

of an indirect band gap is depicted in Figure 9.10. Here the lowest state in the conduction band occurs at a different Bloch vector than the Bloch vector of the highest occupied state in the valence band. This situation occurs, for example, in silicon.

In a direct-band-gap semiconductor, the bottom of the conduction band and the top of the valence band are located at the same Bloch vector, as depicted in Figure 9.11. This situation occurs, for example, in gallium arsenide (GaAs).

In a direct-band-gap semiconductor, a photon can directly excite an electron in the state at the top of the valence band to the state at the bottom of the conduction band, since the process can conserve energy and Bloch wave vector. The material has good optical properties compared to those of indirect-band-gap materials, where absorption and emission of photons must be accompanied by phonon emission in order for the process to conserve energy and crystal momentum. The optical properties of an indirect-band-gap semiconductor are thus accompanied by heat production.

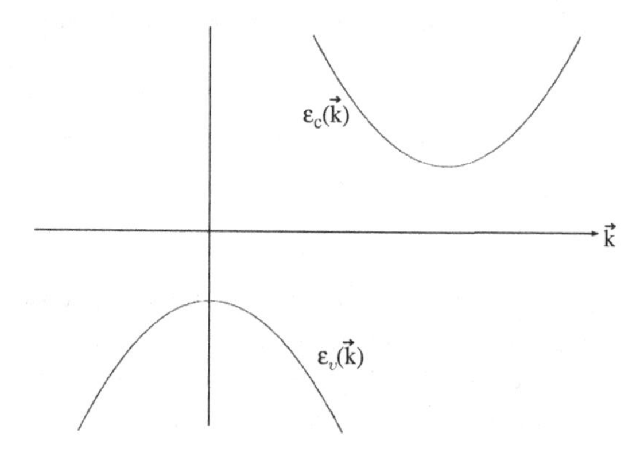

Figure 9.10 Indirect-band-gap case.

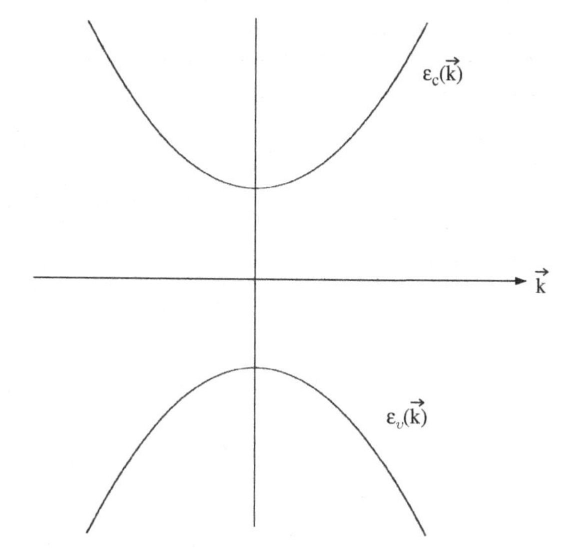

Figure 9.11 Direct-band-gap case.

9.9 Intrinsic Semiconductor

Consider a ground state consisting of completely filled valence bands, and empty conduction bands, as typical of a semiconductor. For simplicity, assume only one relevant conduction and valence band, say, the situation depicted in Figure 9.11. Generalization is straightforward, as seen by the example.

At finite temperatures, a fraction of the electrons in the valence band are thermally excited to the conduction band. The density of electrons in the conduction band is, as discussed in Appendix K, specified by the Fermi–Dirac distribution, Eq. (K.18),

$$n_c(T) = \frac{2}{\Omega}\sum_{\mathbf{k}\in \mathrm{Bz}_1} f(\epsilon_c(\mathbf{k})) = 2\int_{\mathbf{k}\in \mathrm{Bz}_1} \frac{d\mathbf{k}}{(2\pi)^3}\frac{1}{e^{(\epsilon_c(\mathbf{k})-\mu)/kT}+1}$$
$$= 2\int_{\epsilon_c}^{w_c} d\epsilon\, N_c(\epsilon)\,\frac{1}{e^{(\epsilon-\mu)/kT}+1}, \tag{9.76}$$

where the upper integration limit is at the top of the conduction band.

If the electrons in the conduction band only originate from the valence band, a so-called intrinsic semiconductor, there is an equal average concentration of unoccupied electron levels, $p_v(T) = n_c(T)$, in the valence band,

$$p_v(T) = \frac{2}{\Omega}\sum_{\mathbf{k}\in \mathrm{Bz}_1}[1 - f(\epsilon_v(\mathbf{k}))] = 2\int_{w_v}^{\epsilon_v} d\epsilon\, N_v(\epsilon)\,\frac{1}{e^{(\mu-\epsilon)/kT}+1}, \tag{9.77}$$

where w_v denotes the energy value at the bottom of the valence band.

The integrand in Eq. (9.76), the product of the density of states and the probability of occupation, $N_c(\epsilon)f(\epsilon)$, is thus the density of electrons per spin at the energy value in question, i.e. $2N_c(\epsilon)f(\epsilon)\,\Delta\epsilon$ is the number of electrons in the specified energy range per unit volume, the energy specific density. It has the form shown in Figure 9.12. At the conduction band edge, the density of electrons vanishes due to the vanishing of the density of states, and at high energies the function decays rapidly due to the vanishing probability of population at high energies.

The chemical potential is a measure of the energy it costs to add an electron to the system, so adding one in the valence band and one in the conduction band gives the estimate $\mu \sim (\epsilon_v + \epsilon_v + \Delta)/2 = \epsilon_v + \Delta/2$, i.e. the chemical potential is estimated to be located in the middle of the gap, the index indicating the considered case of an intrinsic semiconductor, $\mu = \mu_i$. Assume in the following a non-degenerate semiconductor, $\epsilon_c - \mu,\ \mu - \epsilon_v \gg kT$. In the conduction band, energies are larger than the band edge energy, $\epsilon > \epsilon_c$, and $\epsilon - \mu > \epsilon_c - \mu \gg kT$. The exponential term in the denominator of the Fermi function is therefore much larger than one, and the Fermi–Dirac distribution reduces to the Maxwell–Boltzmann distribution,

$$\frac{1}{e^{(\epsilon-\mu)/kT}+1} \simeq e^{-(\epsilon-\mu)/kT} = e^{-(\epsilon-\epsilon_c)/kT}\, e^{(\mu-\epsilon_c)/kT}. \tag{9.78}$$

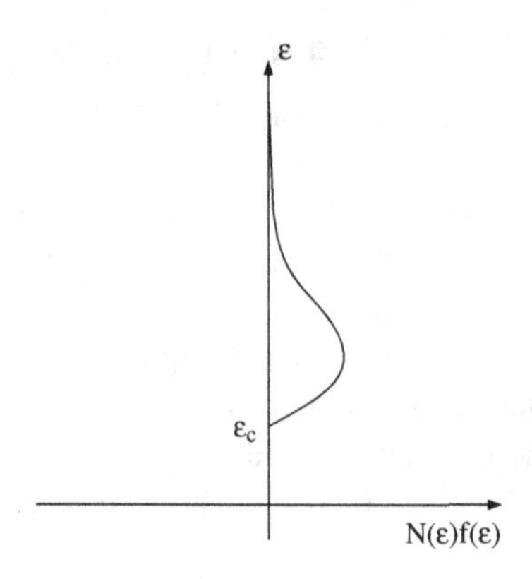

Figure 9.12 Density of conduction electrons as a function of energy.

In the valence band, electron energies are largest at the valence band edge, $\epsilon < \epsilon_v$, and $\mu - \epsilon = \mu - \epsilon_v + (\epsilon_v - \epsilon) > \mu - \epsilon_v \gg kT$. The distribution of the absence of electrons in the valence band therefore becomes

$$\frac{1}{e^{(\mu-\epsilon)/kT}+1} \simeq e^{-(\mu-\epsilon)/kT} = e^{-(\mu-\epsilon_v)/kT}\, e^{-(\epsilon_v-\epsilon)/kT}. \tag{9.79}$$

The expressions in Eqs. (9.76) and (9.77) then become

$$n_c(T) = N_c(T) e^{(\mu-\epsilon_c)/kT} \tag{9.80}$$

and

$$p_v(T) = P_v(T) e^{(\epsilon_v-\mu)/kT}, \tag{9.81}$$

where

$$N_c(T) = 2\int_{\epsilon_c}^{w_c} d\epsilon\, N_c(\epsilon)\, e^{-(\epsilon-\epsilon_c)/kT} \tag{9.82}$$

and

$$P_v(T) = 2\int_{w_v}^{\epsilon_v} d\epsilon\, N_v(\epsilon)\, e^{-(\epsilon_v-\epsilon)/kT} \tag{9.83}$$

are slowly varying functions of the temperature (as the integrations start or end where the arguments of the exponential vanishes) compared to the exponentials in Eqs. (9.80) and (9.81).

Multiplying the expressions for the two types of densities gives the *law of mass action*,

$$n_c(T)\, p_v(T) = N_c(T) P_v(T)\, e^{-\Delta/kT}. \tag{9.84}$$

Since the occupation of states in the conduction band is due to the depopulation of the valence band, $n_c(T) = p_v(T)$, and Eq. (9.84) gives for the intrinsic carrier density

$$n_c(T) = (N_c(T) P_v(T))^{1/2}\, e^{-\Delta/2kT} \equiv n_i. \tag{9.85}$$

Inserting Eq. (9.85) into Eq. (9.80) gives

$$e^{(\mu-\epsilon_c)/kT} = \left(\frac{P_v(T)}{N_c(T)}\right)^{1/2} e^{\Delta/kT}, \tag{9.86}$$

and taking the logarithm, the expression for the chemical potential is

$$\mu = \epsilon_c - \frac{\Delta}{2} + \frac{kT}{2} \ln\left(\frac{P_v(T)}{N_c(T)}\right). \tag{9.87}$$

Likewise, inserting

$$p_v(T) = n_c(T) = (N_c(T)P_v(T))^{1/2}\, e^{-\Delta/2kT} \tag{9.88}$$

into Eq. (9.81) similarly gives the chemical potential expressed in terms of the valence band edge,

$$\mu = \epsilon_v + \frac{\Delta}{2} + \frac{kT}{2} \ln\left(\frac{P_v(T)}{N_c(T)}\right). \tag{9.89}$$

To estimate the logarithmic term, we note that the main contribution to the integrals determining $N_c(T)$ and $P_v(T)$ comes from the energies near the band edges due to the presence of the damping exponential factor, and since band widths are typically much larger than kT at room temperature, $kT_{300\text{K}} \simeq 25$ meV, the energy integrations can be extended to infinity. The spectrum typically has an extremum at the band edge, and assuming for simplicity an isotropic band structure,

$$\epsilon_c(\mathbf{k}) = \epsilon_c + \frac{\hbar^2 k^2}{2m_c}, \tag{9.90}$$

where the curvature of the band dispersion is described by the effective mass m_c of the Bloch electron. The density of states is therefore the same as in the Sommerfeld model, Eq. (3.26), except for the spectrum starting at the band edge,

$$N_c(\epsilon) = \frac{m_c\sqrt{2m_c(\epsilon - \epsilon_c)}}{2\pi^2\hbar^3}, \tag{9.91}$$

i.e. has the form depicted in Figure 9.13, and similarly for the density of states in the valence band.

The integral in Eq. (9.82) (which by the substitution $y = x^{1/2}$ becomes a trivial Gaussian integral) gives

$$N_c(T) = \frac{\sqrt{2m_c^3}}{\pi^2\hbar^3}(kT)^{3/2}\int_0^\infty dx\, x^{1/2} e^{-x} = \frac{\sqrt{2m_c^3}}{\pi^2\hbar^3}(kT)^{3/2}\sqrt{\pi} = 2\left(\frac{m_c kT}{2\pi\hbar^2}\right)^{3/2}. \tag{9.92}$$

Similarly, assuming for the valence band

$$\epsilon_v(\mathbf{k}) = \epsilon_v - \frac{\hbar^2 k^2}{2m_v} \tag{9.93}$$

gives

$$P_v(T) = 2\left(\frac{m_v kT}{2\pi\hbar^2}\right)^{3/2}. \tag{9.94}$$

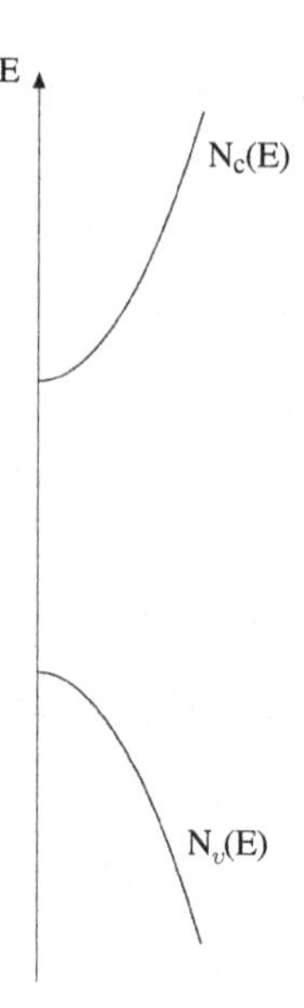

Figure 9.13 Density of states near valence and conduction band edges.

The chemical potential, Eq. (9.89), for an intrinsic semiconductor then reduces to

$$\mu_i = \epsilon_v + \frac{1}{2}\Delta + \frac{1}{2}kT \ln\left(\frac{m_v}{m_c}\right). \tag{9.95}$$

At room temperature, $kT_{300\text{K}} \sim 10^{-2}$ eV, and even if the effective masses differ by two orders of magnitude, the logarithmic term can be neglected for band gaps as small as on the order of 0.1 eV. The intrinsic chemical potential thus typically has a value in the middle of the band gap, and the assumption of non-degeneracy, $\epsilon_c - \mu,\ \mu - \epsilon_v \gg kT$, is justified.

The presence of electrons in the conduction band as well as the absence of electrons in the valence band (the presence of so-called holes, as discussed in Chapter 10) leads to a finite conductivity. The magnitude can be estimated by the Drude expression, the conductivity being proportional to the density of charge carriers. The density of electrons in the conduction band, Eq. (9.85), becomes, according to Eqs. (9.92) and (9.94),

$$n_c(T) = 2(m_c m_v)^{3/4} \left(\frac{kT}{2\pi\hbar^2}\right)^{3/2} e^{-\Delta/2kT} \tag{9.96}$$

and thus has a strong temperature dependence due to the exponential function.

The electron density at room temperature is according to Eq. (9.96) extremely sensitive to the gap size relative to $kT_{300\text{K}}$, the average thermal energy at room temperature. To estimate the electron density, assume $m_c m_v \sim m_e^2$, whereby

$$n_c(T) \sim \left(\frac{m_e kT}{2\pi\hbar^2}\right)^{3/2} e^{-\Delta/2kT} \sim 10^{19}\, e^{-\Delta/2kT}\ \text{cm}^{-3}. \tag{9.97}$$

The density of charge carriers is determined by the activation energy exponential factor in Eq. (9.96). At room temperature, the activation factor is approximately

$$e^{-\Delta/2kT} \sim 10^{-35},\ 10^{-19},\ 0.6 \qquad \text{for } \Delta = 4\,\text{eV},\ 2\,\text{eV},\ 25\,\text{meV}, \tag{9.98}$$

the respective values of the energy gap Δ.

Thus materials with large energy gaps, such as, for example, diamond, for which $\Delta = 5.4\,\text{eV}$, have negligible carrier concentrations and are unable to conduct current – they are insulators. The typical resistivity of insulators is on the order of $\rho \simeq 10^{22}\,\Omega\,\text{cm}$, which should be contrasted to the typical resistivity of a metal, $\rho \simeq 10^{-6}\,\Omega\,\text{cm}$. When thermal excitation at room temperature leads to observable conductivity, the material is called a semiconductor, corresponding, according to the above estimate, to energy gaps less than on the order of two electronvolts, $\Delta_{sc} \lesssim 2\,\text{eV}$. Intrinsic semiconductors have resistivities in the enormous range, $\rho \sim 10^{-3}$–$10^{9}\,\Omega\,\text{cm}$.

Exercise 9.4 The energy gap of a semiconductor has a weak temperature dependence. For example, for temperatures above 250 K, the energy gap of silicon has the linear dependence

$$\Delta(T) = \Delta(0) - aT, \qquad a \simeq 2.8 \times 10^{-4}\,\text{eV/K}. \tag{9.99}$$

Show that this does not change the exponential temperature dependence of the carrier concentrations, say, Eq. (9.96), only the prefactor.

In a metal, the conductivity decreases with temperature since the inelastic scattering decreases and the density of electrons is fixed. In a semiconductor, the conductivity at first strongly increases with increasing temperature due to an increase in the concentration of conduction electrons, before eventually decreasing due to electron–phonon scattering as in a metal. The typical conductivity of a semiconductor as a function of temperature, as depicted in Figure 9.14, is thus quite different from that of a metal. The sensitivity to temperature of the conductivity of semiconductors can be used to make a thermometer.

Illumination of a semiconductor with visible light (consisting of photons with energies on the order of one electronvolt) excites electrons into the conduction band. The conductivity of semiconductors is therefore sensitive to whether the crystal is illuminated or not,

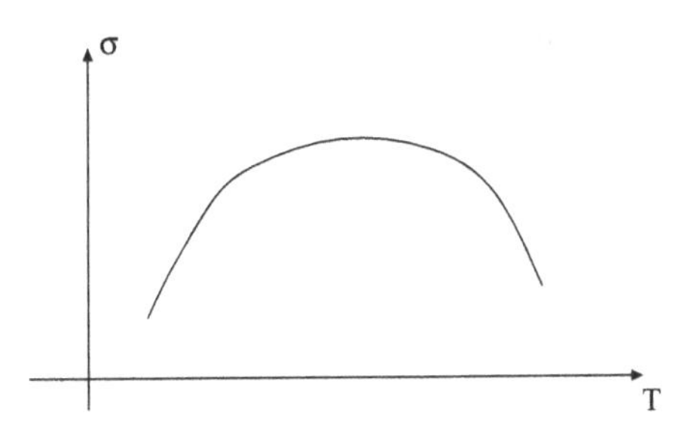

Figure 9.14 Temperature dependence of the conductivity of a semiconductor.

the photoconductivity effect, a commonly exploited effect as used, for instance, in alarm triggers. The photoconductivity effect is also used to produce temperature maps of the surface of the Earth as measured from satellites, and in many other situations of temperature measurement where measuring thermal radiation is of interest, say, in the metallurgical industry.

9.10 Bloch Dynamics

The analysis of the response of an electron in a Bloch state to an external force for the three-dimensional case can be taken directly over from the one-dimensional case discussed in Section 8.4. For typical band gaps and not extreme field strengths, interband transitions can be neglected, and a Bloch state evolves as described by the Bloch vector evolution

$$\mathbf{k}(t) = \mathbf{k}_0 + \frac{-e\mathbf{E}}{\hbar}\, t. \tag{9.100}$$

Compared to the one-dimensional case, aperiodic dynamics can occur, but the motion is still confined to the same band as dictated by Bragg reflection. A filled band is thus inert to an electric field, as illustrated in Figure 8.1.

9.11 Summary

The crystal translation operators commute with the Hamiltonian for the particle in the corresponding periodic potential, and if $\psi(\mathbf{r})$ is a solution to the time-independent Schrödinger equation for a certain energy, so is the function $\psi(\mathbf{r}+\mathbf{R}) = \hat{T}(\mathbf{R})\,\psi(\mathbf{r})$, and $\psi(\mathbf{r}+\mathbf{R})$ is proportional to $\psi(\mathbf{r})$, as described by the eigenvalue for the translation operator. In order for the wave function not to diverge, the proportionality factor must have modulus one, as the translation operator can be repeatedly applied, and the eigenvalue equation for the translation operator takes the form

$$\psi(\mathbf{r}+\mathbf{R}) = e^{i\mathbf{k}\cdot\mathbf{R}}\psi(\mathbf{r}). \tag{9.101}$$

An energy eigenfunction for a particle in a periodic potential is thus characterized by a real vector, $\mathbf{k}$, and satisfies the Bloch condition. Bloch's theorem provides the understanding of electron transport.

10 Semiconductor Doping

The ubiquity of semiconductors in electronic equipment rests on the controlled manipulation that can turn a bad conductor into a conductor of either so-called n- or p-type. The conductivity of a semiconductor is highly sensitive to its content of impurities, and the first step in producing an n- or p-type semiconductor consists in material purification, minimizing unintentional (background) impurities, and then adding specific foreign atoms into the semiconductor, the doping process.

10.1 Donor Impurities

Imagine substituting one of its atoms in a pure semiconductor by an atom with an extra proton in its nucleus, $Z_{\text{imp}} = Z_{\text{host}} + 1$, and its accompanying extra electron for charge neutrality. The crystal then has one extra plus charge fixed in space at the site of the so-called impurity (the extra proton in the nucleus of the substituted atom), as depicted in Figure 10.1, and one excess electron in the otherwise ideal crystal. An example would be the case where a pentavalent atom (five valence electrons) is substituted in a crystal of tetravalent atoms (four valence electrons), such as is the case for a phosphorus atom substituted into a crystal of silicon atoms (being of almost the same size, lattice imperfection is negligible). The impurity atom is then called a donor impurity, since the substituted atom donates an electron to the system. Of interest is the quantum state available for the excess electron.

A single impurity atom in the huge host crystal is not expected to change the global properties of the system. The energy eigenfunctions and energy spectrum, the band structure, of the macroscopic host crystal indeed turn out to be robust, i.e. the Bloch functions of the host lattice are to a good approximation still energy eigenfunctions. In the ground state, the valence bands are therefore still filled just as for the pure semiconductor (a dielectric material, i.e. no free charges at zero temperature), the electrons being tied down in their covalent bonds. Of interest is the accommodation of the extra electron experiencing the periodic potential throughout the crystal and in addition an attractive potential due to the impurity ion, partly screened by the valence electrons, thus weakening the attraction. The energy levels available for the extra electron are thus described by the Hamiltonian of the pure semiconductor, $\hat{H}_0$, and an additional attractive potential, $V(\mathbf{r})$, describing the deviation from the periodic potential near the site of the impurity. The valence band states are occupied and thus unavailable according to the exclusion principle, but the extra electron

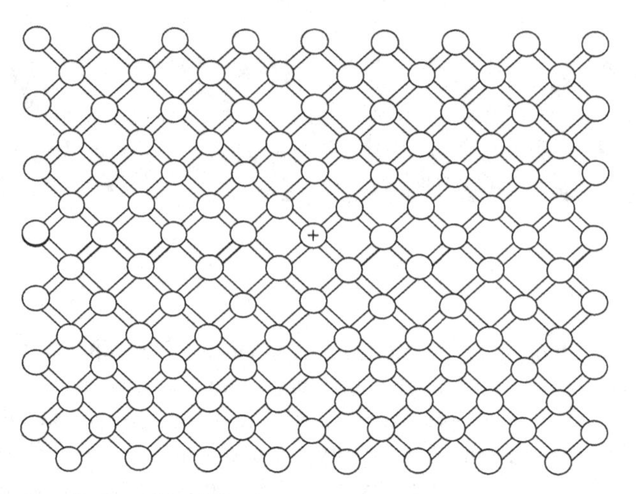

Figure 10.1 Donor impurity ion indicated by its additional plus charge in its nucleus.

is not simply accommodated in the unoccupied state of lowest energy in the conduction band, since a Bloch state is extended and has insignificant probability (inversely proportional to the volume, Ω, of the semiconductor) of being close to the impurity. Instead, its wave function is spatially localized near the impurity ion in whose attractive potential the energy is lowered.

The exclusion principle suppresses the tendency for the additional impurity electron to share common locations with the valence electrons (the bonds cannot accommodate much extra charge before being saturated). The excess electron is thus forced away from the attracting impurity (the electronic bonds near the impurity receive only a small additional electronic charge from the presence of the additional electron). The wave function of this so-called impurity state is thus expected to extend over many unit cells (but localized near the impurity ion on the scale of the size of the system). Such an extended wave function is not very sensitive to the crystal structure, the variations of which are on the scale of the lattice constant, its large extension averaging out such effects. We shall therefore assume that the impurity potential is described by a centrally symmetric potential of Coulombic form (we choose the origin of the reference frame as the location of the impurity atom)

$$V(\mathbf{r}) = -\frac{e^2}{4\pi\epsilon\,|\mathbf{r}|}, \tag{10.1}$$

where the strength of the bare positive impurity charge is screened by the valence (bonded) electrons as expressed by the dielectric constant, ϵ, of the pure semiconductor.

The impurity states for the excess electron are thus determined by the time-independent Schrödinger equation

$$\hat{H}\,\psi(\mathbf{r}) = E\,\psi(\mathbf{r}), \qquad \hat{H} = \hat{H}_0 + V(\mathbf{r}), \tag{10.2}$$

where, besides the periodic potential included in the pure semiconductor Hamiltonian, $\hat{H}_0$, the Hamiltonian contains the impurity potential, Eq. (10.1). The problem is that of an attractive Coulomb potential with the additional feature that the motion is not in free space, as for the hydrogen atom, but in a periodic potential, the eigenstates of $\hat{H}_0$ being the pure semiconductor Bloch states

$$\hat{H}_0\,\psi_{n,\mathbf{k}}(\mathbf{r}) = \epsilon_n(\mathbf{k})\,\psi_{n,\mathbf{k}}(\mathbf{r}), \qquad \psi_{n,\mathbf{k}}(\mathbf{r}) = \frac{1}{\sqrt{\Omega}}\,e^{i\mathbf{k}\cdot\mathbf{r}}u_{n,\mathbf{k}}(\mathbf{r}), \tag{10.3}$$

the Bloch functions being chosen in the form in Eq. (9.45) to have the small inverse volume parameter explicit.

10.2 Impurity States

A solution to the time-independent Schrödinger equation, Eq. (10.2), a localized impurity state, is sought as a superposition of allowed states in the crystal,

$$\psi(\mathbf{r}) = \sum_{n,\mathbf{k}\in \mathrm{Bz}_1}{}' \; a_n(\mathbf{k})\,\psi_{n\mathbf{k}}(\mathbf{r}), \tag{10.4}$$

where the prime indicates that the summation is only over unoccupied states, i.e. conduction bands, the valence band states of the crystal already occupied, and the Bloch functions, $\psi_{n\mathbf{k}}(\mathbf{r})$, are determined by Eq. (10.3). In order for the superposition, Eq. (10.4), to have the lowest possible energy, it should be a superposition of states at the bottom of the, typically degenerate, lowest conduction bands, as they have lowest energy. The coefficients, $a_n(\mathbf{k})$, are thus only non-zero in those regions, and the sought after positions of the energy levels in the attractive impurity potential are therefore expected to be situated below the conduction band edge. For definiteness, the bottom of the conduction bands is located at $\mathbf{k} = \mathbf{0}$ (the so-called Γ-point, as is the case for the materials of interest).[1] The functions $a_n(\mathbf{k})$ therefore vanish outside a small region near $\mathbf{k} = \mathbf{0}$. This energy-based requirement is in concordance with the basic assumption that the wave function, Eq. (10.4), should extend over a range large compared to the lattice constant, since the extension of a function and the range of k values in its Fourier transform obey the relationship $\Delta k\,\Delta r \sim 1$, giving $\Delta k \ll 1/a$.

Inserting the superposition, Eq. (10.4), into Eq. (10.2), and taking the scalar product with a conduction band Bloch function $\psi_{n\mathbf{k}}$ (i.e. multiplying both sides of Eq. (10.2) by the complex conjugate of $\psi_{n\mathbf{k}}$ and integrating over the volume of the crystal), and using the orthonormality of the Bloch functions, gives the equation for the expansion coefficients,

$$(\epsilon_n(\mathbf{k}) - E)a_n(\mathbf{k}) + \sum_{n',\mathbf{k}'\in \mathrm{Bz}_1} V_{nn'}(\mathbf{k},\mathbf{k}')\,a_{n'}(\mathbf{k}') = 0. \tag{10.5}$$

Use has been made of the property of the wave function, Eq. (10.4),

$$\hat{H}_0\,\psi(\mathbf{r}) = \sum_{n,\mathbf{k}\in \mathrm{Bz}_1}{}' \; a_n(\mathbf{k})\,\epsilon(\mathbf{k})\,\psi_{n\mathbf{k}}(\mathbf{r}) \tag{10.6}$$

[1] The case where the bottom of the conduction band is at point $\mathbf{k}_0$ is treated with equal ease by taking out the fast oscillations explicitly in the factor $e^{i\mathbf{r}\cdot\mathbf{k}_0}$ in the Bloch function. An example of this procedure is given in the envelope treatment of heterostructures in Section 12.2.

and the matrix elements of the potential are

$$V_{nn'}(\mathbf{k},\mathbf{k}') = \int_{\Omega} d\mathbf{r}\,\psi^*_{n,\mathbf{k}}(\mathbf{r})\,V(\mathbf{r})\,\psi_{n',\mathbf{k}'}(\mathbf{r}) = \frac{1}{\Omega}\int_{\Omega} d\mathbf{r}\,e^{-i\mathbf{r}\cdot(\mathbf{k}-\mathbf{k}')}\,V(\mathbf{r})\,u^*_{n,\mathbf{k}}(\mathbf{r})\,u_{n',\mathbf{k}'}(\mathbf{r}). \tag{10.7}$$

Since $a_n(\mathbf{k})$ vanishes outside a small region near $\mathbf{k} = \mathbf{0}$, only small values, $\mathbf{k}' \sim \mathbf{0}\ (\sim \mathbf{k})$, contribute in Eq. (10.5) to the sum. For such small wave vectors, the function $e^{-i(\mathbf{k}-\mathbf{k}')\cdot\mathbf{r}}V(\mathbf{r})$ in the integrand in Eq. (10.7) is a slowly varying function on the scale of the unit cell lattice constant. The function $u^*_{n\mathbf{k}}(\mathbf{r})\,u_{n'\mathbf{k}'}(\mathbf{r})$ is a periodic function of the position variable as specified by the Bravais lattice of the crystal in question. The integral, a product of a slowly varying function and a periodic function, is therefore according to formula Eq. (T.1) approximately

$$\begin{aligned} V_{nn'}(\mathbf{k},\mathbf{k}') &\simeq \frac{1}{\Omega^2}\int_{\Omega} d\mathbf{r}\,e^{-i(\mathbf{k}-\mathbf{k}')\cdot\mathbf{r}}V(\mathbf{r})\int_{\Omega} d\mathbf{r}\,u^*_{n\mathbf{k}}(\mathbf{r})\,u_{n'\mathbf{k}'}(\mathbf{r}) \\ &= \delta_{n,n'}\frac{1}{\Omega^2}\int_{\Omega} d\mathbf{r}\,e^{-i(\mathbf{k}-\mathbf{k}')\cdot\mathbf{r}}V(\mathbf{r})\int_{\Omega} d\mathbf{r}\,u^*_{n\mathbf{k}}(\mathbf{r})\,u_{n\mathbf{k}'}(\mathbf{r}), \end{aligned} \tag{10.8}$$

where orthogonality of the u functions in the band index has been used.

Since the Bloch vectors in question are small, $\mathbf{k} \sim \mathbf{0} \sim \mathbf{k}'$, and $u_{n\mathbf{k}}(\mathbf{r})$ has a weak dependence on the Bloch vector (recall that $\mathbf{k}$ is a parameter in Eq. (9.36) determining $u_{n\mathbf{k}}$), then $u^*_{n\mathbf{k}}(\mathbf{r})\,u_{n\mathbf{k}'}(\mathbf{r}) \simeq u^*_{n\mathbf{0}}(\mathbf{r})\,u_{n\mathbf{0}}(\mathbf{r})$ and

$$V_{nn'}(\mathbf{k},\mathbf{k}') \simeq \delta_{n,n'}\frac{1}{\Omega^2}\int_{\Omega} d\mathbf{r}\,e^{-i(\mathbf{k}-\mathbf{k}')\cdot\mathbf{r}}V(\mathbf{r})\int_{\Omega} d\mathbf{r}\,u^*_{n\mathbf{0}}(\mathbf{r})\,u_{n\mathbf{0}}(\mathbf{r}). \tag{10.9}$$

According to the normalization of the u functions, Eq. (9.50), $V_{nn'}(\mathbf{k},\mathbf{k}')$ becomes

$$V_{nn'}(\mathbf{k},\mathbf{k}') \simeq \frac{\Omega}{\Omega^2}\,\delta_{nn'}\int_{\Omega} d\mathbf{r}\,e^{-i(\mathbf{k}-\mathbf{k}')\cdot\mathbf{r}}V(\mathbf{r}) \equiv \frac{1}{\Omega}\,\delta_{nn'}\,V(\mathbf{k}-\mathbf{k}'). \tag{10.10}$$

Since the volume of the system is large, the Fourier transform can be obtained by integrating over spherical coordinates (inserting a convergence factor, $1/r \to e^{-\kappa r}/r$, and eventually letting $\kappa \to 0$, makes the integrations elementary, or simply recall Section 4.6.2), giving

$$V(\mathbf{k}-\mathbf{k}') = \int d\mathbf{r}\,e^{-i\mathbf{r}\cdot(\mathbf{k}-\mathbf{k}')}V(\mathbf{r}) = \frac{-e^2}{\epsilon(\mathbf{k}-\mathbf{k}')^2}. \tag{10.11}$$

Since $V_{nn'}(\mathbf{k},\mathbf{k}')$ is diagonal in the band indices, there is no mixing of the degenerate conduction bands, and we can therefore concentrate on each band separately. For simplicity, we consider the case of a single lowest conduction band, $n = c$, and Eq. (10.5) becomes

$$(\epsilon_n(\mathbf{k}) - E)\,a_n(\mathbf{k}) + \frac{1}{\Omega}\sum_{\mathbf{k}'\in \mathrm{Bz}_1} V(\mathbf{k}-\mathbf{k}')\,a_n(\mathbf{k}') = 0. \tag{10.12}$$

Since $a_n(\mathbf{k})$ is anticipated to vanish outside a small region near $\mathbf{k} = \mathbf{0}$, the conduction band energy is given by the effective mass approximation,

$$\epsilon_n(\mathbf{k}) \simeq \epsilon_n(\mathbf{0}) + \frac{1}{2} \sum_{\alpha,\beta=x,y,z} k_\alpha \frac{\partial^2 \epsilon_n(\mathbf{0})}{\partial k_\alpha \, \partial k_\beta} k_\beta = \epsilon_c + \frac{\hbar^2 k^2}{2m_c}, \tag{10.13}$$

where in the second equality, for simplicity, an isotropic dispersion is assumed, and the notation ϵ_c for the conduction band edge introduced. Equation (10.12) thus becomes

$$\left(\epsilon_c + \frac{\hbar^2 \mathbf{k}^2}{2m_c} - E\right) a_n(\mathbf{k}) - \frac{e^2}{\epsilon \Omega} \sum_{\mathbf{k}' \in \mathrm{Bz}_1} \frac{a_n(\mathbf{k}')}{(\mathbf{k} - \mathbf{k}')^2} = 0. \tag{10.14}$$

This equation is actually a familiar Schrödinger equation written in an unfamiliar form. Since $u_{n\mathbf{k}}$ is a slowly varying function of $\mathbf{k}$ compared to the peaked function $a_n(\mathbf{k})$, the impurity state wave function becomes

$$\begin{aligned} \psi(\mathbf{r}) &= \frac{1}{\sqrt{\Omega}} \sum_{\mathbf{k} \in \mathrm{Bz}_1} a_n(\mathbf{k}) e^{i\mathbf{k}\cdot\mathbf{r}} u_{n\mathbf{k}}(\mathbf{r}) \simeq u_{n\mathbf{0}}(\mathbf{r}) \frac{1}{\sqrt{\Omega}} \sum_{\mathbf{k} \in \mathrm{Bz}_1} a_n(\mathbf{k}) e^{i\mathbf{k}\cdot\mathbf{r}} \\ &= u_{n\mathbf{0}}(\mathbf{r}) \, \chi(\mathbf{r}), \end{aligned} \tag{10.15}$$

where

$$\chi(\mathbf{r}) \equiv \frac{1}{\sqrt{\Omega}} \sum_{\mathbf{k} \in \mathrm{Bz}_1} a_n(\mathbf{k}) \, e^{i\mathbf{k}\cdot\mathbf{r}} \tag{10.16}$$

is an envelope function modulating the Bloch function of the Bloch state at the bottom of the conduction band. Observing that

$$\Delta_{\mathbf{r}} \, \chi(\mathbf{r}) = \frac{1}{\sqrt{\Omega}} \sum_{\mathbf{k} \in \mathrm{Bz}_1} (-\mathbf{k}^2) \, a_n(\mathbf{k}) \, e^{i\mathbf{k}\cdot\mathbf{r}} \tag{10.17}$$

and multiplying Eq. (10.14) by $e^{i\mathbf{k}\cdot\mathbf{r}}/\sqrt{\Omega}$ and summing over the $\mathbf{k}$ values gives

$$\left(-\frac{\hbar^2}{2m_c} \Delta_{\mathbf{r}} - \tilde{E}\right) \chi(\mathbf{r}) - \frac{e^2}{\epsilon \Omega^{3/2}} \sum_{\mathbf{k},\mathbf{k}'} \frac{e^{i\mathbf{k}\cdot\mathbf{r}}}{(\mathbf{k} - \mathbf{k}')^2} a_n(\mathbf{k}') = 0, \tag{10.18}$$

where $\tilde{E} \equiv E - \epsilon_c$, the energy of the impurity state measured relative to the conduction band edge. In the large volume limit, the summation is an integral (recall that $\Delta\mathbf{k} = (2\pi)^3/\Omega$),

$$\frac{1}{\Omega^{3/2}} \sum_{\mathbf{k},\mathbf{k}' \in \mathrm{Bz}_1} \frac{e^{i\mathbf{k}\cdot\mathbf{r}}}{(\mathbf{k} - \mathbf{k}')^2} a_n(\mathbf{k}') = \frac{1}{\Omega^{1/2}} \sum_{\mathbf{k}' \in \mathrm{Bz}_1} a_n(\mathbf{k}') \int_{\mathrm{Bz}_1} \frac{d\mathbf{k}}{(2\pi)^3} \frac{e^{i\mathbf{k}\cdot\mathbf{r}}}{(\mathbf{k} - \mathbf{k}')^2}. \tag{10.19}$$

Since $a_n(\mathbf{k}')$ vanishes outside a small region near $\mathbf{k}' = \mathbf{0}$, the factor $(\mathbf{k} - \mathbf{k}')^{-2} \simeq \mathbf{k}^{-2}$ is small for large values of $\mathbf{k}$, and as the $\mathbf{k}$ integrand in addition has an oscillating exponential function, the $\mathbf{k}$ integration can be extended to all of $\mathbf{k}$-space, giving (shifting the integration variable to $\mathbf{q} = \mathbf{k} - \mathbf{k}'$)

$$\frac{1}{\Omega^{3/2}} \sum_{\mathbf{k},\mathbf{k}' \in \mathrm{Bz}_1} \frac{e^{i\mathbf{k}\cdot\mathbf{r}}}{(\mathbf{k}-\mathbf{k}')^2} a_n(\mathbf{k}') \simeq \frac{1}{\Omega^{1/2}} \sum_{\mathbf{k}'} a_n(\mathbf{k}')\, e^{i\mathbf{k}'\cdot\mathbf{r}} \frac{1}{(2\pi)^3} \int d\mathbf{q}\, \frac{e^{i\mathbf{q}\cdot\mathbf{r}}}{\mathbf{q}^2}$$

$$= \frac{1}{4\pi\,|\mathbf{r}|} \frac{1}{\Omega^{1/2}} \sum_{\mathbf{k}'} a_n(\mathbf{k}')\, e^{i\mathbf{k}'\cdot\mathbf{r}}, \tag{10.20}$$

where performing the Fourier transform to get the $1/|\mathbf{r}|$ potential in fact is superfluous since it is the inverse of the previous transformation, Eq. (10.11). Equation (10.20) thus reads[2]

$$\frac{1}{\Omega^{3/2}} \sum_{\mathbf{k},\mathbf{k}' \in \mathrm{Bz}_1} \frac{e^{i\mathbf{k}\cdot\mathbf{r}}}{(\mathbf{k}-\mathbf{k}')^2} a_n(\mathbf{k}') \simeq \frac{1}{4\pi\,|\mathbf{r}|} \frac{1}{\Omega^{1/2}} \chi(\mathbf{r}) \tag{10.21}$$

and Eq. (10.18) becomes the envelope function equation

$$\left(-\frac{\hbar^2}{2m_c} \Delta_{\mathbf{r}} - \frac{e^2}{4\pi\epsilon\,|\mathbf{r}|} \right) \chi(\mathbf{r}) = \tilde{E}\,\chi(\mathbf{r}). \tag{10.22}$$

Normalization of the impurity state, Eq. (10.15), again using the fact that the integral of a product of a slowly varying function, $|\chi_i(\mathbf{r})|^2$, and a periodic function, $|u_{c\mathbf{0}}(\mathbf{r})|^2$, can be approximated according to Eq. (T.1), gives

$$1 = \int_\Omega d\mathbf{r}\, |\chi(\mathbf{r})|^2\, |u_{c\mathbf{0}}(\mathbf{r})|^2 \simeq \frac{1}{\Omega} \int_\Omega d\mathbf{r}\, |\chi(\mathbf{r})|^2 \int_\Omega d\mathbf{r}\, |u_{c\mathbf{0}}(\mathbf{r})|^2 = \int_\Omega d\mathbf{r}\, |\chi(\mathbf{r})|^2. \tag{10.23}$$

The envelope function is thus constructed to be normalized. From knowledge of the ground-state hydrogen wave function, the envelope function corresponding to lowest energy is

$$\chi_1(\mathbf{r}) = R_{10}(r)\, Y_{00}(\theta,\varphi) = \frac{2}{a_*^{3/2}} \sqrt{\frac{1}{4\pi}}\, e^{-r/a_*}. \tag{10.24}$$

The problem of finding the impurity levels has thus reduced to the hydrogen atom problem, except that different parameters appear: the permittivity of vacuum, the dielectric constant ϵ_0, has been replaced by the dielectric constant of the semiconductor; and the free electron mass, m_e, has been replaced by the conduction band mass. The impurity energy levels (the principal quantum number is denoted by j since n already denotes the band index) can therefore immediately be written down as

$$\tilde{E}_j = -\left(\frac{e^2}{4\pi\epsilon\hbar} \right)^2 \frac{m_c}{2j^2} = -\frac{\hbar^2}{m_c a_*^2 j^2} \simeq -\frac{m_e}{m_c} \left(\frac{a_0}{a_*} \right)^2 \frac{13.6\,\mathrm{eV}}{j^2}, \qquad j = 1, 2, \ldots, \tag{10.25}$$

where a_0 is the Bohr radius and a_* is the *Bohr radius* of the impurity state,

$$a_* = \frac{\hbar^2 4\pi\epsilon}{m_c e^2}. \tag{10.26}$$

[2] The left-hand side is the Fourier transform of the convolution of $V(\mathbf{k}-\mathbf{k}')$ and $a_n(\mathbf{k}')$, which is the product of their Fourier transforms.

The impurity energy levels in a semiconductor are thus, as expected, located below the conduction band edge, as depicted in Figure 10.2, the lowest level having energy

$$E_1 = \epsilon_c + \tilde{E}_1 \simeq \epsilon_c - \frac{m_e}{m_c}\left(\frac{a_0}{a_*}\right)^2 \times 13.6\,\text{eV}. \tag{10.27}$$

At zero temperature, the electron donated by the impurity atom will thus occupy the lowest impurity level, as indicated by the black dot in Figure 10.2. Of interest is how far below the band edge the level is.

The binding energy of an impurity state,

$$\tilde{E}_j = E_j^{\text{H}}\left(\frac{\epsilon_0}{\epsilon}\right)^2 \frac{m_c}{m_e}, \tag{10.28}$$

is typically three orders of magnitude smaller than that of hydrogen, and the difference between the donor levels and the bottom of the conduction band (the conduction band edge) is thus typically in the milli-electronvolt range,

$$\epsilon_c - \epsilon_d = \left(\frac{e^2}{4\pi\epsilon}\right)^2 \frac{m_c}{2\hbar^2} = \left(\frac{\epsilon_0}{\epsilon}\right)^2 \frac{m_c}{m_e} \times 13.6\,\text{eV} \sim 10^{-3}\,\text{eV} \equiv 1\,\text{meV}, \tag{10.29}$$

which corresponds to room temperature thermal energy, $kT_{300\text{K}} \simeq 25$ meV.

To get numbers into play, consider, for example, the gallium arsenide, GaAs, III–V compound semiconductor. This material has dielectric constant 13 times larger than that of vacuum, $\epsilon = 13\epsilon_0$ and the conduction band mass $m_c \simeq 0.07 m_e$. In the case of GaAs, the impurity level is thus, according to the presented theory, situated below the conduction band by the amount

$$\tilde{E}_1 \simeq 5.67\,\text{meV}. \tag{10.30}$$

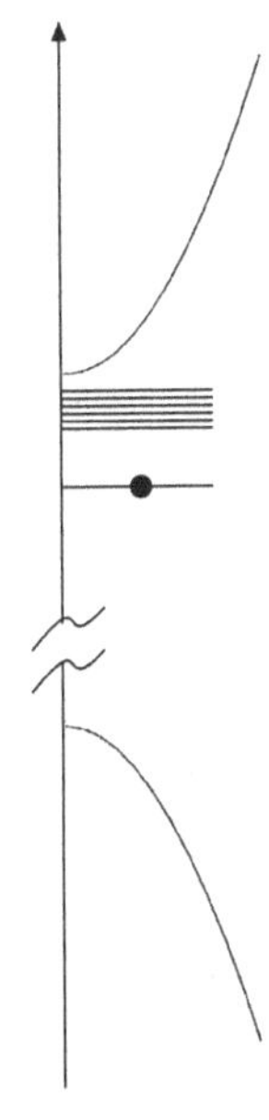

Figure 10.2 Donor energy levels are located just below the conduction band edge.

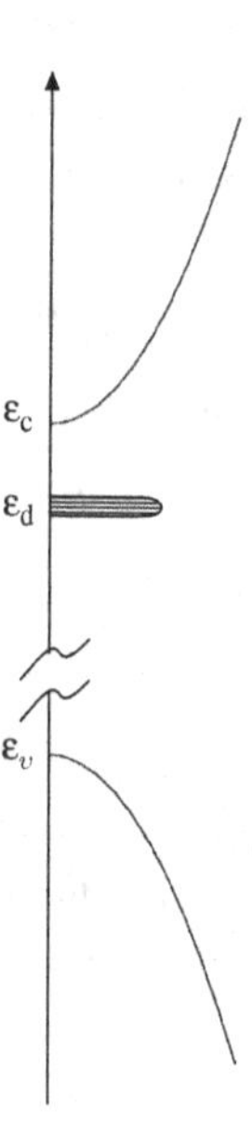

Figure 10.3 Donor levels located a few meV below the conduction band edge.

If the semiconductor is doped with many such impurities, it will at zero temperature have many electrons in impurity states. The lowest level for each impurity is thus due to slight level splitting located, as depicted in Figure 10.3, in a small pocket just below the conduction band. So, for example, doping silicon or germanium with many pentavalent atoms, such as arsenic or phosphorus, at zero temperature, electrons occupy these impurity levels, situated just a few meV below the conduction band edge.

At room temperature, all the impurity state electrons are excited to the conduction band, or, as we say, donated. Such a doped semiconductor is called an n-type semiconductor, a negative charge carrier semiconductor, the electrons in the conduction band acting like electrons in a metal. Doping significantly increases the conductivity by orders of magnitude. If the concentration ratio of donor atoms to host atoms is on the order of 10^{-7}, $n_D/n_H \sim 10^{-7}$, as for say n_{As}/n_{Ge}, this increases the concentration of conduction electrons 2000 times. In this case, the density of electrons in the conduction band is dominated by the donated electrons, not the thermally excited electrons as in the intrinsic case, and we have a so-called extrinsic semiconductor.

Donor impurities, being vital for turning the semiconductor into a good conductor, however, will also be a source of scattering. When an electron is excited from an impurity state into the conduction band, an ionized impurity appears and its corresponding potential leads to scattering of the conduction electrons.

Exercise 10.1 Determine the location of the donor level for the case of a phosphorus atom introduced into a silicon crystal.

10.3 Model Consistency

Confronted with a complicated many-body problem, approximations were made based on qualitative arguments, and not founded on the existence of a small parameter upon which a controlled approximation could be built. The self-consistency of the produced results should therefore be checked, i.e. output quantities exhibit the input assumptions. For gallium arsenide, GaAs, the impurity *Bohr radius* has the magnitude $a_* = \epsilon m_e a_0 / \epsilon_0 m_c \simeq 186 a_0 \simeq 98$ Å. The extension of the impurity energy level wave function indeed is large compared to the lattice constant, which for GaAs is approximately 6 Å. The envelope wave function of the impurity state is thus a hydrogen wave function of gigantic extension on the scale of the lattice constant of the host semiconductor, as depicted in Figure 10.4. The assumption that the extension of the impurity state is large compared to the lattice constant is thus self-consistently confirmed.

The orthogonality of the impurity states to those of the host system is clearly true for the valence states, since by construction the impurity state is built by only conduction band states. For the state at the bottom of the conduction band, $\psi_{c\mathbf{0}}(\mathbf{r}) = u_{c\mathbf{0}}(\mathbf{r})/\sqrt{\Omega}$, the overlap with the impurity state is

$$\frac{1}{\sqrt{\Omega}} \int_{\Omega} d\mathbf{r}\; u_{c\mathbf{0}}(\mathbf{r})^* \, u_{c\mathbf{0}}(\mathbf{r}) \, \chi_i(\mathbf{r}) \simeq \frac{1}{\sqrt{\Omega}} \frac{1}{\Omega} \int_{\Omega} d\mathbf{r}\; \chi_i(\mathbf{r}) \int_{\Omega} d\mathbf{r}\; |u_{c\mathbf{0}}(\mathbf{r})|^2$$

$$= \frac{1}{\sqrt{\Omega}} \int_{\Omega} d\mathbf{r}\; \chi_i(\mathbf{r}) = 4\sqrt{4\pi}\, \frac{a_*^{3/2}}{\sqrt{\Omega}}, \qquad (10.31)$$

where the last equality follows from the envelope function being localized on the scale of the size of the system whereupon the integration can be extended to all space. The overlap is thus vanishingly small even for sub-micrometer sized semiconductor samples, on the order of $\sqrt{v/\Omega}$, where v is the volume of the unit cell. A similar estimate follows for the other conduction band states, as $u_{n\mathbf{k}}$, with the normalization choice Eq. (9.50), is on the order of the square root of the unit cell volume.

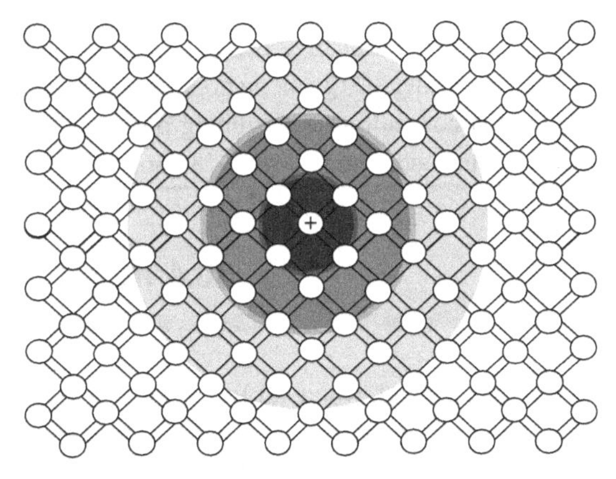

Figure 10.4 Donor level wave function extending over many lattice constants.

From the general relationship between extension in real space and k-space, it follows that the function $a(\mathbf{k})$ is peaked near $\mathbf{k} = \mathbf{0}$, but for explicitness we invert the Fourier series for the expression for the impurity state,

$$\chi_1(\mathbf{r}) = \frac{1}{\sqrt{\Omega}} \sum_{\mathbf{k} \in \mathrm{Bz}_1} a_c^{(1)}(\mathbf{k})\, e^{i\mathbf{k}\cdot\mathbf{r}}, \tag{10.32}$$

giving for the expansion coefficients

$$a_c^{(1)}(\mathbf{k}) = \frac{1}{\sqrt{\Omega}} \int_{\Omega} d\mathbf{r}\, e^{-i\mathbf{k}\cdot\mathbf{r}} \chi_1(\mathbf{r}), \tag{10.33}$$

which is an elementary integral,

$$a_c^{(1)}(\mathbf{k}) = \frac{1}{\sqrt{\Omega \pi a_*^3}} \int d\mathbf{r}\, e^{-i\mathbf{k}\cdot\mathbf{r}}\, e^{-r/a_*} = \frac{8\sqrt{\pi}}{\sqrt{\Omega a_*^5}} \left(\frac{1}{k^2 + a_*^{-2}} \right)^2. \tag{10.34}$$

We note that $a_c^{(1)}(\mathbf{k})$ is a Lorentzian of width a_*^{-2}, the maximum at $\mathbf{k} = \mathbf{0}$, for which $a_c^{(1)}(\mathbf{0}) = 8\sqrt{\pi}\, a_*^{3/2}/\sqrt{\Omega}$. In the case of GaAs, $a_* \sim 98$ Å, giving a width a_*^{-1} of the Lorentzian on the order of 0.01 Å^{-1}, which is only a tiny fraction of the size of the Brillouin zone, which is on the order of $\pi/6$ Å^{-1} ~ 0.2 Å^{-1}. The assumption of the functions $a(\mathbf{k})$ being peaked near $\mathbf{k} = \mathbf{0}$ is thus confirmed.

In the model, the donor energy levels only depend on the dielectric constant and the conduction band effective mass of the host crystal. When comparing with experiment, this turns out not quite to be the case. This is to be expected in view of the assumption for the impurity potential. The impurity potential is not just Coulombic at short distance due to the core electrons, and the presence of the valence (bonded) electrons violates the assumed spherical symmetry of the impurity potential, and the impurity atom does not substitute perfectly for the host atom. However, we note that there is even good quantitative agreement for the impurity energy level. For example, if GaAs is doped with silicon, experiment gives the value 5.8 meV for the binding energy, quite close to the model value in Eq. (10.30); and doping with germanium, the experimental value is 6.1 meV. The obtained result is thus, even for many practical purposes, quantitatively accurate.

Exercise 10.2 Consider 100 g silicon doped homogeneously with 4 μg phosphorus. Calculate the electron concentration, assuming one electron per phosphorus atom. Silicon has density 2.33 g/cm^3 and atomic weight 28.086, and phosphorus has atomic weight 30.974.

10.4 Holes

Consider a valence band, where all Bloch states except one are occupied, i.e. one electron has been removed from the valence band, say, the electron in Bloch state $\mathbf{k}_e$ of given spin.[3]

[3] Exciting an electron to the conduction band by photon absorption being an example. How to remove electrons from the valence band in a permanent way is discussed in the next section.

The total probability current density carried by the valence band with one empty state is the total contribution of the many occupied electron states (the factor of 2 accounts for the spin degeneracy),

$$\mathbf{S}_e = \frac{2}{\Omega} \sum_{\mathbf{k} \in Bz_1} \mathbf{v}_v(\mathbf{k}) - \frac{1}{\Omega} \mathbf{v}_v(\mathbf{k}_e) = -\frac{1}{\Omega} \mathbf{v}_v(\mathbf{k}_e), \tag{10.35}$$

where the second equality follows because the total current for a filled band is zero. Compared to the filled band case, the flux of the empty state is subtracted, because a particle flux opposite to that of the Bloch state $\mathbf{k}_e$ is present in the crystal. The electric current density for the situation is, setting for simplicity $\Omega = 1$,

$$\mathbf{j}_e = -e\mathbf{S}_e = e\mathbf{v}_v(\mathbf{k}_e). \tag{10.36}$$

Instead of considering the electrical properties of the valence band in terms of the occupied electron states, we seek a description referring only to the empty state, the absence of an electron, a so-called hole.

The absence of electron probability due to the absent Bloch electron is given by $|\psi_{v\mathbf{k}_e}(\mathbf{x})|^2$. Define correspondingly the *hole* Bloch vector as minus the Bloch vector of the empty state, $\mathbf{k}_h \equiv -\mathbf{k}_e$, and associate a hole wave function according to $\psi^{(h)}_{v\mathbf{k}_h}(\mathbf{x}) \equiv \psi_{v\mathbf{k}_e}(\mathbf{x}) = \psi_{v-\mathbf{k}_h}(\mathbf{x}) = \psi^*_{v\mathbf{k}_h}(\mathbf{x})$. The probability current density calculated in the hole state, $\psi^{(h)}_{n\,\mathbf{k}_h}(\mathbf{x}) = \psi^*_{v\mathbf{k}_h}(\mathbf{x})$, carries, due to the complex conjugation, a minus sign relative to Bloch state $\mathbf{k}_h$,

$$\mathbf{S}_h = -\mathbf{v}_v(\mathbf{k}_h) = \mathbf{v}_v(-\mathbf{k}_h) = \mathbf{v}_v(\mathbf{k}_e). \tag{10.37}$$

The electric current, Eq. (10.36), can then be represented as a hole state current as

$$\mathbf{j}_h \equiv \mathbf{j}_e = -e\mathbf{S}_e = e\mathbf{v}_v(\mathbf{k}_e) = e_h\mathbf{S}_h, \tag{10.38}$$

where the hole charge is defined as minus the electron charge, the positive charge $e_h \equiv e = |e|$. The electric current is now expressed in terms of the occupation of a single hole instead of in terms of the huge number of occupied electron states.

In the presence of an electric field, electron Bloch states change in time according to the Bloch vector dynamics, $\hbar\dot{\mathbf{k}} = -e\mathbf{E}$. The occupation dynamics of electrons for the considered valence band situation is therefore as depicted in Figure 10.5, where the unoccupied state is indicated by an open circle. The empty state will thus, in the next instant, be the one indicated in the right-hand diagram in Figure 10.5.

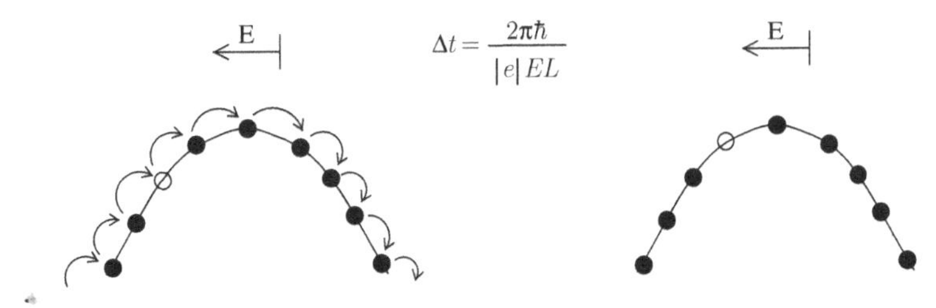

Figure 10.5 Motion of absent Bloch electron.

In the presence of an electric field, the equation of motion for the hole Bloch vector $\mathbf{k}_h \equiv -\mathbf{k}_e$ is thus, according to Eq. (9.100),

$$\hbar \frac{d\mathbf{k}_h(t)}{dt} = -\hbar \frac{d\mathbf{k}_e(t)}{dt} = e\mathbf{E} = e_h\mathbf{E}, \tag{10.39}$$

i.e. the equation of motion corresponding to a Bloch particle of positive charge, as the hole is endowed with.

The energy of a hole is defined as *minus* the energy of the state in which the electron is missing, up to a constant, which is chosen to be the valence band edge,

$$\epsilon_v^{(h)}(\mathbf{k}_h) \equiv \epsilon_v - \epsilon_v(\mathbf{k}_e) = \epsilon_v - \epsilon_v(-\mathbf{k}_h) = \epsilon_v - \epsilon_v(\mathbf{k}_h). \tag{10.40}$$

The deeper down in the valence band the level of the missing electron is, the larger the positive energy of the hole state (reflecting, in electron language, that the energy needed to create such a state is larger, i.e. the larger is the energy needed to remove an electron occupying such a state). The energy spectrum for the hole states is, according to Eq. (10.40), obtained by flipping the electron state dispersion relation around the valence band edge and mirror-reflecting the dispersion relation in $\mathbf{k} = \mathbf{0}$, as depicted in Figure 10.6 (the value of the highest energy level in the valence band assumed at $\mathbf{k} = \mathbf{0}$). An occupied state in the hole spectrum moves in the opposite direction in $\mathbf{k}$-space compared to an electron.

It is convenient to draw the energy axis for holes downwards, as done in Figure 10.7. The electric current carried by a valence band is, instead of being described in terms of a huge number of filled electron states, described in terms of an occupied hole. In the case of many empty electron states, instead, the description is in terms of the corresponding occupied hole states, their electrodynamics that of positively charged particles.

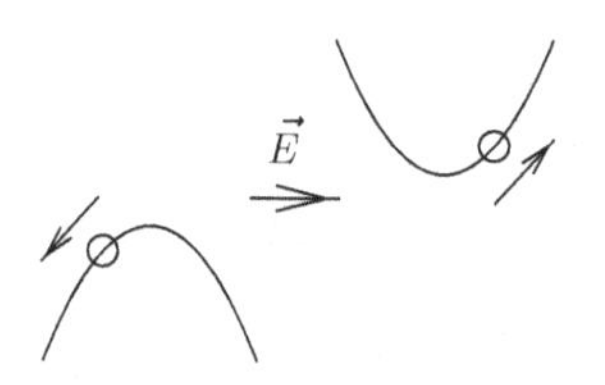

Figure 10.6 Absence of an electron converted to the presence of a hole state.

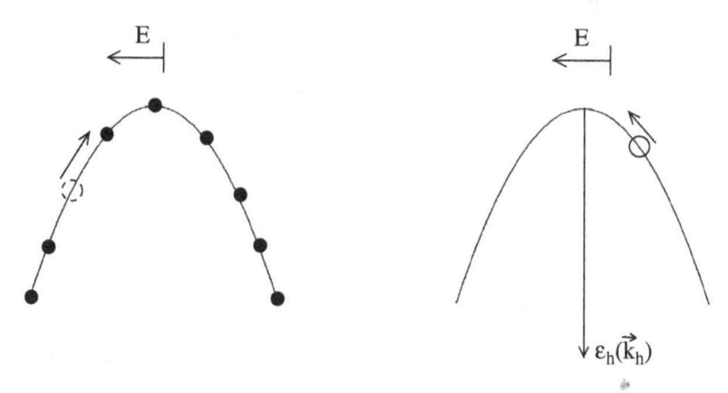

Figure 10.7 Corresponding electron and hole pictures.

The group velocity for a hole is

$$\mathbf{v}_v^{(h)}(\mathbf{k}_h) \equiv \frac{\partial \epsilon_v^{(h)}(\mathbf{k}_h)}{\partial \mathbf{k}_h} = -\frac{\partial \epsilon_v(\mathbf{k}_h)}{\partial \mathbf{k}_h} = \frac{\partial \epsilon_v(\mathbf{k}_e)}{\partial \mathbf{k}_e} = \mathbf{v}_v(\mathbf{k}_e) = -\mathbf{v}_v(\mathbf{k}_h). \tag{10.41}$$

The group velocity of a hole in hole state $\mathbf{k}$ is thus the opposite of the group velocity of an electron in electron state $\mathbf{k}$, $\mathbf{v}_n^{(h)}(\mathbf{k}) = -\mathbf{v}_n(\mathbf{k})$. Electron and hole states with the same Bloch vector thus move in opposite directions in real space.

At finite temperatures, electron states in the valence band will only be partially occupied in accordance with the Fermi–Dirac distribution. The probability for a hole state to be occupied represents the absence of electron occupation of the corresponding electron state, and equals in this case the probability that the electron state is unoccupied, i.e. one minus the probability that the electron state is occupied. In thermal equilibrium, the probability that a hole state is occupied is thus

$$f_{(h)}(\mathbf{k}_h) \equiv 1 - f(\epsilon_v(\mathbf{k}_e)) = 1 - f(\epsilon_v(\mathbf{k}_h)) = \frac{1}{e^{(\mu-\epsilon_v(\mathbf{k}_h))/kT} + 1}. \tag{10.42}$$

At room temperature, a valence band in a semiconductor is only slightly thermally depopulated of its electrons, as discussed in Section 9.9. A more efficient way of creating holes is called for.

10.5 Acceptor Impurities

Instead of providing conduction electrons by donor impurity doping, as discussed in Section 10.2, a semiconductor can also be made a good conductor by making the valence band electrically active. To achieve this, a doping procedure is needed that extracts electrons from the valence band, thereby providing holes to carry electric current. This is achieved if an impurity atom whose valence is one less than that of the host atoms of the semiconductor is substituted, an acceptor impurity, for example, introducing, say, a gallium atom on one of the sites of an otherwise perfect germanium crystal, or boron into a silicon crystal. The charge situation is then that of a fixed negative charge $-e$ placed on top of a host atom at the impurity site, representing the missing proton in the gallium nucleus compared to the germanium nucleus, and the crystal has one less valence electron to accommodate. A two-dimensional rendition is depicted in Figure 10.8.

It is not energetically favorable to just remove an electron from the top of the valence band, since such an absence of electronic charge, i.e. the presence of a hole, is extended over the whole crystal and does not lead to a lowering of the energy of the hole from the attraction by the negative charge on the impurity atom. Instead, a superposition of holes (absence of electrons) in states near the top of the valence band is localized near the impurity charge, and can take advantage of the attractive negative charge representing the missing nuclear charge.[4] In the covalent bonding picture of the missing electron,

[4] In practice, there will often be several degenerate valence bands from which the localized state should be made. But since they are dealt with independently, just like the degenerate conduction bands in the donor case, only one valence band is considered.

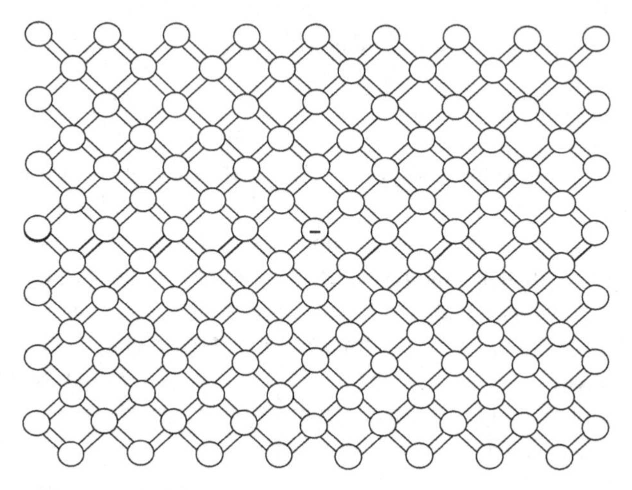

Figure 10.8 Semiconductor doped with an acceptor impurity.

the localized state corresponds to partially depleted bonds in the vicinity of the impurity. Instead of describing the state in terms of the absence of electrons (depleted covalent bonds), the hole description is adopted. The charge configuration of the state is thus perceived as a spread-out positive hole charge distribution (neutralizing the missing electronic bond charge) bound to the negative impurity charge at the impurity site, a bound hole state. The model Hamiltonian for the bound hole state is thus (analogous to the one in the donor impurity case)

$$\hat{H}_h = \hat{H}_0^{(h)} - \frac{e^2}{4\pi\epsilon\,|\mathbf{r}|}, \tag{10.43}$$

where the pure semiconductor *hole* Hamiltonian, $\hat{H}_0^{(h)}$, according to Eq. (10.40), is given by

$$\hat{H}_0^{(h)} = \epsilon_v - \hat{H}_0, \tag{10.44}$$

i.e.

$$\hat{H}_0^{(h)}\,\psi_{v\mathbf{k}}^{(h)}(\mathbf{r}) = (\epsilon_v - \epsilon_v(\mathbf{k}))\psi_{v\mathbf{k}}^*(\mathbf{r}) = \epsilon_v^{(h)}(\mathbf{k})\psi_{v\mathbf{k}}^{(h)}(\mathbf{r}), \tag{10.45}$$

the maximum of the valence band edge assumed at $\mathbf{k} = \mathbf{0}$.[5]

Of interest are the bound hole solutions, i.e. the solutions of the time-independent Schrödinger equation,

$$\hat{H}_h\,\psi_h(\mathbf{r}) = E_h\,\psi_h(\mathbf{r}). \tag{10.46}$$

The localized hole state is a superposition of hole states, which, in order to give lowest energy due to the attractive potential, are those of lowest hole energy, i.e. close to the top of the valence band edge (the complex conjugation of hole functions, $\psi_{v\mathbf{k}}^*(\mathbf{r})$, can be kept

[5] The case where the top of the valence band is at a different point, $\mathbf{k}_0$, is treated with equal ease, amounting to taking out fast spatial variation in the phase factor, as exemplified by the envelope treatment of heterostructures in Section 12.2.

or as in the following left out, the results being unchanged, as $a_v(\mathbf{k})$ will be symmetric around $\mathbf{k} = \mathbf{0}$),

$$\psi_h(\mathbf{r}) = \sum_{\mathbf{k} \in \mathrm{Bz}_1} a_v(\mathbf{k})\, \psi_{v\mathbf{k}}(\mathbf{r}), \tag{10.47}$$

where the absolute square of a coefficient, $|a_v(\mathbf{k})|^2$, is the probability that the hole state $\mathbf{k}$ in the valence band is occupied.

Proceeding analogously to the case of the donor impurity problem, i.e. taking the scalar product of Eq. (10.46) with the Bloch state $\psi_{v\mathbf{k}}$, the energy of the bound hole state is determined by the Schrödinger equation,

$$(\epsilon_v - \epsilon_v(\mathbf{k}) - E_h) a_v(\mathbf{k}) + \frac{1}{\Omega} \sum_{\mathbf{k}' \in \mathrm{Bz}_1} V(\mathbf{k} - \mathbf{k}')\, a_v(\mathbf{k}') = 0. \tag{10.48}$$

Since only hole states near the band edge contribute, the effective mass approximation can be adopted and, employing the isotropic model,

$$\epsilon_v(\mathbf{k}) = \epsilon_v - \frac{\hbar^2 \mathbf{k}^2}{2m_h}, \tag{10.49}$$

where, since the curvature of the valence band is negative, the hole mass m_h is positive. The hole state energy spectrum near the top of the valence band is therefore

$$\epsilon_v^{(h)}(\mathbf{k}) = \epsilon_v - \epsilon_v(\mathbf{k}) = \frac{\hbar^2 \mathbf{k}^2}{2m_h}. \tag{10.50}$$

The bound hole state

$$\begin{aligned} \psi_h(\mathbf{r}) &= \frac{1}{\sqrt{\Omega}} \sum_{\mathbf{k} \in \mathrm{Bz}_1} a_v(\mathbf{k})\, u_{v\mathbf{k}}(\mathbf{r})\, e^{i\mathbf{k}\cdot\mathbf{r}} \simeq u_{v\mathbf{0}}(\mathbf{r}) \frac{1}{\sqrt{\Omega}} \sum_{\mathbf{k} \in \mathrm{Bz}_1} a_v(\mathbf{k})\, e^{i\mathbf{k}\cdot\mathbf{r}} \\ &= u_{v\mathbf{0}}(\mathbf{r})\, \chi_h(\mathbf{r}) \end{aligned} \tag{10.51}$$

is specified by the hole envelope function,

$$\chi_h(\mathbf{r}) \equiv \frac{1}{\sqrt{\Omega}} \sum_{\mathbf{k} \in \mathrm{Bz}_1} a_v(\mathbf{k})\, e^{i\mathbf{k}\cdot\mathbf{r}}, \tag{10.52}$$

extending over many unit cells of the crystal, as depicted in Figure 10.9.

Quite analogous to the donor impurity case, Eq. (10.48) is turned into an equation for the hole envelope function determining the bound hole energies,

$$\left(-\frac{\hbar^2}{2m_h} \Delta_{\mathbf{r}} - \frac{e^2}{4\pi\epsilon|\mathbf{r}|} \right) \chi_h(\mathbf{r}) = E_h\, \chi_h(\mathbf{r}), \tag{10.53}$$

and the lowest bound hole state has the energy

$$E_h = E_1^{\mathrm{H}} \times \left(\frac{\epsilon_0}{\epsilon} \right)^2 \frac{m_h}{m_e} \equiv \epsilon_a, \tag{10.54}$$

where E_1^{H} is the ground-state energy of hydrogen, i.e. $E_1^{\mathrm{H}} \simeq -13.6\,\mathrm{eV}$. Recalling the convention of downward increasing hole energy, the bound hole level is typically a

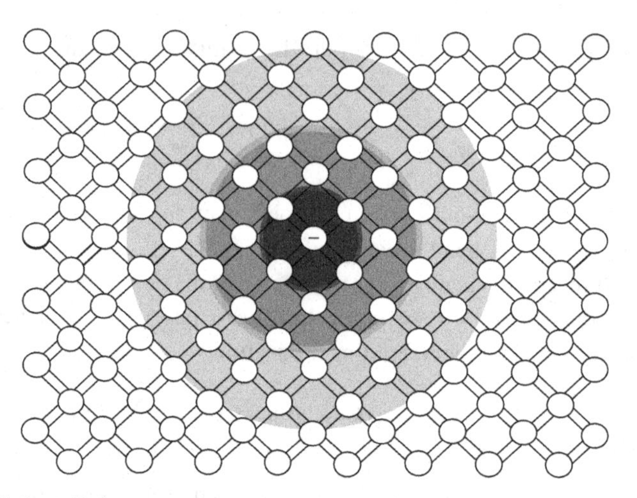

Figure 10.9 Wave function for the bound hole state.

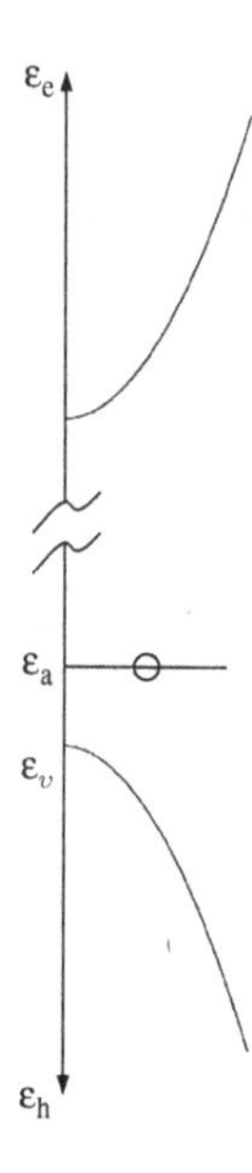

Figure 10.10 Bound hole level.

few milli-electronvolts lower in energy compared to valence band holes, as depicted in Figure 10.10.

In the electron picture, the bound hole state corresponds to the absence of electron probability spread out as described by the opposite charge of the neutralizing bound hole describing the depletion of the stronger electron bonds of the pure semiconductor – a gigantic hydrogenic wave function, a state of absence of electrons, describing an available electron energy level situated a few meV above the valence band edge. At zero temperature, this electron state is empty; but at room temperature, the state will be filled by electron probability corresponding to partially depleting electrons from the top of the valence band, i.e. creating partially filled hole states in the valence band. Or, in *hole* language, the bound hole, at finite temperature, is excited to a hole state in the valence band. When a semiconductor is doped with many such impurity atoms, a pocket of bound hole levels appears, so-called acceptor levels. At room temperature, electrons from the valence band are excited to occupy the acceptor levels, i.e. holes appear in the valence band, acting as positively charged particles able to conduct current. Such type of doping, acceptor impurities providing holes, gives a p-type semiconductor, a positive charge carrier semiconductor. The bound holes are now absent (the excited electrons restoring the covalent bonds of the perfect host semiconductor) and the impurity charges no longer screened, which causes scattering of the otherwise free valence band holes.

Exercise 10.3 Determine the location of the acceptor level for the case of a boron atom introduced into a silicon crystal.

Exercise 10.4 Consider 100 g silicon doped homogeneously with 4 μg aluminum. Calculate the hole concentration, assuming one hole per aluminum atom. Silicon has density 2.33 g/cm^3 and atomic weight 28.086, and aluminum has atomic weight 26.982.

10.6 Doping Methods

Doping of semiconductors can be done with varying sophistication and related cost. A way of introducing donor atoms is by diffusion at high temperature from the surface of the semiconductor (say, a manufacturing process of one hour at 1000 K for silicon); this method will have an uneven concentration of impurity atoms, with most of them at the surface. A solid layer of dopants can be deposited (painted) on the surface, and excess material removed after completed diffusion, or the dopants can be supplied to the surface by a gaseous compound containing the wanted dopant (for example, AsH_3, BH_3, $AlCl_3$). A third way involves nuclear reactions, for example, bombarding a large piece of silicon with neutrons (provided by a nuclear reactor) will transmute some of the silicon nuclei into phosphorus nuclei – the nuclear reaction is ${}^{30}_{14}\mathrm{Si}(\mathrm{n},\gamma){}^{31}_{14}\mathrm{Si}(-,\beta^-){}^{30}_{15}\mathrm{P}$. Transmutation doping provides the most even distribution of dopants and is important for the tolerance in design of integrated circuits. Another way of doping is ion implantation, using an accelerator to shoot in the impurities. Changing the energy of the ions emerging from the accelerator changes the depth of penetration, making the distribution and density of dopants controllable.

10.7 Carrier Densities

A semiconductor can thus be turned into a conductor where electrons are the current carriers, an n-type semiconductor, or one where holes carry the current, a p-type semiconductor. In real semiconductors, both types of impurities are present, and both electrons and holes

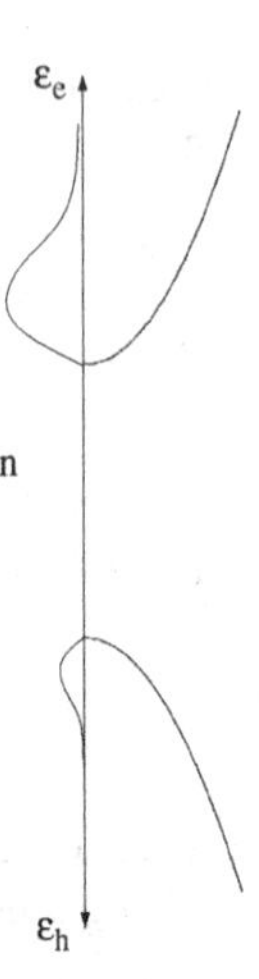

Figure 10.11 Occupation of electrons and holes in an n-type semiconductor.

are generally present. An n-type semiconductor thus refers to the case where mainly electrons are the current carriers, and similarly with p-type.

In an n-type semiconductor at room temperature, the populations of the conduction and valence bands are as depicted on the left side in Figure 10.11. On the right side in Figure 10.11 are shown the density of states of electrons, $N_c(E)$, in the conduction band and the density of states of holes, $N_h(E)$, in the valence band. The density of electrons at a certain energy, ϵ, is thus $N_c(\epsilon)f(\epsilon)$, where f is the Fermi function, and the density of holes in the valence band is $N_c(\epsilon)(1 - f(\epsilon))$. In an n-type semiconductor, at room temperature, there is thus a lot of electrons in the conduction band, located mainly near the conduction band edge, and only a tiny fraction of holes near the valence band edge, as thermal excitation over the band gap is small.

In a p-type semiconductor, the holes are the majority carriers, whereas the electrons are the minority carriers, as depicted in Figure 10.12.

Consider a semiconductor where both donor and acceptor impurities are present. The impurity concentration is assumed not excessive so that its influence on the band structure and thereby the density of states is insignificant. Assuming a non-degenerate semiconductor, $\epsilon_c - \mu,\ \mu - \epsilon_v \gg kT$, Maxwell–Boltzmann statistics applies and the expressions for the electron and hole concentrations are given by the expressions in Eqs. (9.80) and (9.81). However, the chemical potential no longer has the value as in the intrinsic case, and the electron and hole concentrations are different,

$$n_c(T) - p_v(T) \equiv \Delta n. \tag{10.55}$$

Multiplying the expressions for the two types of densities, Eqs. (9.80) and (9.81), the chemical potential drops out, and the *law of mass action*, Eq. (9.84), is still valid,

$$n_c(T)p_v(T) = N_c(T)P_v(T)\,e^{-\Delta/kT} = n_i^2. \tag{10.56}$$

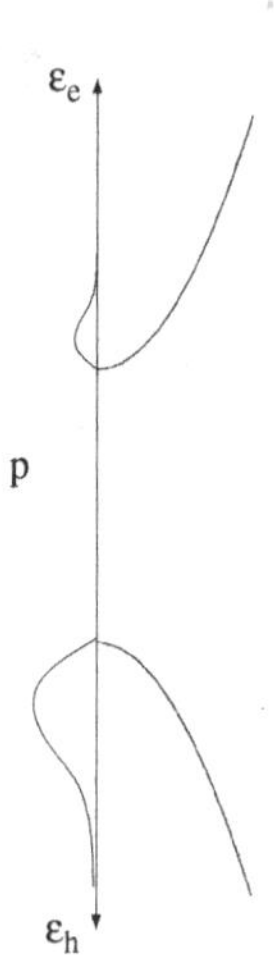

Figure 10.12 Occupation of holes and electrons in a p-type semiconductor.

The product of $n_c(T)$ and $p_v(T)$ still equals the intrinsic carrier concentration squared, since the expressions for $N_c(T)$ and $P_v(T)$, Eqs. (9.82) and (9.83), are the same for the extrinsic and intrinsic cases.

Multiplying Eq. (10.55) by $n_c(T)$, respectively $p_v(T)$, and solving the quadratic equations gives

$$n_c = \frac{1}{2}\sqrt{\Delta n^2 + 4n_i^2} + \frac{1}{2}\,\Delta n, \qquad p_v = \frac{1}{2}\sqrt{\Delta n^2 + 4n_i^2} - \frac{1}{2}\,\Delta n. \tag{10.57}$$

When Δn is large compared to n_i, one of the carrier densities thus dominates and is almost equal to Δn. When impurities are the main source of carriers, i.e. $\max\{n_c, p_v\}$ is much larger than n_i, an extrinsic semiconductor, one of the carrier types is dominant, determining whether the doped semiconductor is n- or p-type.

Using the fact that the expressions for $N_c(T)$ and $P_v(T)$, Eqs. (9.82) and (9.83), are also valid for the extrinsic case, the expressions for the carrier densities can be rewritten as

$$n_c(T) = n_i\, e^{(\mu-\mu_i)/kT}, \qquad p_v(T) = n_i\, e^{-(\mu-\mu_i)/kT}, \tag{10.58}$$

and the measure of the carriers provided by impurities is specified in terms of the deviation of the value of the chemical potential from the intrinsic chemical potential,

$$\frac{\Delta n}{n_i} = 2\sinh\left(\frac{\mu - \mu_i}{kT}\right). \tag{10.59}$$

Exercise 10.5 Consider the temperature dependence of majority carriers, say, $n_c(T)$ for an n-type semiconductor. Sketch the logarithm of $n_c(T)$, $\ln n_c(T)$, as a function of $1/T$. Notice that, at high and low temperatures, $\ln n_c(T)$ depends linearly on $1/T$; whereas at intermediate temperatures, $\ln n_c(T)$ is constant: the intrinsic, extrinsic and freeze-out regions. Obtain the slopes in the two linear regimes.

10.8 Summary

By doping a semiconductor with special atoms, either a p-type or an n-type semiconductor can be created. In the next chapter, it is shown how these can be used to make electronic devices, chief among them the transistor. In the early days of semiconductor production, several transistors would be produced in a piece of semiconductor and then broken off for later insertion into electronic circuits for one's use. However, integrated circuits were soon realized: make the electronic circuit that one is after directly on a single piece of semiconductor. Integrated circuit technology started to develop and has indeed now grown to very large-scale integration (VLSI), silicon chips containing billions of transistors.

11 Transistors

The stage is now set for discussing *classical* electronic devices – classical in the sense that quantum mechanics is needed only in a constitutive sense, i.e. energy band structure and the consequences of the exclusion principle, whereas the charge transport is essentially classical. Cases in question are p–n junctions, metal–semiconductor junctions and p–n–p transistors, where often the motions of electrons and holes are, to sufficient accuracy, simply described by diffusion. In the following chapters we discuss structures where also the transport is governed by quantum mechanics, but before that we discuss the p–n junction, ubiquitous in electronic devices, the workhorse of electronics. Then p–n junctions are combined into three-terminal devices, n–p–n and p–n–p transistors, the first ever-present mass-produced solid-state transistor. Finally, we discus the present-day all-pervasive transistor, the field effect transistor, revolutionizing electronic machinery in providing the integrated electric circuits used in computers.

11.1 The p–n Junction

A semiconductor in which the impurity concentration is non-uniform, i.e. in certain regions there is a vastly greater number of donor impurities compared to acceptor impurities and vice versa in adjacent regions, is referred to as an inhomogeneous semiconductor. The simplest case is where an n-type semiconductor is adjacent to a p-type semiconductor, a p–n junction. The junction must be constructed in one and the same semiconductor. Two pieces of n-type and p-type semiconductor cannot just be *glued* together to form a functioning device, since if not carefully constructed there will be a myriad of electronic interface states, capturing charge, which would be detrimental to device functioning.

Consider a semiconductor in which the impurity concentration is assumed to vary only along one direction. As an idealization, assume that in the left half-space of the semiconductor, $z < 0$, there are only acceptor impurities, and in the right half-space, $z > 0$, only donor impurities. The densities of donor and acceptor impurities are assumed constant, respectively n_d and n_a, and typically different. Understanding this simplified scenario then spells out the modifications to which a deviation from a sharp interface leads.[1]

[1] To have a rather sharp interface separating the n-type and p-type semiconductors is achievable by molecular beam epitaxy (MBE) technology, but certainly not by having dopants diffuse into the semiconductor crystal. Shooting in dopants using an accelerator also allows controlled dopant deposition by tuning the energy of the dopants.

Consider this contact of p- and n-type regions. At zero temperature, all the impurity electrons in the n-type material are bound in their *hydrogen* levels, and in the p-type material there are bound holes to the acceptor impurities, as depicted in Figure 11.1, where only the acceptor and donor impurities in the region near the *interface* are depicted explicitly. For simplicity, the Bohr atomic orbital depiction of the bound impurity electron has been used instead of the proper wave function probability cloud of Figure 10.4, and similarly a bound hole is depicted as in Figure 11.2.

On raising the temperature, the bound electrons in the n-type material are thermally excited to the conduction band of the semiconductor, and at room temperature they are all free carriers. Similarly, there will be free holes in the valence band of the p-type semiconductor, as the bound holes are excited to the valence band (occupied by electrons excited from the valence band). These thermally liberated carriers in Bloch Gaussian states move around, bumping into the ionized impurities, leading to motion of the electrons and holes as described by the Drude theory, or diffusive motion if impurities are plentiful. There are very few thermally excited (over the wide band gap) conduction electrons in the p-part of the semiconductor, and the multitude of conduction band electrons in the n-part are not Pauli-blocked from moving to the p-part. A net flow of electrons will thus pass to the p-side of the p–n junction. Similarly, the abundance of valence band holes in the p-part diffuse to the n-part (the absence of electrons in the p-type material disappears due to electrons flowing from the valence band in the n-type material, leaving behind electron absence, i.e. presence of holes). Electrons and holes have left the region near the interface, the so-called depletion layer. This leaves behind non-neutralized fixed charges, ionized donor impurities in the n-type region and ionized acceptor impurities in the p-type semiconductor. At room temperature, there are thus fixed positive and negative charges in, respectively, the n- and p-type regions near the interface, as depicted in Figure 11.3.

This charge distribution is the main source of the electric field at the p–n interface region, pointing in the direction from the n- to the p-type region, as depicted in Figure 11.4 (the depletion layer is not completely devoid of electrons and holes, which slightly modifies its

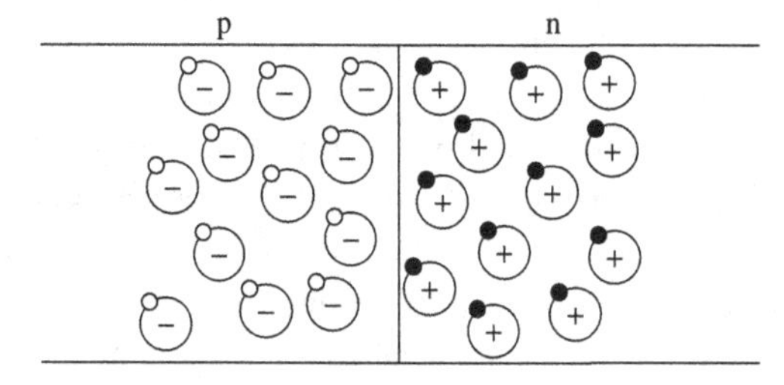

Figure 11.1 The p–n junction at zero temperature; donor electrons and acceptor holes are bound in their hydrogenic ground states.

Figure 11.2 Bound hole.

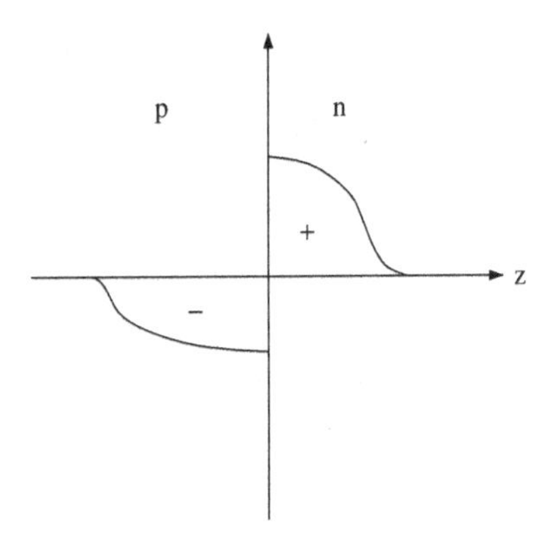

Figure 11.3 Ionized impurity charge distribution.

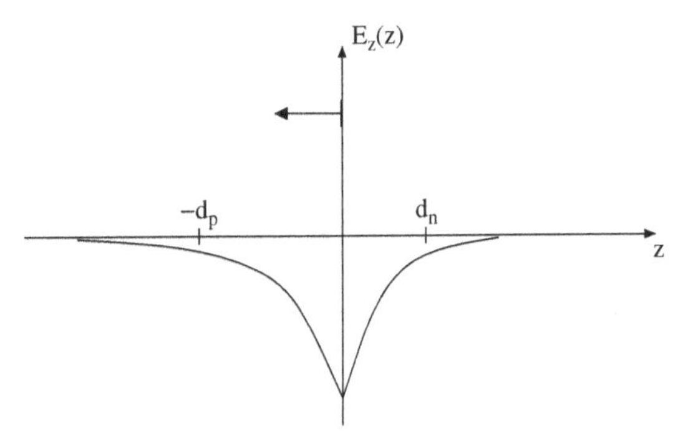

Figure 11.4 Built-in electric field as a function of z.

charge distribution). Outside the depletion layer, the electric field vanishes as the impurity charges are neutralized by the corresponding free electron and hole charges.

The electrostatic potential, $E_z(z) = -\partial_z \varphi(z)$, determining the built-in electric field, will according to Figure 11.4 have a shape as depicted in Figure 11.5. The built-in electric field drives an electric current opposite to the diffusion currents driven by the electron and hole concentration gradients, and in thermal equilibrium the diffusion and driven currents cancel, no net current flows.

Even if the doping configurations are not step-like, and a small density of *wrong* dopants is present, resulting in a situation like that depicted in Figure 11.6, the situation is essentially the same: a built-in electric field is present, the work agent for the nonlinear properties of p–n junctions utilized in electronics.

11.2 Built-in Electric Field

Of interest is the strength of the built-in electric field, the potential drop across the depletion layer, $(\Delta\varphi)_0 \equiv \varphi(\infty) - \varphi(-\infty)$. The built-in electrostatic potential in a p–n junction is, by its construction, slowly varying over the spatial cell, $\Delta\mathbf{x}_c$, and electrons and holes can

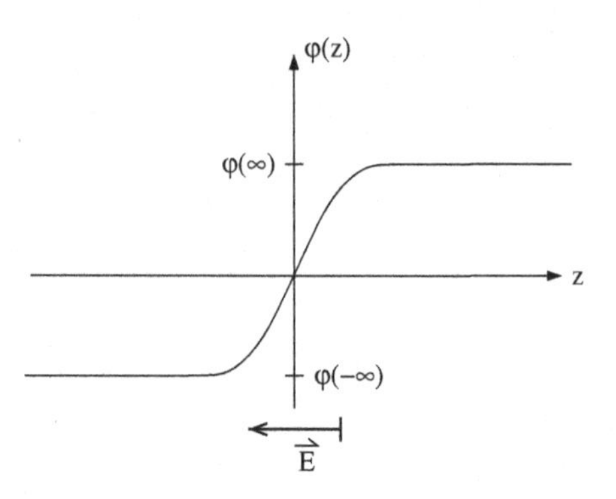

Figure 11.5 Electrostatic potential in a p–n junction.

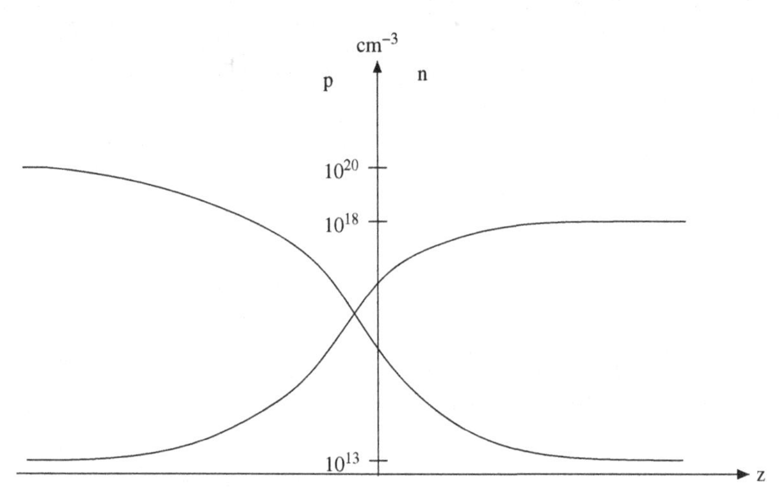

Figure 11.6 Depletion layer formed close to the interface of a p–n junction.

be viewed as occupying Bloch wave packet states. An electron in the conduction band in Bloch Gaussian state $(\mathbf{k}_i, \mathbf{x}_i)$ has energy

$$E_c(\mathbf{k}_i, z_i) = \epsilon_c(\mathbf{k}_i) - e\varphi(z_i). \tag{11.1}$$

In the wave packet representation, the conduction band edge has, according to Figure 11.5, a spatial variation across the p–n junction, the band bending effect of the built-in field as depicted in Figure 11.7. Similarly, the valence band edge has the spatial dependence depicted in Figure 11.8.

The probability of occupation of a Bloch Gaussian state is specified by the local Fermi distribution. The density of electrons, the local electron carrier concentration, is given by the population of the conduction band. Again, assuming a non-degenerate semiconductor, this population is obtained as in Section 9.9, the only difference being the spatial dependence of the energy, and the local electron carrier concentration is, according to Eq. (9.80), for $\epsilon_c - e\varphi(z) - \mu \gg kT$,

$$n_c(z) = N_c(T)\, e^{-(\epsilon_c - e\varphi(z) - \mu)/kT}. \tag{11.2}$$

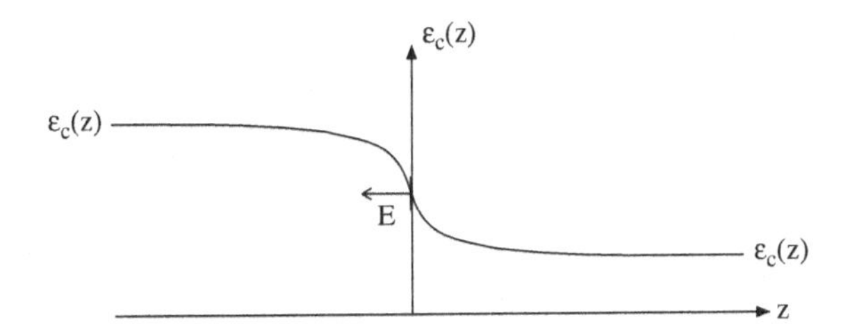

Figure 11.7 Conduction band edge variation in a p–n junction.

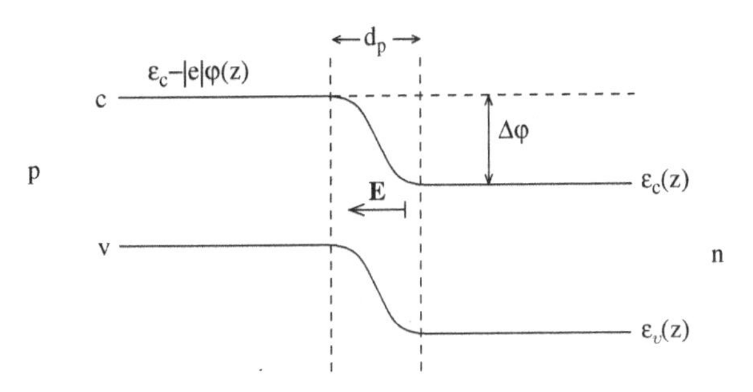

Figure 11.8 Conduction and valence band edges in a p–n junction.

Similarly, according to Eq. (9.81), the local hole carrier concentration is, now for $\mu - \epsilon_v + e\varphi(z) \gg kT$,

$$p_v(z) = P_v(T)\, e^{-(\mu-\epsilon_v+e\varphi(z))/kT}. \tag{11.3}$$

Far from the depletion layer there is no impact of the p–n interface, and, as all impurities are ionized, the electron concentration deep in the n-region is equal to the donor impurity concentration,

$$n_c(\infty) = n_d = N_c(T)\, e^{-(\epsilon_c-e\varphi(\infty)-\mu)/kT}. \tag{11.4}$$

Similarly, deep in the p-type region the density of free holes equals the density of acceptor impurities,

$$p_v(-\infty) = n_a = P_v(T)\, e^{-(\mu-\epsilon_v+e\varphi(-\infty))/kT}. \tag{11.5}$$

Multiplying the previous two equations together and taking the logarithm gives the potential drop $(\Delta\varphi)_0$ and thereby, for the voltage drop across the junction,

$$\begin{aligned} e(\Delta\varphi)_0 \equiv e\varphi(\infty) - e\varphi(-\infty) &= \epsilon_c - \epsilon_v + kT \ln\left(\frac{n_a n_d}{N_c(T)P_v(T)}\right) \\ &= \Delta + kT \ln\left(\frac{n_a n_d}{N_c(T)P_v(T)}\right). \end{aligned} \tag{11.6}$$

The potential drop is determined by the band gap, a large energy compared to the thermal energy, $kT_{\text{room}} \sim 10^{-2}\,\text{eV}$.

The electrostatic potential determining the built-in electric field is determined by Poisson's equation,

$$\frac{d^2\varphi}{dz^2} = -\frac{\rho(z)}{\epsilon}, \tag{11.7}$$

where $\rho(z)$ is the charge distribution, inhomogeneity assumed only in the direction normal to the depletion layer. For our purpose it is enough to estimate the electrostatic potential. A derivative is estimated according to $\varphi'(z) \sim \varphi(z)/z$, giving for the second derivative $\varphi''(d) \sim \varphi'(d)/d$, where $d = d_p + d_n$ is the size of the depletion layer. The derivative of the potential is estimated by the slope in Figure 11.5, $\varphi'(d) \sim (\Delta\varphi)_0/d \simeq \Delta/ed$, and Eq. (11.6) has been used. Poisson's equation then gives the estimate for the size of the depletion layer, using the impurity charge densities as sole source,

$$\frac{\Delta}{ed^2} \sim \frac{en_{a,d}}{\epsilon}, \qquad d = \sqrt{\frac{\epsilon\Delta}{e^2 n_{d,a}}}, \tag{11.8}$$

which for donor concentrations on the order of $n_{d,a} \sim 10^{14}$–$10^{18}\,\mathrm{cm}^{-3}$ and a gap on the order of $\Delta \sim 0.1\,\mathrm{eV}$ gives $d \sim 10^4$–10^2 Å. The magnitude of the built-in electric field $E \sim \Delta/ed$ is thus huge, on the order of 10^6 V/cm.

Anticipating applying voltages across the junction that are small compared to 10^6 V/cm, we therefore do not expect much change in the electric field in the depletion layer in the presence of an applied bias. However, a crucial distinction between current directions has been devised, as we shall elaborate.

Without a voltage applied to the junction, the average current vanishes, but of course a lot of processes go on in this dynamical system of electrons and holes in thermal equilibrium with the lattice vibrations. For example, an electron that has diffused to the p-type material can recombine into one of the many free hole states available in the valence band, with the emission of a phonon (or a photon, like an electron in an excited state in an atom decays into a lower empty state). Similarly, a hole in the n-type material can recombine with one of the many electrons in the conduction band. These processes are referred to as electron–hole recombination processes. Minority carrier densities are thermally generated continuously, tiny amounts of holes in the n-type material and tiny amounts of electrons in the p-type material, or rather electron–hole pairs in both materials.

11.3 Biased p–n Junction

Since the carriers are highly mobile, the carrier densities are low in the region of the built-in field, i.e. the depletion layer. Owing to the greatly reduced carrier densities, the depletion region will have a much higher electrical resistance than outside in the homogeneous regions, the resistance being inversely proportional to the carrier density. Thus the p–n junction can be viewed as an equivalent circuit component where a high resistance is sandwiched in between two low resistances (Figure 11.9).

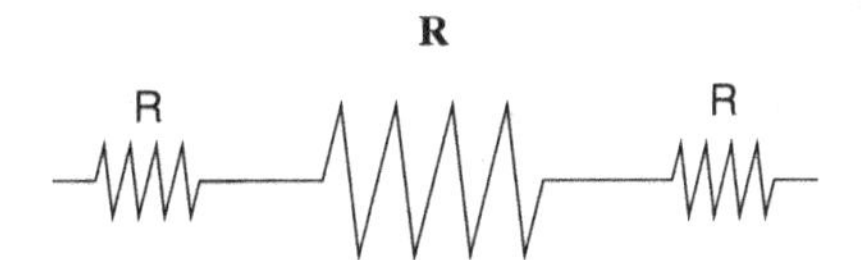

Figure 11.9 Resistance distribution in a p–n junction.

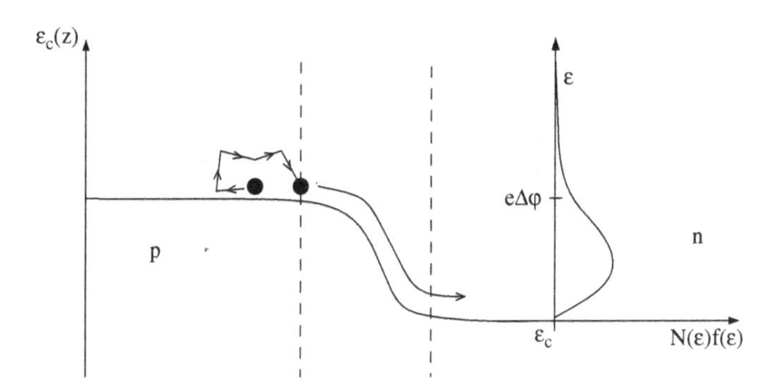

Figure 11.10 Origin of generation and recombination electron currents.

When a potential V is applied across a p–n junction, almost all the voltage drop will therefore be over the high resistance, the depletion layer, and even in the biased case the electrostatic potential will therefore only vary appreciably within the depletion layer. The external potential therefore adds to the internal one, giving a slightly changed potential difference since the built-in field is large,

$$\Delta\phi = (\Delta\phi)_0 - V. \tag{11.9}$$

To understand the properties of a biased p–n junction, the minority currents are important. In the p-part of the semiconductor, in the bulk, as a result of thermal excitation, a minute density of electrons is constantly being generated. Such an electron diffuses around, and if it does not recombine with any of the abundant holes in the p-type region and diffuses into the depletion region, it will be swept across by the built-in electric field to a vacant state in the conduction band in the n-type semiconductor, as depicted in Figure 11.10. In terms of energy, which is conserved, the electron arrives with an energy in the high-energy tail of the Fermi distribution, where it is accommodated in the empty state there.

The resulting current is called the electron generation current density, j_e^{gen}. The electron generation current is generated in the bulk of the p-type semiconductor, and is simply due to the finite temperature (lattice vibrations knocking electrons from the valence band to the conduction band) and is thus, for relevant values of the bias voltage, independent of the applied bias, as all that matters in order to be swept across the junction is to reach the depletion layer. On the contrary, an electron in the n-part (where there is an abundance of such electrons, which are the majority carriers) can only get to the p-side if it can overcome the electric field in the depletion layer, i.e. arrive at the depletion layer with enough energy to overcome the potential energy difference. So only electrons in the high-energy tail on the right in Figure 11.10 have enough energy to make it to the p-side. The density of such high-energy electrons in the conduction band, the high-energy tail of the Maxwell–Boltzmann distribution, is given by, assuming as usual a non-degenerate situation,

$$n = 2 \int_{\epsilon_c + e\Delta\phi}^{w_c} d\epsilon \, N_c(\epsilon)\, e^{-(\epsilon-\mu)/kT} = 2\, e^{-e\Delta\phi/kT} \int_{\epsilon_c}^{w_c - e\Delta\phi} d\epsilon \, N_c(\epsilon + e\Delta\phi)\, e^{-(\epsilon-\mu)/kT}, \tag{11.10}$$

where the last equality simply results from the substitution $\epsilon \to \epsilon - e\Delta\phi$. The integral produces a smooth function of the temperature and the potential difference since the density of states is a smooth function vanishing at the bottom of the conduction band, and the electron current density from the n- to the p-side is proportional to $\exp(-e\Delta\phi/kT)$, i.e. proportional to the density of overcoming high-energy electrons. In terms of the applied voltage, this gives

$$j_e^{\text{rec}} \propto \exp[-e((\Delta\phi)_0 - V)/kT], \tag{11.11}$$

the electron recombination current density, as a high-energy electron injected from the n-side to the p-side eventually recombines with one of the abundant holes on the p-side. An electron is lost from the conduction band, but an electron in that state will be supplied by the external supply at the n-type material. We can easily relate the prefactor in the recombination current to the generation current. In the absence of voltage bias, there is no net electron current across the junction: generation and recombination are in opposite directions but of equal magnitude,

$$j_e^{\text{rec}}\Big|_{V=0} = j_e^{\text{gen}}, \tag{11.12}$$

and we get for the electron recombination current density

$$j_e^{\text{rec}} = j_e^{\text{gen}} \exp(eV/kT). \tag{11.13}$$

The net electric current flow of electrons is therefore

$$j_e = j_e^{\text{rec}} - j_e^{\text{gen}} = j_e^{\text{gen}}(\exp(eV/kT) - 1), \tag{11.14}$$

i.e. the current direction for the p–n junction is chosen in the direction from p to n, opposite to the direction of the built-in electric field.

Next we account for the hole currents. In the n-part of the semiconductor, in the bulk, due to thermal excitation, a minute density of holes is constantly being generated, hole minority carriers. Such a hole diffusing into the depletion region will immediately be swept across by the built-in electric field existing in the region, as depicted in Figure 11.11, resulting in a hole generation current density, j_h^{gen}.

On the contrary, a hole on the p-side can only get to the n-side if it can overcome the electric field in the depletion layer, i.e. arrive at the depletion layer having enough energy to overcome the potential energy difference. The density of such high-energy holes, the

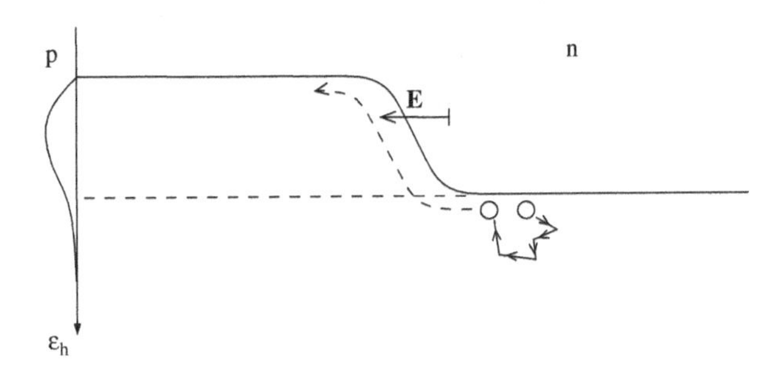

Figure 11.11 Origin of generation and recombination hole currents.

high-energy hole tail depicted on the left in Figure 11.11, is equal to the absence of electrons in states deep down in the valence band, and is given by

$$p = 2 \int_{\epsilon_v - e\Delta\phi}^{w_v} d\epsilon\, N_v(\epsilon)\, e^{(\epsilon-\mu)/kT} = 2\, e^{-e\Delta\phi/kT} \int_{\epsilon_v}^{w_v - e\Delta\phi} d\epsilon\, N_v(\epsilon - e\Delta\phi)\, e^{(\epsilon-\mu)/kT}. \quad (11.15)$$

The hole current density from the p- to the n-side is therefore proportional to $\exp(-e\Delta\phi/kT)$,

$$j_h^{\rm rec} \propto \exp[-e((\Delta\phi)_0 - V)/kT] \quad (11.16)$$

the hole recombination current density, as one of the abundant electrons on the n-side eventually will drop into the empty level that constitutes the arrived hole. An electron in the conduction band has been annihilated, but an electron in that state will be supplied by the external supply at the n-type material. In the absence of voltage, there is no net hole current across the junction: generation and recombination currents are in opposite directions but of equal magnitude,

$$j_h^{\rm rec}\Big|_{V=0} = j_h^{\rm gen}, \quad (11.17)$$

and the hole recombination current density is

$$j_h^{\rm rec} = j_h^{\rm gen} \exp(eV/kT). \quad (11.18)$$

The net flow of holes is therefore

$$j_h = j_h^{\rm rec} + j_h^{\rm gen} = j_h^{\rm gen}(\exp(eV/kT) - 1). \quad (11.19)$$

The various electron and hole contributions flow in opposite directions but their electric currents add, giving for the total current density through a biased p–n junction

$$j = (j_h^{\rm gen} + j_e^{\rm gen})(\exp(eV/kT) - 1) \quad (11.20)$$

or equivalently for the total current through the p–n junction

$$I = I_0(\exp(eV/kT) - 1). \quad (11.21)$$

The current, of course, vanishes in the absence of voltage, $V = 0$, but putting on a small positive voltage generates an exponentially large current. For a small negative voltage, the current saturates at the value $I \simeq I_0$, which is tiny since it is the magnitude of the generation

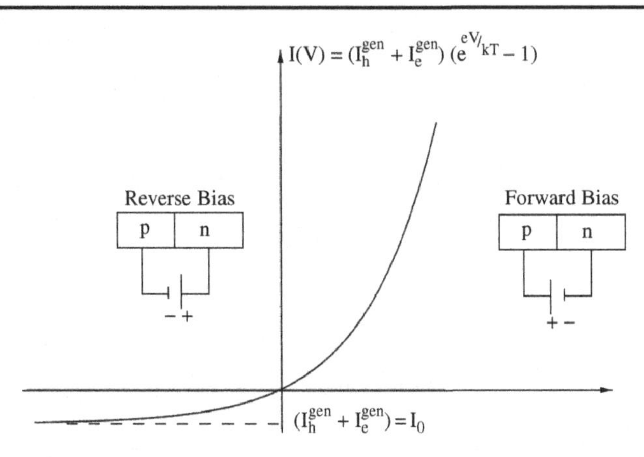

Figure 11.12 The I–V characteristic of a p–n junction.

currents, the minority current. The p–n junction is thus a highly asymmetric device. We emphasize that the exponential dependence of the current on the voltage is due to the exponential sensitivity of the number of majority carriers that have enough energy to make it across the uphill potential, the exponential tail of the Maxwell–Boltzmann distribution.

If the plus terminal of the battery is attached to the p-side of the p–n junction, the conduction band edge is decreased in the p-region and, as the negative terminal is on the n-side, increased there. The potential drop at the p–n interface is thus decreased: the positive voltage, V, in Eq. (11.9) subtracts from the built-in potential, and the built-in electric field is thus reduced. There are thus bigger tails of high-energy electrons and holes, exponentially increasing the recombination currents. If the terminals of the battery are attached oppositely, $V < 0$ in Eq. (11.9), the recombination currents are exponentially suppressed, the reverse bias case (see Figure 11.12).

The p–n junction is highly asymmetric with respect to current flow: in one direction it is hard for any bias voltage to push even the slightest current through, whereas in the other direction the current grows exponentially with the bias. The p–n junction thus acts as a rectifier.

11.4 Photovoltaics

A p–n junction can become a light source, producing light in response to an applied d.c. voltage. If a p–n junction is strongly forward biased, electrons and holes are dragged into the depletion layer, as depicted in Figure 11.13.

An electron in the conduction band of the n-part decays through the coupling to the electromagnetic field, making an energy- and momentum-conserving transition to an empty electron state in the valence band, analogously to an atomic transition of an electron from an excited state to a lower state with the emission of a photon. In the process of electron–hole recombination with the emission of a photon, the p–n junction functions as a light-emitting diode (LED).

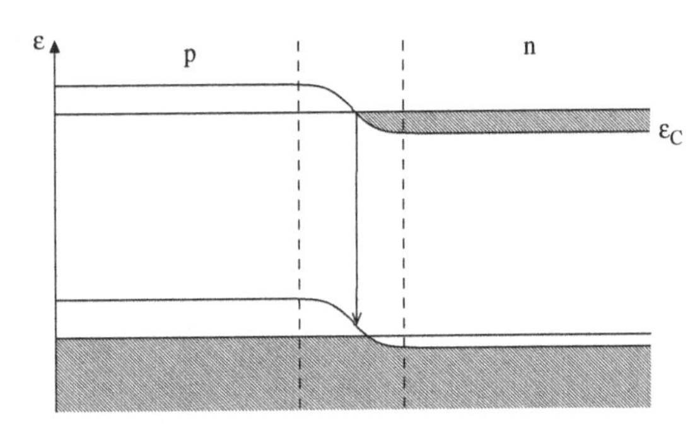

Figure 11.13 Spatial overlap of electrons and holes in a strongly forward biased p–n junction.

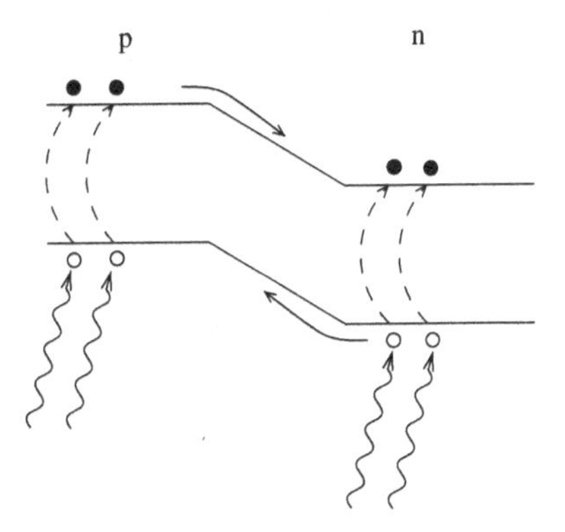

Figure 11.14 Light-produced minority carriers creating current.

The opposite process can be used to make a battery, a photovoltaic cell such as a solar cell. If a p–n junction is illuminated by light, photons excite electrons from the valence band to the conduction band. Minority carriers are thereby created, which leads to a current in the p–n junction (an additional generation current). Photon-supplied electrons to the conduction band on the p-side are flushed to the n-side, and holes created on the n-side are flushed to the p-side, as depicted in Figure 11.14.

This light-generated current, I_L, is thus added to the current through the p–n junction,

$$I = I_0(\exp(eV/kT) - 1) - I_L. \tag{11.22}$$

The I–V characteristic of a p–n junction illuminated by light is thus as depicted in Figure 11.15. At the indicated point, where $IV < 0$, the battery provides its maximum effect to an external circuit.

11.5 Bipolar Transistor

A semiconductor where a p-type material is sandwiched between two n-type materials is a so-called n–p–n transistor. A schematic depiction of such a device is displayed in Figure 11.16.

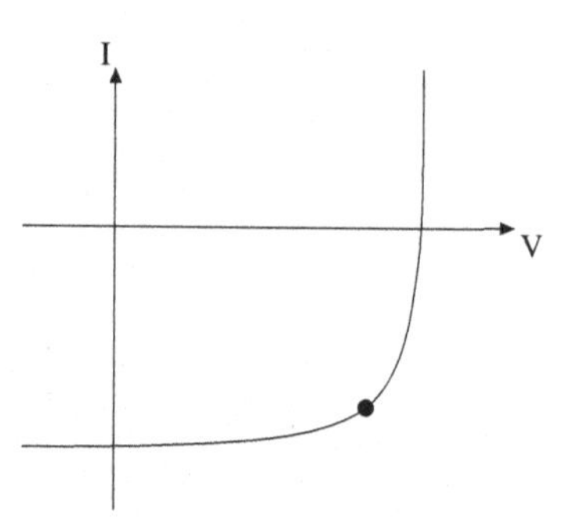

Figure 11.15 The I–V characteristic of a light-illuminated p–n junction.

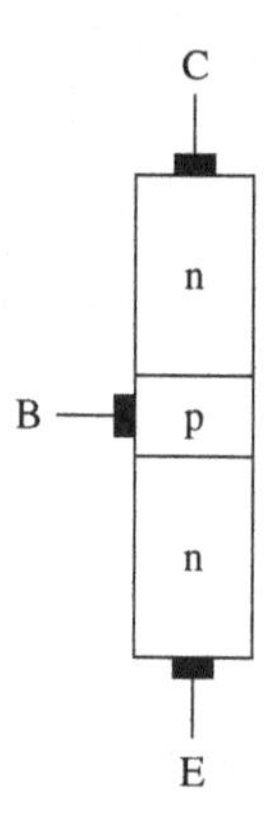

Figure 11.16 Bipolar n–p–n transistor.

The three parts of the device are referred to as the base (B), emitter (E) and collector (C), the latter two referring to analogously functioning parts in old vacuum tube technology, and the word "base" refers to the transistor being built of this material. The built-in electric fields at the junctions give the conduction and valence band edges as shown in Figure 11.17 by the dashed lines.

The base and collector are connected to batteries and can be biased relative to the grounded emitter as depicted in Figure 11.18.

Consider first applying voltages so that the emitter–base n–p junction is forward biased and the base–collector p–n junction strongly reverse biased. The conduction and valence band edges are thus shifted as shown in Figure 11.17 by the solid lines.

Let us first account for the action in the conduction band, i.e. what are the electron currents? At the forward biased emitter–base junction, the high-energy tail of the electrons in the emitter conduction band can overcome the barrier, and the current $I_e \simeq I_0 \exp(eV_b/kT)$ is injected into the base. At the reversed biased base–collector junction, electron injection is strongly suppressed and that electron current is negligible. An electron entering the p-type material through the forward biased junction diffuses around in the p-type material, and has the possibility of recombining with the multitude of free holes in the base, and thus not contribute to the collector current, but instead increase the base current. This is

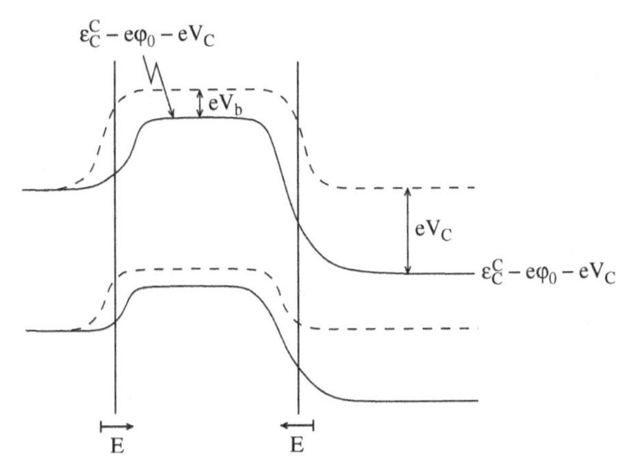

Figure 11.17 Biased and unbiased (dashed lines) band profiles.

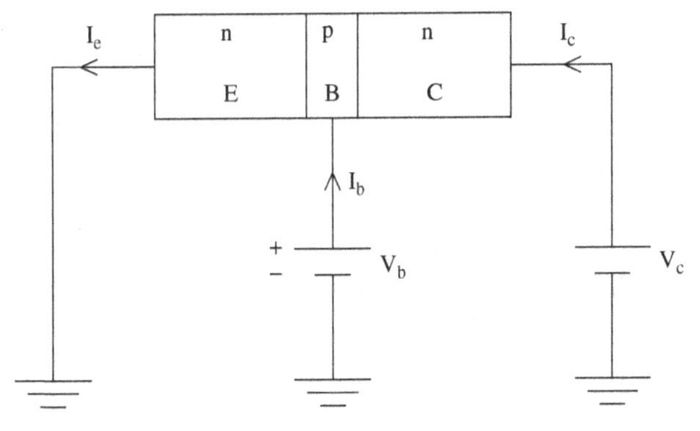

Figure 11.18 Transistor terminals connected to batteries.

to be avoided, and the key ingredient in the bipolar transistor operation is a geometric characteristic, *viz.* to make the p-type material so thin (much narrower than the transverse dimensions) that most of the electrons injected into the p-type material diffuse across to the forward biased p–n junction where the electrons are swept to the n-type material. The n-type material from where the electrons are injected, or emitted, is therefore referred to as the emitter and the other n-type material is referred to as the collector. Owing to the thinness of the base and base electrons being minority carriers, their contribution to the current in the base is negligible. We define β as the ratio of the number of electrons that are lost from the injected electrons (either by recombination in the base or by diffusing through the base and leaving through the base terminal) to the number of electrons captured by the collector, i.e. $I_b = \beta I_c$. This ratio is typically in the few percent range, $\beta \sim 10^{-2}$, depending on the shape and size of the transistor.

Next, let us account for the current carried by the holes, i.e. we consider the action in the valence band. In the collector, which is n-type material, the holes, being minority carriers, make an insignificant contribution to the collector current, $I_c = I_c^{(e)} + I_c^{(h)} \simeq I_c^{(e)}$. The base, which is p-type, has an abundance of free holes. They are prevented from moving from the base to the collector by the strong reverse bias, i.e. the large field in the base–collector junction. On the other hand, the emitter–base junction is assumed only slightly forward biased, and a hole current from the base to the emitter of the same order of magnitude as the electron current from the emitter to the base is to be expected, unless special precautions are taken. This base current of holes into the emitter is detrimental to the supposed functioning of a transistor since it increases the base current, I_b, for a given current of electrons to the collector, and thus decreases the amplification ratio I_c/I_b. To avoid this unwanted effect, the p-type material, the base, is weakly doped in comparison to the more heavily n-doped emitter (however, the doping is still such that a built-in field between emitter and base is maintained). The currents in an n–p–n transistor are thus dominated by the electron currents.

Increasing the base voltage slightly (slightly increasing the base current) will give rise to a large increase, ΔI_e, in the emitter current since the working point of the emitter–base junction is on the steep part of the I–V characteristic, $\Delta I_e \simeq I_e(V_b)\exp(e\Delta V_b/kT)$ (the mechanism being that the barrier for electron injection is lowered). Since $I_e = I_b + I_c$, the result is a large change in the collector current in response to a small change in the base current. Since the base current remains a small fraction of the collector current, $I_b = \beta I_c$, the transistor thus works as an amplifier. A small base current I_b gives a large current (typically two orders of magnitude larger) at the collector electrode $I_c \simeq (1/\beta)I_b$. That the amplification is linear is important in order for an amplified signal not to be distorted. We have only accounted for the main currents, and many other processes, such as additional current decreases due to recombination in depletion layers, etc., may lead to small corrections. However, the basic transistor action is explained: the current in one part of the device (the collector current I_c) is controlled by a voltage (V_{be} or equivalently the current I_b) in another part of the device.

The electric circuit symbol for an n–p–n transistor is depicted in Figure 11.19. An equivalent transistor can be built by swapping the n- and p-type regions, giving a p–n–p transistor. The circuit symbol for the p–n–p transistor has just the arrows reversed, indicating that the injector current is carried by holes.

The situation discussed where the base current is amplified is referred to as the common emitter configuration and is depicted as in Figure 11.20.

The bipolar transistor has three modes of operation. (i) In the *active mode*, which we have just discussed, the emitter–base junction is forward biased and the collector–base junction is reverse biased. (The inverse active mode, where forward and reverse biasing of the junctions are interchanged, is similar but asymmetric due to the asymmetry in doping.)

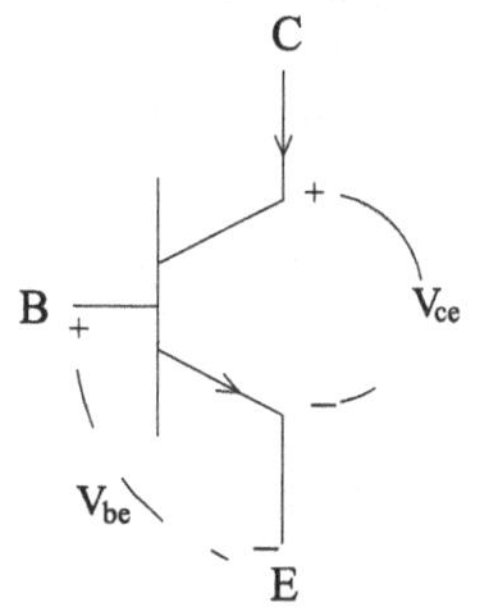

Figure 11.19 Circuit symbol for a bipolar n–p–n transistor.

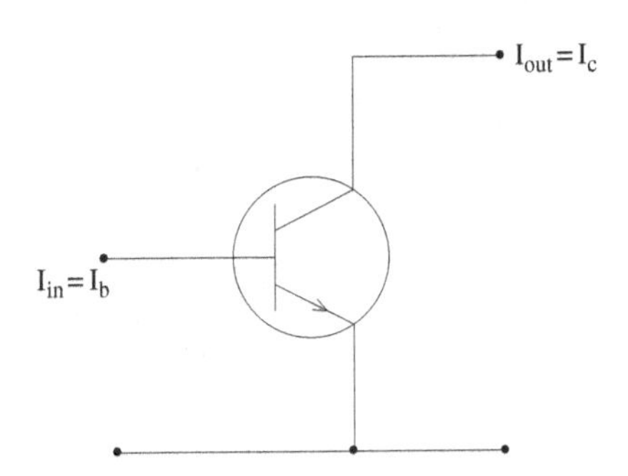

Figure 11.20 Common emitter configuration.

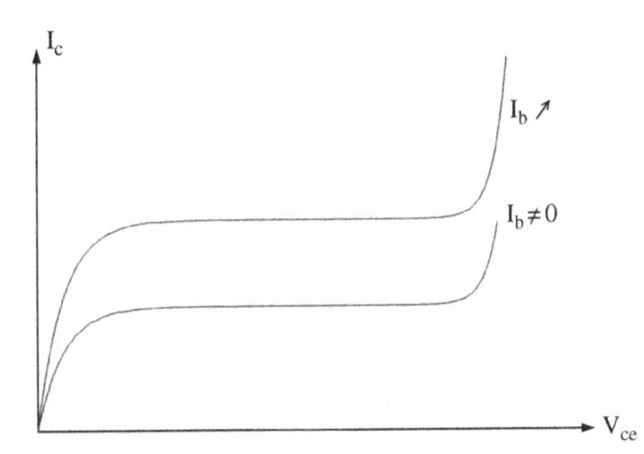

Figure 11.21 The I_C–V_{ce} characteristic.

(ii) In the *cut-off mode*, both junctions are reverse biased and no current flows through the transistor. The transistor is in the *OFF* state. (iii) In the *saturation mode*, both junctions are forward biased and current flows through the transistor with the tiniest voltage across it. The transistor is in the *ON* state.

In the active mode, the collector current is almost independent of the magnitude of the collector voltage V_c ($V_c = V_{ce}$ the voltage drop over the device), as long as the base–collector junction is reverse biased, since electrons are swept across at the base–collector junction irrespective of the variation in V_c as long as there is a built-in field. The I_c–V_{ce} characteristic is thus mostly flat, as depicted in Figure 11.21.

At low voltages, there is an ohmic regime, as a small voltage is needed to get the carriers drifting; and at high enough voltage, both junctions are eventually forward biased, flushing current through. However, in between, the transistor in the common-emitter mode acts as a *current source*, and with the advantage of low dissipation compared to that of a voltage source in series with a resistor with high resistance.

11.6 Field Effect Transistor

The bipolar transistor still finds its uses, but the most common type of transistor is the field effect transistor (FET), which is devised on a different operation principle, and is unipolar, i.e. only transport of one type of carriers is involved, either electrons or holes. The proliferation of field effect transistors, such as in computers, is essentially due to the manufacturing processes of chips. The field effect transistor is ideal for very large-scale integrated (VLSI) circuits.

Consider a wafer of purified silicon that is then p-doped. By subsequently heavily n-doping two regions, the source (S) and drain (D), two back-to-back p–n junctions are created as depicted in Figure 11.22. On top of the p-region is then created an insulator (I), SiO_2, by exposing the Si wafer to air or water, and on top of that is placed a conductor, the gate (G), in practice very heavily doped silicon, polycrystalline silicon that functions as a metallic gate. In practice, the different regions are created in the opposite order by etching and doping techniques; however, the end result of a field effect transistor is as depicted in Figure 11.22.

By charging the gate positively, the band edges respond as depicted in Figure 11.23, and electrons are attracted to reside underneath the gate. With a large enough positive gate voltage, the surface of the p-type semiconductor has been inverted to an n-type semiconductor; an inversion layer of electrons is formed underneath the gate, as depicted in Figure 11.24.

A tiny voltage drop between the source and drain will result in a current flowing between the source and drain. The device is called metal–oxide–semiconductor field effect transistor (MOSFET). The gate voltage controls whether a current easily flows between source and drain. The device acts as a transistor; a voltage in one part of the device controls the current between the source and drain. The transistor thus acts as a switch: if the gate voltage is zero, no current can flow; whereas for a sufficient gate voltage it can.

If the gate voltage is negative, the band bending is opposite and, since a p-type semiconductor is considered, more holes are accumulated near the oxide–semiconductor interface than in the bulk. The interface appears more p-type than the bulk. A layer of holes can, for this gate voltage, conduct a current, as depicted in Figure 11.25.

Swapping n- and p-regions in Figure 11.22, an equivalent device is created. A gate is then separated by an insulator from an n-type semiconductor, and, charging the gate negatively,

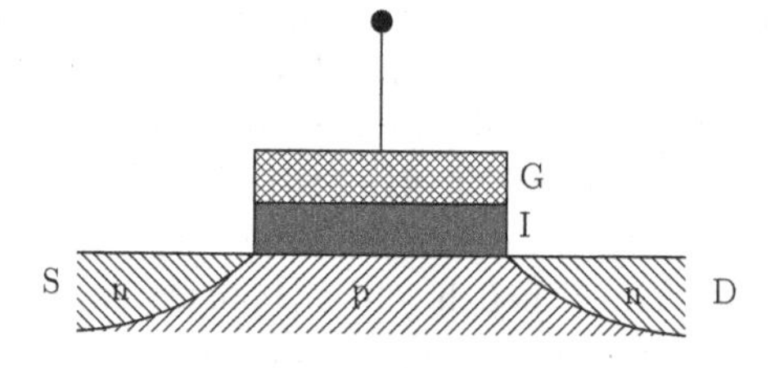

Figure 11.22 Field effect transistor.

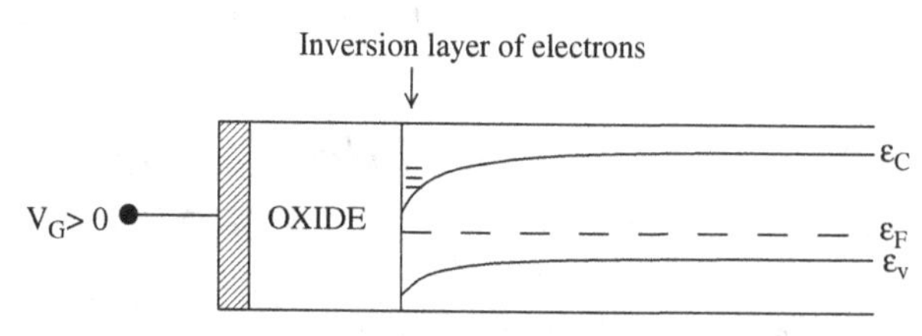

Figure 11.23 Inversion layer of electrons.

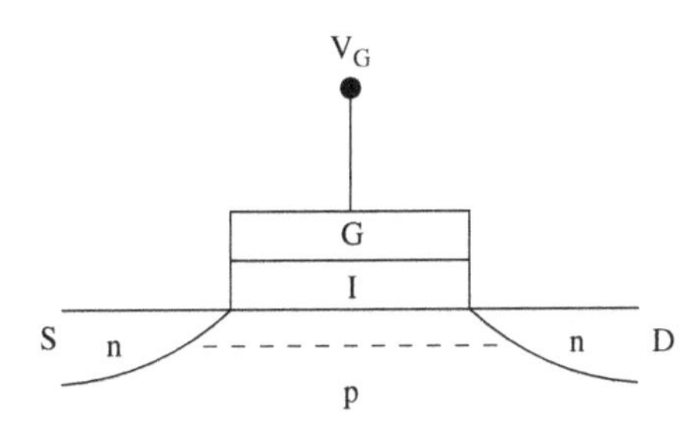

Figure 11.24 Inversion layer of electrons in a field effect transistor.

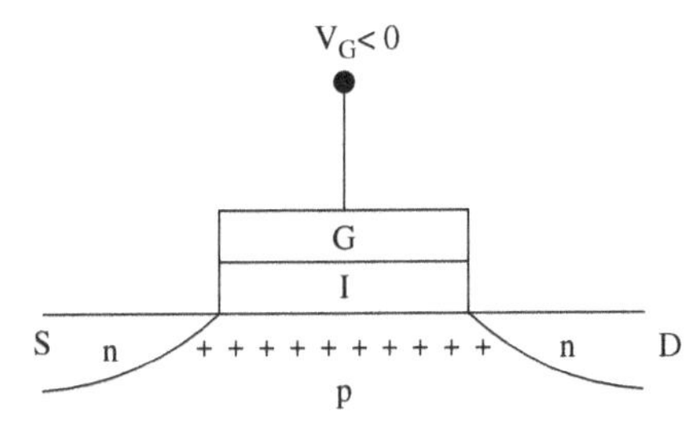

Figure 11.25 Layer of holes in a field effect transistor.

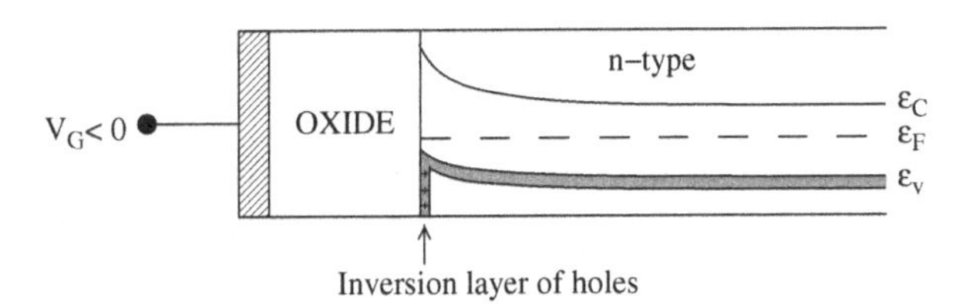

Figure 11.26 Inversion layer of holes.

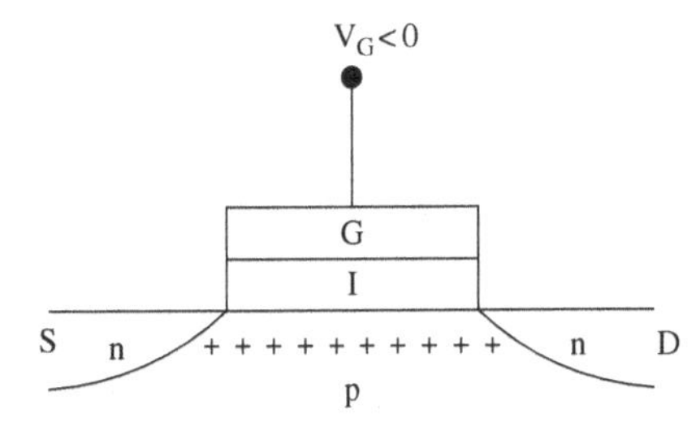

Figure 11.27 Inversion layer of holes in a field effect transistor.

the band edges are bent as depicted in Figure 11.26. By having a large enough positive gate voltage, an inversion layer of holes is attracted to be located underneath the gate. A conducting inversion layer of holes is then created as depicted in Figure 11.27.

Incorporating both types of MOSFETs on the same wafer, having both electron and hole based transistors on the same chip, is the currently used complementary metal–oxide–semiconductor technology, CMOS technology, using these complementary MOSFETs.

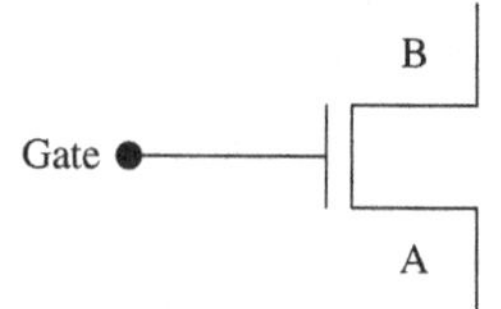

Figure 11.28 Circuit symbol for a field effect transistor.

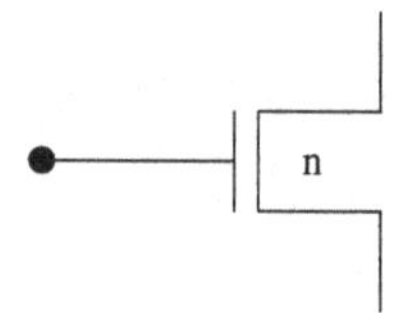

Figure 11.29 Circuit symbol for an n-type field effect transistor.

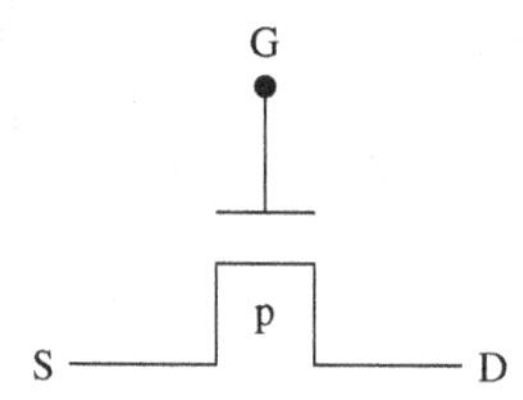

Figure 11.30 Circuit symbol for a p-type field effect transistor.

The electric circuit symbol for the field effect transistor is depicted in Figure 11.28. In fact, we must be more specific, so for an n-type field effect transistor its electric circuit symbol is as depicted in Figure 11.29; and for a p-type field effect transistor we have the diagrammatic symbol depicted in Figure 11.30.

11.7 Summary

The quantitative theory of semiconductor electronics needs detailed consideration of the transport properties of materials, and is the subject of a large electrical engineering literature. However, understanding the basic electric functioning of the considered semiconductor devices is provided by the qualitative considerations presented in this chapter, and is sufficient when we later discuss their use in the construction of logic gates in computers.

G. E. Moore, one of the founders of the chip-maker INTEL, noted the empirical relationship that the number of components (transistors, diodes, resistors and capacitors) on a given size of silicon chip roughly doubled every 18 months. Such a packing rate increase for silicon technology will soon come to an end. For example, tunneling is becoming a nuisance to present-day electronics, since the ever decreasing thinness of the silicon oxide barriers on a chip is reaching the limit where barrier penetration leads to leak currents detrimental to device functioning.

12 Heterostructures

Heterostructures are artificial human-made materials consisting of different compound semiconductors grown on top of each other. Typically they are III–V semiconductor compounds such as GaAs–AlAs, or II–VI compounds such as, for example, CdTe/HgTe. The creation of such structures was achieved due to the advancement of semiconductor production technology. We shall describe in detail one such fabrication technology, molecular beam epitaxy (MBE), conceived in the 1960s and greatly perfected through the 1970s. Other epitaxial techniques widely used in electronics production, such as metal–organic chemical vapor decomposition (MOCVD), and, for example, metal–organic vapor-phase epitaxy (MOVPE) and metal–organic molecular beam epitaxy (MOMBE), as well as others, were developed in the 1980s and later. Each technique has superior features depending on which type of heterostructure is to be created. Using MBE, it is possible to grow compounds of crystals with almost atomic precision, and the technique is important and further promising for nanostructure fabrication. We shall here describe the MBE technique, which stands out due to the simplicity of its operating principle, which in turn is also the basis for its versatility.

12.1 Molecular Beam Epitaxy

In an MBE machine, collision-free thermal beams of atoms or molecules are evaporated from effusion (Knudsen) cells with small orifices (typically with a beam intensity of 10^{11}–10^{16} atoms(molecules)/cm^2 s), and condensed, under ultra-high-vacuum (UHV) conditions, onto a heated substrate. The impinging atoms or molecules can stick to the substrate, and such adsorbed atoms or molecules will, due to their thermal energy provided by the heated substrate, migrate along the substrate surface to energetically favorable lattice sites. The functioning of an MBE machine is conveniently described by referring to the sketch depicted in Figure 12.1. The outer periphery marks the chamber, which has been pumped to the UHV condition before the effusion cells are opened.

For example, having the ovens with gallium and arsenic open, the substrate is bombarded with gallium and arsenic and it is possible to grow almost perfectly the III–V compound semiconductor GaAs. The gallium flux is always monatomic whereas the arsenic flux comprises either As_2 or As_4 molecules. Having to work with a highly toxic substance such as arsenic is delicate. However, compared to MOCVD, which works with gaseous components, a great advantage of the MBE technique is that the hazardous chemicals are handled

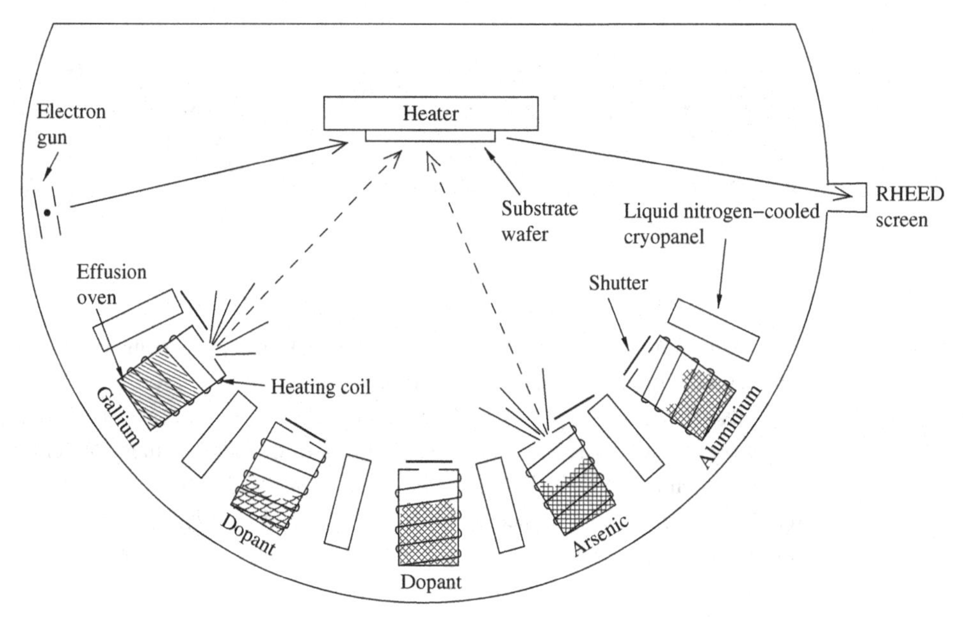

Figure 12.1 Schematics of a molecular beam epitaxy machine.

in solid form until they are processed inside the MBE vacuum vessel. Perfect stoichiometry of the growing crystal can be obtained because the group V atoms adhere less well to the substrate than do the group III atoms. By keeping the substrate sufficiently hot, a group V atom evaporates unless it is next to – or by thermal migration comes next to – and bonds with a group III atom, since a group III atom will stick to the substrate provided the substrate is not too hot. Thus, the formation of layers of III–V semiconductor compounds is controlled by the adsorption of the group V element, whereas the growth rate is determined only by the flux of the group III element. In this way, monolayer by two-dimensional monolayer, a III–V semiconductor compound is built up with an almost perfect crystal lattice structure. The III–V semiconductor compounds have the zincblende (or sphalerite) crystal structure. This is because two such atoms together have eight valence atoms to share to form covalent bonding, as described in Section 6.6. The crystal structure is thus the one where each group III atom is surrounded by four group V atoms (occupying the apexes of a tetrahedron with the group III atom at the center), and vice versa for the group V atom. Thus the crystal lattice is that of diamond, but instead of having carbon ions at each site, the sites are occupied alternately by gallium and arsenide.[1]

Typically, for AlGaAs compounds, the cooking temperature is 600–700°C. At low substrate temperatures, less than 750 K, the sticking coefficient of Ga is unity and the sticking

[1] The zincblende crystal structure can be viewed as consisting of two interpenetrating fcc lattices on which the group III and V atoms reside, respectively. The two cubic lattices, of cube size a, are oriented parallel to each other, and the corner of one cube is displaced relative to the other along the body diagonal by the vector $(a/4, a/4, a/4)$. The Bravais lattice for the zincblende structure may be considered an fcc lattice with a basis containing the two different atoms at positions $(0, 0, 0)$ and $(a/4, a/4, a/4)$, respectively. However, these crystallographic details are of no importance for understanding the properties of the heterostructures of interest here.

coefficient of arsenic, in the chemical form As_2, practically vanishes except if there is an excess surface population of gallium. The growth rate of the heterostructure is therefore determined solely by the arrival rate at the substrate of the group III atoms, here Ga, and, due to the described "delicacies", is a slow process, typically proceeding at one monolayer (of size 2.83 Å for GaAs) per second, corresponding to one micrometer per hour (growth rates can vary from extremely slow, less than 1 Å/s, to fast, 35 Å/s). The growth process is highly sensitive to temperature, and, for example, at 1000°C, even a change of 0.1°C has an effect on the growth process. Good temperature control and stability, provided by liquid-nitrogen-cooled cryo-panels, is thus vital for MBE. Early attempts using MBE to make layered crystal structures date back to the early 1960s. However, lacking a fundamental understanding of the growth mechanisms at crystal surfaces, it has remained more of an art form, of trial and error, than simple engineering. The problems of the GaAlAs technology, for example, were solved empirically in the 1980s.

Consider, first, growing the III–V semiconductor GaAs, and then, by use of shutters, closing the Ga oven and opening the Al oven, in order to grow the III–V semiconductor AlAs on top. Thereby a heterostructure has been created, as schematically depicted in Figure 12.2. Aluminum is chemically similar to gallium, being the neighboring group III element in the periodic table. Thus the lattice constants are almost the same in the two compound semiconductors, and the interface between them is an almost ideal plane interface.

Since MBE takes place in ultra-high vacuum, it allows for monitoring techniques, and most importantly growth can be monitored *in situ*, giving MBE its importance, since it thus allows crystalline layer growth with accurate dimensional control down to the atomic level. The most important characterization technique controlling the growth of the heterostructure is reflection high-energy electron diffraction (RHEED) since it provides atomic-scale resolution and does not disrupt the crystal growth process. As depicted earlier in Figure 12.1, an electron gun sends a beam of 10–30 keV electrons to impinge on the sample at an angle of typically 0.5–2.5 degrees. The diffracted beam is converted to visible light by hitting a phosphor screen (the high energy of the electron beam is needed in order to image a sufficient area of reciprocal lattice space into the small solid angle at the phosphor screen) and analyzed by a charge-coupled device (CCD) camera placed outside the chamber, the RHEED screen depicted in Figure 12.1. The intensity of the diffraction pattern depends on the number of atomic monolayers that have been deposited. In fact, the specularly reflected electron beam shows oscillations in the intensity as a function of diffraction angle. Thus the RHEED oscillations not only give real-time information on the growth rates, layer

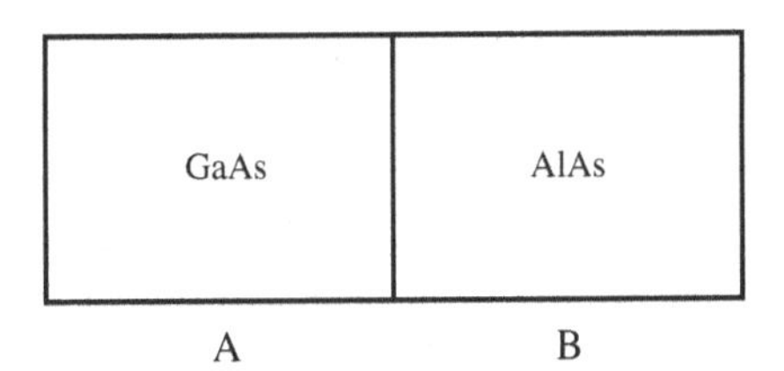

Figure 12.2 Gallium arsenide–aluminum arsenide heterostructure depiction.

thickness and alloy composition, but also can be utilized for feedback control, such as ascertaining the layer thickness. After growth of a definite number of monolayers of GaAs, the Ga oven is closed and the Al oven opened and the process now proceeds with ascertained growth of the AlAs crystal. An MBE chamber has further apparatus such as mass spectrometers (convenient for measuring leaks in the vacuum system or for measuring the water vapor in the residual gas) and ionization gauges for controlling the beam fluxes.

Using MBE it is possible to join two different materials into a heterostructure with an interface that is almost abrupt on atomic scale, as symbolized in Figure 12.2. This requires that the two materials have the same crystal structure, and, more importantly, in order to avoid strain, they should have almost the same lattice constant. To achieve this requires that the substituted atoms in the semiconductor compound have almost identical chemical nature. Therefore, a GaAs–AlAs heterostructure should be a good candidate since Ga and Al are next to each other in group III of the periodic table, and therefore as similar as possible. The lattice constants of the compound semiconductors GaAs and AlAs are very close, and these two compound semiconductors were historically one of the first candidates to be tried. Therefore, one of the first heterostructures to be grown successfully was GaAs–AlAs structures, for which a lattice mismatch, $(a_B - a_A)/a_A$, between the two materials of less than 0.1% is easily obtained.

The MBE machine is also equipped with effusion cells containing dopants. During the growth process, donor and acceptor atoms can thus be substituted for atoms of the heterostructure, and thereby provide the heterostructure with charge carriers, electrons and holes.

12.2 Envelope Function

Our interest lies in understanding the electron states in heterostructures. The MBE technique can make heterostructures with interfaces that are almost abrupt on an atomic scale. In practice, deviations from the case of an ideal interface occur: the bonds between atoms near the interface will deviate from those in the bulk, and the assumption of a perfect monolayer-by-monolayer growth is not possible, leaving the interface region varying in the directions perpendicular to the growth direction. In reality, rather than a sharp surface, one thus encounters an interface *layer*. However, the electronic states of interest in the heterostructure will have relatively little probability for being at the interface and a description employing an ideal interface should work to a first approximation. We shall therefore start out from the idealization that the heterostructure is perfectly lattice-matched, as a consequence of which the structure has an ideal flat interface between the two compound semiconductors, say GaAs and AlAs. Choosing the z direction as the growth direction, the interface is a plane in the x, y directions chosen to be at location $z = 0$.

Consider first the simplest case of a heterostructure consisting of two compound semiconductors labeled A and B, as depicted in Figure 12.2. The Hamiltonian determining the dynamics of an electron in the heterostructure is then

$$\hat{H} = \frac{\hat{\mathbf{p}}^2}{2m} + V_A(\mathbf{r})\theta(-z) + V_B(\mathbf{r})\theta(z), \tag{12.1}$$

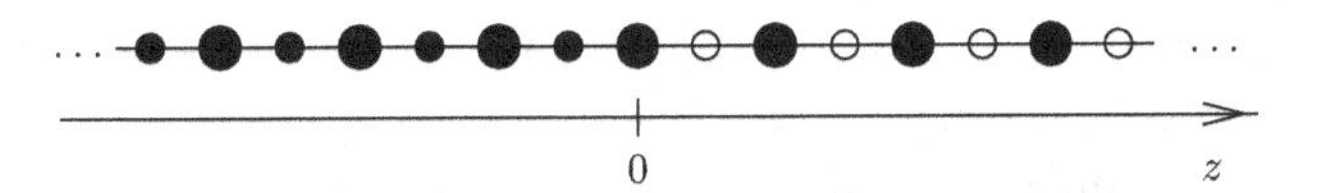

Figure 12.3 One-dimensional analog of a heterostructure.

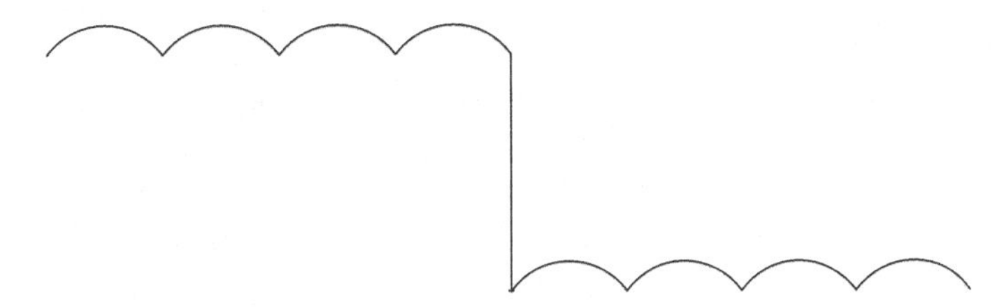

Figure 12.4 Heterostructure potential.

where V_A (V_B) is the periodic potential in material A (B), respectively.[2]

In Figure 12.3 is depicted a one-dimensional analog of the atomic structure of a heterostructure, such as for example GaAs–AlAs. The mean field potential felt by an electron in this idealization of a heterostructure is depicted in Figure 12.4. The atomic difference between the two different atoms characterizing the heterostructure, here gallium and aluminum, leads in the crudest approximation to a jump in potential at the ideal interface. This is the model Hamiltonian to be investigated in the following.

The interest is finding the energy eigenvalues and eigenfunctions for the heterostructure Hamiltonian, i.e. the solutions of the time-independent Schrödinger equation

$$\hat{H}\psi(\mathbf{r}) = \epsilon\, \psi(\mathbf{r}) \tag{12.2}$$

for the Hamiltonian specified in Eq. (12.1).

In a perfect bulk semiconductor, any state of an electron can be expanded into its complete set of Bloch states,

$$\psi(\mathbf{r}) = \sum_{n\mathbf{k}} f_{n\mathbf{k}}\, \psi_{n\mathbf{k}}(\mathbf{r}) = \sum_{n\mathbf{k}} f_{n\mathbf{k}}\, e^{i\mathbf{k}\cdot\mathbf{r}}\, u_{n\mathbf{k}}(\mathbf{r}), \tag{12.3}$$

where the $\mathbf{k}$ summation is over the Bloch vectors in the first Brillouin zone, $\mathbf{k} \in \mathrm{Bz}_1$, and n is the band index. In the case of a heterostructure, the solution of the time-independent Schrödinger equation can therefore be sought in this form in the two different regions,

$$\psi(\mathbf{r}) = \begin{cases} \sum_{n\mathbf{k}} f^A_{n\mathbf{k}}\, e^{i\mathbf{k}\cdot\mathbf{r}}\, u^A_{n\mathbf{k}}(\mathbf{r}), & \mathbf{r} \in A,\ z < 0, \\ \sum_{n\mathbf{k}} f^B_{n\mathbf{k}}\, e^{i\mathbf{k}\cdot\mathbf{r}}\, u^B_{n\mathbf{k}}(\mathbf{r}), & \mathbf{r} \in B,\ z > 0, \end{cases} \tag{12.4}$$

[2] In real III–V semiconductors, the energy also has a spin–orbit part, $(\sigma \times \nabla_{\mathbf{r}} V)\cdot\hat{\mathbf{p}}$, due to the difference between the atoms making up the compound. We neglect this effect since we want to keep the discussion as simple as possible and no qualitative features will change. However, for a quantitatively correct account of, for example, the split-off band from the degenerate valence bands of GaAs and AlGaAs, the so-called light and heavy hole bands, the spin–orbit interaction must be included. For a discussion of such important practical details, we refer the interested reader to the vast technical literature on heterostructures. Here the intent is to convey the essential idea.

leaving the interface condition of continuity of the wave function to determine the expansion coefficients. The two materials making up the A–B heterostructure have identical crystal structures and assumed identical lattice constants, and therefore have the same Brillouin zones, which therefore is not indicated in the summations in Eq. (12.4).

For a perfect lattice-matched heterostructure with an ideal interface, and the chemical nature of the substituted atoms being almost identical, and their size not differing too much, the periodic potentials in the two materials have the same periodicity and symmetry, and they differ only by a constant, a drop taking place on atomic scale at the interface as depicted in Figure 12.4. The periodic parts of the Bloch functions, $u^{B,A}_{n\mathbf{k}}(\mathbf{r})$, satisfy (recall Eq. (9.36)) in the two respective regions the equations

$$\left\{\frac{\hbar^2}{2m}(-i\nabla_{\mathbf{r}}+\mathbf{k})^2+V_A(\mathbf{r})\right\}u^A_{n\mathbf{k}}(\mathbf{r})=\epsilon^A_n(\mathbf{k})\,u^A_{n\mathbf{k}}(\mathbf{r}),\qquad z<0, \tag{12.5}$$

and

$$\left\{\frac{\hbar^2}{2m}(-i\nabla_{\mathbf{r}}+\mathbf{k})^2+V_B(\mathbf{r})\right\}u^B_{n\mathbf{k}}(\mathbf{r})=\epsilon^B_n(\mathbf{k})\,u^B_{n\mathbf{k}}(\mathbf{r}),\qquad z>0. \tag{12.6}$$

The jump in the periodic potential at the interface can be compensated by an equal jump between the values of $\epsilon^A_n(\mathbf{k})$ and $\epsilon^B_n(\mathbf{k})$, as depicted later in Figure 12.5, which is equivalent to assuming that the periodic part of the Bloch function is identical in the two materials,

$$u^A_{n\mathbf{k}}(\mathbf{r})=u^B_{n\mathbf{k}}(\mathbf{r}). \tag{12.7}$$

The sought after wave function for an energy eigenstate in the heterostructure thus has the form

$$\psi(\mathbf{r})=\sum_{n\mathbf{k}}f^{(A,B)}_{n\mathbf{k}}\,e^{i\mathbf{k}\cdot\mathbf{r}}u_{n\mathbf{k}}(\mathbf{r}). \tag{12.8}$$

The left-hand side of the time-independent Schrödinger equation for the state in Eq. (12.4) thus takes the form in the two regions

$$\hat{H}\psi(\mathbf{r})=\begin{cases}\sum_{n\mathbf{k}}f^A_{n\mathbf{k}}\,\epsilon^A_n(\mathbf{k})\,e^{i\mathbf{k}\cdot\mathbf{r}}\,u^A_{n\mathbf{k}}(\mathbf{r}), & \mathbf{r}\in A,\ z<0,\\ \sum_{n\mathbf{k}}f^B_{n\mathbf{k}}\,\epsilon^B_n(\mathbf{k})\,e^{i\mathbf{k}\cdot\mathbf{r}}\,u^B_{n\mathbf{k}}(\mathbf{r}), & \mathbf{r}\in B,\ z>0.\end{cases} \tag{12.9}$$

We shall seek the energy eigenfunctions built from states near band edges since, as usual, the states of lowest energy are those of interest, i.e. the function $f_{n\mathbf{k}}$ is peaked around a band edge Bloch vector $\mathbf{k}_0$. Since the Bloch function, $u_{n\mathbf{k}}$, has a weak dependence on the Bloch vector, and the functions $f_{n\mathbf{k}}$ are assumed to be peaked functions of the Bloch vector, the sought after solution is approximately

$$\psi(\mathbf{r})\simeq\psi^{A(B)}_{\mathbf{k}_0}(\mathbf{r})=e^{i\mathbf{k}_0\cdot\mathbf{r}}\sum_n u_{n\mathbf{k}_0}(\mathbf{r})f^{A(B)}_n(\mathbf{r})=\sum_n\psi_{n\mathbf{k}_0}(\mathbf{r})f^{A(B)}_n(\mathbf{r}). \tag{12.10}$$

Here we have indicated that the heterostructure energy eigenstate depends on the band edge Bloch vector $\mathbf{k}_0$, and, if $\mathbf{k}_0\neq\mathbf{0}$, fast oscillations in the exponential function due to $\mathbf{k}_0$ have been taken out, so that

$$f_n^{A(B)}(\mathbf{r}) = \sum_{\mathbf{k}} f_{n\mathbf{k}}^{A(B)} \, e^{i(\mathbf{k}-\mathbf{k}_0)\cdot\mathbf{r}}. \tag{12.11}$$

By "peaked" is meant that, for each band index n, $f_{n\mathbf{k}}$ as a function of $\mathbf{k}$ is only non-vanishing in a region near $\mathbf{k}_0$ that is small compared to the size of the Brillouin zone. The function $f_n(\mathbf{r})$ thus represents a smooth spatial variation on the scale of the unit cell on top of the Bloch function, $\psi_{n\mathbf{k}_0}(\mathbf{r})$, which has rapid spatial variation, i.e. $f_n(\mathbf{r})$ is an envelope function that hardly varies over the unit cell.

A wave function is continuous, and since the functions $u_{n\mathbf{k}_0}(\mathbf{r})$ for different n values are linearly independent, the envelope functions for each band index must match at the interface, $z = 0$. Assuming that the function $u_{n\mathbf{k}_0}$ does not have a node at the interface, i.e. vanishes at the interface, the linear independence thus demands that the coefficients, the envelope functions, satisfy the interface condition

$$f_n^A(\mathbf{r}_\perp, z = 0) = f_n^B(\mathbf{r}_\perp, z = 0) \tag{12.12}$$

or equivalently, according to Eq. (12.11),

$$\sum_{\mathbf{k}} f_{n\mathbf{k}}^A \, e^{i(\mathbf{k}-\mathbf{k}_0)\cdot\mathbf{r}_\perp} = \sum_{\mathbf{k}} f_{n\mathbf{k}}^B \, e^{i(\mathbf{k}-\mathbf{k}_0)\cdot\mathbf{r}_\perp} \tag{12.13}$$

where $\mathbf{r}_\perp$ is a position vector in the interface plane. Since the two materials are assumed perfectly lattice-matched, with an ideal interface, the envelope function is, according to Eq. (12.13), translationally invariant in the interface coordinate $\mathbf{r}_\perp = (x, y)$ for all values of z; the effects of the periodic potential in these directions are described by the $u_{n\mathbf{k}_0}$ functions. There is therefore inhomogeneity only in the z direction, and the envelope function factorizes to

$$f_n^{A(B)}(\mathbf{r}) = f_n^{A(B)}(\mathbf{r}_\perp, z) = \chi_\perp(\mathbf{r}_\perp) \, \chi_n(z), \tag{12.14}$$

where

$$\chi_n(z) = \begin{cases} \chi_n^A(z), & z > 0, \\ \chi_n^B(z), & z < 0, \end{cases} \tag{12.15}$$

and

$$\chi_\perp(\mathbf{r}_\perp) = \frac{1}{\sqrt{S}} \sum_{\mathbf{k}_\perp} e^{i(\mathbf{k}_\perp - (\mathbf{k}_0)_\perp)\cdot\mathbf{r}_\perp} \tag{12.16}$$

is a sum of the plane wave part of the Bloch functions in the x, y plane, and the area of the interface of the heterostructure is denoted by S. We have, according to Eq. (12.11), expanded $\chi_\perp$ on a range of $\mathbf{k}_\perp$ near $\mathbf{k}_0$, but can in fact treat the z dependence of the coefficient functions separately since the plane waves are linearly independent orthogonal functions. The interface condition, Eq. (12.12), thus becomes a boundary condition on the z direction envelope functions,

$$\chi_n^A(z = 0) = \chi_n^B(z = 0). \tag{12.17}$$

In the following, for definiteness we shall assume that we are dealing with states near the zone center, i.e. the case of band structure where $\mathbf{k}_0 = \mathbf{0}$. (The band edges of the materials in question have precisely this property; for example, it is the relevant case for a GaAs–AlAs heterostructure grown along the (001)-direction.)

To determine the energy eigenvalues and envelope functions, the sought after solution

$$\psi(\mathbf{r}) = \frac{1}{\sqrt{S}} \sum_{n'\mathbf{k}'_\perp} e^{i\mathbf{k}'\cdot\mathbf{r}_\perp}\, \chi_{n'}(z)\, u_{n'\mathbf{0}}(\mathbf{r}) \tag{12.18}$$

is inserted into the time-independent Schrödinger equation, Eq. (12.2), and the scalar product with the wave function $\chi_n(z)\, e^{i\mathbf{k}_\perp\cdot\mathbf{r}_\perp} u_{n\mathbf{0}}(\mathbf{r})$ is taken, obtaining

$$0 = \int d\mathbf{r}\, [\chi_n(z)]^*\, e^{-i\mathbf{k}_\perp\cdot\mathbf{r}_\perp} u^*_{n\mathbf{0}}(\mathbf{r})(\hat{H} - \epsilon)\, \psi(\mathbf{r}). \tag{12.19}$$

Inserting the expression for the wave function, Eq. (12.18), the task is then to simplify the integrand

$$0 = \int_V d\mathbf{r}\, \chi^*_n(z)\, e^{-i\mathbf{k}_\perp\cdot\mathbf{r}_\perp} u^*_{n\mathbf{0}}(\mathbf{r}) \left(-\epsilon + \frac{\hat{\mathbf{p}}^2}{2m} + \theta(-z)V_A(\mathbf{r}) + \theta(z)V_B(\mathbf{r}) \right)$$
$$\times \sum_{n'\mathbf{k}'_\perp} \chi_{n'}(z)\, e^{i\mathbf{k}'_\perp\cdot\mathbf{r}_\perp} u_{n'\mathbf{0}}(\mathbf{r}). \tag{12.20}$$

Recalling Eq. (9.34), we have for the Hamiltonian given by Eq. (12.1)

$$\hat{H} e^{i\mathbf{k}\cdot\mathbf{r}_\perp} u_{n\mathbf{k}}(\mathbf{r})$$
$$= e^{i\mathbf{k}\cdot\mathbf{r}} \left(\frac{\hbar^2}{2m}(-i\nabla_{\mathbf{r}_\perp} + \mathbf{k}_\perp)^2 - \frac{\hbar^2}{2m}\frac{\partial^2}{\partial z^2} + \theta(-z)V_A(\mathbf{r}) + \theta(z)V_B(\mathbf{r}) \right) u_{n\mathbf{k}}(\mathbf{r}) \tag{12.21}$$

and Eq. (12.19) becomes (see the similar calculation in connection with the effective mass approximation in Appendix S)

$$0 = \sum_{\mathbf{k}'_\perp} \int_V d\mathbf{r}\; e^{i(\mathbf{k}'_\perp - \mathbf{k}_\perp)\cdot\mathbf{r}_\perp}\, \chi^*_n(z)\, u^*_{n\mathbf{0}}(\mathbf{r}) \left(-\epsilon + \frac{\hbar^2\mathbf{k}'^2_\perp}{2m} + \frac{\hbar\mathbf{k}'_\perp}{m}\cdot\hat{\mathbf{p}} + \hat{H} \right)$$
$$\times \sum_{n'} \chi_{n'}(z)\, u_{n'\mathbf{0}}(\mathbf{r}). \tag{12.22}$$

Using

$$\hat{H} u_{n\mathbf{0}}(\mathbf{r}) = \begin{cases} \epsilon^A_n(\mathbf{0})\, u_{n\mathbf{0}}(\mathbf{r}), & \mathbf{r} \in A, \\ \epsilon^B_n(\mathbf{0})\, u_{n\mathbf{0}}(\mathbf{r}), & \mathbf{r} \in B, \end{cases} \tag{12.23}$$

turns Eq. (12.22) into

$$0 = \sum_{\mathbf{k}'_\perp} \int_V d\mathbf{r}\, e^{i(\mathbf{k}'_\perp - \mathbf{k}_\perp)\cdot\mathbf{r}_\perp}\, \chi^*_n(z)\, u^*_{n\mathbf{0}}(\mathbf{r})$$
$$\times \left\{ \sum_{n'} u_{n'\mathbf{0}}(\mathbf{r}) \left(-\epsilon - \frac{\hbar^2}{2m}\frac{\partial^2}{\partial z^2} + \frac{\hbar^2\mathbf{k}'^2_\perp}{2m} + \epsilon^A_{n'}(\mathbf{0})\,\theta(-z) + \epsilon^B_{n'}(\mathbf{0})\,\theta(z) \right) \chi_{n'}(z) \right.$$
$$\left. + \sum_{n'} \left(\chi_{n'}(z) \frac{\hbar\,\mathbf{k}'_\perp}{m} \cdot \hat{\mathbf{p}}\, u_{n'\mathbf{0}}(\mathbf{r}) - \frac{\hbar^2}{m} \frac{\partial \chi_{n'}(z)}{\partial z} \frac{\partial u_{n'\mathbf{0}}(\mathbf{r})}{\partial z} \right) \right\}. \tag{12.24}$$

The integrand is a product of a periodic function, *viz.* $u^*_{n\mathbf{0}}(\mathbf{r})\, u_{n'\mathbf{0}}(\mathbf{r})$, or derivatives of this periodic function, which again is periodic, and a slowly varying function (on the scale of the size of the unit cell, i.e. the lattice constant). We can therefore employ the formula (T.6) to turn the integral into the product of integrals,

$$
\begin{aligned}
0 \simeq & \frac{1}{V} \sum_{n'\mathbf{k}'_\perp} \int_V d\mathbf{r}_1\, u^*_{n\mathbf{0}}(\mathbf{r}_1)\, u_{n'\mathbf{0}}(\mathbf{r}_1) \int_V d\mathbf{r}\, e^{i(\mathbf{k}'_\perp - \mathbf{k}_\perp)\cdot \mathbf{r}_\perp}\, \chi^*_n(z) \\
& \times \left(-\epsilon - \frac{\hbar^2}{2m}\frac{\partial^2}{\partial z^2} + \frac{\hbar^2 \mathbf{k}'^2_\perp}{2m} + \epsilon^A_{n'}(\mathbf{0})\,\theta(-z) + \epsilon^B_{n'}(\mathbf{0})\,\theta(z) \right) \chi_{n'}(z) \\
& + \frac{1}{V} \sum_{n'\mathbf{k}'_\perp} \int_V d\mathbf{r}_1\, u^*_{n\mathbf{0}}(\mathbf{r}_1) \int_V d\mathbf{r}\, e^{i(\mathbf{k}'_\perp - \mathbf{k}_\perp)\cdot \mathbf{r}_\perp}\, \chi^*_n(z) \\
& \times \left(\chi_{n'}(z) \frac{\hbar\, \mathbf{k}'_\perp}{m} \cdot \hat{\mathbf{p}}\, u_{n'\mathbf{0}}(\mathbf{r}_1) - \frac{\hbar^2}{m} \frac{\partial \chi_{n'}(z)}{\partial z} \frac{\partial u_{n'\mathbf{0}}(\mathbf{r}_1)}{\partial z_1} \right) .
\end{aligned}
\tag{12.25}
$$

The orthogonality of the periodic part of the Bloch functions, Eq. (9.48), makes the first n' sum collapse, leaving

$$
\begin{aligned}
0 = & \sum_{\mathbf{k}'_\perp} \int_V d\mathbf{r}\, e^{i(\mathbf{k}'_\perp - \mathbf{k}_\perp)\cdot \mathbf{r}_\perp}\, \chi^*_n(z) \\
& \times \left(-\epsilon - \frac{\hbar^2}{2m}\frac{\partial^2}{\partial z^2} + \frac{\hbar^2 \mathbf{k}'^2_\perp}{2m} + \epsilon^A_{n}(\mathbf{0})\,\theta(-z) + \epsilon^B_{n}(\mathbf{0})\,\theta(z) \right) \chi_{n}(z) \\
& + \frac{1}{V} \sum_{n'\mathbf{k}'_\perp} \int_V d\mathbf{r}\, e^{i(\mathbf{k}'_\perp - \mathbf{k}_\perp)\cdot \mathbf{r}_\perp}\, \chi^*_n(z) \left(\chi_{n'}(z) \frac{\hbar\, \mathbf{k}'_\perp}{m} \cdot \int_V d\mathbf{r}_1\, u^*_{n\mathbf{0}}(\mathbf{r}_1) \frac{\hbar}{i} \nabla_{\mathbf{r}_1} u_{n'\mathbf{0}}(\mathbf{r}_1) \right. \\
& \left. - \frac{\hbar^2}{m} \frac{\partial \chi_{n'}(z)}{\partial z} \int_V d\mathbf{r}_1\, u^*_{n\mathbf{0}}(\mathbf{r}_1) \frac{\partial u_{n'\mathbf{0}}(\mathbf{r}_1)}{\partial z_1} \right) .
\end{aligned}
\tag{12.26}
$$

Similarly, the orthogonality of the plane waves makes the $\mathbf{k}'_\perp$ sums collapse, leaving

$$
\begin{aligned}
0 = & S \int dz\, \chi^*_n(z) \left(-\epsilon - \frac{\hbar^2}{2m}\frac{\partial^2}{\partial z^2} + \frac{\hbar^2 k^2_\perp}{2m} + \epsilon^A_{n}(\mathbf{0})\,\theta(-z) + \epsilon^B_{n}(\mathbf{0})\,\theta(z) \right) \chi_{n}(z) \\
& + \frac{1}{L_z} \sum_{n'} \int dz\, \chi^*_n(z) \left(\chi_{n'}(z) \frac{\hbar \mathbf{k}_\perp}{m} \cdot \int_V d\mathbf{r}_1\, u^*_{n\mathbf{0}}(\mathbf{r}_1) \frac{\hbar}{i} \nabla_{\mathbf{r}_1} u_{n'\mathbf{0}}(\mathbf{r}_1) \right. \\
& \left. - \frac{\hbar^2}{m} \frac{\partial \chi_{n'}(z)}{\partial z} \int_V d\mathbf{r}_1\, u^*_{n\mathbf{0}}(\mathbf{r}_1) \frac{\partial u_{n'\mathbf{0}}(\mathbf{r}_1)}{\partial z_1} \right) ,
\end{aligned}
\tag{12.27}
$$

which, by using Dirac notation

$$\langle n, \mathbf{0}|\hat{\mathbf{p}}|n', \mathbf{0}\rangle = \frac{1}{v}\int_v d\mathbf{r}\; u^*_{n\mathbf{0}}(\mathbf{r})\frac{\hbar}{i}\nabla_{\mathbf{r}}\, u_{n',\mathbf{0}}(\mathbf{r}), \tag{12.28}$$

can be compactly written as

$$0 = \int dz\; \chi^*_n(z)\left\{\left(-\epsilon - \frac{\hbar^2}{2m}\frac{\partial^2}{\partial z^2} + \frac{\hbar^2 k_\perp^2}{2m} + \epsilon_n^A(\mathbf{0})\,\theta(-z) + \epsilon_n^B(\mathbf{0})\,\theta(z)\right)\chi_n(z)\right.$$
$$\left. + \sum_{n'}\left(\frac{\hbar\,\mathbf{k}_\perp}{m}\cdot\langle n, \mathbf{0}|\hat{\mathbf{p}}_\perp|n', \mathbf{0}\rangle - \frac{i\hbar}{m}\langle n, \mathbf{0}|\hat{p}_z|n', \mathbf{0}\rangle\frac{\partial}{\partial z}\right)\chi_{n'}(z)\right\} \tag{12.29}$$

or

$$0 = \int dz\; \chi^*_n(z)\sum_{n'}\left(H^{\text{eff}}_{nn'}\left(z, -i\hbar\frac{\partial}{\partial z}\right) - \epsilon\,\delta_{nn'}\right)\chi_{n'}(z), \tag{12.30}$$

where

$$H^{\text{eff}}_{nn'}\left(z, -i\hbar\frac{\partial}{\partial z}\right) = \left(\epsilon_n^A(\mathbf{0}) + V_n(z) + \frac{\hbar^2 k_\perp^2}{2m} - \frac{\hbar^2}{2m}\frac{\partial^2}{\partial z^2}\right)\delta_{nn'}$$
$$+ \frac{\hbar\,\mathbf{k}_\perp}{m}\cdot\langle n, \mathbf{0}|\hat{\mathbf{p}}_\perp|n', \mathbf{0}\rangle - \frac{i\hbar}{m}\langle n, \mathbf{0}|\hat{p}_z|n', \mathbf{0}\rangle\frac{\partial}{\partial z} \tag{12.31}$$

is the effective Hamiltonian for the problem, and the step potential

$$V_n(z) = \begin{cases} 0, & z \in A,\, z > 0, \\ \epsilon_n^B(\mathbf{0}) - \epsilon_n^A(\mathbf{0}), & z \in B,\, z < 0, \end{cases} \tag{12.32}$$

is determined by the band edge offsets in the two materials.[3]

Since the Bloch velocity, $\mathbf{v}_n(\mathbf{0})$, is zero, the last two terms in Eq. (12.31) contain only off-diagonal elements. All the rapid atomic scale details of the wave function for an electron energy eigenstate in a heterostructure now only enter through matrix elements, i.e. have been lumped into effective parameters. Collecting the envelope components into a column vector, χ, it follows from Eq. (12.30) that the envelope function satisfies the matrix equation

$$H^{\text{eff}}\chi = \epsilon\,\chi \tag{12.33}$$

determined by the hermitian matrix operator H^{eff}. The Hamiltonian $H^{\text{eff}}_{nn'}$ in Eq. (12.33) is analogous to the $\mathbf{k}\cdot\mathbf{p}$ Hamiltonian considered in Appendix S, except for the step potential, which depends on the position being in either of the two materials and the occurrence of partial differentiation with respect to z instead of just k_z.

Bands far away in energy can be neglected and closer ones treated perturbatively in the effective mass approximation, as discussed in Appendix S, with the new feature that the

[3] Of course, a step function cannot be taken literally; the transition from one semiconductor to the other typically takes place at best over a few lattice constants (and electron readjustments at the interface shift the bands relative to one another). However, for wave functions that are mostly non-zero away from the interface, say, inside a quantum well, details of the drop may be neglected in a first approximation.

effective mass is now z-dependent due to $V_n(z)$. A complicating feature is that heterostructures typically have energy bands that are degenerate at band edges, and in general a matrix equation has to be solved.[4] For simplicity, we consider the case of non-degenerate bands (except for spin degeneracy), and the simplest case where the envelope function only has a contribution from a single band (an approximation that works reasonably well for the conduction band in GaAs). The equation for the envelope function is then

$$\left\{ -\frac{\partial}{\partial z}\frac{\hbar^2}{2m^*(z)}\frac{\partial}{\partial z} + \epsilon_n^A(\mathbf{0}) + V_n(z) + \frac{\hbar^2 k_\perp^2}{2m^*(z)} \right\} \chi_n(z) = \epsilon\,\chi_n(z), \tag{12.34}$$

with the difference compared to a Schrödinger equation being that the effective masses in the two parts of the heterostructure can be different,

$$m^*(z) = \begin{cases} m_A^*, & z \in A, z > 0, \\ m_B^*, & z \in B, z < 0, \end{cases} \tag{12.35}$$

which leads to the feature that it is not the derivative of the *wave function* but the quantity

$$\frac{1}{m^*(z)} \frac{\partial \chi_n(z)}{\partial z}$$

that should be continuous in order to conserve current, as can be seen by going back to the original Schrödinger equation and integrating it across the interface. Otherwise, the problem of determining the electronic states in a heterostructure is reduced to the one-dimensional problem of a particle in a step potential determined by the band edges of the semiconductors that make up the heterostructure.

Heterostructures provide a new tool for controlling carriers, as the different band edges in the different materials will force carriers into definite spatial regions, in the envelope formulation described by the step potentials, say, electrons to the low band edge spatial region.

12.3 Structure Types

In the simplest of cases, only one conduction band and one valence band in the heterostructure are relevant, a situation for example as depicted in Figure 12.5. The step potentials felt by electrons and holes in the conduction and valence bands are specified by the band edges of the two materials making up the heterostructure.

Tailor-made structures of step potentials can be produced by MBE technology. Sandwiching a narrow-band-gap material in between a wide-band-gap material creates quantum wells as depicted in Figure 12.6.

[4] In the case of a direct-band-gap III–V compound semiconductor heterostructure, such as GaAs–AlAs, four bands need to be taken into account: the conduction band and three valence bands (two heavy hole bands and a light hole band, which are split due to spin–orbit interaction). The relevant equation, Eq. (12.33), is then an 8×8 matrix equation. For these real-life complications, the interested reader is referred to the vast literature.

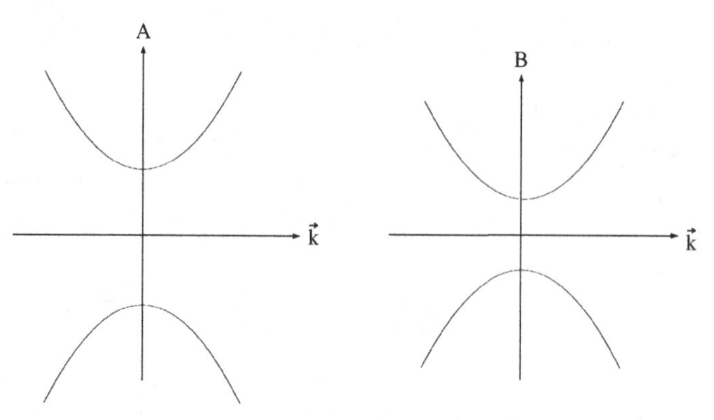

Figure 12.5 Band edges in the two materials constituting the *A*–*B* heterostructure.

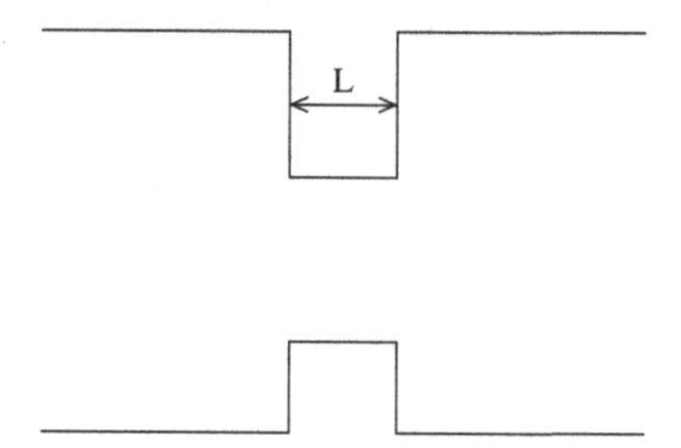

Figure 12.6 Single heterostructure quantum well.

Proper dopant atoms provided during growth of a heterostructure provide electrons experiencing the conduction band edge profile, and holes experiencing the valence band edge profile. The band edge profiles depicted in Figure 12.6 lead to quantum wells for both electrons and holes, a type I heterostructure. The energy levels in the quantum wells are determined by the *Schrödinger* equation for the envelope function. The case where there is only one bound electron state and one bound hole state is depicted in Figure 12.7.

Heterostructures can have electronic and photonic properties superior to those of the bulk materials and alloys. The original motivation for creating heterostructures was to build double heterostructures: sandwiching a narrow-band-gap semiconductor in between two wide-band-gap semiconductors. The conduction band edge will be lower in the narrow-band-gap material (and the valence band edge higher) than in the wide-band-gap material, and electrons and holes will be trapped in the quantum well region, making such structures superior to p–n junctions as lasers. In the case of p–n junction lasers, the wave length of the emitted light is determined by the band gap of the semiconductor (a quantity determined by nature), giving typically red lasers. With the advent of heterostructures, the quantum well confinement can enhance the energy difference between electron and hole states, and therefore the energy of the emitted photons. Owing to the transverse direction of the heterostructure, a large confined inverted electron–hole population can exist (and can be electronically maintained). Just like an electron in an excited state in an atom decays by emission of a photon, the same process takes place in a quantum well, where an electron

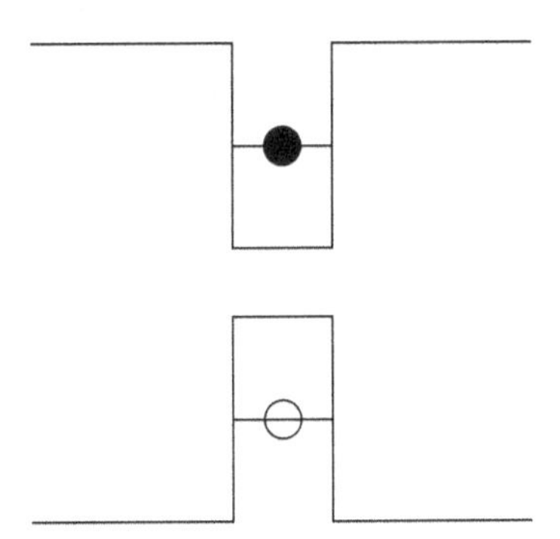

Figure 12.7 Electron and hole quantum well levels.

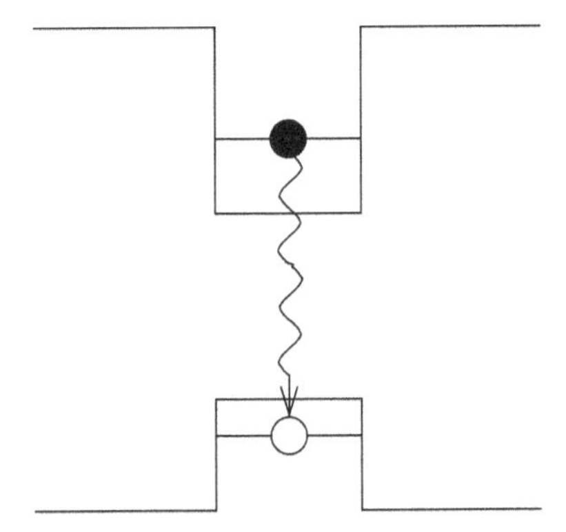

Figure 12.8 Light emission by electron–hole transition.

in the conduction band quantum well makes the transition to a quantum well hole state, as depicted in Figure 12.8.

Heterostructure technology has produced the *blue* laser, with important applications for reading more compressed digital information stored on disks, and energy efficiently providing the third color source for making white light.

The physical properties of a particle in a heterostructure can be expressed in terms of the envelope function. To calculate a *z direction property*, for example, the average velocity of an electron in the z direction,

$$v_z \equiv \langle \psi | \hat{v}_z | \psi \rangle = \int d\mathbf{r}\ \psi^*(\mathbf{r}) \frac{-i\hbar}{m} \frac{\partial}{\partial z} \psi(\mathbf{r}), \tag{12.36}$$

for an electron in a heterostructure state described by the wave function

$$\psi(\mathbf{r}) = \chi_n(z)\, u_{n\mathbf{0}}(\mathbf{r}), \tag{12.37}$$

the slowly varying character of the envelope function and the periodicity of the Bloch function lead, by employing formula (T.6), to

$$v_z \simeq \langle \chi | \hat{v}_z | \chi \rangle + \langle n, \mathbf{0} | \hat{v}_z | n, \mathbf{0} \rangle = \langle \chi | \hat{v}_z | \chi \rangle \tag{12.38}$$

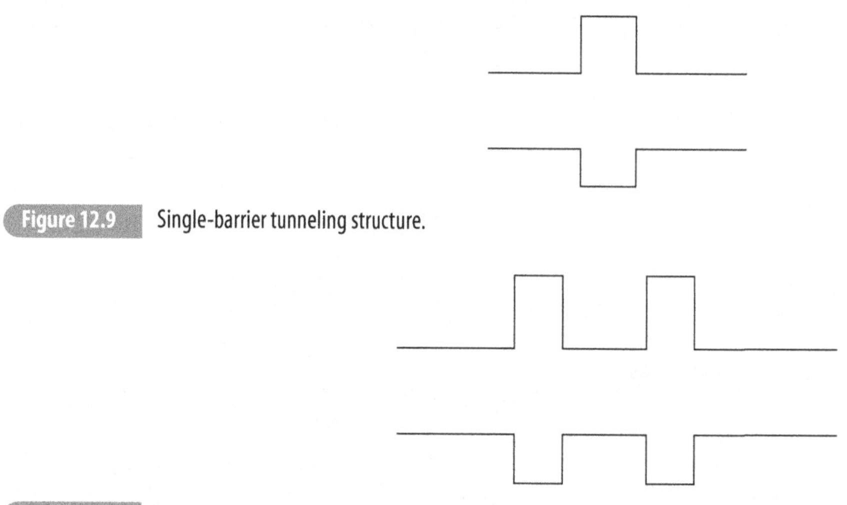

Figure 12.9 Single-barrier tunneling structure.

Figure 12.10 Double-barrier tunneling structure.

as

$$\left. \frac{\partial \epsilon_n(\mathbf{k})}{\partial k_z} \right|_{\mathbf{k}=\mathbf{0}} = 0. \tag{12.39}$$

The average velocity is thus determined only by the envelope function.

If a wide-band-gap material is sandwiched in between a narrow-band-gap material, a heterostructure with square barrier potentials for both electrons in the conduction band and holes in the valence band is constructed, as depicted in Figure 12.9.

Repeating such a construction a double-barrier tunneling structure is formed as depicted in Figure 12.10. Such a structure has interesting electronic properties as discussed in Chapter 13.

By periodically repeating a quantum well structure, multiple quantum wells are formed, an artificial crystal structure or superlattice. The properties of an electron in such a conduction band profile will then be governed by the envelope function satisfying the equation for a particle in a periodic step function potential,

$$\left\{ -\frac{\hbar^2}{2m^*}\frac{\partial^2}{\partial z^2} + \frac{\hbar^2 k_\perp^2}{2m^*} + V_{\text{per}}(z) \right\} \chi(z) = \epsilon \chi(z), \tag{12.40}$$

and as regards the z direction kinematics a band structure appears, so-called mini-bands. The *lattice* constant, a, of a superlattice can be tailored at will, giving a parameter for changing the Bloch period $T_B = 2\pi\hbar/eEa$. For example, making the superlattice constant a thousand times larger than that of a semiconductor lowers the Bloch period by three orders of magnitude. Bloch oscillations should thus be easier to observe in superlattices, as the amount of impurities in heterostructures can be kept low, and lower electric field strengths can be used, for which Zener tunneling is weaker. Modest field strengths $F = eE = 10^3$ V/cm and a lattice constant $a = 100$ Å give a high-frequency oscillator, $\nu = 250$ GHz, and would thereby enhance the capabilities of microwave devices.

12.4 Band-Gap Engineering

In order to control electrons in a heterostructure, wide-band-gap materials are preferable. For the sandwich GaAs–AlAs, this cannot be achieved because the materials have a common anion. For device applications, in, say, a III–V semiconductor such as GaAs, one should substitute some fraction x of the gallium atoms with another type of atom from group III, say aluminum, so that certain lattice sites will be occupied by Al ions instead of Ga ions. This substitution is easily achieved in an MBE machine by having all three ovens with Ga, As and Al open at the same time, and $Al_xGa_{1-x}As$ can be grown on top of GaAs. However, the site where an aluminum atom substitutes cannot be controlled, only the fraction, x, so an electron in such a structure does not experience a perfect periodic potential, but a disordered solid. For many purposes, however, the average potential, where the disordered potential is replaced by the average periodic one, has proved its practical utility, the virtual crystal approximation,

$$\begin{aligned} V_{\mathrm{Al}_x\mathrm{Ga}_{1-x}\mathrm{As}}(\mathbf{r}) &\equiv V_{\mathrm{As}}(\mathbf{r}) + xV_{\mathrm{Al}}(\mathbf{r}) + (1-x)V_{\mathrm{Ga}}(\mathbf{r}) \\ &= (1-x)V_{\mathrm{GaAs}}(\mathbf{r}) + xV_{\mathrm{AlAs}}(\mathbf{r}) \\ &= V_{\mathrm{GaAs}}(\mathbf{r}) + x(V_{\mathrm{Al}}(\mathbf{r}) - V_{\mathrm{Ga}}(\mathbf{r})). \end{aligned} \tag{12.41}$$

In this approximation, the periodic potential machinery can be applied, and the envelope function description appears, but now with the substitutional fraction, x, as a parameter with which the band edge offsets can be monitored. It is thus possible to monitor the band gap between the conduction and valence band edges as a function of the substitutional fraction, $\Delta = \Delta(x)$, and one speaks of band-gap engineering.

12.5 Modulation Doping

Dopants not only provide carriers but also act as scatterers of electrons and holes, and are thus an inherent limitation to high mobilities of the carriers. With the advent of MBE to create heterostructures, the strength of dopant scattering can be reduced by using so-called modulation doping, and thereby provide high-mobility structures.

A heterostructure made of an $Al_xGa_{1-x}As$ part next to a GaAs part will have a step potential with the lowest potential in the GaAs part of the $Al_xGa_{1-x}As$–GaAs heterostructure. Consider now that during the growth process the $Al_xGa_{1-x}As$ is heavily n-doped. In such a structure, the donor electrons are thus transferred to the narrow-band-gap material, GaAs, as the potential energy is lower there. The charge distribution of the fixed positive ionized donor charges in the AlGaAs part and the negative charge of the transferred electrons will give rise to the modified conduction band edge potential profile depicted in Figure 12.11.

By proper doping, the potential felt by the transferred electrons can have bound states below the Fermi energy, as symbolized by the horizontal lines in the well in Figure 12.11.

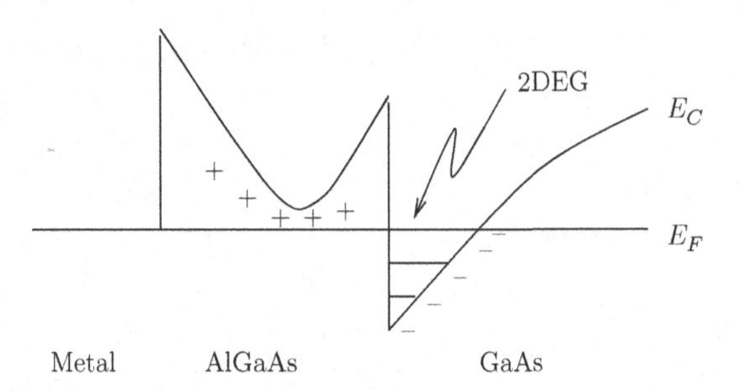

Figure 12.11 Conduction band edge for a modulation-doped GaAs–AlGaAs heterostructure.

At sufficiently low temperatures, only the lowest bound state is populated, and the electrons are trapped in one channel state at the interface, in an inversion layer similar to the situation discussed in Section 11.6. The electrons can therefore move only along the interface, their transverse degree of freedom having been frozen out, and a two-dimensional electron gas (2DEG) has been created. The trick in modulation doping is to create a so-called spacer layer in the AlGaAs part just next to the GaAs, a layer where there is no n-doping. As a result, the positively charged dopant impurities are spatially separated from the trapped electrons, thereby drastically reducing their scattering potential effect on the electrons in the two-dimensional electron gas. The electrons can therefore move almost freely in this so-called inversion layer formed in the gallium arsenide just below the interface. Mobilities a thousand times larger than in semiconductors can easily be achieved.

Besides being used in electronic devices, the two-dimensional electron gas is also of fundamental importance, as unexpected physics takes place in them due to the two-dimensional nature of the system. The electrons truly experience a two-dimensional world. For instance, at low temperatures and high transverse magnetic fields ($\sim$ 10 T), the Hall resistance exhibits quantum features, the so-called (integer) quantum Hall steps. The position of the plateaus are with incredible precision (better than 10^{-7}) positioned at the universal values

$$\rho_{xy} = \frac{2\pi\hbar}{ie^2}, \qquad i = 1, 2, 3 \ldots . \tag{12.42}$$

The fact that the quantum Hall plateaus are determined solely by constants of nature is of very practical importance, as the value of the first plateau is now used as reference for the resistance standard. Together with the Josephson effect, which is used as the basis for the voltage standard, solid-state systems nowadays are used for fundamental practical purposes as standards. The precision of these solid-state electronic standards are much better than the previous mechanical ones, and, curiously enough, quite easy to maintain due to the rigidity and universal features of the quantum effects upon which they are based. This was a quite unexpected development since one would *a priori* not expect such systems to exhibit universal high-precision effects, as large systems are messy on the atomic level.

In Section 13.8, a two-dimensional electron gas is studied in the presence of a constriction, the quantum point contact, which is shown to exhibit conductance quantization, and thereby provide a candidate for the third electronic standard, the current standard.

12.6 Summary

Molecular beam epitaxy has led to the realization of material structures of incredibly small dimensions in the growth direction, and to the development of new device concepts where quantum mechanics enters at the transport level, in contrast to that of usual semiconductor and silicon technology, where quantum mechanics is only important in the constitutive sense, the electronic transport being classical, e.g. simple diffusion. The MBE technique used to construct heterostructures is therefore an important tool in the emerging nanotechnology. The envelope function description of the electronic states in heterostructures has been presented. What it lacks in rigor seems to be outshone by its physically appealing simplicity and applicability.

13 Mesoscopic Physics

A mesoscopic sample is a small structure through which the motion of an electron is quantum mechanically coherent. The sample has sufficiently low temperature so that inelastic collisions with other electrons or lattice vibrations can be neglected, i.e. dissipation can be neglected inside the sample. To study currents through mesoscopic samples, a scattering approach is thus appropriate, the sample constituting the obstacle.

13.1 Scattering Approach to Conductance

The simplest mesoscopic transport situation is depicted in Figure 13.1. Two electron *reservoirs*, kept at certain electrochemical potentials, populate states in metal leads connected to the sample (S), a *black box* giving rise to transmission and reflection.

The electron reservoirs are macroscopic, so electrons fed into states in the left and right leads are uncorrelated, and the currents generated by the left and right reservoirs are added, as depicted in Figure 13.2. Electrons are not coherently reflected back into the leads. In this way the reservoirs give rise to breaking of time reversal symmetry, establishing a non-equilibrium situation with dissipation, inelastic scattering, taking place in the reservoirs.

The considered transport situation is analogous to the case of the tunnel junction. Consider first the one-dimensional case; the argument is thus identical to that for obtaining the tunneling current discussed in Chapter 3. The average current flowing through the system is obtained by counting the contributions from each of the states in the leads (though assuming perfect leads, so that the lead states are labeled by momentum or wave number, is not essential; any quantum labeling leads to an analogous result). Consider counting the current in the *left* lead. The average current carried by momentum states $\pm p$ on the left is specified by their current and population (L is the length of the left lead, the region in which the current is carried by the plane wave of periodic boundary condition, say)

$$I_p^L = \frac{ev_p}{L}(f_p^L - R_p f_p^L - T_p f_p^R), \tag{13.1}$$

where $f_p^{L(R)}$ is the probability with which the left (right) reservoir populates the $\pm p$ states in the left (right) lead, respectively, and R_p and T_p are the reflection and transmission probabilities of the structure for the energy in question, $\epsilon_p = p^2/2m$. The first term is the current fed into the p state by the left reservoir, and the second term is the current carried in the $-p$ state due to reflection. The third term is the transmitted part of the current

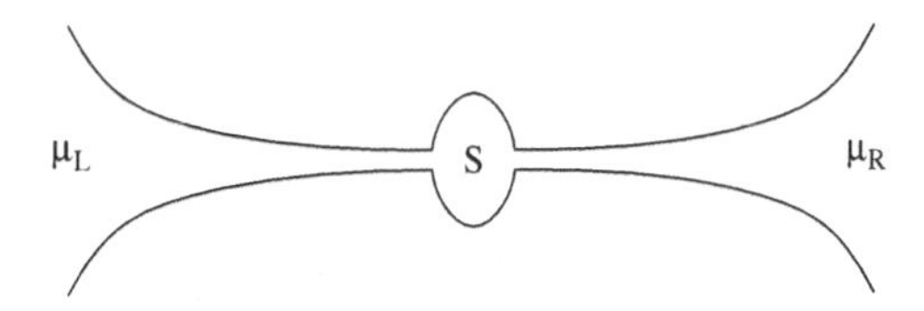

Figure 13.1 Electron reservoirs connected by leads to a sample.

Figure 13.2 Addition of left and right currents.

originating from the current fed by the right reservoir into lead state $-p$ (see the general depiction in Figure 13.2). The populations of p states are assumed to equal the probabilities of population of corresponding energy states in the reservoir, i.e.

$$f_p^{L(R)} = \frac{1}{e^{(\epsilon_p - \mu_{L(R)})/kT} + 1} \tag{13.2}$$

is the Fermi function, and the μ are the electrochemical potentials of the reservoirs. The reservoir connections to the leads are thus not assumed to cause additional scattering, associated contact resistance, and scattering is due only to sample properties.

Using the fact that an impinging particle is either transmitted or reflected, $R_p + T_p = 1$, the average current in the left lead carried by momentum states $\pm p$ is rewritten as

$$I_p^L = \frac{ev_p}{L} T_p (f_p^L - f_p^R). \tag{13.3}$$

Analogously the counting can be done in the right lead and would establish the consequence of current conservation, $I_p^L = I_p^R$.

The total average current is then (the factor of 2 accounting for spin degeneracy)

$$\begin{aligned} I = 2 \sum_{p>0} I_p^L &= 2 \int_0^\infty d\epsilon \, \frac{1}{2} N_1(\epsilon)\, v_\epsilon T(\epsilon)(f^L(\epsilon) - f^R(\epsilon)) \\ &= \frac{2e}{2\pi\hbar} \int_0^\infty d\epsilon \; T(\epsilon)(f(\epsilon) - f(\epsilon + \Delta\mu)), \end{aligned} \tag{13.4}$$

where the energy dependence of the density of states and the velocity cancel, $N_1(\epsilon)\, v_\epsilon = 2/2\pi\hbar$, and $\Delta\mu \equiv \mu_R - \mu_L$ is the electrochemical potential difference between the reservoirs, specified by, say, a voltage difference sustained by a battery, $\Delta\mu = eV$.

Assuming that the transmission coefficient is smooth on the scale of eV, or interest lies only in sufficiently small voltages, the current, $I = GV$, is expressed in terms of the conductance of the mesoscopic structure,

$$G = \frac{2e^2}{2\pi\hbar} T(\epsilon_F). \tag{13.5}$$

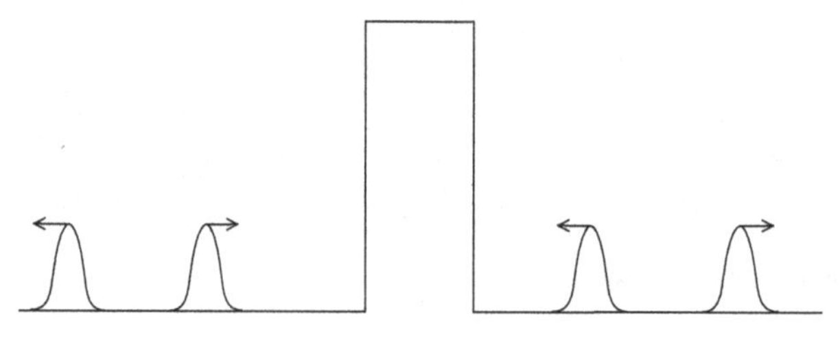

Figure 13.3 Transmitted and reflected wave packets.

The transport properties of the mesoscopic system are determined by a scattering property of the sample, the transmission coefficient for an electron with the Fermi energy, i.e. identical to the tunnel current result, Eq. (3.50).[1]

Instead of using plane waves, the current can also be viewed as being carried by the complete set of Gaussian wave packets, a few of them depicted in Figure 13.3.

Counting the current at a position x in, say, the left lead gives, analogous to the argument leading to Eq. (13.1) (recall also Eqs. (4.106) and (4.107)),

$$I = 2e \sum_{x_i, p_i > 0} v_{p_i} \, |\psi_{x_i,p_i}(x)|^2 \, T_{p_i}(f^L_{p_i} - f^R_{p_i}), \tag{13.6}$$

the analogous formula to Eq. (13.1) but occupations interpreted in terms of Gaussian wave packets. The current is due to left- and right-moving wave packet states being populated to different p_i state heights in the local Fermi seas by the left and right reservoirs. Only spatial cells near the position x effectively contribute to the sum, and since the summation is equivalent for any x, the current expression, Eq. (13.6), is, as it should be, independent of the location. Integrating the expression over the length of contributing cells, L_c, gives (inserting a factor of unity)

$$I = 2e \sum_{x_i, p_i > 0} \frac{\Delta x_c \, \Delta p_c}{2\pi\hbar} \, v_{p_i} \, \frac{1}{L_c} \int_{x - L_c/2}^{x + L_c/2} d_x \, |\psi_{x_i,p_i}(x)|^2 \, T_{p_i}(f^L_{p_i} - f^R_{p_i}). \tag{13.7}$$

By normalization of the Gaussian functions, and since the summation over Δx_c, by definition of L_c, sums up to L_c, the current is

$$I = \frac{2e}{2\pi\hbar} \sum_{p_i > 0} \Delta p_i T_{p_i}(f^L_{p_i} - f^R_{p_i}), \tag{13.8}$$

and thereby the same expression as in Eq. (13.4) is obtained, as the contributions to the sum over Gaussian momentum state labels are only from the region near the Fermi energy, and the sum can be converted into an integral.

[1] The scattering approach to the transport of confined waves was introduced in the context of electromagnetic wave guides by Julian Schwinger. In the context of electron scattering, the conductance formula, Eq. (13.5), is referred to as a Landauer formula (for a two-terminal device), after Rolf Landauer, who championed the scattering approach to conductance.

In the Gaussian wave packet picture emerges the scattering scenario of the reservoirs "throwing" electrons at the sample, leading to a current determined by the scattering properties of the sample.

13.2 Resonant Tunnel Diode

As discussed in Chapter 12, molecular beam epitaxy (MBE) makes it possible to produce heterostructures where the electrons experience square potential barriers. For example, alternating layers of, say, GaAs and GaAlAs can be grown with atomic precision. An example of the potential profile of a double-barrier resonant tunneling heterostructure is depicted to the left in Figure 13.4.

For the case of a double-barrier potential, the phenomenon of resonant tunneling is encountered. As discussed in Section 2.8, depending on parameters, for special energy values the double barrier can have transmission coefficient equal to one, as shown to the right in Figure 13.4, where the transmission probability as a function of energy is displayed. The tunneling probability is thus a highly non-monotonic function of energy. The energies for which perfect transmission occurs correspond to the resonance energies in the quantum well, as indicated in Figure 13.4. Our extensive discussion in Section 2.8 of the transmission properties of the double barrier was thus not purely academic, as such structures can be manufactured (as discussed in Section 12.2). Double-barrier resonant tunneling structures (DBRTS) have potential for use as oscillators and transistors, and for functioning as a new type of memory device in computers: multi-state memory, since there can be more than two resonant levels.

In Section 13.1 the current through a mesoscopic sample was shown to be proportional to the tunneling probability for an energy value equal to the Fermi energy of the electron

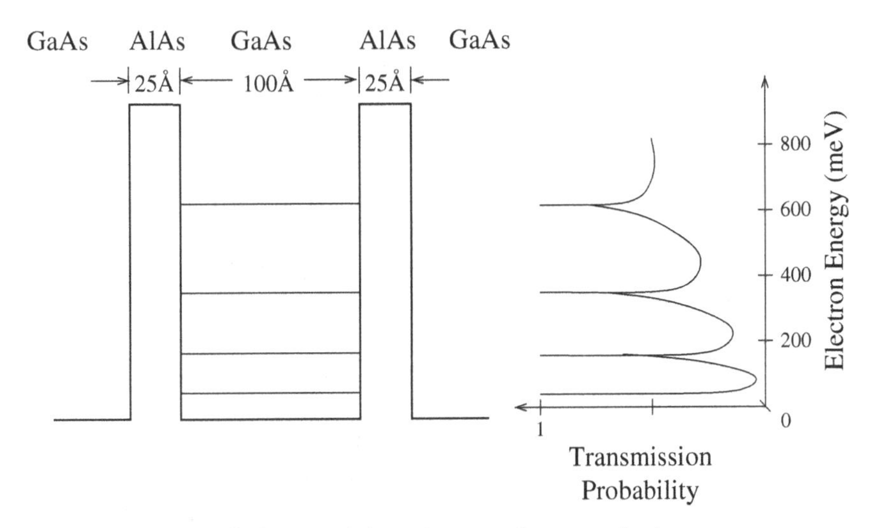

Figure 13.4 Double-barrier heterostructure. Perfect transmission and corresponding resonant levels.

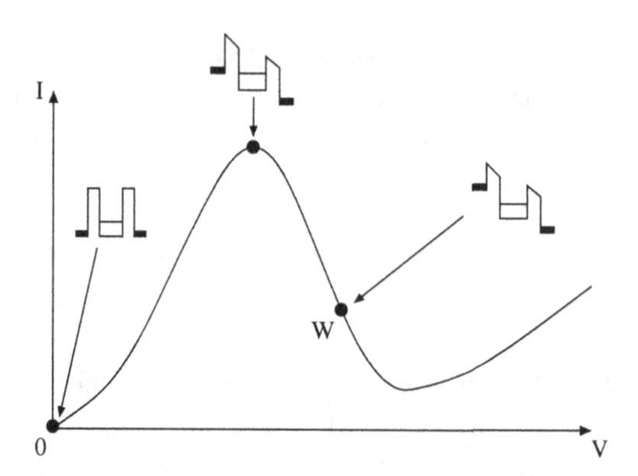

Figure 13.5 The I–V characteristic of a double-barrier resonant tunneling structure.

reservoirs connected to the sample. The current through the double-barrier structure therefore has a maximum when the Fermi energy equals a resonant energy. The double-barrier structure therefore has the current–voltage characteristic depicted in Figure 13.5.

At low voltages the tunnel diode exhibits normal resistive behavior: raising the voltage raises the amount of current through the device. However, once the bias voltage has raised the electrons beyond the resonance, the current decreases for increased voltage, and the device exhibits negative differential resistance. Such a device finds numerous applications in electronics; for example, it can be used to extract energy from a d.c. battery such as to counter the dissipative damping characteristic of circuits built with passive elements. A device with such a property is referred to as an active circuit element. We now show that the tunnel diode can be used for current amplification. The negative differential resistance of the resonant tunnel diode can be used to compensate the positive resistance in an electric circuit, and generate amplified current oscillations.

13.3 Current Amplification

The transport of charge in electronic circuits meets resistance as represented by resistors, and with only the passive elements of capacitors and inductors at hand current oscillations will damp out. We now investigate the consequences of including in a circuit a device with an I–V characteristic as depicted in Figure 13.5, a device with a negative differential resistance branch, for example, the resonant tunnel diode discussed in the previous section. Consider, for example, the electronic circuit depicted in Figure 13.6.

For the resonant tunnel diode, we use the diagrammatic circuit symbol depicted in Figure 13.7. If the diode is disconnected from the circuit, the circuit will exhibit damping. Owing to dissipation in the resistor, the circuit built by passive elements and a d.c. battery will have a current behavior that vanishes exponentially on the time scale L/R as discussed in Section 5.10. If the tunnel diode is present, the circuit can, instead of damping, exhibit *amplification* of the current.

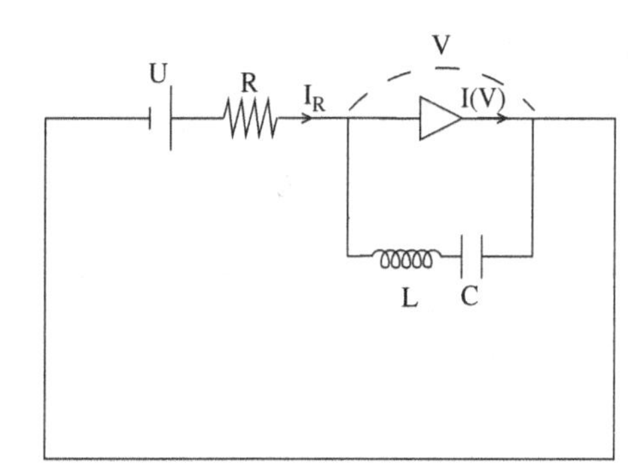

Figure 13.6 Oscillator circuit with tunnel diode.

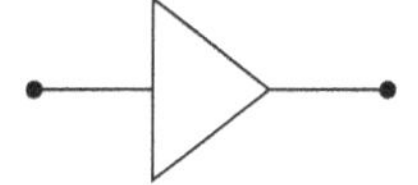

Figure 13.7 Diagrammatic circuit symbol for the tunnel diode.

The electronic circuit diagram depicted in Figure 13.6 consists of two loops, and its dynamical behavior is obtained by applying the two *Kirchhoff's laws* to them. A d.c. battery provides the voltage U, which is dropped over the resistor and the tunnel diode; the latter voltage, denoted V, is also the voltage over the in-parallel capacitor and inductor series element. Applying Kirchhoff's law to the large loop gives

$$U = RI_{\mathrm{R}} + V. \tag{13.9}$$

The time-dependent quantities, here $I_{\mathrm{R}}(t)$ and $V(t)$, have their time dependence suppressed. Current conservation at the nodes of the circuit gives

$$I_{\mathrm{R}} = I_1 + I(V), \tag{13.10}$$

where $I(V)$ denotes the current through the tunnel diode at the voltage V across it, and I_1 is the current through the inductor and capacitor, again with the instant of time t in question suppressed. Equation (13.10) can in view of Eq. (13.9) be rewritten as

$$I(V) + I_1 = \frac{U - V}{R}. \tag{13.11}$$

The voltage drop over the active component equals the voltage drop over the capacitor in series with the inductor, giving the second circuit equation,

$$V(t) = V_C(t) + L\frac{dI_1(t)}{dt}, \tag{13.12}$$

where V_C is the voltage drop over the capacitor.

Introducing the notation

$$V_C(t) = W + g_1(t), \qquad V(t) = W + g_2(t) \tag{13.13}$$

i.e. measuring the two voltages from a common fixed, i.e. time-independent, reference voltage W, Eq. (13.12) becomes

$$L\frac{dI_1(t)}{dt} = g_2(t) - g_1(t). \tag{13.14}$$

Assume that, when the circuit is switched on, the voltage, V, over the tunnel diode is close to the voltage on the middle of the negative differential resistance branch, i.e. the voltage is close to the value W depicted in Figure 13.5. For voltages V close to the value W, the tunnel diode current can be Taylor-expanded and only low-order terms need be kept,

$$I(V) \simeq I(W) + \frac{dI(W)}{dV}(V - W). \tag{13.15}$$

Introducing the notation

$$k = \left.\frac{dI(V)}{dV}\right|_{V=W} \equiv \frac{dI(W)}{dV}, \tag{13.16}$$

the derivative of the current is negative, $k < 0$, since W is assumed to be on the negative differential resistance branch. Since $g_2(t) \equiv V(t) - W$, Eq. (13.15) becomes

$$I(V) \simeq I(W) + kg_2(t). \tag{13.17}$$

Inserting into Eq. (13.11) gives

$$I(W) + kg_2(t) + I_1(t) \simeq \frac{U - (W + g_2(t))}{R}. \tag{13.18}$$

The capacitor is charged by the current I_1, so its voltage is

$$V_C(t) = \frac{Q(t)}{C} = \frac{1}{C}\int^t dt'\, I_1(t'). \tag{13.19}$$

To remove the charge history of the capacitor, we differentiate, as usually done to get rid of initial information, obtaining

$$\frac{dV_C(t)}{dt} = \frac{I_1(t)}{C} \tag{13.20}$$

or, in view of Eq. (13.13),

$$\frac{dg_1(t)}{dt} = \frac{I_1(t)}{C}. \tag{13.21}$$

Similarly, to remove the initial conditions in Eq. (13.14), we differentiate the equation, obtaining

$$L\frac{d^2I_1(t)}{dt^2} = \frac{dg_2(t)}{dt} - \frac{dg_1(t)}{dt}. \tag{13.22}$$

Finally, differentiating Eq. (13.18) gives

$$\left(k + \frac{1}{R}\right)\frac{dg_2(t)}{dt} = -\frac{dI_1(t)}{dt}. \tag{13.23}$$

Inserting Eqs. (13.23) and (13.21) into Eq. (13.22) gives the equation for the current through the series element of inductor and capacitor,

$$L\frac{d^2I_1(t)}{dt^2} + r\frac{dI_1(t)}{dt} + \frac{1}{C}I_1(t) = 0, \tag{13.24}$$

where the notation

$$\frac{1}{r} \equiv k + \frac{1}{R} \tag{13.25}$$

has been introduced. The solution to Eq. (13.25) in, for example, the parameter regime $r^2 < 4L/C$ is (compare with the solution in Eq. (5.85) in Chapter 5)

$$I_1(t) = A\,e^{-(r/4L)t}\sin(\omega t) \tag{13.26}$$

where the oscillation frequency is

$$\omega = \sqrt{\frac{1}{LC} - \frac{r^2}{4L^2}}. \tag{13.27}$$

When the circuit is switched on, the initial current, $I_1(t = 0)$, vanishes due to the inertial effect of an inductor. Choosing the resistance R large enough so that $r < 0$, the amplitude in the current oscillations $I_1(t)$ *increases* in time. Owing to the presence of the tunnel diode, the current amplitude is exponentially *increasing* in time, i.e. the circuit exhibits amplification. The tunnel diode acts as an active device.

Rewriting Eq. (13.23) gives

$$\frac{dV(t)}{dt} = -r\frac{dI_1(t)}{dt}, \tag{13.28}$$

which states that the voltage over the diode is proportional to the current I_1, up to a constant. Choosing the initial voltage over the diode equal to W, $V(t = 0) = W$, the voltage over the diode is

$$V(t) = -rA\,e^{-(r/4L)t}\sin(\omega t) + W. \tag{13.29}$$

The current through the tunnel diode, $I_{\rm TD}(t) \equiv I(V(t))$, is found by differentiating Eq. (13.11) and using Eq. (13.23),

$$\frac{dI_{\rm TD}(t)}{dt} = \left(\frac{r}{R} - 1\right)\frac{dI_1(t)}{dt}. \tag{13.30}$$

When the current through the diode increases, the current in the parallel connection decreases and vice versa. Equation (13.30) states that, up to an additive constant, $I_{\rm TD}$ is proportional to I_1. According to Eq. (13.10), the initial current through the resistor equals the current through the diode and the constant is determined by Eq. (13.11), giving

$$I_{\rm TD}(t) = \left(\frac{r}{R} - 1\right)I_1(t) + \frac{U - W}{R}. \tag{13.31}$$

Current conservation, Eq. (13.10) or (13.9), gives for the current through the resistor

$$I_R(t) = \frac{r}{R}I_1(t) + \frac{U - W}{R} = \frac{r}{R}A\,e^{-(r/4L)t}\sin(\omega t) + \frac{U - W}{R}. \tag{13.32}$$

If we assume that the capacitor initially is uncharged, Eq. (13.12) determines the amplitude to be

$$A = \frac{W}{\omega}. \tag{13.33}$$

The active element, the tunnel diode, is capable of exploiting the energy source, the battery, for amplification. Of course, nothing can behave exponentially for all times, which here is reflected by the linear approximation used to derive the results at some point no longer being valid. The voltage starts pulsating around the work voltage, W, with increasing amplitude and the nonlinearity of the I–V characteristic must be taken into account. The obtained formulas are therefore only applicable to a finite time interval. However, the point was for the moment just to exemplify the feature of amplification.

Exercise 13.1 Calculate from Eq. (13.24) the current for the cases $r^2 = 4L/C$ and $r^2 > 4L/C$.

13.4 Oscillator

We now illustrate how negative differential resistance can be used to compensate the positive resistance in an electric circuit, and generate current oscillations. Consider the circuit depicted in Figure 13.8.

When the switch is closed, oscillations appear in such a circuit with an LC oscillator, oscillations which in the absence of the diode would damp out due to the resistance R (recall Section 5.9). At each half-period the voltage polarity in the oscillatory circuit will change, in which case the voltage across the oscillator part of the circuit is added or subtracted from the supply voltage of the battery. If the diode is operated in the negative differential resistance regime at the working point (W), a decrease in the forward bias

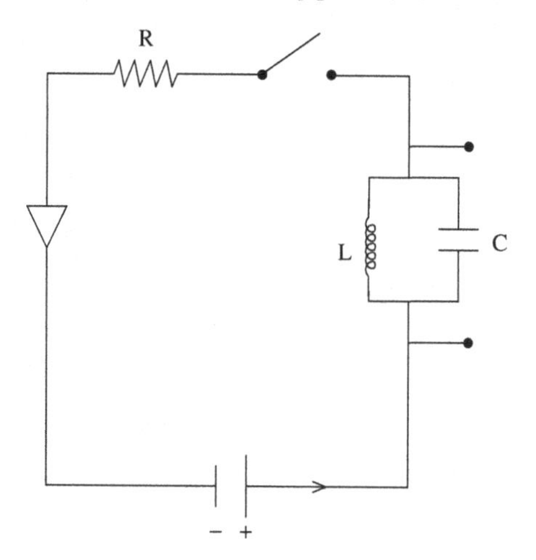

Figure 13.8 Diode-oscillator circuit.

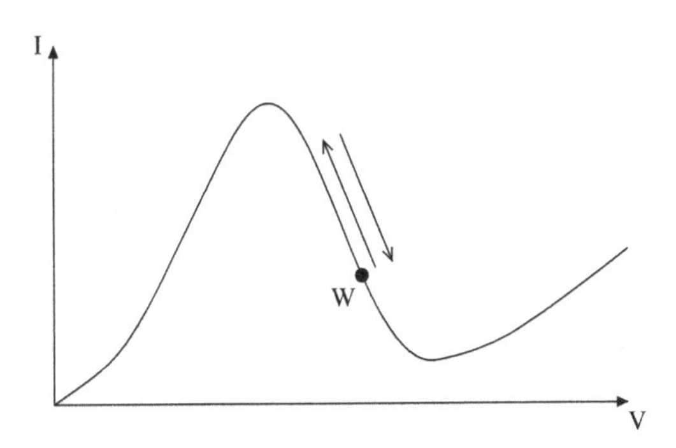

Figure 13.9 Voltage and current oscillations.

will cause an increase in the current through the diode (and hence in the entire circuit). This then changes the polarity of the oscillator, increasing the forward bias, and thereby decreasing the current. As a result, the current and voltage pulsate in the circuit, as depicted in Figure 13.9.

13.5 Bistability and Memory Cell

Consider a device with an I–V characteristic that has a branch of negative differential resistance in parallel with a resistor, as depicted in Figure 13.10.

The voltage on the external terminal of the resistor, U, is dropped over the resistor and the diode, $RI + V = U$, the circuit equation. The load curve, $(U - V)/R$, specifies the possible solutions of the circuit equation, and, for the chosen value of U, has three intersections with the diode I–V characteristic, as shown in Figure 13.11. For the given external voltage, the diode can be in three different voltage states.

For the case of the diode having voltage V_A and the current through the diode is decreased slightly, the circuit equation gives that the voltage is slightly increased and, being on the positive resistance part of the I–V characteristic, the current is pushed back up to its value at voltage V_A. If, in the opposite situation, the current through the diode is increased slightly, the voltage is slightly decreased and the current is pushed back up to its value at voltage V_A. The voltage state V_A is stable. A similar situation holds for voltage state V_C, but for state V_B a slight deviation will drive the voltage state of the diode further away from V_B, i.e. it is unstable, until ending up in one of the stable states. The diode exhibits bistability. The device can thus be used as a memory cell to store binary information, and one can switch between the two states. Reading corresponds to sensing the voltage and writing corresponds to changing between the high and low voltage states.

13.6 Quantum Wires

Leads in real life are not one-dimensional, and each lead state will be characterized by a longitudinal quantum number, the wave number $k = p/\hbar$, and discrete quantum numbers

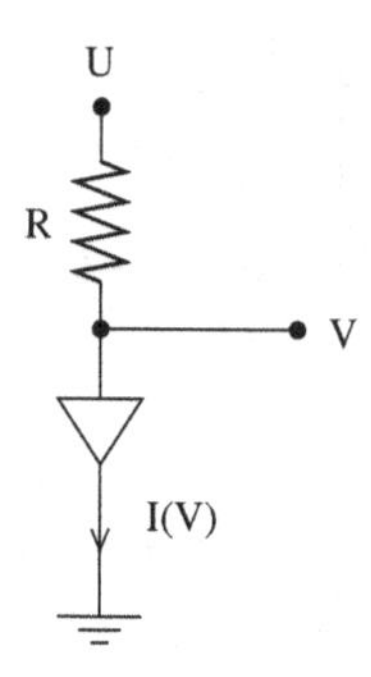

Figure 13.10 Diode in series with a resistor.

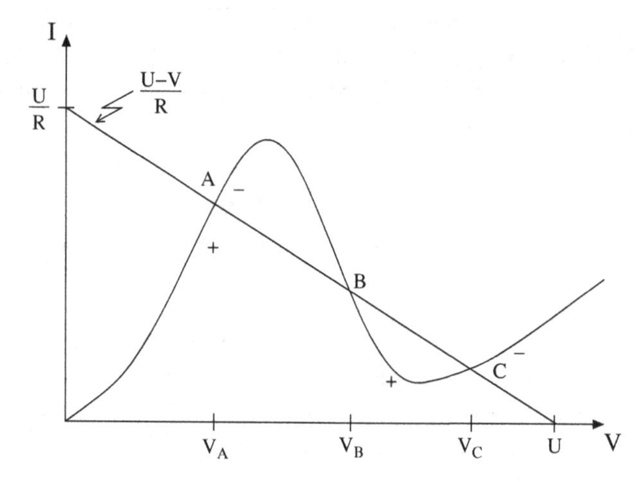

Figure 13.11 Load curve with three intersections.

for the transverse directions. Consider first a two-dimensional wire of width d, and assume the electrons are confined in a box in the transverse direction, corresponding to the box potential

$$V(x,y) = \begin{cases} \infty, & |y| > d/2, \\ 0, & |y| < d/2, \end{cases} \tag{13.34}$$

the zero of the y axis thus having been chosen in the middle of the wire. In Section 13.8 we discuss how to realize a two-dimensional situation making such a potential relevant.

The time-independent Schrödinger equation for an electron in such a potential landscape is

$$-\frac{\hbar^2}{2m}\left(\frac{\partial^2}{\partial x^2} + \frac{\partial^2}{\partial y^2}\right)\psi_E(x,y) = E\,\psi_E(x,y) \tag{13.35}$$

inside the wave guide, and the box boundary condition makes the wave function vanish outside, i.e.[2]

$$\psi_E(x, y = \pm d/2) = 0. \tag{13.36}$$

The problem separates, $\psi(x,y) = \chi(x)\,\phi(y)$, and the Schrödinger equation becomes

$$-\frac{\hbar^2}{2m}\left(\frac{\chi''(x)}{\chi(x)} + \frac{\phi''(y)}{\phi(y)}\right) = E, \tag{13.37}$$

leaving the two terms constant,

$$-\frac{\hbar^2}{2m}\frac{\chi''(x)}{\chi(x)} = E_\parallel, \qquad -\frac{\hbar^2}{2m}\frac{\phi''(y)}{\phi(y)} = E_\perp, \qquad E = E_\parallel + E_\perp. \tag{13.38}$$

In the x direction we have a free problem, so that the longitudinal part of the wave function is a plane wave

$$\chi_k(x) = \frac{1}{\sqrt{L}}\,e^{ikx}, \qquad k \in \mathbb{R}, \tag{13.39}$$

corresponding to the longitudinal energy

$$E_\parallel = \frac{\hbar^2 k^2}{2m}. \tag{13.40}$$

For the transverse problem corresponding to the longitudinal quantum number k, i.e. for the wave function

$$\psi(x,y) = \chi_k(x)\,\phi(y), \tag{13.41}$$

the eigenvalue equation for the transverse part of the wave function is

$$-\frac{\hbar^2}{2m}\,\phi''(y) = \left(E - \frac{\hbar^2 k^2}{2m}\right)\phi(y), \tag{13.42}$$

with the general solution

$$\phi_q(y) = A_u\,e^{iqy} + A_d\,e^{-iqy}, \qquad q^2 + k^2 = \frac{2mE}{\hbar^2}. \tag{13.43}$$

The box boundary condition, the two equations in Eq. (13.36), gives the matrix equation

$$\begin{pmatrix} e^{iqd/2} & e^{-iqd/2} \\ e^{-iqd/2} & e^{iqd/2} \end{pmatrix}\begin{pmatrix} A_u \\ A_d \end{pmatrix} = \begin{pmatrix} 0 \\ 0 \end{pmatrix}, \tag{13.44}$$

and the physically acceptable solutions ($(A_u, A_d) \neq (0,0)$) require that the determinant vanishes,

$$0 = \begin{vmatrix} e^{iqd/2} & e^{-iqd/2} \\ e^{-iqd/2} & e^{iqd/2} \end{vmatrix} = 2i\sin(qd). \tag{13.45}$$

[2] Clearly, such an electronic wave guide has many formal features in common with an electromagnetic wave guide. One noticeable difference, however, is the difference in boundary conditions, where typically, for the electromagnetic wave function, a component of the electric field, the derivative would be zero on a lossless metallic wave guide surface.

The boundary condition thus quantizes the possible transverse wave vectors to the values

$$q_n = \frac{\pi}{d}\, n, \qquad n = 1, 2, 3, \ldots, \tag{13.46}$$

as the $n = 0$ mode does not produce a proper wave function, since $A_u = -A_d$ renders the function equal to zero.

Noting that, according to Eq. (13.44), A_u and A_d are proportional, $A_u = -e^{iqd} A_d$ (or $A_u = -e^{-iqd} A_d$, up and down amplitudes only differing by a phase factor, as nothing distinguishes up from down), the corresponding normalized transverse wave functions are

$$\phi_{q_n}(y) = \sqrt{\frac{2}{d}}\, \sin\!\left(\frac{\pi n}{2d}(2y + d)\right), \qquad n = 1, 2, 3, \ldots, \tag{13.47}$$

and referred to as channel or sub-band wave functions, of which the two of lowest energy are depicted in Figure 13.12.

The considered two-dimensional quantum wire, a straight wave guide, thus has the energy eigenstates

$$\psi_{nk}(x, y) = \frac{1}{\sqrt{L}}\, e^{\pm ikx} \sqrt{\frac{2}{d}}\, \sin\!\left(\frac{\pi n}{2d}(2y + d)\right), \tag{13.48}$$

with corresponding energies

$$E_n(k) = \frac{\hbar^2 k^2}{2m} + \frac{\pi^2 \hbar^2}{2md^2}\, n^2, \qquad k \in \mathbb{R}, \quad n = 1, 2, 3, \ldots. \tag{13.49}$$

The energy spectra for the channels, the sub-band energies, are thus shifted parabolas, as depicted in Figure 13.13.

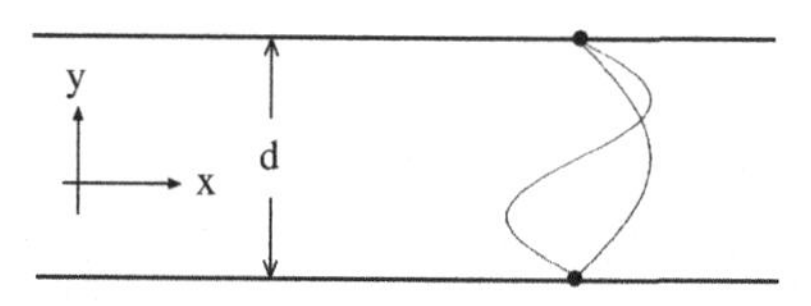

Figure 13.12 Depiction of the two lowest energy channel wave functions.

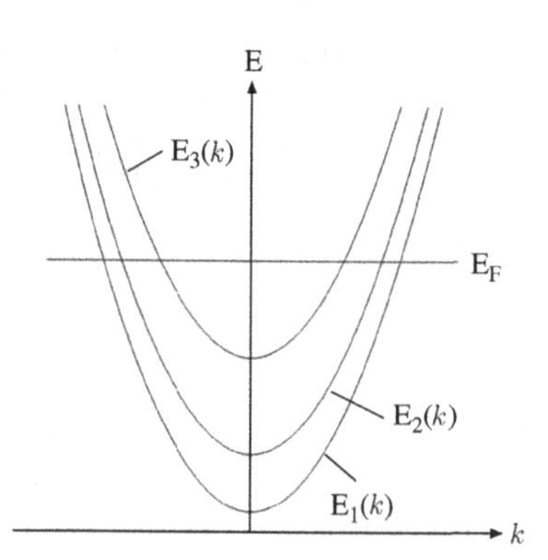

Figure 13.13 Channel energy spectra.

In a quasi-three-dimensional structure, a three-dimensional quantum wire, where two directions are confined, similarly the eigenstates for the two confined transverse directions are, with the label $i = (n_y, n_z)$,

$$\psi_{i,k}(\mathbf{r}) = \frac{1}{\sqrt{L}} e^{ikx} \chi_i(y, z), \qquad E_i(k) = E_i + \frac{\hbar^2 k^2}{2m}, \qquad E_i = \frac{\hbar^2 \pi^2}{2m} \left(\frac{n_y^2}{d_y^2} + \frac{n_z^2}{d_z^2} \right), \tag{13.50}$$

where, as usual, the longitudinal direction is chosen to be the x direction, and the transverse dimensions are $d_{y,z}$. The transverse states are given by (with the mid-point choice of origin)

$$\chi_i(y, z) = \sqrt{\frac{2}{d_y}} \sin\left(\frac{\pi n_y}{2d_y}(2y + d_y) \right) \sqrt{\frac{2}{d_z}} \sin\left(\frac{\pi n_z}{2d_z}(2z + d_z) \right), \qquad n_{y,z} = 1, 2, 3, \ldots. \tag{13.51}$$

In many cases the transverse part of the eigenfunctions need not be specified, since only their orthonormality properties are needed. For example, we have for the longitudinal velocity

$$\begin{aligned} \langle k, i | \hat{v}_x | k, i' \rangle &= \frac{1}{L} \frac{\hbar k}{m} \delta_{ii'} = \frac{v_k}{L} \delta_{ii'} = \frac{1}{L} \frac{1}{\hbar} \frac{\partial E_i(k)}{\partial k} \delta_{ii'} \\ &\equiv \frac{1}{L} v_k^i \delta_{ii'}. \end{aligned} \tag{13.52}$$

Once the transverse quantum numbers are specified for an allowed energy state of energy E, the longitudinal k value is specified by

$$\frac{\hbar^2 k^2}{2m} = E - E_i. \tag{13.53}$$

13.7 Multichannel Scattering

In the multichannel case, a mesoscopic system is represented by a *black box* with incoming and outgoing attached channel states, as depicted in Figure 13.14. Consider the case of two-dimensional (or three-dimensional) constrained leads. The states in the leads are thus labeled by (i, k), where i is the transverse quantum number. An incoming particle in a channel state can be transmitted to states in the right lead or reflected to states in the left lead, provided these states have the same energy as the incoming channel.

Accounting for the average current in such a device is analogous to the discussion for the tunnel junction. The average current carried by lead states $(i_l, \pm k)$ on the left is obtained, just as in the one-dimensional case, by counting

$$I_{i_l,k}^L = \frac{e v_k^{i_l}}{L} \left(f_{i_l,k}^L - \sum_{j_l} R_{i_l,-k;j_l,k'} f_{j_l,k'}^L - \sum_{i_r} T_{i_l,-k;i_r,k''} f_{i_r,k''}^R \right). \tag{13.54}$$

Here the first term is the current fed into the k lead state in question by the left reservoir; the second term is the reflection into the $-k$ lead state in question arising from the left reservoir

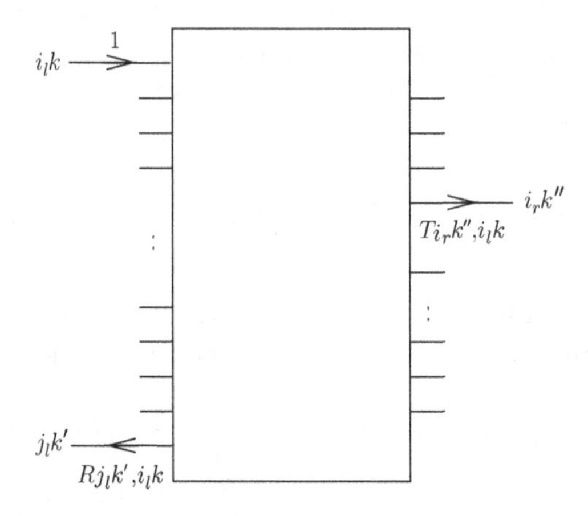

Figure 13.14 Multichannel device reflecting and transmitting incoming electrons.

feeding into any of the left lead states, (j_l, k'); and the third term is the transmission into the $-k$ lead state in question arising from the right reservoir feeding into any of the right lead states, (i_r, k'').

The transmission and reflection coefficients vanish unless they satisfy the energy conservation requirement

$$E_{j_l}(k') = E_{i_l}(-k), \qquad E_{i_r}(k'') = E_{i_l}(-k) \tag{13.55}$$

or

$$E_{j_l}(k') = E_{i_l}(k) = E_{i_r}(k'') \tag{13.56}$$

as, according to Eq. (13.49),

$$E_{i_l}(-k) = E_{i_l}(k) \tag{13.57}$$

as demanded by time reversal symmetry. The redundancy in notation can therefore be pruned,

$$R_{i_l j_l}(E_{i_l}(k)) \equiv R_{i_l,-k;\,j_l,k'}, \qquad T_{i_l,i_r}(E_{i_l}(k)) \equiv T_{i_l,-k;\,i_r,k''}, \tag{13.58}$$

and the total average current takes the form

$$I = \frac{2e}{L} \sum_{i_l,k>0} v_k^{i_l} \left\{ f^L(E_{i_l}(k)) - \sum_{j_l} R_{i_l j_l}(E_{i_l}(k)) f^L(E_{i_l}(k)) \right.$$
$$\left. - \sum_{i_r} T_{i_l,i_r}(E_{i_l}(k)) f^R(E_{i_l}(k)) \right\}. \tag{13.59}$$

In each channel there is again energy dependence cancelation between the density of states and the velocity, $N_i(\epsilon)\, v_\epsilon^i = 2/2\pi\hbar$, and the average current is

$$I = \frac{2e}{2\pi\hbar} \sum_{i_l} \int_{E_{i_l}^{\min}}^{\infty} dE \left\{ f^L(E) - \sum_{j_l} R_{i_l j_l}(E) f^L(E) - \sum_{i_r} T_{i_l,i_r}(E) f^R(E) \right\}. \tag{13.60}$$

Since an incoming particle in channel j_l with total energy E has to be either reflected or transmitted, the reflection and transmission coefficients satisfy the normalization condition,

$$\sum_{i_r} T_{i_r j_l}(E) + \sum_{i_l} R_{i_l j_l}(E) = 1, \tag{13.61}$$

and the average current can be rewritten as

$$I = \frac{2e}{2\pi\hbar} \sum_{i_l} \int_{E_{i_l}^{\min}}^{\infty} dE \left\{ f^L(E) \sum_{i_r} T_{i_r,j_l}(E) - \sum_{i_r} T_{j_l,i_r}(E) f^R(E) \right\}. \tag{13.62}$$

Using the fact that, as a consequence of time reversal symmetry, the transmission properties from left to right are identical to those from right to left,

$$T_{i_l,i_r}(E) = T_{i_r,i_l}(E), \tag{13.63}$$

the average current is specified by

$$I = \frac{2e}{2\pi\hbar} \sum_{i_l,i_r} \int_{E_{i_l}}^{\infty} dE \; T_{i_r,i_l}(E) \{ f^L(E) - f^R(E) \}. \tag{13.64}$$

For small bias, $\Delta\mu \ll kT$, the conductance becomes

$$G \equiv \frac{I}{\Delta\mu/e} = \frac{2e^2}{2\pi\hbar} \int_{E_{i_l}}^{\infty} dE \left(-\frac{\partial f(\epsilon_F)}{\partial E} \right) \sum_{i_l,i_r} T_{i_r,i_l}(E). \tag{13.65}$$

In the degenerate case, for example for metallic contacts, where the derivative of the Fermi function is sharply peaked at the Fermi level, $\epsilon_F \equiv (\mu_l + \mu_r)/2$, or for the case where the transmission coefficient is a smooth function, the conductance reduces to

$$G = \frac{2e^2}{2\pi\hbar} \sum_{i_l,i_r} T_{i_r,i_l}(\epsilon_F). \tag{13.66}$$

The conductance of a mesoscopic sample is expressed by the scattering properties of the sample.

13.8 Quantum Point Contact

In Section 12.5 we discussed how electrons can be confined in a two-dimensional inversion layer, as illustrated in Figure 13.15, in which, by modulation doping, the strength of the donor impurity potential is diminished, and the electrons have high mobility. To a first approximation, the motion of the electrons is thus assumed free: the inversion layer contains a two-dimensional electron gas (2DEG).

If two metallic gates are placed at some distance, and electrically insulated, from the two-dimensional electron gas (for instance, on top of an insulating Si–AlGaAs layer), by connection to a battery, the two gates can be negatively charged. In this way high potential

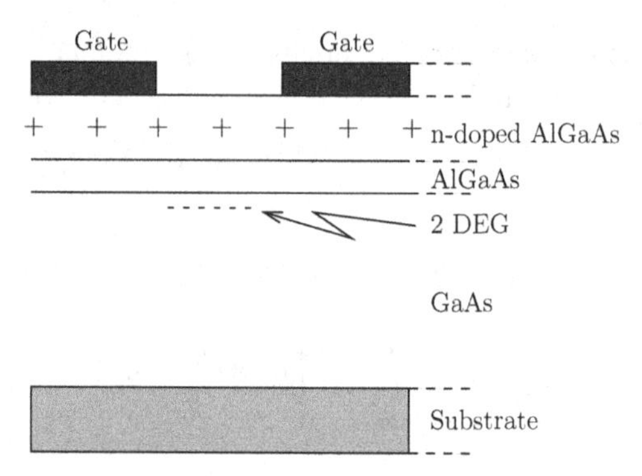

Figure 13.15 Cross-section of a modulation-doped heterostructure.

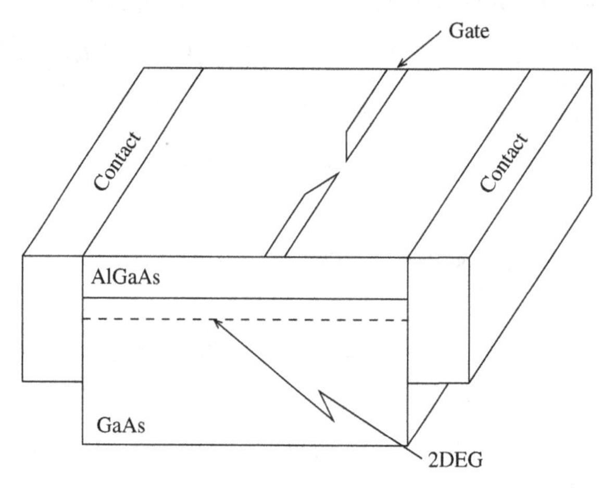

Figure 13.16 Schematics of a split gate.

energy for the two-dimensional electrons is created in the region below the gates, which thereby is depleted of electrons. A top view of such a split-gate device is depicted in Figure 13.16.

Such a split gate leads to a constriction of the channel located below the gate position. When the split gate is highly negatively charged, the potential energy profile for the two-dimensional electron gas, as indicated in Figure 13.17, is at a high potential below the split gate.

In that case, wave function leakage into the depletion region can be neglected, i.e. box boundary conditions can be assumed, as indicated in Figure 13.17. The two-dimensional channel is described by the varying profile distance $d(x)$. The potential landscape indicated in Figure 13.17 for the electrons in the two-dimensional layer is thus approximated by

$$V(x,y) = \begin{cases} \infty, & |y| > d(x)/2, \quad \text{depletion regions,} \\ 0, & |y| < d(x)/2, \quad \text{wave guide,} \end{cases} \tag{13.67}$$

assuming the potential profile to be mirror-symmetric with respect to the chosen x axis (the leads on the two sides are assumed identical for simplicity). If the constriction is very

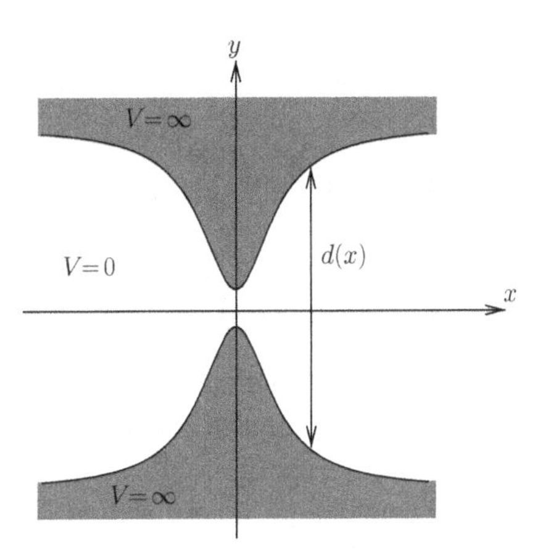

Figure 13.17 Constriction.

narrow, we speak of a point contact. The width of the opening, $d_0 \equiv d(x = 0)$, can be varied by changing the split-gate voltage.

We are interested in the conductance of the device. Far away from the constriction, the electrons occupy the quantum states of a two-dimensional quantum wire or lead, i.e. described by longitudinal and transverse quantum numbers, as discussed in Section 13.6. Reservoirs on the left and right feed electrons into the quantum states in the leads and they are transmitted and reflected into other transverse modes by the constriction, as discussed in Section 13.7. To have a strictly two-dimensional electron gas, the temperature is assumed to be very low, and we speak of a quantum point contact. According to Eq. (13.66), the conductance of the device is

$$G = \frac{2e^2}{2\pi\hbar} \sum_{n,n'=1}^{\infty} T_{n,n'}(\epsilon_F), \tag{13.68}$$

where ϵ_F is the Fermi energy of the electron reservoirs. The conductance is determined by the transmission and reflection properties of the geometric constriction, and the time-independent Schrödinger equation for an electron in such a potential landscape is

$$-\frac{\hbar^2}{2m}\left(\frac{\partial^2}{\partial x^2} + \frac{\partial^2}{\partial y^2}\right)\psi(x,y) = E\,\psi(x,y) \tag{13.69}$$

inside the wave guide, and for simplicity a box boundary condition is assumed,[3]

$$\psi(x, y = \pm d(x)/2) = 0. \tag{13.70}$$

Since we expect the electrostatically produced split-gate potential to be smooth, the profile $d(x)$ is a smooth function of position, and the solution for the constriction is sought in *adiabatic* form,

[3] Mathematically, the problem is quite analogous to that of not *matter* waves but electromagnetic waves propagating inside a tube made of conducting material, say, a microwave guide. The only formal change is that the boundary condition would then be that the derivative of the electric field should vanish on the surface of the conductor (see reference [3]).

$$\psi_E(x,y) = \sum_{n=1}^{\infty} \chi_n^E(x)\, \phi_n(x,y), \qquad n = 1,2,3,\ldots, \tag{13.71}$$

where the transverse functions[4]

$$\phi_n(x,y) \equiv \sqrt{\frac{2}{d(x)}}\, \sin\!\left(\frac{\pi n}{2d(x)}(2y + d(x))\right) \equiv \phi_n^x(y), \qquad n = 1,2,3,\ldots \tag{13.72}$$

guarantee that the box boundary condition is satisfied. For fixed x, the transverse channel functions in Eq. (13.72) constitute a complete set, and any electron state in the leads can be expanded on them. The expansion coefficients are functions of the longitudinal variable, giving rise to the longitudinal wave function, $\chi_n^E(x)$, to be determined. Away from the constriction, say, of longitudinal size L, where the channel width is constant, $d(x \gg L) = d$, the quantum wire states of Section 13.6 are recovered. The scattering due to the constriction will in general mix these states. In the adiabatic case, it is now shown that the non-mixing channel approximation can be implemented.

The transverse direction kinetic energy operator in Eq. (13.69) introduces the channel profile explicitly, whereas the longitudinal one gives

$$\frac{\partial^2(\chi_n^E(x)\,\phi_n(x,y))}{\partial x^2} = \phi_n(x,y)\,\frac{\partial^2 \chi_n^E(x)}{\partial x^2} + \chi_n^E(x)\,\frac{\partial^2 \phi_n(x,y)}{\partial x^2} + 2\,\frac{\partial \chi_n^E(x)}{\partial x}\,\frac{\partial \phi_n(x,y)}{\partial x}. \tag{13.73}$$

In order to determine the longitudinal part of the wave function, $\chi_n^E(x)$, corresponding to the total energy E and transverse channel state n, the scalar product of the Schrödinger equation, Eq. (13.69), with $\phi_n^x(y)$ is taken (integrating out the transverse degree of freedom), giving

$$\left\{-\frac{\hbar^2}{2m}\frac{\partial^2}{\partial x^2} - E\right\} \chi_n^E(x) + \sum_{n'=1}^{\infty} V_{n,n'}(x)\chi_{n'}^E(x) + \sum_{n'=1}^{\infty} U_{n,n'}(x)\frac{\hbar}{i}\,\frac{\partial \chi_{n'}^E(x)}{\partial x} = 0. \tag{13.74}$$

Here

$$U_{n,n'}(x) = \frac{\hbar}{im}\left\langle \phi_n^x \,\middle|\, \frac{\partial \phi_{n'}^x}{\partial x} \right\rangle \tag{13.75}$$

and

$$V_{n,n'}(x) = \frac{(\hbar\pi n)^2}{2m(d(x))^2}\,\delta_{n,n'} + \tilde{V}_{n,n'}(x), \tag{13.76}$$

the first term being the transverse direction kinetic energy operator contribution, with

$$\tilde{V}_{n,n'}(x) = \frac{\hbar^2}{2m}\left\langle \phi_n^x \,\middle|\, \frac{\partial^2 \phi_{n'}^x}{\partial x^2} \right\rangle. \tag{13.77}$$

[4] Recall Section 13.6, where an electronic wave guide without a constriction, i.e. a straight wave guide, was considered.

In the above, the angle bracket notation indicates the scalar product over the transverse direction, for example,

$$\left\langle \phi_n^x \left| \frac{\partial^2 \phi_{n'}^x}{\partial x^2} \right. \right\rangle \equiv \int_{-d(x)/2}^{d(x)/2} dy\, [\phi_n^x(y)]^* \frac{\partial^2 \phi_{n'}^x(y)}{\partial x^2}, \tag{13.78}$$

the complex conjugation in fact being inoperative because the transverse functions are real. Using the normalization condition of the transverse functions,

$$\langle \phi_n^x \mid \phi_n^x \rangle = 1, \tag{13.79}$$

or by direct calculation, the diagonal elements of the matrix $U_{n,n'}(x)$ are seen to vanish, $U_{n,n}(x) = 0$.

Collecting the longitudinal parts of the wave function, labeled by the channel numbers, into the column vector,

$$\boldsymbol{\chi}^E(x) = \begin{pmatrix} \chi_1^E(x) \\ \chi_2^E(x) \\ \vdots \end{pmatrix}, \tag{13.80}$$

the set of equations determining the channel states can be rewritten in matrix form as

$$\mathbf{H}\left(x, \frac{\hbar}{i}\frac{\partial}{\partial x}\right) \boldsymbol{\chi}^E(x) = E\, \boldsymbol{\chi}^E(x), \tag{13.81}$$

where the *longitudinal* matrix Hamiltonian is

$$H_{n,n'}\left(x, \frac{\hbar}{i}\frac{\partial}{\partial x}\right) = \frac{1}{2m}\left(\frac{\hbar}{i}\frac{\partial}{\partial x}\right)^2 \delta_{nn'} + \tilde{V}_{n,n'}(x) + U_{n,n'}(x)\frac{\hbar}{i}\frac{\partial}{\partial x}. \tag{13.82}$$

By integrating out the transverse degree of freedom, the two-dimensional restriction geometry problem is traded for an infinite set of coupled *one*-dimensional equations.

Since

$$\frac{\partial \phi_n^x(y)}{\partial x} = -\frac{d'(x)}{d(x)}\left(\frac{1}{2}\phi_n^x(y) + y\frac{\partial \phi_n^x(y)}{\partial y}\right) \tag{13.83}$$

and

$$\begin{aligned}\frac{\partial^2 \phi_n^x(y)}{\partial x^2} &= -\frac{d''(x)}{d(x)}\left(\frac{1}{2}\phi_n^x(y) + y\frac{\partial \phi_n^x(y)}{\partial y}\right) \\ &\quad + \left(\frac{d'(x)}{d(x)}\right)^2\left(\frac{3}{4}\phi_n^x(y) + 3y\frac{\partial \phi_n^x(y)}{\partial y} + y^2\frac{\partial^2 \phi_n^x(y)}{\partial y^2}\right),\end{aligned} \tag{13.84}$$

the inter-channel mixing elements $U_{n,n'}(x)$ and $\tilde{V}_{n,n'}$ are small for a smooth profile, being at least on the order of $d'(x)/d(x)$ or $d''(x)/d(x)$, and smaller than the diagonal term in Eq. (13.76). Though not necessary, since an order-of-magnitude estimate is sufficient, calculation of the trigonometric integrals gives

$$\left\langle \phi_n^x \left| \frac{\partial \phi_{n'}^x}{\partial x} \right. \right\rangle = \frac{d'(x)}{d(x)} \begin{cases} 2nn'/(n^2 - n'^2), & n \neq n',\ n + n' \text{ even}, \\ 0, & n \neq n',\ n + n' \text{ odd}, \\ 0, & n = n', \end{cases} \tag{13.85}$$

and

$$\left\langle \phi_n^x \left| \frac{\partial^2 \phi_{n'}^x}{\partial x^2} \right. \right\rangle = \begin{cases} -\left(\frac{d'(x)}{2d(x)}\right)^2 \left(1 + \frac{\pi^2}{3} n^2\right), & n = n', \\ -\left(\frac{d'(x)}{d(x)}\right)^2 \frac{2nn'^3 + 6nn'}{(n^2 - n'^2)^2} + \frac{d''(x)}{d(x)} \frac{2nn'}{n^2 - n'^2}, & n \neq n',\ n + n' \text{ even}, \\ 0, & n \neq n',\ n + n' \text{ odd}. \end{cases} \tag{13.86}$$

In order for $\tilde{V}_{n,n}(x)$ to be much smaller than $V_{n,n}(x)$, the first term in the expression in Eq. (13.76) should dominate and the profile must satisfy $|d'(x)|^2 \ll 1$, i.e. the function $d(x)$ should vary only slightly over distances on the order of $d(x)$. In that case, all terms in $U_{n,n'}(x)$ and $\tilde{V}_{n,n'}$ are small, and for a smooth profile the $U_{n,n'}(x)$ and $\tilde{V}_{n,n'}$ inter-channel mixing terms can be neglected, the channels decouple, and the longitudinal part of the wave function is determined by

$$\left\{ -\frac{\hbar^2}{2m} \frac{\partial^2}{\partial x^2} + V_n(x) \right\} \chi_n^E(x) = E\, \chi_n^E(x), \tag{13.87}$$

where

$$V_n(x) = \frac{1}{2m} \left(\frac{\hbar\pi}{d(x)}\right)^2 n^2. \tag{13.88}$$

In the adiabatic limit the constriction causes no channel mixing, but a potential barrier in the longitudinal direction appears. The channel potentials have the form shown in Figure 13.18 for various channel indices ($V_n(x)$ is largest when $d(x)$ is smallest, i.e. at the location of the point contact). The maxima of the potentials increase with the channel number squared according to Eq. (13.88).

In the adiabatic limit, a set of decoupled one-dimensional scattering problems is encountered. An incoming electron in a definite channel state is, by energy conservation, only transmitted into the same channel state in the other lead. The transmission matrix is therefore diagonal in the channel indices,

$$T_{n,n'}(\epsilon_F) = T_n(\epsilon_F)\, \delta_{n,n'}, \tag{13.89}$$

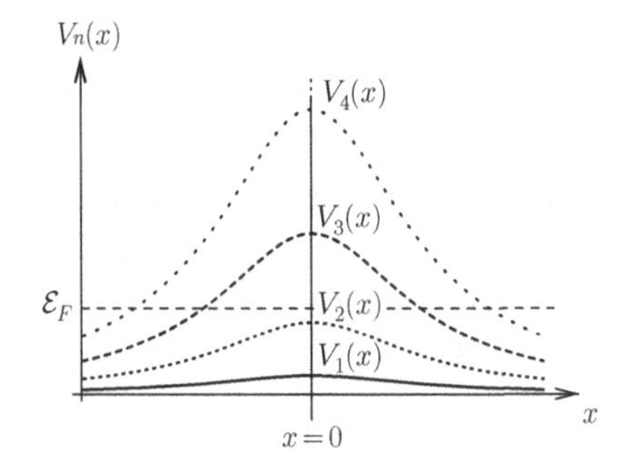

Figure 13.18 Effective channel potentials.

and $T_n(\epsilon_F)$ is the transmission coefficient for transmission from the channel n in one lead of the constriction to the channel n in the other lead for the Fermi energy of the electron gas in question. The conductance formula (13.68) thus reduces to the sum of the diagonal elements. In the adiabatic limit, the conductance of a quantum point contact is therefore

$$G^{\text{adia}} = \frac{2e^2}{2\pi\hbar}\sum_{n=1}^{\infty} T_n(\epsilon_F), \tag{13.90}$$

the sum of the conductance contributions from each channel,

$$G^{\text{adia}} = \sum_{n=1}^{\infty} G_n, \qquad G_n = \frac{2e^2}{2\pi\hbar} T_n(\epsilon_F). \tag{13.91}$$

For each channel index, n, a barrier potential, $V_n(x)$, determines the transmission probability at the Fermi energy.

The Fermi energy has some position relative to the channel potentials as indicated in Figure 13.18. The top of the potentials depends, according to Eq. (13.88), on the constriction minimum, d_0, and the channel index,

$$V_n(x) = n^2 V_1(x), \qquad V_1(x) = \frac{1}{2m}\left(\frac{\pi\hbar}{d(x)}\right)^2. \tag{13.92}$$

When the Fermi energy is larger than $V_n(0)$, an electron at the Fermi energy will have a transmission coefficient for that channel corresponding to predominantly being transmitted, i.e. equal almost to unity, whereas the transmission coefficient for a channel index where the Fermi energy is smaller than $V_n(0)$ predominantly vanishes due to almost total reflection.

The conductance of the constriction can be changed by either changing the Fermi energy, i.e. the density of the electrons, by a gate underneath the 2DEG, or changing the constriction, i.e. the size of the opening d_0. By charging the split-gate electrode less negatively, the constriction opening increases and more channel potentials dip below the Fermi energy, i.e. more channels become conducting. Experimentally it is easier to let the top of the potentials V_n cross the fixed Fermi energy by increasing d_0 than to move the Fermi energy up through the barriers.

If the Fermi energy is not close to any of the potential tops, transmission is almost total for the channels with tops below, and reflection is almost total for tops above the Fermi energy. We therefore first treat the transmission problem classically, i.e. under-barrier tunneling and over-barrier reflection are neglected. In that case, the transmission coefficient is one for all channels where $V_n(0)$ is below the Fermi energy, and zero for all channels where $V_n(0)$ is above, i.e.[5]

$$T_n^{\text{cl}}(\epsilon_F) = \begin{cases} 1, & n < d_0 k_F/\pi, \\ 0, & n > d_0 k_F/\pi, \end{cases} \tag{13.93}$$

[5] This shows explicitly that the transmission coefficient, and thereby the conductance, can be viewed as a function of either the Fermi energy (or Fermi wave vector) or the contact opening size d_0.

giving the *classical* expression for the conductance,

$$G^{\mathrm{cl}} = \frac{2e^2}{2\pi\hbar} \sum_{n=1}^{n_{\max}(d_0)} 1, \tag{13.94}$$

where

$$n_{\max}(d_0) = \left[\frac{k_F d_0}{\pi} \right] \tag{13.95}$$

is the highest channel number for which there is transmission, i.e. the square brackets denote *the whole part of*, giving, in the *classical* approximation,

$$G^{\mathrm{cl}} = \frac{2e^2}{2\pi\hbar} \left[\frac{k_F d_0}{\pi} \right]. \tag{13.96}$$

Plotting the conductance G^{cl} as a function of $k_F d_0/\pi$, the step structure as depicted by the thin line in Figure 13.19 is obtained.

By charging the split-gate electrode less negatively, d_0 increases and more channel potentials dip below the Fermi energy, i.e. more channels become conducting and the current is stepwise increased. Increasing the opening of the quantum point contact, d_0, by half a Fermi wave length, λ_F, adds one more conducting channel, and increases the conductance of the point contact by the amount $e^2/\pi\hbar$, a quantum of conductance. The conductance is quantized in units of the universal value $2e^2/2\pi\hbar$, i.e. it does not depend on, for example, the effective mass of the electrons, or any other material specifics.[6] When $k_F d_0/\pi < 1$, the quantum point contact is pinched off; it does not conduct any current.

The quantum effects of traversing potential barriers lead to the feature that, when a barrier potential top passes through the Fermi level, there occurs under-barrier tunneling and over-barrier reflection. We therefore expect that, in the quantum mechanical treatment of the conductance of a constriction, the sharp steps will be smeared out when a potential

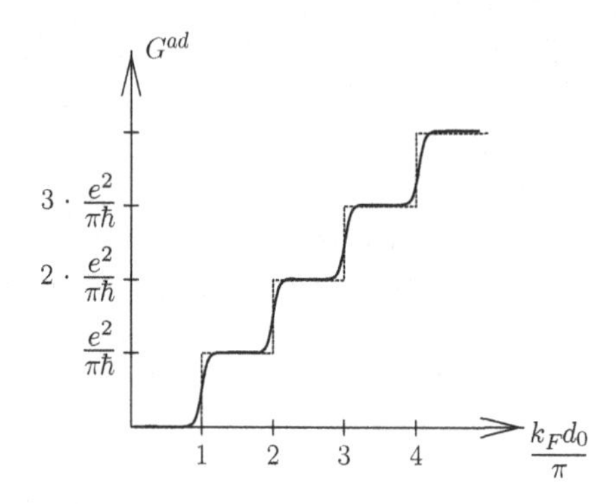

Figure 13.19 Quantized conductance of a point contact.

[6] It has therefore been suggested to base the current standard on this effect, and in this way, together with the voltage standard based on the Josephson effect and the resistance standard based on the quantum Hall effect, to close the electromagnetic *metrological* triangle.

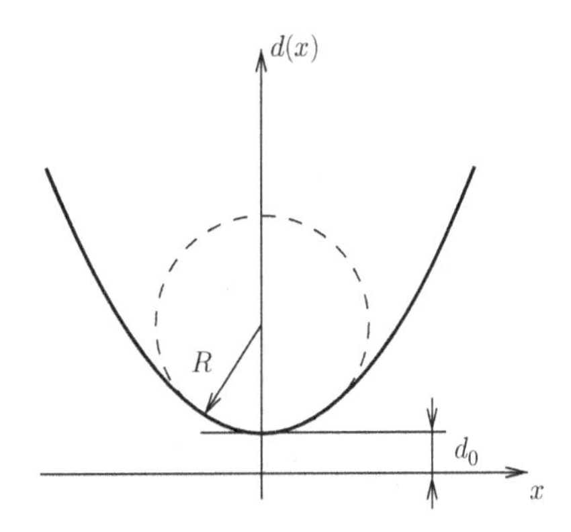

Figure 13.20 Constriction profile.

passes through the Fermi level. We are thus interested in solving the barrier transmission problem for energies close to the top of the barriers.

The constriction profile is depicted in Figure 13.20. Close to the constriction minimum, the wave guide profile has the Taylor expansion

$$d(x) = d_0 + \frac{x^2}{2R} + \cdots \tag{13.97}$$

expressed in terms of its curvature, $R^{-1} \equiv d''(x=0)$ and the constriction opening.

Close to the constriction minimum, the quantity of interest becomes

$$\left(\frac{1}{d(x)}\right)^2 \cong \frac{1}{d_0^2}\left(1 + \frac{x^2}{2Rd_0}\right)^{-2} \cong \frac{1}{d_0^2}\left(1 - \frac{x^2}{Rd_0}\right), \qquad x^2 \ll Rd_0, \tag{13.98}$$

and the potential near the constriction becomes

$$V_n(x) = \frac{(\hbar\pi n)^2}{2md_0^2}\left(1 - \frac{x^2}{Rd_0}\right), \qquad x^2 \ll Rd_0. \tag{13.99}$$

The length $L = \sqrt{Rd_0}$ thus describes the size of the point contact; as the constriction potential vanishes at this distance, the profile has healed to its size in the leads. Since the main correction to the classical result, the smearing effect, is for energies close to the top of the barrier, the potential can be approximated by the harmonic approximation,

$$V_n(x) = V_n(0) - \tfrac{1}{2}k_n x^2, \tag{13.100}$$

where the *force* constant is

$$k_n = \frac{(\hbar\pi)^2}{mRd_0^3}\, n^2. \tag{13.101}$$

The problem can thus be solved by referring to a known one, as the transmission coefficient for the inverted parabola is

$$T(E) = \frac{1}{1 + e^{-2\pi(E/\hbar)\sqrt{m/k}}}, \qquad V(x) = -\tfrac{1}{2}kx^2. \tag{13.102}$$

The case of interest can thus be inferred by including the offset in energy and the different curvature for the channel potentials,

$$T_n(E) = \frac{1}{1 + e^{-2\pi[(E - V_n(0))/\hbar]\sqrt{m/k_n}}}. \tag{13.103}$$

The translation between "parabola" notation and the present channel transmission problem being

$$\frac{2\pi}{\hbar}\sqrt{\frac{m}{k_n}} \leftrightarrow \frac{2md_0\sqrt{Rd_0}}{n\hbar^2}, \tag{13.104}$$

for the transmission coefficient at the Fermi energy we obtain

$$T_n(\epsilon_F) = \frac{1}{1 + e^{-(\epsilon_F - V_n(0))/\Delta_n}}, \tag{13.105}$$

where $\Delta_n = n\Delta_1$, and

$$\Delta_1 = \frac{\hbar^2}{2md_0\sqrt{Rd_0}}. \tag{13.106}$$

For k_F close to $n\pi/d_0$, i.e. for the Fermi energy near $V_n(0)$,

$$\epsilon_F - V_n(0) = \frac{\hbar^2 k_F^2}{2m} - \frac{(\hbar\pi n)^2}{2md_0^2} \simeq n\frac{(\hbar\pi)^2}{md_0^2}\left(\frac{k_F d_0}{\pi} - n\right), \qquad \hbar k_F \cong n\frac{\pi\hbar}{d_0}, \tag{13.107}$$

and in the regime of interest the transmission coefficient is

$$T_n(\epsilon_F) \simeq \frac{1}{1 + e^{-2\pi^2\sqrt{R/d_0}\,(k_F d_0/\pi - n)}}. \tag{13.108}$$

The transmission coefficient thus depends on $k_F d_0/\pi$, as depicted in Figure 13.21.

Not only is there a large prefactor, $\sqrt{R/d_0}$, in the exponential in Eq. (13.108), but the smoothness of the potential gives in addition a numerically large prefactor $2\pi^2 \sim 20$ (a general quasi-classical feature). Tunneling and over-barrier reflection therefore give rise to very little smearing of the otherwise sharp conductance quantum steps. The conductance quantization in a ballistic point contact was originally observed in the AlGaAs structure at low temperatures of a few kelvins [4], requiring sophisticated technology, as

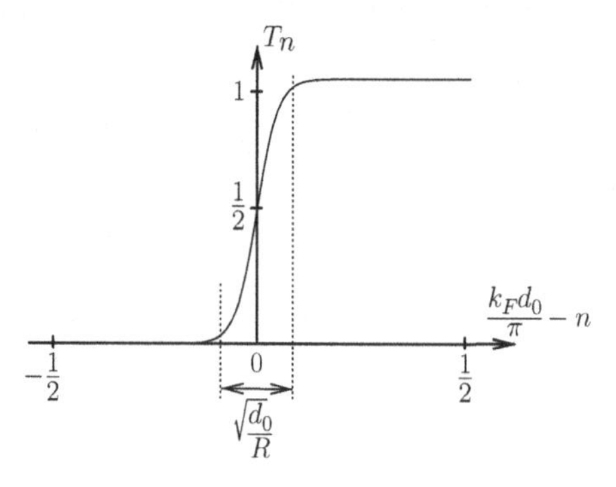

Figure 13.21 Broadening of conductance step.

discussed in Chapter 12. The conductance was measured as a function of the voltage on the split gate controlling the opening in the constriction, which could be varied in the region $0 \leq d_0 \leq 250\,\text{nm}$.

Conductance quantization can also be observed in table-top experiments with an electrical contact. When a light is turned on or off in a room by operating an old type of electrical switch contact, the switch is a metal tip that is plugged in to or pulled out of another piece of metal. When, for example, a light is turned off in this way, thin filaments, channels, of the metal are formed and when one breaks, a channel for conduction disappears and the conductance of the switch decreases by a quantum unit of conductance. The conductance of the switch thus does not continuously or abruptly decrease to zero, but drops in steps [5, 6].

The observation of quantized plateaus in the conductance of high-mobility electron gases has generated interest in these devices, since the strong sensitivity dependence of the source–drain conductance of a quantum point contact on electrostatic fields can provide detection of small electronic signals. The quantum point contact has detected charge motion and monitored quantum decoherence in an electron interferometer [7, 8].

If the transverse dimension becomes macroscopic, then the summation over the channels approaches the value of the integral,

$$G^{\text{cl}} = \frac{2e^2}{2\pi\hbar} \sum_{n=1}^{n_{\max}(d_0)} 1 \to \frac{2e^2}{2\pi\hbar} \int_0^{k_F d_0/\pi} dq = \frac{2e^2}{2\pi\hbar} \frac{k_F d_0}{\pi}, \tag{13.109}$$

and thus we obtain the classical conductance of a point contact in two dimensions.

Exercise 13.2 Consider two Fermi gases in three dimensions, separated by a wall with an orifice of radius d, the gases being at different chemical potentials. There will thus be a net flow of fermions through the orifice. Use classical kinetics to obtain that the flow is described by the conductance, the three-dimensional analog of Eq. (13.109),

$$G = \frac{2e^2}{2\pi\hbar} \frac{k_F d_0}{\pi}. \tag{13.110}$$

A point contact can thus be used for probing Fermi surface properties.

13.9 Summary

When a transport situation is specified by an obstacle intersecting leads, its properties are determined by the scattering properties of the obstacle. Typical such mesoscopic systems were shown to have characteristic nonlinear current–voltage relationships. They therefore find interesting applications as electronic devices and detectors.

14 Arithmetic, Logic and Machines

In this chapter, basic elements of computers are described. We first remind ourselves that arithmetic rules, such as addition and multiplication, are mechanical operations. They can therefore be handled by machines executing the rules. Recall the thousand-year-old Chinese invention of the abacus, where beads moved around represent arithmetic calculations (with amazing speed in the hands of a skilled person), and present-day technology based on pushing electrons around (objects of much smaller mass and higher velocities).

A computing machine does logic operations – it does so nowadays even to the extent that an interaction with a machine can fool you into believing that it acts as an intelligent human being that you start talking to! But still a machine is an idiot, a mechanical device executing programmed operations about whose content it has no comprehension. It does not know *what* it is doing.

14.1 Counting

The human brain has the capacity to make distinctions in the photonic information of the *world* that our eyes provide its electric circuitry, the pictures of vision. It has the ability to quantify the macroscopic features provided. I see *one* or *two* sheep in the meadow, or *two* cows and *three* horses. I can *count*: point at objects and say one, two, three, etc. This reflects the important property of the brain that it has a good memory – in fact, for simple objects, a perfect one. Being able to quantify the number of objects is the basis of arithmetic as represented by the *natural* numbers.

The natural numbers, *one*, *two*, *three*, etc., the basic symbols of quantification, are represented below by three different types of symbols:

	|	||	|||	||||	|||||	...
	I	II	III	IV	V	...
0	1	2	3	4	5	...

In the first row, a stick or line represents the number *one*. The next number, *two*, has an added stick, etc. The Indo-Arab civilization provided the total abstraction where representation (symbol) and meaning are no longer correlated, the third row. This represents a complete compression of the meaning into a single symbol representing a large amount of sticks. Here the *perfect* memory of our brain is vital for this representation to exist and to be of practical use. The Roman numerals (second row) also have compressed information, e.g. the number *ten* being represented by the symbol X, and *fifty*, *hundred* and *thousand* by the symbols (capital Roman letters) L, C and M, respectively. The number *nine* is represented by IX, expressing the rule that when a number (symbol) precedes that of a symbol representing a larger number, it should be subtracted (similarly XL symbolizes thus the number *forty*), whereas the opposite order of the symbols, XI, represents the number *eleven* and LX the number *sixty*.

However, when it comes to capturing the essence of the natural numbers and their rules of compositions, it is better to go back to the representation where meaning and symbol are one and the same, as depicted in the first row above. The simplest and most direct quantification of the natural numbers is to represent them by sticks, or fingers, recalling that the word *digit* is Latin for the English word *finger* (two different symbols for the same *thing*). Note that using identically looking sticks is just for convenience. The symbol

$$\triangle \ \triangleright \ \bigcirc$$

is just as valid a representation of the quantity *three* as are three sticks. We count *three* animals in the meadow whether it is *one* cow and *two* horses or *two* cows and *one* horse. The number *three* labels all collections of three members.

For convenience, the row of natural numbers, *one*, *two*, *three*, etc., is represented by lines or *sticks*. We therefore have an arsenal of sets of sticks representing the natural numbers as depicted in the first row above. We note that each subsequent number in the line of stick arrangements is obtained by placing yet one more stick next to the previous stick collection. Take one stick from the collection of single sticks and place it next to one other stick and you get the two-stick representation of the natural number *two*; then place a further single stick next to the two sticks and you have the three-stick representation of *three*. Repeating the procedure produces the *follower* in the line-up of natural numbers. The infinite set of natural numbers can thus be obtained by introducing the operation of *addition*: here in the form of adding one stick at a time.

14.2 Arithmetic

Arithmetic is the study of quantity, the art or science of computation: the composition of numbers into other numbers. The operation of one-stick addition can be extended to a composition for any pair of numbers. Adding sticks, addition of numbers, can be stated as follows. Place, for example, two sticks next to another set of two sticks (both copied from the two-stick collection in the row of sticks). Then move the combined collection along the arsenal of stick arrangements. The stick arrangement matches exactly one in the line-up of sticks, which, for the present example, will be at the number, stick collection, we call *four*. Asking why *two* plus *two* is equal to *four* is thus simply a sign of ignorance, since this is simply the meaning endowed in the names of the numbers and the composition *addition*, putting sticks next to each other. In view of the above mechanical procedure, addition can be done by a machine. The composition, or process, of addition is mathematically represented by the symbol $+$, and for the above example we have, using Arabic number notation, $2+2=4$, the symbol $=$ thus having the meaning of identity, *equal to*.

Once we can *add* sticks, we can also do the reverse process, *subtraction* of sticks. Denote this process, or composition, by $-$, then what is meant by, for example, $4-2$ is the number obtained by moving the two-stick configuration below the four-stick configuration and removing two sticks from the four-stick configuration, which leaves us with a two-stick configuration, i.e. $4-2=2$. In general, $N_1 - N_2$ is the number obtained by moving the N_2-stick configuration below the N_1-stick configuration and removing N_2 sticks from the N_1-stick configuration. We note that this leaves three possibilities: if the N_1-stick configuration consists of more sticks than the N_2-stick configuration, corresponding to being further out in the line-up of natural numbers, subtraction constructs another natural number. If $N_2 = N_1$, removing, subtracting, an equal amount of sticks produces the absence of sticks, and the *value* it represents, or lack of value, is called *zero*. The statement of absence of a quantity is represented by a circle, 0, in honor of Babylonians, $N_1 - N_1 = 0$. In the situation where the N_2-stick pile contains more sticks than the N_1-stick pile, a number of sticks cannot be *subtracted*, a deficit of sticks is produced, and we shall say that a negative number is produced. A deficit of sticks, for example, represented by hollow sticks, symbolize the so-called *negative* whole numbers progressing to the left of the line-up of natural numbers. A negative number is symbolized by the sign $-$, called *minus*, in front of the natural number expressing the number of hollow sticks, for example, $2-4=-2$.

Another composition of numbers is *multiplication*. Representing the multiplication operation by a centered dot, $\cdot$, for example three multiplied by two, $3 \cdot 2$, is endowed with the following meaning: *take three times a two-stick configuration and place them next to each other*, i.e. add them together

$$|||\cdot|| \quad \rightarrow \quad ||+||+|| \quad \rightarrow \quad ||||||$$

Multiplication thus expresses multiple additions (a subroutine of additions). Instead of the above mechanical prescription, we can write the meaning of multiplication in Arabic mode, $3\cdot 2 \equiv 2+2+2=6$. Multiplication is *commutative*, i.e. $3\cdot 2 = 2\cdot 3$; both operations give the same number, the one represented by six sticks (as an exercise, prove in general the commutation property at the stick level).

For large numbers, the stick representation is not convenient. Multiplication and addition give a different way of representing numbers, different from the fundamental one where any number is reached by *adding*, one upon one, one upon one upon one, etc. Instead, numbers are expressed in terms of powers of a number larger than one, in a *base* larger than one. Say, choose base *ten*: then $1 \equiv 10^0$, $10 \equiv 10^1$, and one hundred, 100, which denotes the number 10 *times* 10, $100 = 10\cdot 10$, is written as ten to the power two, $100 \equiv 10^2$, etc. All natural numbers are then expressed in terms of powers of ten with the help of multiplication and addition. The number *one hundred and fifty six* is then written 156 and represents the number $1\cdot 10^2 + 5\cdot 10^1 + 6\cdot 10^0$. In terms of sticks, it means: take one bunch from the hundred stockpile of sticks and add to that five bunches of sticks from the ten stockpile of sticks and finally add six sticks from the one stockpile. Using a base is a convenient way of compressing information, representing numbers in a coded form by powers of the base. Any number expressed in base ten, say $a_3a_2a_1a_0$, thus uniquely specifies the number $a_3 \cdot 10^3 + a_2 \cdot 10^2 + a_1 \cdot 10^1 + a_0 \cdot 10^0$, where any a_n can be any of the ten numbers between zero and nine. To add numbers expressed in base ten, say $156 + 7$, adding seven sticks to the 156 pile, we note that when adding seven sticks to six sticks gives 10 plus 3, i.e. the ten stockpile number is increased by one and we have three sticks from the one stockpile, $156 + 7 = 163$. The basic rule of addition of numbers in base representation, in digital form, is thus that of a *carry*. So six plus seven gives three plus a carry, $6+7 = 1 \cdot 10 + 3$ $(= 13)$; *one* is added, carried, to the left neighboring digital place representing the ten produced by addition. Similarly, if we added 60 to 156, adding 60 to 50 would produce a carry, but now to the hundred bunch, i.e. $156 + 60 = 216$. Adding 67 to 156 produces two carries as $156 + 67 = 223$.

We do not attempt the impossible task of memorizing multiplication of large numbers. Instead, a list of steps are memorized together with memorizing the multiplication table of multiplying any two single-digit numbers. Multiplication of large numbers can then be accomplished.

We can ask the question: which numbers can be obtained as the product of other numbers? Going through the natural numbers starts out fairly easy, analyzing a finite number of product possibilities. Products of two, even numbers, $4, 6, 8, \ldots$, are clearly such a type of number, but there are others, say, $9 = 3 \cdot 3$. The numbers that cannot be represented by a product of other numbers, except in the trivial way $n = 1 \cdot n$, the so-called *prime* numbers, are at first readily identified as $2, 3, 5, 7, 11, 13, 17, 19, 23, 29, 31, 37, 41, \ldots$.[1] The non-prime numbers are thus those that can be factorized into a product of primes, any number only having one set of prime factors. The ancient Greeks proved that there are infinitely many prime numbers (the biggest known prime has more than four million digits). There exists no usable method for factorizing large numbers, i.e. it is difficult to factor. So, given a large number, N, which is the product of two primes, $N = p \cdot q$, both more than a hundred digits each, it is very difficult, perhaps impossible, for a computer on a practical time scale to find out what p and q are. On the other hand, given the prime number, say p, it is not as hard to find out that it is a prime. This is the basis for a practical matter, since large prime numbers then can be used for cryptography.[2]

Given two natural numbers, n and m, one can ask for their greatest common divisor (g.c.d.). Since the g.c.d. is smaller than the smaller of the two, there is a finite number of possibilities, and the g.c.d. can be found by brute force. Try dividing n and m by $1, 2, 3, \ldots$ in succession and choose the greatest number that divides both. However, the process can be speeded up by using another algorithm, the Euclidean algorithm. For example, if $n = 153$ and $m = 68$, then $n = 2 \cdot 68 + 17$ and $m = 4 \cdot 17$, and the g.c.d. is 17. The Euclidean algorithm runs as follows. Assuming $n > m$ (or determine which is greater), divide n by m, giving $n = k_1 \cdot m + r_1$. If the remainder r_1 equals zero, $r_1 = 0$, the g.c.d. is m. Otherwise,

[1] Numbers have many properties. For example, if a number is divisible by 9, the sum of the digits making up the number is also a number divisible by 9, as exemplified by the numbers 18, 36, 72, 162. Numbers have the flavor of magic, as testified by *lucky* numbers.

[2] The digital computer is an input/output machine that can be represented by the classical mechanics of, for example, scattering of balls. Quantum mechanics offers the possibility for far more capable machines, e.g. finding the prime factors of a number much faster than present computers. On the other hand, it also offers the new possibility of unbreakable kinds of cryptography.

divide m by r_1, giving $m = k_2 \cdot r_1 + r_2$. If the remainder r_2 equals zero, $r_2 = 0$, the g.c.d. is r_1. If r_2 is different from zero, then divide r_1 by r_2, giving $r_1 = k_3 \cdot r_2 + r_3$. If the remainder r_3 equals zero, $r_3 = 0$, the g.c.d. is r_2. Repeating this process, eventually one obtains $r_{n-1} = k_{n+1} \cdot r_n$ and r_n is the g.c.d.

An algorithm is a finite set of rules stating, step by step, precisely what to do to solve a given class of problems (i.e. a mechanical method). Anybody can solve the problems provided that they have the ability to perform the operations required by the algorithm and that they follow the rules exactly. A school child can learn the Euclidean algorithm, without knowing why it gives the desired answers. However, the above description of the Euclidean algorithm is too informal for an ordinary computer to handle, it needs to be more mechanical. Such a description makes use of labeled boxes (technically, registers with addresses) to store questions, answers, intermediate results and instructions (acts to be performed). For example, store the natural numbers a and b in registers A and B, the answer in register D, and intermediate results in registers C and A and B. Using the notation $\langle X \rangle$ to represent the content of register X, a computer program to find the g.c.d. could look as follows:

Address	Content	Explanation
A	a	
B	b	
C		
D		
1	STO (REM (A, B), C)	Store remainder of $\langle A \rangle / \langle B \rangle$ in C
2	TZE (C, 6)	Transfer to instruction 6 if $\langle C \rangle = 0$, otherwise continue with instruction 3
3	STO (B, A)	Store $\langle B \rangle$ in A
4	STO (C, B)	Store $\langle C \rangle$ in B
5	TRA (1)	Go to 1
6	STO (B, D)	Store $\langle B \rangle$ in D
7	Print (D)	Print out $\langle D \rangle$ as answer

The procedure defines a function g.c.d.$(a, b) = \langle D \rangle$. If $a = b$, the instructions 1, 2 and 6 give $\langle D \rangle = b$ as the g.c.d. of a and b. If $a < b$, rem$(a, b) = b$, instructions 3 and 4 interchange a and b. If $\langle A \rangle > \langle B \rangle$, the loop from 1 to 5 repeats divisions of $\langle A \rangle$ by $\langle B \rangle$ until remainder 0 is reached. Then instruction 2 exits to give the answer, $\langle D \rangle = \langle B \rangle$.

14.3 Binary Numbers

A transistor has two states (like an ordinary electric switch), so to represent numbers physically by states of transistors they must be expressed in base *two*, the binary numbers, where a number expressed in digital form as $\cdots a_3 a_2 a_1 a_0$ represents the number

$$N = \cdots + a_3 \cdot 2^3 + a_2 \cdot 2^2 + a_1 \cdot 2^1 + a_0 \cdot 2^0. \qquad (14.1)$$

That is, we pick numbers from the one-stick pile $1 = 2^0$, the two-stick pile $2 = 2^1$, the four-stick pile $4 = 2^2$, the eight-stick pile $8 = 2^3$, etc. So zero and one are denoted by 0 and 1 (as in any base) and any a_n can take on the values 0 and 1. The binary expression for the decimal number two is 10, as $2 = 1 \cdot 2^1 + 0$; and for example the decimal number five is 101 and the decimal number ten is 1010.

Exercise 14.1 Represent the decimal numbers from zero to twenty in base two.

The basic rules for binary addition are

$$0+0=0, \qquad 0+1=1+0=1, \qquad 1+1=0 \text{ and a carry.} \qquad (14.2)$$

The last statement means that one *plus* one is two, i.e. that the number *two* in binary code is represented by the binary string 10. When two ones are added, the result consists of two digits, the higher significant bit, 1, is called a carry. The basic rule for addition in binary number notation is thus:

$$\begin{array}{r} 1 \\ + \quad 1 \\ \hline 1\ 0 \end{array}$$

For larger numbers, the rule is just repeated. For example, addition of two plus three, which in base ten goes like $2+3 = 5$, is in binary achieved by the basic binary rules of Eq. (14.2):

$$\begin{array}{r} 1\ 0 \\ + \quad 1\ 1 \\ \hline 1\ 0\ 1 \end{array}$$

To mechanize the process of addition in the binary code, we thus want a device that has two input terminals and two output terminals, as depicted in Figure 14.1.

The single-bit adder, or half adder, should for given inputs A and B, being either 0 or 1, have as output S and C representing the sum of the numbers A and B in binary code, i.e. the input/output sequence as given in the table below.

A	B	S	C
0	0	0	0
0	1	1	0
1	0	1	0
1	1	0	1

This looks like the Boolean algebraic representation of the truth tables of formal logic, where *true* and *false*, T and F, are represented by 1 and 0.

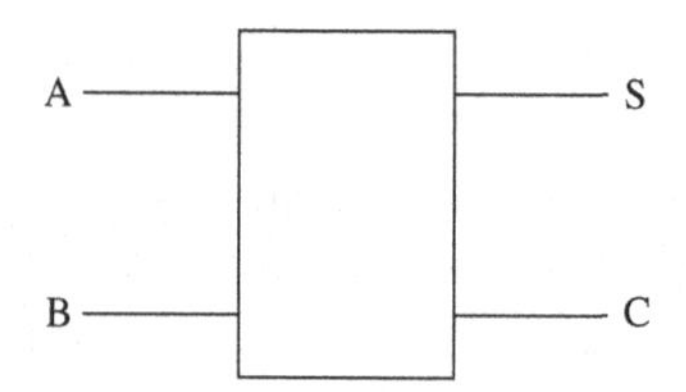

Figure 14.1 The half adder.

14.4 Logic

In his *Syllogisms* (483 BC), Aristotle laid down rules for correct inferences. Consider the two propositions (the technical term a logician uses to stress the *meaning* of an informative sentence (a linguistic complex)):

p_1: *All gentlemen are polite.*
p_2: *No bankers are polite.*

From these two propositions, we logically infer the conclusion:

c: *No gentlemen are bankers.*

The correctness of the inference seems obvious, ingrained in our brains as if unable to operate by the rule making existence unlikely.

A simple way of expressing the correctness of the inference is by using a pictorial representation of the involved objects, as depicted in Figure 14.2.

The first statement, p_1, places gentlemen in the camp of polite people, as depicted on the left-hand side of Figure 14.2. The second assertion, p_2, says that there is no overlap between the set of bankers and polite people. Bankers are a subset of the set of impolite people, as also depicted on the left-hand side of Figure 14.2. Viewing this content from the discriminating point of view of dividing people into bankers and non-bankers, as depicted on the right-hand side of Figure 14.2, the set of gentlemen, having no overlap with the set of bankers, is a subset of the non-bankers, and the figure shows that no gentlemen are bankers. By assigning truth to the two propositions, we arrive at another truth, the conclusion.[3] For logic inference to get going, assumptions are needed, and discriminations about the objects of the world. We divided, classified, the individuals of the world as to whether they possess a property or not, say, to be polite or not.

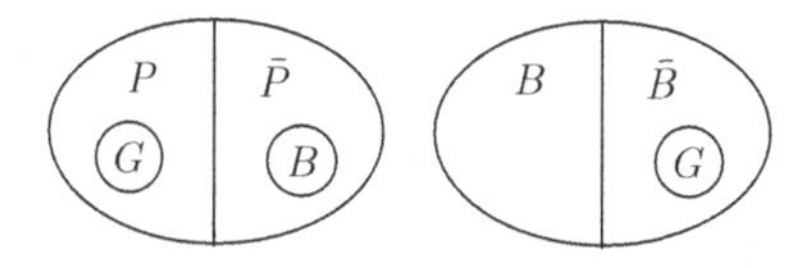

Figure 14.2 Logic inference based on set representation.

[3] The inference can also be expressed using set theory notation, the symbol $\subset$ endowed with the meaning *contained in*: $G \subset P$ and $B \subset \bar{P}$ therefore $G \subset \bar{B}$.

Exercise 14.2 Asserting the truth value of the above considered categorical syllogism is efficiently done by drawing a Venn diagram, i.e. three overlapping circles representing the sets of gentlemen, bankers and polite people. Do this. The propositions will then make parts of the circles empty (mark this, for example, by hatching the empty parts). Draw the conclusion by inspecting the diagram.

From propositions, new propositions can be formed by the three basic logic operations: *and* $\equiv \wedge$; *or* $\equiv \vee$; and *negation* $\equiv \neg$. A proposition can be assigned either true or false, the value T or F (the logic principle of the excluded middle). The logic operations can then be specified by their truth tables.

For example, the logic operation of *negation* is represented by the following truth table. It shows that, if proposition A is assigned true, its negation, ¬A, is false, and vice versa.

A	¬A
T	F
F	T

Of importance for reasoning is logic *and*, which has the following truth table. The logic *and* of two propositions is only true if both statements are true.

A	B	A ∧ B
F	F	F
F	T	F
T	F	F
T	T	T

Logic implication $\Rightarrow$ (*if* ... *then* ... in common usage) has the following truth table. Logic, or material, implication rules out the possibility of the *antecedent*, p_1, being true but the *consequent*, p_2, being false. The first and third entries in the last row therefore follow from the common usage of *implication*, and the remaining entries by the considerations in the following exercise.

p_1	p_2	$p_1 \Rightarrow p_2$
T	T	T
F	T	T
T	F	F
F	F	T

Exercise 14.3 Consider the correct, true, implication for all numbers: if $n > 5$ then $n > 3$. Consider the choices $n = 4$ and $n = 2$ with corresponding antecedents and consequents. Realize that suggests the remaining entries, so that implication is always true except for the case where the antecedent is true and the consequent false.

A tautology is a statement schemata that comes out to be true regardless of the truth values assigned to the propositions, *variables*, involved. An example of such a formally true schemata, true on logical grounds alone (due to the logical form), is $(p \wedge q) \Rightarrow q$ as the following truth table shows. The column to the right follows from the truth table for implication.

p	q	$p \wedge q$	$(p \wedge q) \Rightarrow q$
T	T	T	T
F	T	F	T
T	F	F	T
F	F	F	T

A conclusion from two premises, p_1 and p_2 therefore c, is a formally valid argument (or not) according to whether $(p_1 \wedge p_2) \Rightarrow c$ is (or is not) a tautology. If it is a tautology, the premises cannot be true and the conclusion false, and thus the argument, or inference, (p_1 and p_2 therefore c) is valid. If it is not a tautology, then it is possible for the premises to be true and the argument (p_1 and p_2 therefore c) to be false, so the inference scheme cannot be valid. The concept of a *tautology* can therefore be used to settle the validity of deductive inferences. With the help of truth tables, correct inference can be mechanized.

Exercise 14.4 Show that, for the categorical syllogism considered in Exercise 14.2, the schemata $(p_2 \wedge p_2) \Rightarrow c$ is a tautology. For example, use the Venn diagram to determine the truth values of the propositions, and show that $(p_2 \wedge p_2) \Rightarrow c$ comes out with only T entries.

14.5 Logic Gates

Instead of assigning true or false values to a proposition by letters, numbers can be used. A function maps a statement onto the set $\{0, 1\}$: proposition $\rightarrow$ $\{0, 1\}$. If assigned value 0, the proposition is considered false; and if assigned value 1, it is considered true. We shall now move on to represent logic operations by binary electronic gates.

The logic operation AND is thus specified by the following binary table. The logic operation AND has the entries corresponding to multiplication of the entries A and B, i.e. $A \cdot B$.

A	B	A AND B
0	0	0
0	1	0
1	0	0
1	1	1

The logic operations are beneficially (for their later combination) represented pictorially by gates, logical gates which have input terminals and an output terminal. Logic gates are switching functions. Their input is binary-valued variables and they compute a function. The logic gate AND is represented by the symbol in Figure 14.3. The AND gate has the prescribed functioning that, if the possible Boolean values are attained at the input terminals A and B, then the corresponding value for A AND B appears at the output terminal.

Another logic gate is the OR gate, which operates according to the Boolean prescription in the following table (in accordance with logic *or*, ∨, considered true if at least one of the statements are true). The logic OR gate symbol is depicted in Figure 14.4.

A	B	A OR B
0	0	0
0	1	1
1	0	1
1	1	1

A more distinct *or* is the exclusive-or, the XOR gate, which has the following Boolean table. Logic exclusive-or is only true if exactly one of the propositions is true. The XOR gate is represented by the symbol in Figure 14.5.

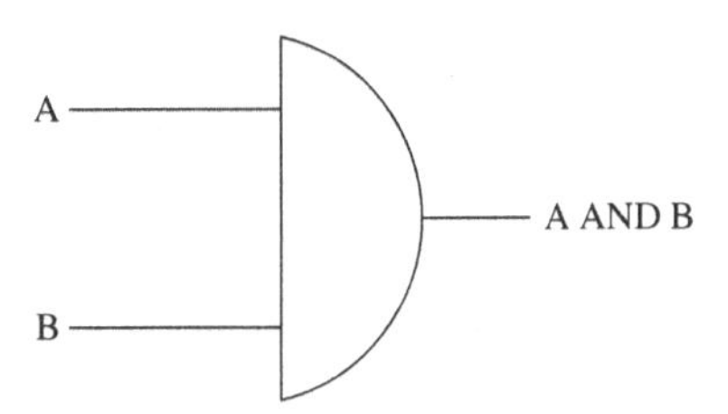

Figure 14.3 The diagrammatic symbol for the AND gate.

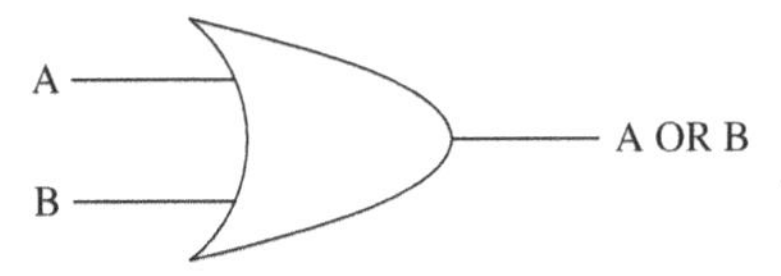

Figure 14.4 The diagrammatic symbol for the OR gate.

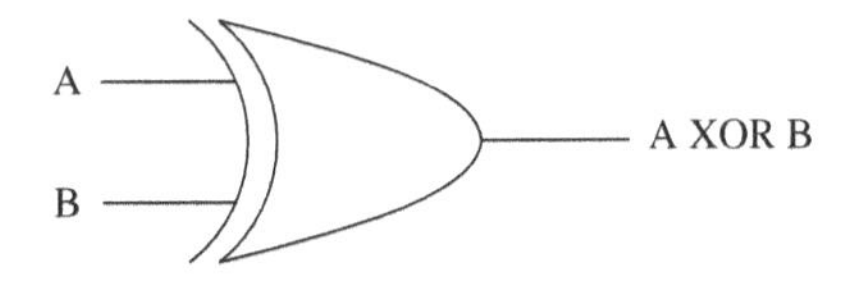

Figure 14.5 The diagrammatic symbol for the exclusive-or gate, the XOR gate.

A	B	A XOR B
0	0	0
0	1	1
1	0	1
1	1	0

In the true/false scenario of logic, the logic operation of negation is convenient. Similarly, in binary representation, the NOT gate is central, swapping the inputs, i.e. in binary code the NOT gate operates according to the following Boolean table. The symbol for the NOT gate is displayed in Figure 14.6.

A	NOT A
0	1
1	0

The NOT gate has many uses. For example, consider a wire branching into two, fanning out, and place a NOT gate on one of the wires. Then, for logic input 1, the wire with the NOT gate has output 0 and the other one is still logic 1, and vice versa for input 0. With two such gates and AND gates at hand, a circuit is simply built where the binary numbers 00, 01, 10 and 11 are coded to different output lines.

The logic operations are not independent but can be built from each other. As an example, Figure 14.7 shows how an OR gate can be built from an AND and NOT gates.

To see this, let us look at the output on the various terminals for given inputs on terminals A and B, and collect them in the table below. The inputs on terminals A and B in Figure 14.7 enter both NOT gates, and for the inputs A and B in the first two columns the outputs of the NOT gates are those listed in the third and fourth columns. These outputs enter as inputs of the AND gate, and according to its working the corresponding outputs of the AND gate are those listed in the fifth column. Finally, this output is negated by the last NOT gate, and the output of the total gate, i.e. the output on terminal C, is listed in the last column and is seen to be identical to that of the OR gate.

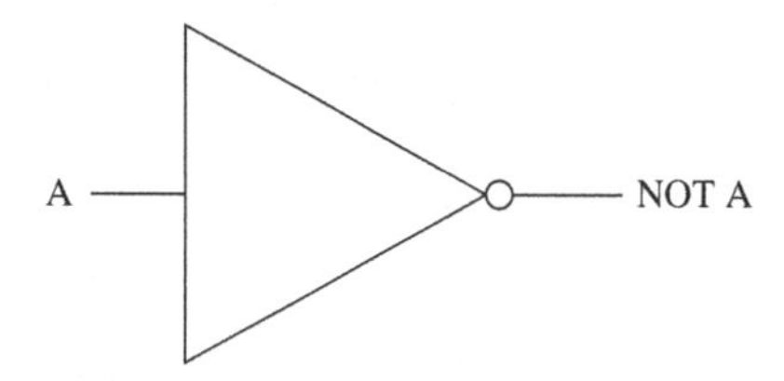

Figure 14.6 The diagrammatic symbol representing the NOT gate.

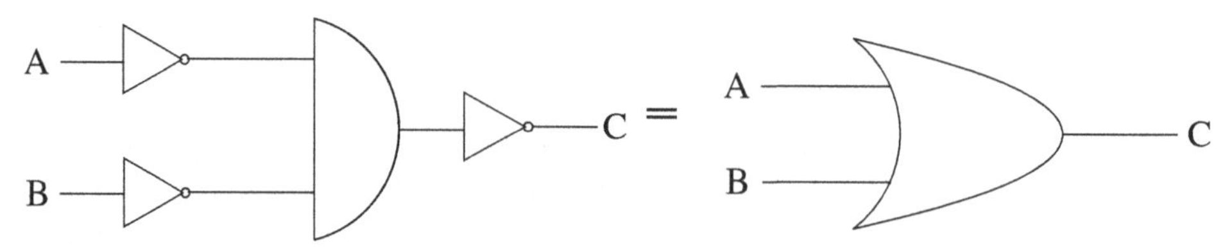

Figure 14.7 OR gate built by an AND gate and three NOT gates.

A	B	¬A	¬B	¬A AND ¬B	C
0	0	1	1	1	0
0	1	1	0	0	1
1	0	0	1	0	1
1	1	0	0	0	1

Indeed, if we combine logic gates as depicted in Figure 14.7, the OR gate is represented in terms of an AND gate and three NOT gates. Thus to build logic circuits not all logic gates are needed. In fact it can be shown that AND, OR and NOT is a complete set of gates, i.e. from these all logical functions can be built. Any black box with an assigned output state for each possible input state can be built from these three gates.

The AND gate followed by the NOT gate, the NOT{AND} gate, or negated AND, is called the NAND gate. The Boolean table for the NAND gate is therefore as shown in the table below. It is depicted in Figure 14.8.

A	B	A NAND B
0	0	1
0	1	1
1	0	1
1	1	0

As discussed in the next section, the physics of present technology favors the NAND gate as the fundamental logic gate, as it is easier to implement than the AND gate, which then is obtained by a NAND gate followed by a NOT gate. A diagrammatic symbol is introduced for the NAND gate as depicted in Figure 14.9.

The NOT gate is easily constructed from a NAND gate. Just let an input line, A, fan out into the two inputs of a NAND gate. Then the truth table for the device is as follows, i.e. the truth table for a NOT gate.

A	A	A NAND A
0	0	1
1	1	0

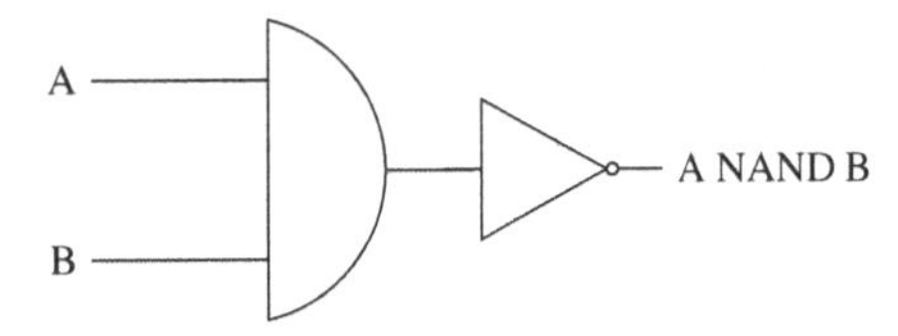

Figure 14.8 The NAND gate is the negated AND gate.

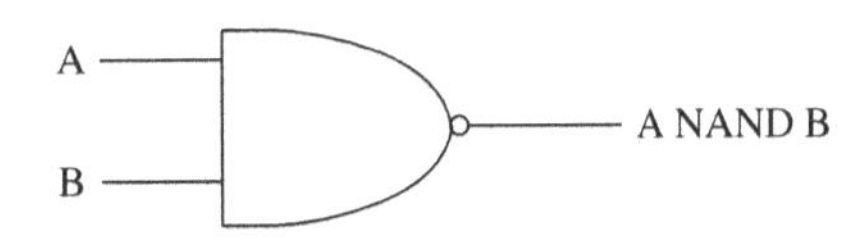

Figure 14.9 Diagrammatic symbol for the NAND gate.

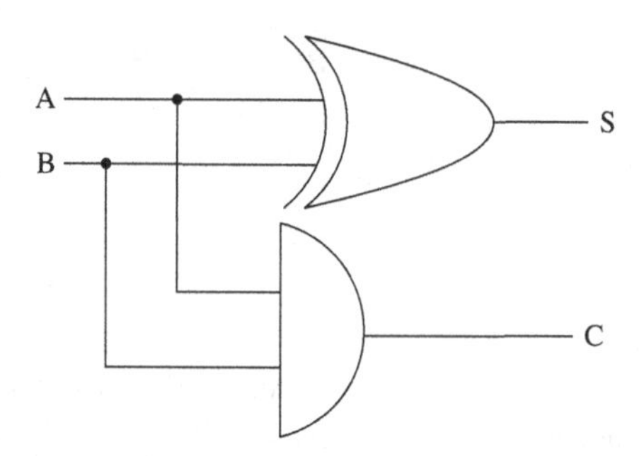

Figure 14.10 The half adder built by the logical gates XOR and AND.

Exercise 14.5 Show that a NAND gate where one of the inputs is kept fixed at value one acts as a NOT gate. Show that, if the output of a NAND gate is fed into another NAND gate where the other input is kept fixed at one, it acts as an AND gate. The NAND gate is thus logically complete, i.e. all gates can be built from NAND gates.

Returning to arithmetic, the half adder is represented by the combination of the logic gates AND and XOR, as depicted in Figure 14.10, since the sum output, S, is equivalent to the output on the XOR gate, and the carry output, C, is equivalent to the output on the AND gate.

The dots on the A and B lines indicate that the signal is transferred along both lines (the lines are connected or joined), the signal fans out; whereas the crossing of two lines without a dot indicates that the two lines are not connected. The input/output of the gate in Figure 14.10 is therefore as in the following table, and thus S = A XOR B and C = A AND B.

A	B	A XOR B	A AND B
0	0	0	0
0	1	1	0
1	0	1	0
1	1	0	1

Having half adders at hand, arbitrary binary numbers can be added. The next exercise should demonstrate that, to be able to add two arbitrary n-bit binary numbers, $2n-1$ simple adders are needed.

Exercise 14.6 Consider adding the two-bit binary numbers 10 and 11. Feed the *first* numbers 0 and 1 into an adder, as respectively inputs A and B, which then gives the S output 1, which is stored on that line, whereas the output C = 0 is fed into the next adder together with 1, the *one* from binary 10, producing C = 0 and S = 1, the latter then entering a last adder together with 1, the second *one* from binary 11. Make a drawing of the above scenario, and end up with three wires, one C terminal and two S terminals with output values 101.

A full adder is a circuit that can add three binary digits. It consists of three inputs, the three bits, and two outputs, sum and carry. The full adder operates according to the following truth table. A full adder adds binary numbers accounting for values carried in and out.

A	B	C_{in}	C_{out}	S
0	0	0	0	0
1	0	0	0	1
0	1	0	0	1
0	1	1	1	0
0	0	1	0	1
1	0	1	1	0
1	1	0	1	0
1	1	1	1	1

Exercise 14.7 Show that the full adder can be implemented by two half adders and an OR gate. The output of the first XOR gate is fed as inputs into the next XOR and AND gates, whereas the output of the first AND gate is fed into the OR gate. The carry input, C_{in}, is fed into the second XOR and AND gates, and the XOR output is the sum, S, and the AND output is fed as the other input of the OR gate whose output is the carry output, C_{out}.

Show that the switch function for the full adder is S = (A XOR B) XOR C_{in} and C_{out} = ((A XOR B) AND C_{in}) OR (A AND B).

Exercise 14.8 With AND gates at hand, multiple AND gates can be constructed. For example, consider four input terminals, A, B, C and D, entering a black box with a single output terminal. Then the multiple AND gate should only give output 1 if all the inputs are 1, and otherwise 0. Show that this is achieved by three AND gates where the outputs of the two AND gates, A AND B and C AND D, are fed into the third AND gate.

14.6 Physical Gates

We now turn to the physical implementation of the logic gates by electronic devices. Consider first the case where the binary code is represented physically by voltages on terminals according to the table below, with bits represented by voltages.

Bit	Voltage
1	$V > 0$
0	$V = 0$

We start with the NOT gate. We should thus construct an electronic device with the input/output relationship specified in the following table. An electronic implementation

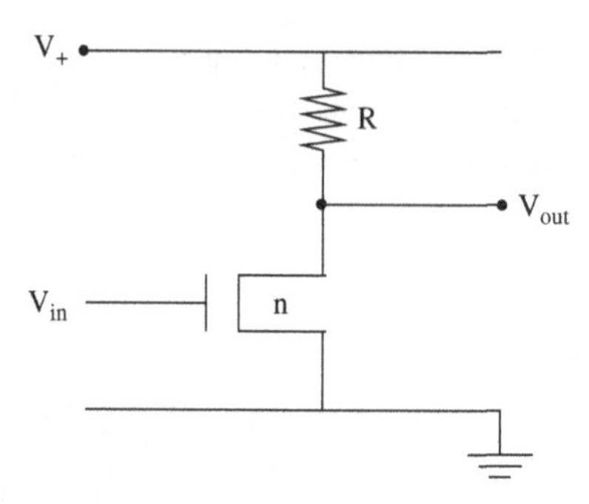

Figure 14.11 NOT gate realized by transistor and resistor.

of the NOT gate is provided by the circuit in Figure 14.11. Here the n-type field effect transistor (n-type MOS, or simply nMOS) is employed.

Bit	V_{in}	V_{out}	Bit
1	$V > 0$	0	0
0	0	$V > 0$	1

To realize this, let us check how the device works electrically. The input voltage, V_{in}, is taken as the voltage on the gate of the field effect transistor, $V_{\text{in}} = V_G$. The voltage on the gate determines whether the plus voltage terminal, V_+, is electrically connected or not to the ground terminal. If the voltage on the gate is zero, or in practice slightly negative, $V_G = 0^-$, no current can flow through the field effect transistor, the output terminal is electrically cut off from the ground terminal as there is no inversion layer to conduct, and the transistor is an insulator. Since no current in that case flows through the circuit, and thereby no current through the resistor, the voltage drop over the resistor is zero, i.e. the voltage on the output terminal is the same as the voltage at the other terminal of the resistor, $V_{\text{out}} = V_+$, and the output terminal is representing the bit 1. Next, since the transistor is an n-type field effect transistor, it is conducting for a positive gate voltage due to the inversion layer of electrons that then exists. Introducing a positive voltage on the gate, the output terminal is electrically connected to ground by a good conductor and the device cannot sustain a voltage difference between its source and drain terminals. The voltage on the output terminal will after a short delay time thus become the same as that of the other terminal of the transistor, i.e. the zero voltage of ground, $V_{\text{out}} = 0$, representing the bit 0. There is then no voltage drop between the source and drain terminals of the field effect transistor and the current through it has ceased to flow. The response of the device is thus precisely as in the table above; the output bit is the inversion of the input bit.

Next we want to physically realize the NAND gate, i.e. a device that in response to two input voltages, V_A and V_B, has the output voltage as described in the table below. In Figure 14.12 is a NAND gate realized by two transistors and a resistor.

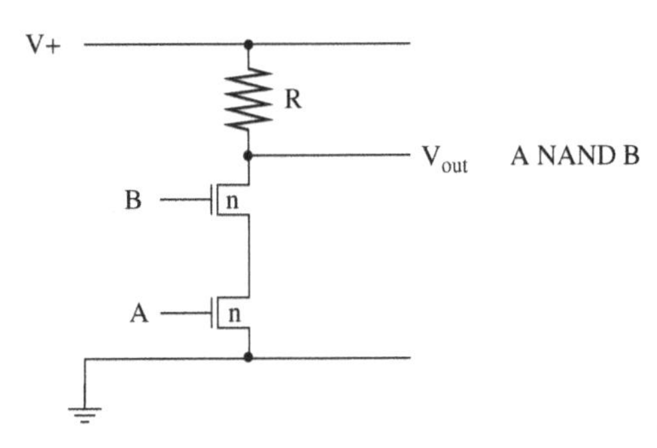

Figure 14.12 NAND gate realized by transistors and resistor.

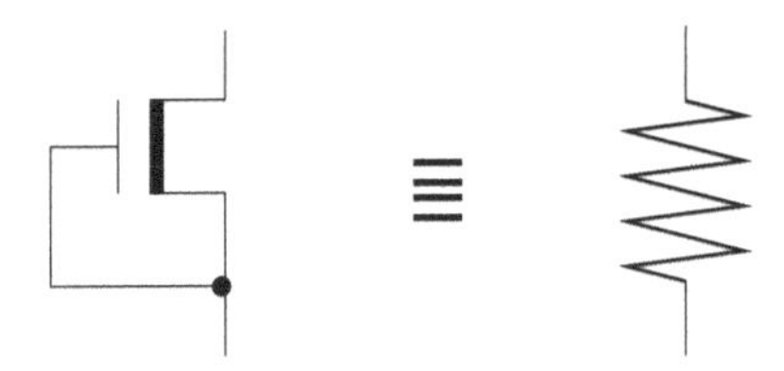

Figure 14.13 Depletion mode transistor acts as a resistor.

V_A	A	V_B	B	A NAND B	V_{out}
0	0	0	0	1	V
0	0	V	1	1	V
V	1	0	0	1	V
V	1	V	1	0	0

The input voltages, V_A and V_B, are the gate voltages of the two field effect transistors. To realize that the device operates as a NAND gate, we note that only if both the input voltages are positive, $V_A \simeq V_B > 0$, are both transistors conducting and the circuit is electronically connected and the output voltage vanishes, $V_{\text{out}} = 0$ representing the bit 0. For the other three possible voltage inputs, at least one of the transistors is not conducting and the ground terminal is cut off from the rest of the circuit. No current flows and the other terminal potential, V_+, is pulled down to the output terminal, leaving the output voltage equal to V_+, $V_{\text{out}} = V_+$, representing the bit 1.

To create standard resistors on a silicon chip is inconvenient, for space and cost reasons, and they are simply implemented instead by field effect transistors. If the gate and source are connected, the gate and source voltages are the same $V_G = V_S$, the device always conducts and the transistor, in this so-called depletion mode, acts as a resistor, as displayed in Figure 14.13.

In reality, on a silicon chip, the NOT gate is thus realized as depicted in Figure 14.14.

Exercise 14.9 Consider the device depicted in Figure 14.15.

Show that it represents logic identity, i.e. the logic state on the output is the same as that on the input. Though logically inoperative, it can be used for delay and amplifying purposes, countering the inevitable power loss in transistors.

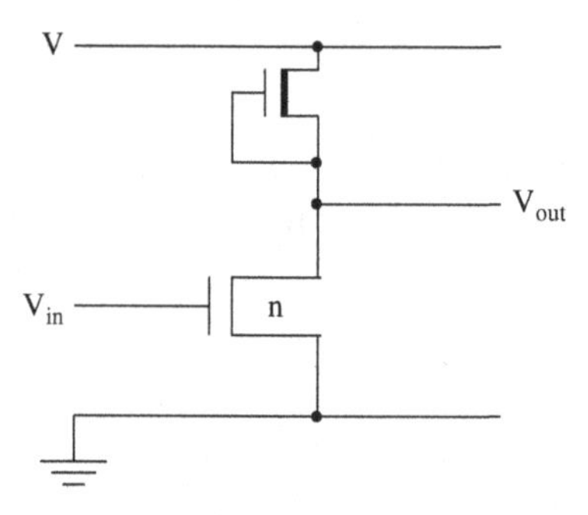

Figure 14.14 NOT gate or inverter realized by only transistors.

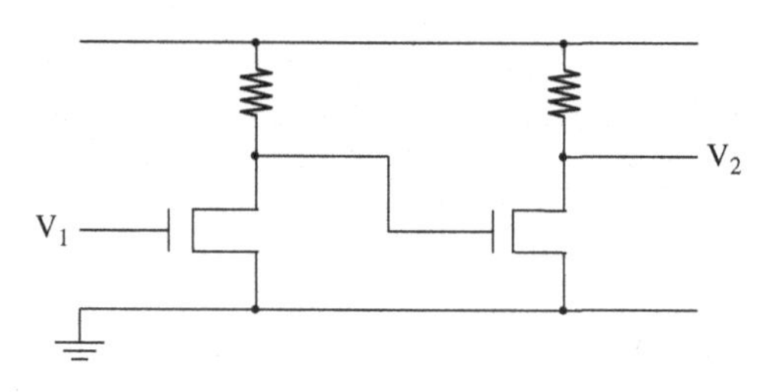

Figure 14.15 Identity or follower circuit.

On a silicon chip, the NAND gate is realized as depicted in Figure 14.16, realized by using both n-type and p-type field effect transistors, the CMOS technology.

As in the previous nMOS case, logic one is held by V, but now logic zero corresponds to voltage $-V$. To see that the device in Figure 14.16 acts as a NAND gate, we note that the output voltage can only be pulled down to $-V$, i.e. logical 0, if both the n-transistors are conducting, which requires both input voltages to be positive, V, i.e. both representing logical 1.

Early logic gate technologies were based on the bipolar transistor, where one of the logic states, the ON state, corresponded to a current-carrying state, thus involving heating. The advantage of the complementary MOS (CMOS) technology is that the logic states correspond to physical states without any currents flowing, thus vastly reducing heat production in the operation of a machine built by such devices. Only the change of a bit involves current flow. This is, of course, an idealization: there will always be small currents flowing as a result of voltage differences due to the finite, though large, resistances of the components, and power loss is unavoidable.

14.7 Feedback

Of importance for the efficiency of operation of a network of logic gates is the ability to re-use logic gates in later steps. After a simple adder has done its addition, it can do another one. This entails feedback. The simplest type, to be considered here, is where the output

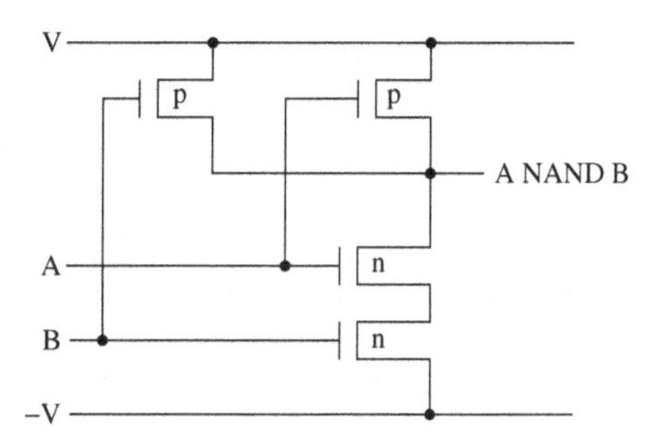

Figure 14.16 The NAND gate realized in CMOS technology.

wire of a logic gate is fed in as one of the input wires on the same gate. If we think of the gates as operating instantaneously, a delay component, the trivial identity gate, should in practice be inserted in the *loop* line.

Let us consider what feedback entails. Take, for example, the OR gate and feed the output C back as input (as an exercise, you could draw the gate picture), assuming a next C is produced according to the OR gate operation, i.e. as in the table below.

A	C	next C
0	0	0
1	0	1
0	1	1
1	1	1

If A = 0, then C will repeatedly give whatever value it was initially, i.e. C stays stable. If A = 1 and C = 0, then C will change to 1 and repeatedly give 1. If A = 1 and C = 1, then C also repeatedly gives 1. The OR gate with feedback thus remembers whether the input has ever been ON, i.e. carried input signal A = 1. These devices can be used as a *recognizer*: accepting every finite string of "0"s and "1"s that contain at least one 1.

If we consider feedback with the XOR gate, the next output, next C, is produced according to the following table. The feedback XOR gate looks like a memory cell where the memory C stays the same if A = 0, and the memory is switched if A = 1. However, the next C will again switch and the C value oscillates, so the device cannot be used as a memory cell. When using feedback, one must thus take care and choose logic operations to avoid such oscillations.

A	C	next C
0	0	0
0	1	1
1	0	1
1	1	0

When playing with logic gates, we can assume that the output occurs at the same instant as the inputs. This is, of course, not true for physical gates. A physical process takes a finite amount of time; it takes time for the voltages to achieve the values representing logic 1 or 0. Since signals traveling through different components take different times, mismatch of signals will occur, and the machine goes haywire, not operating according to plan. Only if signals can be put on halt can operation according to design provide for proper current flows and controlled operation. This is achieved by ensuring that signals are only sent through a device at regular intervals. Thus a clock is introduced to pace sequential operations. For example, consider a stepwise voltage, being alternately 0 and V, fed simultaneously as one of the inputs of two AND gates. The inputs on the two other input wires of the two AND gates are then only transferred when the clock wire corresponds to logic 1. Accounting for the finite delay times of devices makes designing the arrangement of components of a computing machine, its architecture, an art form in itself.

Feedback is also essential when it comes to creating essential machine components. A desirable device is one that can hold a value, 0 or 1, until it is decided that it should be reset by a signal on a wire, a so-called flip-flop. To achieve this, consider a gate with two input lines, S and R, and two output lines, Q and ¬Q, the latter notation signifying that the signals on the two output wires should always be their logic complements. The device should operate according to the following truth table.

Q	S	R	next Q
0	0	0	0
0	0	1	0
0	1	0	1
1	0	0	1
1	1	0	1
1	0	1	0

Clearly, when built by logic gates, it must entail feedback, Q (and ¬Q) being fed back. Note that the S and R lines are assumed never to have value 1 simultaneously, a feature that must be ensured by the device environment, the machine architecture. Let us see what the truth table gives for the behavior of the device.

The Q-line signal represents the contents of the flip-flop, and, according to the truth table, stays constant, at whatever its value, 0 or 1, when both S and R are 0, i.e. the flip-flop has two stable states. Consider first the case where the R wire carries no signal, R = 0. If Q is initially zero, Q = 0, then setting the S wire to one, S = 1 (S = 0 → 1), changes the Q value to 1, and any future change of S does not change this. If Q is initially one, Q = 1, it stays that way no matter how S is changed. So through the S wire the value of the flip-flop can be set to 1 (the set wire). Next consider what happens if the R line is changed, R = 0 → 1 (then, by construction, S has value zero, S = 0). If Q = 0 and R = 1, the Q value will, according to the truth table, stay at value 0, but if Q = 1 then R = 1 resets it to zero, Q = 0. Thus the R wire resets, or clears, the contents of the flip-flop.

Exercise 14.10 A flip-flop can be constructed by two NOR gates (NOT(OR)) where S and R are inputs on different NOR gates and Q is the output on the gate with R input and fed back to the other gate as input (i.e. together with S), and ¬Q is the output on the gate with

S input and fed back to the other gate as input (i.e. together with R). Show that indeed the above truth table follows and that the output on the ¬Q line indeed always is the opposite of that on the Q line.

14.8 Summary

Providing voltages to terminals of connected electric devices, the flow of electric charge then gives the correct voltage drops over components, giving intentional outputs for given voltage inputs. In this fashion, a mechanical machine is realized by the pushing of electrons. The voltages are determined by physical gates realizing logic operations. A digital computer is a mechanical machine where any wire can hold two different voltage values, corresponding to two discrete states, thereby representing the binary numbers 0 and 1. The flow of electricity in the wire determines output voltages representing the outcome of, for example, arithmetic calculations. The computer is a programmable machine designed to automatically carry out a sequence of arithmetic or logic operations. For the processor to perform instructions, they must be encoded as binary numbers. A computer has different parts: a memory; element(s) doing arithmetic or logic operations; sequencing and control units that can change the order of operations based on information stored; and a processing unit that executes a series of instructions that make it read, manipulate and then store data.

Even though each terminal wire holds only one of two voltages, V or 0, representing the bits 1 and 0, simple logic circuits can encode so that a voltage on a wire can represent any number expressed in binary code. By providing voltage to input wires, another *coded* wire, representing a binary number, triggers a third specific terminal, has *its* voltage turned on, now representing the binary number of interest, which is then the input voltage to a gate for further manipulation. Such multiplexers can thus make the computer transport any coded binary number (and thereby sequences of such) to other device terminals after whose operation it represents another number specified by the logic operations that have been carried out. In this way data are transmitted within the machine. A programming language must thus be translated, coded, into a representation by binary numbers in order for the computer to perform the program's instructions. The computer language's *words* are binary numbers, the machine language. The computer language sentences, in the rigid machine language, consist of sequences of the binary numbers 0 and 1, i.e. far from our natural language. The development of programming languages, capable of the precision of machine language, evolves toward easier use.

Computers do arithmetic calculations and are widely used in the sciences and beyond for numerical calculations and simulations. Besides that, non-numerical algorithms can also be realized on computers. For example, the letters of the alphabet can be coded, i.e. represented by numbers, and be manipulated in the computer according to the logic operations.

A computer has five basic units: the central processing (control) unit (CPU); the memory (store); the arithmetic–logic unit; the input (I); and the output (O) (the latter two communicating with the outside world). They perform the following tasks: the input transmits data and instructions to the store; the control unit locates in the store the (next) instruction to be carried out (decodes it) and, depending on the result of the decoding, either carries out a *computation* in the arithmetic–logic unit, or puts the result of a *computation* at an appropriate location in the store, or determines the next instruction to be carried out by examining the content of some register, or inputs a word to the store, or outputs a word from the store. Except for the latter case, the control unit carries out the next instruction in the instruction list in the store.

The topic of mechanizability of arithmetic and logic inference has a long history of invention of various mechanical machines, real or imagined. An example of such is Charles Babbage's *analytical engine* (1830), which, conceived but only started to be built, could function like today's digital computers. Having no other efficient way to represent and manipulate the digits, awkward mechanical means like the positions of cogwheels were envisioned to represent the discrete states. The technological breakthrough came with electric switches. The first practical digital computer, built in 1944 by H. H. Aiken, was an electro-mechanical machine employing some three thousand telephone relays. It was a slowly operating machine. Vacuum tubes were first employed in the subsequent generation of computers, stimulating the emergence of the modern era of the electronic computer, finding its present form based on the solid-state transistor. The development of the mechanical processing of information since the 1950s has been phenomenal, by any standards, and computers have come to utterly determine human activities on the planet and beyond.

Appendix A: Principles of Quantum Mechanics

To appreciate how profoundly different from classical mechanics, not to say utterly strange, quantum mechanics is, it is useful to break the theory down into its basic constructs. Instead of postulating quantum dynamics in terms of a differential equation, the Schrödinger equation, an integral characterization can be obtained, engineered solely by the use of a few principles, in particular the basic principle of quantum mechanics, the superposition principle. The Schrödinger equation then follows as a corollary, as shown in this appendix. The elaboration testifies to the fact that the principles behind an equation can be more robust as regards their range of validity than the equation itself. In fact, the Schrödinger equation does not encompass all phenomena: it describes non-relativistic quantum phenomena. In relativistic quantum theory, other equations take over, but the quantum principles behind them are the same. From a practical point of view, such as understanding electronic devices, the topic of this book, non-relativistic quantum theory suffices.

In order to elucidate the conceptual structure of quantum mechanics, we consider the space-time approach, since it offers the intuitive formulation of quantum mechanics. The content and consequences of the principles of quantum mechanics are illustrated at first by considering the simplest example of a physical system, *viz.* that of a single particle.[1]

A.1 Quantum Kinematics

The first question to deal with is how to describe quantum mechanical systems, i.e. the kinematics of a quantum system. To accomplish this, a new concept, foreign to common sense, must be introduced, *viz.* probability *amplitudes*. We must learn to speak the language of quantum mechanics.

The quantitative description of a physical system is provided by the Born rule or the kinematic principle, as follows.

[1] The reader is at this point assumed to be ignorant about physics. Quantum mechanics is assumed the fundamental theory of how the world works, and classical mechanics is at first given no role. Only common sense knowledge is assumed, in particular, simple counting understanding of probability calculus. The following two principles indeed earn the prestige of being called principles; a hundred years of search has found no gears and wheels beyond them.

Kinematic principle

To any possible event is attributed a complex number ψ called the probability amplitude. The probability P of the event taking place is given by the absolute square of the amplitude

$$P = |\psi|^2. \tag{A.1}$$

In the description of a single particle, the events are the different locations of the particle, and the probability amplitude associated with the event of the particle at position $\mathbf{x}$ is denoted $\psi(\mathbf{x})$. The probability for the event "particle at position $\mathbf{x}$" to take place is thus $P(\mathbf{x}) = |\psi(\mathbf{x})|^2$. Denoting a probability, the number $P(\mathbf{x})$ of course has the practical meaning that on repeating many times a measurement revealing the location of the particle, $P(\mathbf{x})$ equals the relative frequency for the outcome to observe the particle at position $\mathbf{x}$. Measurement is here understood in its intuitive way; position measurement is inferred from, say, blackening of photographic emulsion. The discourse of establishing these *facts* being fairly straightforward, a few finger points to describe the experimental set-up should lead two inquiring minds to agree on hypothesizing on what is going on.

Considering space to be a continuum, we must speak of the probability *density* for a particle's whereabouts. The absolute square of the probability amplitude gives the probability density $P(\mathbf{x}) = |\psi(\mathbf{x})|^2$ for the event of the particle at position $\mathbf{x}$, i.e. the probability for the event of the particle located in a small volume $\Delta\mathbf{x}$ around position $\mathbf{x}$ is given by $|\psi(\mathbf{x})|^2\Delta\mathbf{x}$. Since $|\psi(\mathbf{x})|^2$ is a probability density for a continuum space, the probability amplitude function, $\psi(\mathbf{x})$, is a *continuous* function of space.[2]

A.2 Time and Quantum Dynamics

In quantum mechanics, time has its usual common sense meaning. Time describes the continuous progression from past through the present to the future. Time is an external parameter, i.e. not a property of a considered physical system. The continuous parameter, t, describes the progression of time, which in practice is quantified by the *time* we read off our watches (or any clockwork motion such as a pendulum or that of the Moon or planets in the sky). An event, in accordance with our usual use of this concept as referring to a straightforwardly agreed physical manifestation, is thus understood as also referring to a certain instant of time. We can thus readily introduce the time dependence of probability amplitudes. The possible events of a single particle are the particle's possible different positions at a time in question, and the probability amplitude associated with the event "particle at position $\mathbf{x}$ at time t" (space-time point $(\mathbf{x}, t)$ for short) is denoted $\psi(\mathbf{x}, t)$. In accordance with tradition, we shall also refer to this probability amplitude as the wave function for the particle. Taking the absolute square of the probability amplitude gives the probability density, $P(\mathbf{x}, t) = |\psi(\mathbf{x}, t)|^2$, for the event of the particle at position $\mathbf{x}$ at time t, i.e. the probability for the event of the particle at time t in a small volume $\Delta\mathbf{x}$ around position $\mathbf{x}$ is given by $|\psi(\mathbf{x}, t)|^2\Delta\mathbf{x}$.

[2] For ease of language we shall often refer to probability *densities* as simply probabilities.

The temporal ordering of events that the concept of time introduces allows us to consider a sequence of two events. For example, the kinematic principle allows as a possible scenario that at time t' the particle is definitely at position $\mathbf{x}'$. The said principle then attests to the whereabouts of the particle at a later time t, as described by the probability amplitude function $\psi_{(\mathbf{x}',t')}(\mathbf{x},t)$, the indices indicating the initial condition. This probability amplitude function predicts possible outcomes at time t *given* the initial event "particle at space-time point $(\mathbf{x}',t')$". According to the kinematic principle, the whereabouts of the particle later in time has the associated probability distribution $|\psi_{(\mathbf{x}',t')}(\mathbf{x},t)|^2$. Instead of using indices, we introduce standard notation from probability calculus for this conditional probability (density), i.e.

$$P(\mathbf{x},t;\mathbf{x}',t') \equiv |\psi_{(\mathbf{x}',t')}(\mathbf{x},t)|^2, \tag{A.2}$$

the probability for the event of the particle at position $\mathbf{x}$ at time t *given* it was at position $\mathbf{x}'$ at time t'.

Analogously we introduce the notation for the conditioned probability amplitude function,

$$K(\mathbf{x},t;\mathbf{x}',t') \equiv \psi_{(\mathbf{x}',t')}(\mathbf{x},t), \tag{A.3}$$

and refer to it as the conditional probability amplitude, the amplitude for the event of the particle at time t at position $\mathbf{x}$ *given* (with certainty the event of) the particle at time t' was at position $\mathbf{x}'$. The conditional probability is thus expressed in the standard quantum way, i.e. in terms of the absolute square of the corresponding conditional probability amplitude,

$$P(\mathbf{x},t;\mathbf{x}',t') = |K(\mathbf{x},t;\mathbf{x}',t')|^2. \tag{A.4}$$

If, instead, the general situation reigns, where the state of the particle at time t' is specified by an arbitrary wave function, $\psi(\mathbf{x}',t')$, the whereabouts of the particle at time t' is specified by the corresponding probability distribution $P(\mathbf{x}',t') = |\psi(\mathbf{x}',t')|^2$. We then inquire about the joint probability distribution, $P(\mathbf{x},t,\mathbf{x}',t')$, i.e. the probability for the event of the particle at position $\mathbf{x}'$ at time t' *and* at position $\mathbf{x}$ at the later time t. This joint probability distribution is related to the conditional probability and the initial probability according to

$$P(\mathbf{x},t,\mathbf{x}',t') = P(\mathbf{x},t;\mathbf{x}',t')\,P(\mathbf{x}',t'). \tag{A.5}$$

This is the simple probability calculus statement that the probability for the two events to take place equals the probability for the first event to take place *multiplied by* the probability for the second event to take place *given* the first has taken place.

Introducing the joint probability amplitude according to

$$|A(\mathbf{x},t,\mathbf{x}',t')|^2 \equiv P(\mathbf{x},t,\mathbf{x}',t'), \tag{A.6}$$

the relationship (A.5) can then be expressed in terms of the corresponding probability *amplitudes*,

$$|A(\mathbf{x},t,\mathbf{x}',t')|^2 = |K(\mathbf{x},t;\mathbf{x}',t')|^2\,|\psi(\mathbf{x}',t')|^2 = |K(\mathbf{x},t;\mathbf{x}',t')\,\psi(\mathbf{x}',t')|^2, \tag{A.7}$$

and thereby (as always, leaving amplitudes defined only up to a phase factor)

$$A(\mathbf{x},t,\mathbf{x}',t') = K(\mathbf{x},t;\mathbf{x}',t')\,\psi(\mathbf{x}',t'). \tag{A.8}$$

The joint probability amplitude, $A(\mathbf{x},t,\mathbf{x}',t')$, for the considered sequence of particle position events "at time t' particle position is $\mathbf{x}'$" *and* "at time t particle position is $\mathbf{x}$" equals the amplitude for the initial event multiplied by the conditional probability amplitude for the two events in question. The relationship, Eq. (A.8), is solely a consequence of the powers vested in the kinematic principle and time being a parameter describing the evolution of the wave function, and refers therefore only to the particle. It describes the isolated particle quantum dynamics for a sequence of two events, stating that the joint probability amplitude equals the product of the corresponding conditional probability amplitude multiplied by the amplitude for the initial event. We refer to this observation as the two-event multiplication rule.

So far we have learned a new language, how to talk about quantum mechanical systems, the language of assigning probability *amplitudes* to events, and how to ascribe probability amplitudes to a temporal sequence of events. Combining this observation with the superposition principle, the structural form of quantum dynamics can now be established, i.e. the relationship between the wave functions at different times t and t': the specification of $\psi(\mathbf{x},t)$ in terms of $\psi(\mathbf{x},t')$.

A.3 Superposition Principle

The true core of quantum mechanics is the rule for assigning the amplitude to an event that can be effected in indistinguishable ways. More precisely, if an event is realized under conditions without distinction to the alternative ways it can be effected, we must for asserting its probability use the superposition principle as follows.

Superposition principle

If an event can be effected in indistinguishable ways, the amplitude for the event is the sum of the individual amplitudes ψ_i for the alternative ways the event can be effected,

$$\psi = \sum_i \psi_i, \tag{A.9}$$

and the associated probability for the event is therefore

$$P = |\psi|^2 = \left|\sum_i \psi_i\right|^2 = \sum_i |\psi_i|^2 + \sum_{i\neq j} \psi_i\psi_j^*. \tag{A.10}$$

The superposition principle is enforced by the characteristic interference phenomena exhibited in nature, as embodied in the last sum of terms on the right-hand side of Eq. (A.10). In order to illustrate the physical implication of the superposition principle, we consider the double-slit experiment, as illustrated in Figure A.1, where, after being emitted from a source, a particle can arrive at space points (the *detector* screen) in only two ways, *viz.* through either of two holes of an otherwise impenetrable wall.

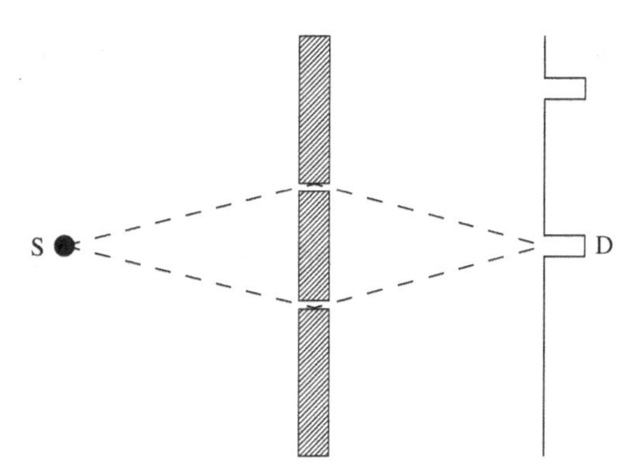

Figure A.1 A particle emitted from the source has two alternatives to reach a location behind the double slit.

The two ways of passing through the double slit are indistinguishable if there is no signature that in principle distinguishes which one of the holes the particle passed through. Arrivals through either hole are thus indistinguishable alternatives, and the individual amplitudes, ψ_1 and ψ_2, for each of the two ways of possible arrival should, according to the superposition principle, be added before squaring to get the probability for the event of arrival. We note that the so-called quantum interference terms in Eq. (A.10), for the present two-alternative case, can be rewritten

$$\psi_1\psi_2^* + \psi_2\psi_1^* = 2\,\Re(\psi_1\psi_2^*) = 2|\psi_1\psi_2|\cos(\varphi_1 - \varphi_2), \qquad \psi_i = |\psi_i|\, e^{i\varphi_i}, \tag{A.11}$$

where modulus and phase for the arrival amplitudes have been introduced. Since the phases, $\varphi_{1,2}$, depend on the alternative paths in question (a dependence on the path length is to be expected), the quantum interference term oscillates, positive and negative, as a function of the position on the detector screen, thereby producing the characteristic oscillating quantum interference arrival pattern.[3] The point of arrival of the particle on the detector screen is unpredictable *in principle*, and a confirmation of the probabilistic assertions of quantum mechanics, that the relative frequency of arrival at places is so and so, requires repeating the experiment. Erratically but surely the interference pattern in accordance with the probabilities predicted by quantum mechanics emerges from such a series of identical experiments. We emphasize the counter-intuitive feature of quantum interference: opening up the second hole through which particles can reach the detector screen thus does not for all detector points increase the number of arriving particles; it can, in fact, according to Eq. (A.11), decrease the number, even to zero. Quantum dynamics thus has wave character, and constructive and destructive interference takes place. In quantum mechanics we are dealing with a new and fundamental type of waves, matter or particle waves described by wave functions. The exotic probabilistic concept of matter waves will have very practical implications, for example, when we consider the motion of an electron in a crystal structure. There the scattering of matter waves by the regularly spaced ions gives rise to propagation for certain states and the absence of propagation for other states of an electron. This gives the all-important distinction between metals and insulators, and so-called

[3] The double-slit experiment has been performed with electrons, neutrons and even large molecules such as C_{60}, and much earlier with photons, unknowingly and in fact in an attempt to disprove the particle nature of light.

semiconductors, the latter having special engineerable properties, the topic discussed in Chapter 10.

A.4 Fundamental Dynamic Equation

Employing the superposition principle of quantum mechanics, we can now establish the *form* of the fundamental dynamic equation for a single quantum particle. At times t' and t the particle is somewhere in space, and according to the kinematic principle has associated the probability amplitude functions $\psi(\mathbf{x}', t')$ and $\psi(\mathbf{x}, t)$ for the events of being at respective but arbitrary places at the times in question. Application of the multiplication rule and the superposition principle specifies the relationship between the conditional probability amplitude $K(\mathbf{x}, t; \mathbf{x}', t')$ and the amplitude functions at different times according to

$$\psi(\mathbf{x}, t) = \int d\mathbf{x}'\; K(\mathbf{x}, t; \mathbf{x}', t')\, \psi(\mathbf{x}', t'). \tag{A.12}$$

According to the multiplication rule, Eq. (A.8), the amplitude to arrive at the position $\mathbf{x}$ at time t via the space-time point $(\mathbf{x}', t')$ equals the product of amplitudes $K(\mathbf{x}, t; \mathbf{x}', t')\, \psi(\mathbf{x}', t')$ for the two events in sequence: event $(\mathbf{x}', t')$ followed by the event of arriving at $(\mathbf{x}, t)$ *given* at $(\mathbf{x}', t')$. Since the particle can arrive at $(\mathbf{x}, t)$ coming via any alternative position $\mathbf{x}'$ at time t', we have to sum over all these indistinguishable alternatives according to the superposition principle. For a continuum space, we have to sum over all the alternative small volumes $\Delta\mathbf{x}'$ via which the particle can arrive, i.e. we must integrate. We thus arrive at the fundamental dynamic equation (A.12). The quantum dynamics of a particle is thus specified in terms of the kernel of the integral equation (A.12), the conditional probability amplitude $K(\mathbf{x}, t; \mathbf{x}', t')$, which is also referred to as the propagator, since it propagates the wave function. The propagator is thus the quantity that completely describes the quantum dynamics of a particle. Once the propagator is known and the wave function at a given moment in time is specified, the future probability amplitudes, and thereby the probabilities for the later whereabouts of the particle, can, according to the fundamental equation specifying the quantum dynamics, Eq. (A.12), be predicted.

At equal times, the propagator, according to Eq. (A.12), is the Dirac delta function (discussed in Appendix B),

$$K(\mathbf{x}, t'; \mathbf{x}', t') = \delta(\mathbf{x} - \mathbf{x}'). \tag{A.13}$$

The physical interpretation of the equal-time conditional probability amplitude, Eq. (A.13), is simply that *given* the particle at time t' is at position $\mathbf{x}'$, then the probability amplitudes for it, at that time, to be at any other position vanish, which is precisely what is expressed by the delta function. The wave function for a particle definitely at position $\mathbf{x}'$ is therefore

$$\psi_{\mathbf{x}'}(\mathbf{x}) \equiv K(\mathbf{x}, t'; \mathbf{x}', t') = \delta(\mathbf{x} - \mathbf{x}'), \tag{A.14}$$

the initial condition for the wave function introduced in Eqs. (A.2) and (A.3), for which the particle at time t' definitely is at position $\mathbf{x}'$.[4]

A.5 Multiplication Rule

Using the dynamical equation (A.12) repeatedly for subsequent times, we obtain the equation (for $t > t_1 > t'$)

$$\psi(\mathbf{x},t) = \int d\mathbf{x}_1 \int d\mathbf{x}' \; K(\mathbf{x},t;\mathbf{x}_1,t_1)\, K(\mathbf{x}_1,t_1;\mathbf{x}',t')\, \psi(\mathbf{x}',t'), \tag{A.15}$$

showing, according to Eq. (A.12), that the propagator has the following property ($t > t_1 > t'$):

$$K(\mathbf{x},t;\mathbf{x}',t') = \int d\mathbf{x}_1 \; K(\mathbf{x},t;\mathbf{x}_1,t_1)\, K(\mathbf{x}_1,t_1;\mathbf{x}',t'). \tag{A.16}$$

This property, *Markovian* dynamics at the amplitude level, reflects the superposition principle, as propagation between the space-time points in question can proceed via any position at some intermediate chosen time.

Repeated use of Eq. (A.12) thus gives for the propagator the equation

$$\begin{aligned} K(\mathbf{x},t;\mathbf{x}',t') = \int d\mathbf{x}_N \cdots \int d\mathbf{x}_2 \int d\mathbf{x}_1 \; & K(\mathbf{x},t;\mathbf{x}_N,t_N)\, K(\mathbf{x}_N,t_N;\mathbf{x}_{N-1},t_{N-1}) \cdots \\ & \cdots K(\mathbf{x}_{N-1},t_{N-1};\mathbf{x}_{N-2},t_{N-2}) \cdots K(\mathbf{x}_1,t_1;\mathbf{x}',t') \end{aligned} \tag{A.17}$$

for a set of N intermediate times $t' < t_1 < t_2 < \cdots < t_N < t$. The propagator between position $\mathbf{x}'$ at time t' and position $\mathbf{x}$ at time t is thus expressed as a sequence of propagations through positions at specified intermediate times, and all possible intermediate positions are then summed over in accordance with the superposition principle, as they constitute indistinguishable alternatives.

According to Eq. (A.17), the sequence of $N+2$ events "particle at subsequent space-time points $(\mathbf{x}',t') \to (\mathbf{x}_1,t_1) \to \cdots \to (\mathbf{x}_N,t_N) \to (\mathbf{x},t)$" can be assigned the amplitude

$$\begin{aligned} &\mathcal{K}(\mathbf{x},t,\mathbf{x}_N,t_N,\ldots,\mathbf{x}_1,t_1,\mathbf{x}',t') \\ &\quad = K(\mathbf{x},t;\mathbf{x}_N,t_N)\, K(\mathbf{x}_N,t_N;\mathbf{x}_{N-1},t_{N-1}) \cdots K(\mathbf{x}_1,t_1;\mathbf{x}',t'), \end{aligned} \tag{A.18}$$

[4] This function is not normalizable, but we shall nevertheless call it a wave function. In fact, the square of a delta *function* cannot be defined in any meaningful way. However, the meaning of Eq. (A.14) as an initial condition is clear. Such mathematical complications, due to considering space a continuum, are really not of importance from a practical point of view, since in fact position cannot be measured with arbitrary accuracy, say, a limit of the size of an atom or usually much larger, as in the case of a grain in a photographic silver emulsion. If space is accordingly discretized into a lattice with this size as lattice constant, we would encounter the Kronecker function instead of the delta function, $K(\mathbf{x},t';\mathbf{x}',t') = \delta_{\mathbf{x},\mathbf{x}'}$, a properly normalized probability amplitude lattice function. For details, see, for example, Chapter One of reference [1].

the multiplication rule for a sequences of events. This assignment we refer to in general as the multiplication rule.[5]

A.6 The Propagator

By employing the superposition principle and the multiplication rule, a suggestive formula for the propagator can be obtained. We can obtain a *path integral* expression for the propagator by the following consideration. At any moment in time, any position in space corresponds to a possible event: the event of the particle at the location in question. Specifying positions at moments in time thus specifies a possible sequence of events. We illustrate each such possible sequence of events pictorially by dots in a space-time plot, as shown in Figure A.2.

The amplitude for such a sequence of events is, according to the multiplication rule, specified by the corresponding product of propagators, the expression in Eq. (A.18). The path in space, $\mathbf{x}_t$, connecting the space points corresponding to the space-time path in Figure A.2 (its projection onto space), is thus the erratic path, which has definite directions $\mathbf{x}_{n+1} - \mathbf{x}_n$ (or *velocities* $(\mathbf{x}_{n+1} - \mathbf{x}_n)/(t_{n+1} - t_n)$) on each path segment, jumping discontinuously to another velocity on the next straight-line segment, etc. At the intermediate times, where the event corresponds to that of a definite particle position, the velocity of the particle is thus undefined. Each such alternative space-time event sequence thus corresponds to a continuous path in space, $\mathbf{x}_t$, having an associated probability (density) amplitude, the expression in Eq. (A.18), which for short we denote $A_{\mathbf{x},t;\mathbf{x}',t'}[\mathbf{x}_t]$ (the parametric dependence on the fixed starting and end points is suppressed in the following).

The particle is left unobserved and could equally well have propagated along any other alternatives of paths, two of which are depicted in Figure A.3.

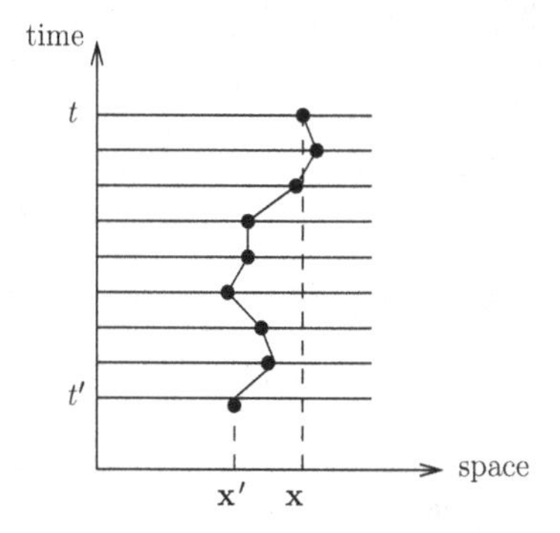

Figure A.2 A sequence of space-time points, possible events, and their associated straight-line space-time path.

[5] Taking the absolute square of Eq. (A.18) gives the probability for this sequence of events. However, on the whole set of such *histories*, one cannot consistently ascribe a probability measure due to quantum interference. See, for example, Chapter One in reference [1].

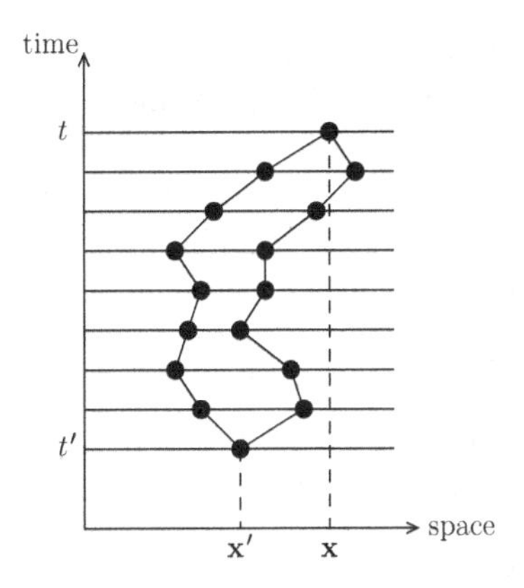

Figure A.3 Two alternative paths connecting space-time points $(\mathbf{x}', t')$ and $(\mathbf{x}, t)$.

According to the superposition principle, the conditional probability amplitude for the particle, $K(\mathbf{x}, t; \mathbf{x}', t')$, is the sum of the amplitudes for all the alternative indistinguishable possible paths connecting the space-time points $(\mathbf{x}', t')$ and $(\mathbf{x}, t)$,[6]

$$K(\mathbf{x}, t; \mathbf{x}', t') = \sum_{\mathbf{x}_t} A[\mathbf{x}_t]. \tag{A.19}$$

A path $\mathbf{x}_t$ is represented by the sequence of its successive straight-line paths, and each such sub-path $\mathbf{x}_t^{(i)}$ has associated a conditional amplitude $A[\mathbf{x}_t^{(i)}]$ for propagation between the two space-time points in question, *viz.* the corresponding propagator. According to its successive construction, a path $\mathbf{x}_t$ is the sequence or *sum* of its sub-paths $\mathbf{x}_t^{(i)}$, i.e. $\mathbf{x}_t$"="$\sum_i \mathbf{x}_t^{(i)}$, and the amplitude for the path $\mathbf{x}_t$ is, according to the multiplication rule, or explicitly the expression in Eq. (A.18), the product of the sequential amplitudes of its constituent sub-paths,

$$A[\mathbf{x}_t] = A\left[\sum_i \mathbf{x}_t^{(i)}\right] = \prod_i A[\mathbf{x}_t^{(i)}]. \tag{A.20}$$

The amplitude for a path is therefore of the exponential form

$$A[\mathbf{x}_t] = \exp\left\{\frac{i}{[S]} S_{\mathbf{x}t\mathbf{x}'t'}[\mathbf{x}_t]\right\}, \tag{A.21}$$

where $S_{\mathbf{x}t\mathbf{x}'t'}[\mathbf{x}_t]$ is a functional of the path,[7] and, according to Eq. (A.20), equals the sum of the contributions from its constituent sub-paths,

$$S_{\mathbf{x}t\mathbf{x}'t'}[\mathbf{x}_t] = \sum_i S_{\mathbf{x}_{i+1}t_{i+1}\mathbf{x}_i t_i}[\mathbf{x}_t^{(i)}]. \tag{A.22}$$

We call this functional the action. The presence of the imaginary unit, demanded by the conservation of the total probability – the particle must be somewhere in space at all

[6] Summing over paths is an infinite-dimensional continuum summation, called a path integral, and should be approached with some care. However, for the following discussion, an intuitive understanding of a sum over paths will suffice.

[7] A *functional* maps a function onto a number, $\mathbf{x}_t \to S[\mathbf{x}_t]$, explaining the use of the square bracket.

times – then requires the action to be a real functional. If the action had an imaginary part, for any path we would be multiplying the complex numbers for each sub-path each having modulus different from one. For space-time points separated by a large time difference, we would end up with a wave function that either vanishes or becomes infinity, in either case violating the requirement of normalization. The quantity $[S]$ is inserted in order to make the exponent dimensionless, i.e. to account for an eventual dimension of the action (we are doing *physics* and physical quantities carry dimensions). We immediately introduce the short-hand notation $\hbar \equiv [S]$ to denote the possible new constant of nature specifying quantum dynamics.

We have thus arrived at the expression

$$\begin{aligned} A_{\mathbf{x},t;\mathbf{x}',t'}[\mathbf{x}_t] &\equiv \mathcal{K}(\mathbf{x},t,\mathbf{x}_N,t_N,\ldots,\mathbf{x}_1,t_1,\mathbf{x}',t') \\ &= K(\mathbf{x},t;\mathbf{x}_N,t_N)\,K(\mathbf{x}_N,t_N;\mathbf{x}_{N-1},t_{N-1})\cdots K(\mathbf{x}_1,t_1;\mathbf{x}',t') \\ &= \exp\{S_{\mathbf{x}t\mathbf{x}_Nt_N}[\mathbf{x}_t^{(N)}]\}\,\exp\{S_{\mathbf{x}_Nt_N\mathbf{x}_{N-1}t_{N-1}}[\mathbf{x}_t^{(N-1)}]\}\cdots \\ &\quad\cdots\exp\{S_{\mathbf{x}_1t_1\mathbf{x}'t'}[\mathbf{x}_t^{(1)}]\} \end{aligned} \tag{A.23}$$

for propagation along a path.

Since the actions of sub-paths are additive, Eq. (A.22), the action for an infinitesimal straight-line segment is proportional to its time step Δt and is further uniquely characterized by the segment's position $\mathbf{x}_t$ and direction $\dot{\mathbf{x}}_t$. The action for an isolated particle can therefore be written as an integral over a function $L(\mathbf{x}_t, \dot{\mathbf{x}}_t)$ we call the Lagrangian,

$$S[\mathbf{x}_t] = \sum_i S_i = \sum_i \Delta t_i L(\mathbf{x}_{t_i}, \dot{\mathbf{x}}_{t_i}) = \int_{t'}^{t} d\bar{t}\; L(\mathbf{x}_{\bar{t}}, \dot{\mathbf{x}}_{\bar{t}}). \tag{A.24}$$

We have thus arrived at the following expression for the conditional probability amplitude:[8]

$$K(\mathbf{x},t;\mathbf{x}',t') = \sum_{\mathbf{x}_t} \exp\left\{\frac{i}{\hbar} S_{\mathbf{x}t\mathbf{x}'t'}[\mathbf{x}_t]\right\} = \sum_{\mathbf{x}_{\bar{t}}} \exp\left\{\frac{i}{\hbar}\int_{t'}^{t} d\bar{t}\; L(\mathbf{x}_{\bar{t}}, \dot{\mathbf{x}}_{\bar{t}}, \bar{t})\right\}. \tag{A.25}$$

Each alternative path contributes to the propagator through a phase factor determined by the action of the path, an as yet unknown functional of the path, however. The contribution from each alternative path to the propagator can thus be represented by a two-dimensional unit arrow, a complex number of unit length, equivalent to the position of the dial on a stop watch. The assigned probability amplitude to a given path can thus be viewed as if the particle has associated an internal clock winding around as it propagates along the path, and the final direction of the associated arrow is specified by the action of the path in question.

This is how far the principles for ascribing probability amplitudes can take us. To get a handle on the analytical form of the action or equivalently the Lagrangian, we can appeal to the empirical fact that the massive objects we see around us follow the trajectory dictated by Newton's equation. Under circumstances of a heavy mass, quantum dynamics must

[8] The situation where the particle is not isolated but interacting with a time-dependent classical field is handled with equal ease, simply leaving the Lagrangian explicitly dependent on time, $L(\mathbf{x}_t, \dot{\mathbf{x}}_t, t)$, say, due to the presence of a classical time-dependent potential $V(\mathbf{x}, t)$. A classical field does not represent any observation of the particle; the field is not influenced by the whereabouts of the particle.

thus reduce to classical motion of certainty along the classical path, say, for the motion in a potential $V(\mathbf{x})$.

A.7 Action and Correspondence Limit

Evidently, the motion of the center of mass of a large object, say, a billiard ball dropped from a tree, follows the trajectory dictated by Newton's equation. Such a state of affairs results from the path integral expression, Eq. (A.25), if only the classical path contributes to the path integral. This, in turn, is achieved if we choose the action for any path as that of the classical action expression offered by classical mechanics, and it so happens that the quantum of action is small compared to the action of the classical path, $S_{\mathbf{x}t\mathbf{x}'t'}[\mathbf{x}_t^{\mathrm{cl}}] \gg \hbar$. We recall that the classical path is determined by stationarity of the classical action,[9]

$$\left.\frac{\delta S}{\delta \mathbf{x}_t}\right|_{\mathbf{x}_t=\mathbf{x}_t^{\mathrm{cl}}} = \mathbf{0}. \tag{A.26}$$

This equation states that a path deviating slightly from the classical path has the same action as the classical path to lowest order in the deviation. Therefore, only paths very close to the stationary one will in that case contribute to the sum over paths in Eq. (A.25). Contributions from paths deviating from the classical path cancel each other on account of the rapidly oscillating phase factor since, according to the stationarity condition, $S_{\mathbf{x}t\mathbf{x}'t'}[\mathbf{x}_t] > S_{\mathbf{x}t\mathbf{x}'t'}[\mathbf{x}_t^{\mathrm{cl}}] \gg \hbar$ (assuming here that the classical path minimizes the action, as is typically the case, otherwise the magnitudes of the actions enter the inequality). Consider the paths in a space-time plot, such as Figure A.3, specified by passing a certain volume at a certain time, and assume that the classical path does not cross this chosen volume. Then the multitude of these paths are assigned phases $S_{\mathbf{x}t\mathbf{x}'t'}[\mathbf{x}_t]/\hbar \gg 1$, and for these paths we are adding up arrows on the unit circle pointing in all directions and therefore adding up to zero. The contributions from paths away from the classical path thus, by interfering destructively, make no contribution to the sum over paths. In this way it is ascertained that probabilities for the particle to be found at places other than those dictated by classical mechanics are vanishingly small. The conditional probability amplitude is then, in the quasi-classical case, $S_{\mathbf{x}t\mathbf{x}'t'}[\mathbf{x}_t^{\mathrm{cl}}] \gg \hbar$, specified by the contribution from the classical path,

$$K_{\mathrm{cl}}(\mathbf{x}, t; \mathbf{x}', t') = A\, e^{(i/\hbar)S_{\mathrm{cl}}(\mathbf{x},t;\mathbf{x}',t')}, \tag{A.27}$$

where S_{cl} denotes the action for the classical path, $S_{\mathrm{cl}}(\mathbf{x}, t; \mathbf{x}', t') \equiv S_{\mathbf{x}t\mathbf{x}'t'}[\,\mathbf{x}_t^{\mathrm{cl}}]$, and the prefactor A reflects that contributions from a tube of paths near the stationary classical path add up constructively.[10]

[9] If a quick reminder of classical mechanics, Newton's law and its equivalent Lagrange and Hamilton formulations is needed, we refer to Appendix D.

[10] For the reader not satisfied with this heuristic argument, we note that when $S_{\mathbf{x}t\mathbf{x}'t'}[\mathbf{x}_t^{\mathrm{cl}}] \gg \hbar$, the path integral can with sufficient accuracy be evaluated in the stationary phase approximation, which amounts to having only the classical path contribute to the path integral. See, for example, Appendix A in reference [1].

From correspondence with the classical limit, it thus follows that the action appearing in Eq. (A.25) is the classical action, and consequently that $\hbar$ has the dimension of action, i.e. has the dimension of energy multiplied by time. This constant of nature characterizing quantum mechanics is called the quantum of action or Planck's constant. Quantum theory does not fix its value; it has to be taken from comparison of theoretical predictions with experimental data. As follows from the next two sections, we are in fact now empowered with an explicit equation for the quantum dynamics of particles and could start calculating physical properties, say, calculate energy levels for the hydrogen atom. Then, comparing the properties of the spectroscopic data, and using measured values of the mass and charge of the electron, would give us the ability to determine the value of $\hbar$. Experiments of highest accuracy (employing the Josephson and quantum Hall effects) set the value at $\hbar = 1.054571596(82) \times 10^{-34}$ J s. This is, indeed, a small value compared to common experienced actions: moving a mass of 1 g the distance 1 cm in 1 s requires the action 5×10^{-8} J s. The correspondence argument is thus consistent.

In accordance with the correspondence principle, we have thus established that the quantum dynamics of a particle is determined by the propagator

$$K(\mathbf{x},t;\mathbf{x}',t') = \sum_{\mathbf{x}_t} e^{(i/\hbar)S_{\mathbf{x}t\mathbf{x}'t'}[\mathbf{x}_t]} = \sum_{\mathbf{x}_{\bar{t}}} e^{(i/\hbar)\int_{t'}^{t} d\bar{t}\, L(\mathbf{x}_{\bar{t}},\dot{\mathbf{x}}_{\bar{t}})}, \tag{A.28}$$

where S is the functional specified by the classical action, i.e. the Lagrangian, L, is Lagrange's function of classical mechanics. Each alternative path contributes to the propagator by a phase factor determined by inserting the path in the expression specified by the classical action. We stress that in quantum mechanics, where actions are on the order of $\hbar$, not only the classical but *all* paths in general contribute to the propagator, the *superposition principle* of quantum mechanics.

A.8 Infinitesimal Time Step Propagator

From the path integral expression for the propagator, Eq. (A.28), a differential characterization of the dynamics can be obtained, the Schrödinger equation, even without explicit knowledge of *how to sum over paths*. We only need to know the propagator for an infinitesimal time step. For an infinitesimal time step, only the straight-line path connecting the two space points in question (denote them $\mathbf{x}'$ and $\mathbf{x}$) contributes to the path integral, since no intermediate events are to be considered, and we have for the propagator for an infinitesimal time

$$K(\mathbf{x}, t+\Delta t; \mathbf{x}', t) \propto \exp\left\{\frac{i}{\hbar}\,\Delta t\, L\left(\mathbf{x}, \frac{\mathbf{x}-\mathbf{x}'}{\Delta t}, t\right)\right\}, \tag{A.29}$$

where the proportionality sign again reflects our failure regarding how to assign the measure for the density of paths involved in the path integral. Appealing to the correspondence principle has already given us knowledge of the Lagrangian, and considering for definiteness the case of a particle of mass m moving in a potential V for which the Lagrange function is

$$L(\mathbf{x}_t, \dot{\mathbf{x}}_t, t) = \tfrac{1}{2} m \dot{\mathbf{x}}_t^2 - V(\mathbf{x}_t, t), \tag{A.30}$$

we have

$$K(\mathbf{x}, t + \Delta t; \mathbf{x}', t) \propto \exp\left\{ \frac{i}{\hbar} \frac{m(\mathbf{x} - \mathbf{x}')^2}{2\Delta t} - \frac{i}{\hbar} \Delta t \, V(\mathbf{x}, t) \right\}. \tag{A.31}$$

We can now justify our laxity regarding how to *sum over paths*, because the prefactor, turning the proportionality sign into equality in Eq. (A.31), is easily assessed. The potential term smoothly produces the factor unity as the time step vanishes, $\Delta t \to 0$, but the kinetic energy term becomes pathological, a unit arrow winding around like a crazy clock. However, we can arrest this behavior by recalling that at equal times the propagator, according to Eq. (A.13), should reduce to the Dirac delta function. The prefactor in Eq. (A.31) is therefore immediately asserted (see the representation of the delta function, Eq. (B.18) of Appendix B), and the propagator for an infinitesimal time step is (in three spatial dimensions)

$$K(\mathbf{x}, t + \Delta t; \mathbf{x}', t) = \left(\frac{m}{2\pi i \hbar \Delta t} \right)^{3/2} \exp\left\{ \frac{i}{\hbar} \frac{m(\mathbf{x} - \mathbf{x}')^2}{2\Delta t} - \frac{i}{\hbar} \Delta t \, V(\mathbf{x}, t) \right\}. \tag{A.32}$$

Exercise A.1 The Lagrange function for two particles in external fields and interacting through a scalar potential is

$$\begin{aligned} &L(\mathbf{x}_2(t), \dot{\mathbf{x}}_2(t), \mathbf{x}_1(t), \dot{\mathbf{x}}_1(t), t) \\ &= \tfrac{1}{2} m_1 \dot{\mathbf{x}}_1^2(t) - V_1(\mathbf{x}_1(t), t) - V(\mathbf{x}_2(t), \mathbf{x}_1(t)) + \tfrac{1}{2} m_2 \dot{\mathbf{x}}_2^2(t) - V_2(\mathbf{x}_2(t), t), \end{aligned} \tag{A.33}$$

as Lagrange's equations are Newton's equations for the system in question.

Show that the infinitesimal two-particle propagator for the system is

$$\begin{aligned} &K(\mathbf{x}_2, \mathbf{x}_1, t + \Delta t; \mathbf{x}_2', \mathbf{x}_1', t) \\ &= \left(\frac{m_2}{2\pi i \hbar \Delta t} \right)^{3/2} \exp\left\{ \frac{i}{\hbar} \frac{m_2(\mathbf{x}_2 - \mathbf{x}_2')^2}{2\Delta t} - \frac{i}{\hbar} \Delta t \, V_2(\mathbf{x}_2, t) \right\} \\ &\times \left(\frac{m_1}{2\pi i \hbar \Delta t} \right)^{3/2} \exp\left\{ \frac{i}{\hbar} \frac{m_1(\mathbf{x}_1 - \mathbf{x}_1')^2}{2\Delta t} - \frac{i}{\hbar} \Delta t \, V_1(\mathbf{x}_1, t) \right\} \\ &\times \exp\left\{ -\frac{i}{\hbar} \Delta t \, V(\mathbf{x}_2, \mathbf{x}_1, t) \right\}. \end{aligned} \tag{A.34}$$

A.9 The Schrödinger Equation

With the knowledge of the infinitesimal time step propagator, the Schrödinger equation can now be derived from the fundamental equation for the dynamics, Eq. (A.12), considering an infinitesimal time step. The wave function is propagated an infinitesimal time step by inserting into Eq. (A.12) the propagator for an infinitesimal time step, Eq. (A.32), whereby we obtain

$$\psi(\mathbf{x}, t+\Delta t) = \left(\frac{m}{2\pi i\hbar\Delta t}\right)^{d/2} \int d\mathbf{x}' \exp\left\{\frac{i}{\hbar}\frac{m(\mathbf{x}-\mathbf{x}')^2}{2\Delta t} - \frac{i}{\hbar}\Delta t\, V(\mathbf{x}, t)\right\} \psi(\mathbf{x}', t). \quad \text{(A.35)}$$

Introducing the variable $\bar{\mathbf{x}} = \mathbf{x} - \mathbf{x}'$ we obtain[11]

$$\psi(\mathbf{x}, t+\Delta t) = e^{-(i/\hbar)\Delta t\, V(\mathbf{x},t)} \left(\frac{m}{2\pi i\hbar\Delta t}\right)^{d/2} \int d\bar{\mathbf{x}} \exp\left\{\frac{i}{\hbar}\frac{m\bar{\mathbf{x}}^2}{2\Delta t}\right\} \psi(\mathbf{x} - \bar{\mathbf{x}}, t). \quad \text{(A.36)}$$

The exponential part of the integrand is wildly oscillating for values of $|\bar{\mathbf{x}}|$ much larger than the length scale $\sqrt{\Delta t\hbar/m}$, and in the limit $\Delta t \to 0$ this therefore suggests that we evaluate the integral by Taylor expansion of the wave function. Taylor-expanding on both sides, on the left with respect to time and on the right with respect to the spatial coordinate, gives (with only terms resulting in linear order terms in Δt displayed)

$$\begin{aligned} &\psi(\mathbf{x}, t) + \Delta t \frac{\partial \psi(\mathbf{x}, t)}{\partial t} + \cdots \\ &= \left(\frac{m}{2\pi i\hbar\Delta t}\right)^{d/2} \left[1 - \frac{i}{\hbar}\Delta t V(\mathbf{x}, t) + \cdots\right] \\ &\times \int d\bar{\mathbf{x}} \exp\left\{\frac{i}{\hbar}\frac{m\bar{\mathbf{x}}^2}{2\Delta t}\right\} \left[\psi(\mathbf{x}, t) - \bar{\mathbf{x}} \cdot \frac{\partial \psi(\mathbf{x}, t)}{\partial \mathbf{x}} + \frac{1}{2}\bar{x}_\alpha \frac{\partial^2 \psi(\mathbf{x}, t)}{\partial x_\alpha \partial x_\beta}\bar{x}_\beta + \cdots\right], \end{aligned} \quad \text{(A.37)}$$

where in the last factor summation over repeated Cartesian indices is implied.

Consider the term generated when the factor one is chosen in the first brackets and the function $\psi(\mathbf{x}, t)$ in the second brackets. Both factors can be taken outside the integral, which is with respect to $\bar{\mathbf{x}}$, and the integral simply produces the inverse of the first factor on the right-hand side of the equation, and this term thus cancels the first term on the left-hand side of Eq. (A.37). An exact equation for the time derivative of the wave function can now be obtained by evaluating all terms to linear order in Δt on the right-hand side of Eq. (A.37). The term proportional to $\psi(\mathbf{x}, t)\, V(\mathbf{x}, t)$ is proportional to Δt. The term containing the first spatial derivative of the wave function ψ is multiplied by an integral having an integrand that is an odd function in the integration variable $\bar{\mathbf{x}}$ and therefore vanishes. The

[11] It is useful to do the following calculation first for the one-dimensional case, and then observe that additional spatial dimensions are just repetitive.

term containing the second spatial derivative of the wave function is proportional to the easily evaluated integral

$$\int d\bar{\mathbf{x}}\, \bar{x}_\alpha\, \bar{x}_\beta \exp\left\{\frac{i}{\hbar}\frac{m\bar{\mathbf{x}}^2}{2\Delta t}\right\} = \delta_{\alpha,\beta}\int d\bar{\mathbf{x}}\, x_\alpha^2 \exp\left\{\frac{i}{\hbar}\frac{m\bar{\mathbf{x}}^2}{2\Delta t}\right\}$$

$$= \left(\frac{\hbar\Delta t}{m}\left(\frac{2\pi i\hbar\Delta t}{m}\right)^{1/2}\right)\left(\frac{2\pi i\hbar\Delta t}{m}\right)^{(d-1)/2}, \tag{A.38}$$

where the first factor in parentheses results from the α direction integral and the other factor is from the other Cartesian directions. The term containing the second spatial derivative of the wave function ψ thus also generates, in view of the first factor on the right-hand side of Eq. (A.37), a term linear in Δt, and the further terms generated by the Taylor expansion of the wave function are of order $(\Delta t)^2$ and higher.[12] Thus performing the Gaussian integrals, and letting Δt approach zero after dividing by it, we obtain the Schrödinger equation for a particle in a potential,

$$i\hbar\frac{\partial\psi(\mathbf{x},t)}{\partial t} = \left(-\frac{\hbar^2}{2m}\frac{\partial^2}{\partial\mathbf{x}^2} + V(\mathbf{x},t)\right)\psi(\mathbf{x},t). \tag{A.39}$$

We note that the Hamiltonian is obtained by substituting in Hamilton's function (see Appendix D) for the momentum the differential operator

$$\hat{H}(t) = H\left(\mathbf{x}, \frac{\hbar}{i}\nabla_{\mathbf{x}}, t\right). \tag{A.40}$$

This is a very efficient formulation of the correspondence between classical and quantum mechanics: performing in Hamilton's function the substitution

$$\mathbf{p} \to \frac{\hbar}{i}\nabla_{\mathbf{x}}, \tag{A.41}$$

the canonical quantization rule.

Exercise A.2 In the presence of both a scalar and vector potential, $\phi(\mathbf{x},t)$ and $\mathbf{A}(\mathbf{x},t)$, the Hamiltonian is, according to Eq. (D.20) and the canonical quantization rule, given by

$$\hat{H} = \frac{1}{2m}\left(\frac{\hbar}{i}\nabla_{\mathbf{x}} - q\mathbf{A}(\mathbf{x},t)\right)^2 + q\phi(\mathbf{x},t). \tag{A.42}$$

Show that the probability current density entering the continuity equation for a particle of charge q becomes

[12] All the integrals are performed by recalling the Gaussian integral $\int_{-\infty}^{\infty} dx\, x^{2n}\, e^{-ax^2} = \sqrt{\pi}\, d^n a^{-1/2}/da^n \propto a^{-1/2-n} \propto \Delta t^{n+1/2}$. The term with $n = 1$ is thus of order $\Delta t^{1+1/2}$ where $\Delta t^{1/2}$ cancels the prefactor associated with each Cartesian direction, and the term is of order Δt. This is the order at which we need to evaluate the terms in Eq. (A.37), since we eventually divide by Δt on both sides of the equation and let Δt approach zero, $\Delta t \to 0$, making second- and higher-order terms in Δt disappear in this limit.

$$\mathbf{S}(\mathbf{x},t) = \frac{\hbar}{2mi}[\psi^*(\mathbf{x},t)\nabla_{\mathbf{x}}\psi(\mathbf{x},t) - \psi(\mathbf{x},t)\nabla_{\mathbf{x}}\psi^*(\mathbf{x},t)] - \frac{q}{m}\mathbf{A}(\mathbf{x},t)\,|\psi(\mathbf{x},t)|^2. \tag{A.43}$$

The presence of the vector potential leads to the appearance of the so-called diamagnetic current term where the vector potential enters explicitly.

Exercise A.3 Show that the infinitesimal time step two-particle propagator considered in Exercise A.1 leads to the two-particle Schrödinger equation

$$i\hbar\frac{\partial\psi(\mathbf{x}_2,\mathbf{x}_1,t)}{\partial t} = \hat{H}\psi(\mathbf{x}_2,\mathbf{x}_1,t), \tag{A.44}$$

where the Hamiltonian is

$$\begin{aligned}\hat{H} = &-\frac{\hbar^2}{2m_1}\frac{\partial^2}{\partial\mathbf{x}_1^2} + V_1(\mathbf{x}_1,t) + V(\mathbf{x}_1,\mathbf{x}_2)\\ &-\frac{\hbar^2}{2m_2}\frac{\partial^2}{\partial\mathbf{x}_2^2} + V_2(\mathbf{x}_2,t).\end{aligned} \tag{A.45}$$

When one is considering a single-particle situation, such as when interpreting the double-slit experiment, an interpretation in terms of a matter wave in space seems a natural description. However, one should bear in mind that, for a many-particle system, one is dealing with a description in a configuration space usually of incredibly many dimensions.

Consider a normalized two-particle state, $\psi(\mathbf{x}_1,\mathbf{x}_2,t)$. Using the Schrödinger equation and noting that only the Laplacian terms are uncanceled gives, analogous to Eq. (1.36) for $P(\mathbf{x}_1,\mathbf{x}_2,t) = |\psi(\mathbf{x}_1,\mathbf{x}_2,t)|^2$, that

$$\begin{aligned}\frac{\partial P(\mathbf{x}_1,\mathbf{x}_2,t)}{\partial t} = &-\frac{\hbar}{im}\nabla_{\mathbf{x}_1}\cdot\left(\psi^*(\mathbf{x}_1,\mathbf{x}_2,t)\ \overleftrightarrow{\nabla}_{\mathbf{x}_1}\ \psi(\mathbf{x}_1,\mathbf{x}_2,t)\right)\\ &-\frac{\hbar}{im}\nabla_{\mathbf{x}_2}\cdot\left(\psi^*(\mathbf{x}_1,\mathbf{x}_2,t)\ \overleftrightarrow{\nabla}_{\mathbf{x}_2}\ \psi(\mathbf{x}_1,\mathbf{x}_2,t)\right),\end{aligned} \tag{A.46}$$

where we have introduced the abbreviated notation

$$\overleftrightarrow{\nabla}_{\mathbf{x}} = \frac{1}{2}\left(\frac{\overrightarrow{\partial}}{\partial\mathbf{x}} - \frac{\overleftarrow{\partial}}{\partial\mathbf{x}}\right) \tag{A.47}$$

for the differential operator associated with the current operator. The average particle density

$$n(\mathbf{x},t) = \int d\mathbf{x}_2\ P(\mathbf{x},\mathbf{x}_2,t) + \int d\mathbf{x}_1\ P(\mathbf{x}_1,\mathbf{x},t) \tag{A.48}$$

satisfies

$$\frac{\partial n(\mathbf{x},t)}{\partial t} = -\frac{\hbar}{im}\int d\mathbf{x}_2\, \nabla_{\mathbf{x}} \cdot \left(\psi^*(\mathbf{x},\mathbf{x}_2,t)\ \overleftrightarrow{\nabla}_{\mathbf{x}}\ \psi(\mathbf{x},\mathbf{x}_2,t)\right)$$
$$-\frac{\hbar}{im}\int_S d\mathbf{s}_2 \cdot \psi^*(\mathbf{x},\mathbf{x}_2,t)\ \overleftrightarrow{\nabla}_{\mathbf{x}_2}\ \psi(\mathbf{x},\mathbf{x}_2,t)$$
$$-\frac{\hbar}{im}\int_S d\mathbf{s}_1 \cdot \psi^*(\mathbf{x}_1,\mathbf{x},t)\ \overleftrightarrow{\nabla}_{\mathbf{x}_1}\ \psi(\mathbf{x}_1,\mathbf{x},t)$$
$$-\frac{\hbar}{im}\int d\mathbf{x}_1\, \nabla_{\mathbf{x}} \cdot \left(\psi^*(\mathbf{x}_1,\mathbf{x},t)\ \overleftrightarrow{\nabla}_{\mathbf{x}}\ \psi(\mathbf{x}_1,\mathbf{x},t)\right), \tag{A.49}$$

where the two middle terms on the right-hand side have been rewritten by using Gauss's theorem, and are seen to vanish.

If $\psi(\mathbf{x}_1,\mathbf{x}_2,t) = \psi_1(\mathbf{x}_1,t)\,\psi_2(\mathbf{x}_2,t)$, the total current density is then simply the sum of the two single-state currents,

$$\mathbf{j}(\mathbf{x},t) = \frac{\hbar}{im}\,\psi_1^*(\mathbf{x},t)\ \overleftrightarrow{\nabla}_{\mathbf{x}}\ \psi_1(\mathbf{x},t) + \frac{\hbar}{im}\,\psi_2^*(\mathbf{x},t)\ \overleftrightarrow{\nabla}_{\mathbf{x}}\ \psi_2(\mathbf{x},t). \tag{A.50}$$

A.10 Evolution Operator

Let us obtain the formal solution of the Schrödinger equation for the case when the Hamiltonian is time-independent, i.e. for an isolated system or a system in time-independent fields. For a small time step $t = t' + \Delta t$, we have from the Schrödinger equation

$$\psi(\mathbf{x},t) = \psi(\mathbf{x},t') + \frac{\Delta t}{i\hbar}\hat{H}\,\psi(\mathbf{x},t') + \mathcal{O}(\Delta t^2) = \left(1 + \frac{\Delta t}{i\hbar}\hat{H}\right)\psi(\mathbf{x},t') + \mathcal{O}(\Delta t^2), \tag{A.51}$$

where $\mathcal{O}(\Delta t^2)$ signifies all terms beyond linear order. To generate a finite displacement in time, we use the tactic of small steps: stepping repeatedly back in time, we obtain successively using the Schrödinger equation, with $\Delta t = (t - t')/N$,

$$\psi(\mathbf{x},t) = \left(\hat{I} + \frac{\Delta t}{i\hbar}\hat{H}\right)\psi(\mathbf{x},t_{N-1}) = \left(1 + \frac{\Delta t}{i\hbar}\hat{H}\right)^N \psi(\mathbf{x},t') \equiv \hat{U}_N(t,t')\,\psi(\mathbf{x},t'), \tag{A.52}$$

where higher-order terms in Δt have been dropped, since we eventually take the limit of large N, the limit $\Delta t \to 0$ *and* $N \to \infty$ in such a way that their product is constant, $\Delta t N = t - t'$. Introducing in this way still shorter time steps, we obtain for the time evolution operator

$$\psi(\mathbf{x},t) = \hat{U}(t,t')\,\psi(\mathbf{x},t') \tag{A.53}$$

the expression

$$\hat{U}(t,t') = \lim_{N\to\infty}\hat{U}_N(t,t') = \lim_{N\to\infty}\left(1 + \frac{\Delta t}{i\hbar}\hat{H}\right)^N = \lim_{N\to\infty}\left(1 + \frac{t-t'}{i\hbar N}\hat{H}\right)^N. \tag{A.54}$$

Recalling the binomial formula[13]

$$(a+b)^N = \sum_{n=0}^{N} \binom{N}{n} a^{N-n}\, b^n \tag{A.55}$$

gives

$$\hat{U}_N(t,t') = \sum_{n=0}^{\infty} \binom{N}{n} 1^{N-n} \left(\frac{\Delta t}{i\hbar}\hat{H}\right)^n . \tag{A.56}$$

Considering the nth term in the sum

$$\begin{aligned}\binom{N}{n}\left(\frac{\Delta t}{i\hbar}\hat{H}\right)^n &= \frac{N(N-1)\cdots(N-(n-1))}{n!}\left(\frac{\Delta t}{i\hbar}\hat{H}\right)^n \\ &= \frac{1}{n!}\frac{N(N-1)\cdots(N-(n-1))}{N^n}\left(\frac{t-t'}{i\hbar}\hat{H}\right)^n ,\end{aligned} \tag{A.57}$$

in the limit $N \to \infty$ it becomes

$$\lim_{N\to\infty}\frac{1}{n!}\frac{N(N-1)\cdots(N-(n-1))}{N^n}\left(\frac{t-t'}{i\hbar}\hat{H}\right)^n = \frac{1}{n!}\left(\frac{t-t'}{i\hbar}\hat{H}\right)^n \tag{A.58}$$

and we have

$$\begin{aligned}\hat{U}(t,t') &= \lim_{N\to\infty}\hat{U}_N(t,t') = \lim_{N\to\infty}\left(1+\frac{\Delta t}{i\hbar}\hat{H}\right)^N \\ &= 1+\frac{t-t'}{i\hbar}\hat{H}+\frac{1}{2!}\left(\frac{t-t'}{i\hbar}\right)^2\hat{H}^2+\frac{1}{3!}\left(\frac{t-t'}{i\hbar}\right)^3\hat{H}^3+\cdots \\ &= e^{-(i/\hbar)\hat{H}(t-t')},\end{aligned} \tag{A.59}$$

where in the last line the exponential of the Hamiltonian is defined in terms of its series expansion. The energy operator is the generator of displacements in time.

[13] Since only one operator is involved, the algebra is identical to that of numbers and results can be taken over.

Appendix B: Dirac Delta Function

The Dirac delta function, δ, is customarily defined according to

$$\int d\mathbf{x}\, \delta(\mathbf{x}) f(\mathbf{x}) = f(\mathbf{0}) \tag{B.1}$$

for arbitrary continuous functions $f(\mathbf{x})$. However, it is a caricature of a function, zero everywhere except at $\mathbf{x} = \mathbf{0}$, where it is infinite. Intuitively, the Dirac delta function is a very peaked function and we shall represent it as a limiting function of a series of functions being progressively peaked.

A sequence of functions, f_n, that becomes progressively peaked at a certain value, say, at zero, so that, for any smooth function ϕ,

$$\lim_{n\to\infty} \int_{-\infty}^{\infty} dx\, f_n(x)\phi(x) = \phi(0), \tag{B.2}$$

is called a Dirac sequence.

A Dirac sequence f_n can be made from any normalized function f,

$$\int_{-\infty}^{\infty} dx\, f(x) = 1, \tag{B.3}$$

by contracting a normalized function and making it progressively peaked,

$$f_n(x) = nf(nx). \tag{B.4}$$

The sequence will then have the weak convergence property, Eq. (B.2), for any smooth function ϕ, since for sufficiently large n we have that the integral is completely dominated by the *peakedness* of f_n at the origin, or, by change of variable, we can run the test function to its value at zero,

$$\begin{aligned}\lim_{n\to\infty} \int_{-\infty}^{\infty} dx\, \phi(x) f_n(x) &= \lim_{n\to\infty} \int_{-\infty}^{\infty} dx\, \phi\left(\frac{x}{n}\right) f(x) \\ &= \phi(0) \int_{-\infty}^{\infty} dx\, f(x) = \phi(0).\end{aligned} \tag{B.5}$$

We shall cavalierly take the limiting procedure before the integration,[1] and use for the limiting function the notation

$$\lim_{n\to\infty} f_n(x) = \delta(x), \tag{B.6}$$

[1] Rigorous justification of this step is provided by the theory of distributions. The delta *function* is of course a functional, $\delta[\phi] = \phi(0)$.

defining in this way the Dirac delta function as a *function* having the property

$$\int_{-\infty}^{\infty} dx\, \phi(x)\, \delta(x) = \phi(0). \tag{B.7}$$

As a first example, we can compress the gate function

$$f(x) = \tfrac{1}{2}\theta(1 - |x|), \tag{B.8}$$

where θ is the step function

$$\theta(x) \equiv \begin{cases} 1, & \text{for } x > 1, \\ 0, & \text{for } x < 1, \end{cases} \tag{B.9}$$

which will then become a consecutively narrower and higher gate.

We can also have an oscillatory suppression by choosing the normalized function

$$f(x) = \frac{\sin x}{\pi x} \tag{B.10}$$

and thereby the Dirac sequence

$$f_n(x) = \frac{\sin nx}{\pi x}, \tag{B.11}$$

since the integral in Eq. (B.5) will again be dominated by the peak at zero, and away from zero the rapid oscillation of the function renders no contribution to the integral, and we have the representation of the delta function,

$$\delta(x) = \lim_{n\to\infty} \frac{\sin nx}{\pi x}. \tag{B.12}$$

For a continuous label, one speaks of Dirac families. For example, from the normalized function

$$f(x) = \frac{e^{-x^2}}{\sqrt{\pi}} \tag{B.13}$$

we get the Dirac family

$$f_t(x) = \frac{1}{\sqrt{t}} f\left(\frac{x}{\sqrt{t}}\right) \tag{B.14}$$

and obtain the expression for the delta function,

$$\delta(x) = \lim_{t\to 0} f_t(x) = \lim_{t\to 0} \frac{e^{-(x/\sqrt{t})^2}}{\sqrt{t\pi}}. \tag{B.15}$$

Another useful representation of the delta function is

$$\delta(x) = \frac{1}{\pi} \lim_{a\to 0} \frac{a}{x^2 + a^2}. \tag{B.16}$$

For the oscillatory function

$$f(x) = \frac{e^{ix^2}}{\sqrt{i\pi}} \tag{B.17}$$

we get

$$\delta(x) = \lim_{t\to 0} f_t(x) = \lim_{t\to 0} \frac{e^{i(x/\sqrt{t})^2}}{\sqrt{i\pi t}}. \tag{B.18}$$

We will also encounter functions whose derivatives are delta functions, such as the derivative of the step function, $\theta(x)$, which can be obtained from the Dirac family of smoothed-out step functions. For example, choosing

$$\theta_a(x) = \frac{1}{2}\left(1 + \tanh\frac{x}{a}\right), \tag{B.19}$$

we have for

$$\theta(x) \equiv \lim_{a\to 0} \theta_a(x) \tag{B.20}$$

that

$$\theta'(x) = \lim_{a\to 0} \theta_a'(x) = \lim_{a\to 0} \frac{1}{2a}\frac{1}{\cosh^2(x/a)}, \tag{B.21}$$

and the normalization property of the function $\cosh^{-2}$ then gives

$$\theta'(x) = -\delta(x). \tag{B.22}$$

The derivative of the Fermi function

$$f_0(\epsilon) \equiv \frac{1}{e^{(\epsilon-\epsilon_F)/kT}+1} = \frac{1}{2}\left(1 - \tanh\left(\frac{\epsilon-\epsilon_F}{kT}\right)\right) = (1 - \theta_{kT}(\epsilon - \epsilon_F)) \tag{B.23}$$

will therefore at zero temperature become a delta function[2]

$$\lim_{T\to 0} \frac{\partial f_0}{\partial \epsilon} = -\delta(\epsilon - \epsilon_F). \tag{B.24}$$

The delta function is now shown to have the integral representation

$$\int_{-\infty}^{\infty} \frac{dk}{2\pi}\, e^{\pm ik(x-x')} = \delta(x - x'). \tag{B.25}$$

First the integrand is modified, since otherwise the integral is not convergent; a convergence factor is introduced

$$\int_{-\infty}^{\infty} \frac{dp}{2\pi}\, e^{\pm ikx} \equiv \lim_{\epsilon\to 0} \int_{-\infty}^{\infty} \frac{dp}{2\pi}\, e^{\pm ikx-\epsilon p^2}. \tag{B.26}$$

Now the integral is meaningful for any positive value of ϵ. This gives a Dirac family

$$\delta_\epsilon(x - x') = \int_{-\infty}^{\infty} \frac{dk}{2\pi}\, e^{\pm ikx-\epsilon p^2}, \tag{B.27}$$

[2] Since our energies are measured from the zero level, we only encounter the Fermi function for positive energies. Mathematically, there is nothing preventing us from defining the Fermi function also for negative values of the energy, and one immediately gets that the weight, the area, of the derivative of the Fermi function is unity, $f(-\infty) - f(\infty) = 1 - 0 = 1$.

since, by completing the square, we obtain (see below)

$$\lim_{\epsilon\to 0}\int_{-\infty}^{\infty}\frac{dk}{2\pi}\,e^{\pm ikx-\epsilon p^2}=\lim_{\epsilon\to 0}\sqrt{\frac{1}{4\pi\epsilon}}\,e^{-x^2/(4\epsilon)},\tag{B.28}$$

which we earlier, Eq. (B.15), have shown to be a representation of the delta function.[3]

The basic Gaussian integral, a non-negative,

$$I(a,b)=\int_{-\infty}^{\infty}dx\;e^{-ax^2\pm bx}\tag{B.29}$$

is evaluated by *completing the square*, $-ax^2+bx=-a(x-b/2a)^2+b^2/4a$, and shifting the integration variable, $x'=x-b/2a$, whereby the integral reduces to the simple Gaussian

$$I(a,b)=e^{b^2/4a}\int dx'\;e^{-ax'^2}=\sqrt{\frac{\pi}{a}}\;e^{b^2/4a}.\tag{B.30}$$

If $b/2a$ is complex, one should note that the contour of integration is shifted into the complex plane. A short course in analytic functions, however, shows the result stated is still correct. Otherwise, use the Taylor expansion form of $\exp\{bx\}$, and the resulting elementary integrals are performed and provide the above result.

Exercise B.1 Show that the plane waves, Eq. (1.6), are so-called delta-normalized

$$\int d\mathbf{x}\;A^*_{\mathbf{k}}A_{\mathbf{k}'}\,e^{i\mathbf{x}\cdot(\mathbf{k}'-\mathbf{k})}=(2\pi)^3\,|A_{\mathbf{k}}|^2\,\delta(\mathbf{k}-\mathbf{k}').\tag{B.31}$$

Intuitively, the delta function corresponds to concentrating the total *weight* of a physical quantity in a point. A familiar example from electrostatics of a delta function is the charge distribution of a point charge, which is the singular limit of a, say, spherical, extended charge distribution

$$\rho(\mathbf{x})=\begin{cases}\rho_0, & |\mathbf{x}|\le a,\\ 0, & |\mathbf{x}|>a.\end{cases}\tag{B.32}$$

Letting the size of the spherical charge distribution, a, shrink to zero and the charge density, ρ_0, approach infinity in such a way that $3\pi a^3\rho_0/4\to q$ (we assume for definiteness three dimensions), we get for the charge distribution of a point particle with charge q the delta distribution

$$\rho(\mathbf{x})=q\,\delta(\mathbf{x}).\tag{B.33}$$

[3] If ϵ is much smaller than any other length scale of a physical problem of interest, the result of integrating a function together with the smoothed delta function will have a limit for vanishing ϵ, and the appearance of the delta function corresponds to leaving out the convergence factor in explicit expressions, demonstrating the meaning of the formula (B.25).

The electrostatic potential of a point particle

$$\phi(\mathbf{x}) = \frac{q}{4\pi\epsilon_0|\mathbf{x}|} \tag{B.34}$$

satisfies the Poisson equation

$$\Delta\phi(\mathbf{x}) = -\frac{\rho(\mathbf{x})}{\epsilon_0} = -\frac{q}{\epsilon_0}\,\delta(\mathbf{x}) \tag{B.35}$$

in accordance with the formula

$$\Delta\frac{1}{|\mathbf{x}|} = -4\pi\,\delta(\mathbf{x}). \tag{B.36}$$

Exercise B.2 Show by direct differentiation that $\Delta\, 1/|\mathbf{x}| = 0$ for $\mathbf{x} \neq 0$, and that by integration of $\Delta\, 1/|\mathbf{x}|$ over, say, a sphere, and using Gauss's theorem, that $\Delta\, 1/|\mathbf{x}|$ is delta-singular at the origin with strength -4π.

Appendix C: Fourier Analysis

Given the series of numbers f_n, $n = 0, \pm 1, \pm 2, \ldots$, construct the function of real x,

$$f(x) \equiv \frac{1}{L} \sum_{n=-\infty}^{\infty} f_n \, e^{2\pi i n x/L}, \tag{C.1}$$

which is seen to be periodic with period L (real), $f(x + L) = f(x)$. Next, note the representation of the Kronecker function,

$$\begin{aligned} \frac{1}{L} \int_{-L/2}^{L/2} dx \, e^{2\pi i(n-n')x/L} &= \frac{1}{L} \left(\frac{e^{\pi i(n-n')} - e^{-\pi i(n-n')}}{2\pi i(n-n')/L} \right) \\ &= \begin{cases} 1, & \text{for } n = n', \\ 0, & \text{for } n \neq n', \end{cases} \\ &\equiv \delta_{n,n'}. \end{aligned} \tag{C.2}$$

Multiplying the right-hand side of Eq. (C.1) by $\exp\{(-2\pi i n' x)/L\}$ and integrating gives the expression

$$\frac{1}{L} \int_{-L/2}^{L/2} dx \sum_{n=-\infty}^{\infty} f_n \, e^{2\pi i(n-n')x/L} = \sum_{n=-\infty}^{\infty} f_n \, \delta_{n,n'} = f_{n'}, \tag{C.3}$$

where the first equality follows from interchange of integration and summation (trivially valid if the sum only has finite many terms, which can often be assumed, as the relevant functions can be assumed to vanish for large numbers n). Treating the left-hand side of Eq. (C.1) similarly gives the identity, the inverse Fourier transform,

$$f_{n'} = \int_{-L/2}^{L/2} dx \, e^{-2\pi i n' x/L} f(x). \tag{C.4}$$

Inserting Eq. (C.4) into Eq. (C.1) (requiring a shift of dummy variable $x \to y$) and interchanging integration and summation gives

$$f(x) = \int_{-L/2}^{L/2} dy \, f(y) \frac{1}{L} \sum_{n=-\infty}^{\infty} e^{(2\pi i n/L)(x-y)} \tag{C.5}$$

and thereby the delta function representation

$$\frac{1}{L}\sum_{n=-\infty}^{\infty} e^{(2\pi i n/L)(x-y)} = \sum_{p=-\infty}^{\infty} \delta(x-y+pL). \tag{C.6}$$

On introducing

$$k_n \equiv \frac{2\pi}{L}n, \qquad \Delta k \equiv \frac{2\pi}{L}, \qquad \tilde{f}(k_n) \equiv f_n,$$

Eq. (C.1) can be rewritten as

$$f(x) = \sum_{n=-\infty}^{\infty} \frac{\Delta k}{2\pi}\tilde{f}(k_n)\, e^{ik_n x}, \tag{C.7}$$

which in the limit $L \to \infty$ becomes the Fourier integral

$$f(x) = \int_{-\infty}^{\infty} \frac{dk}{2\pi}\tilde{f}(k)\, e^{ikx}, \tag{C.8}$$

and the inversion formula, Eq. (C.4), becomes

$$\tilde{f}(k) = \int_{-\infty}^{\infty} dx\, e^{-ixk} f(x). \tag{C.9}$$

The Fourier integration formulas also follow immediately by use of the identity (B.25).

Exercise C.1 Show that the Fourier transform of a convolution becomes simply a product, i.e. for

$$\psi_3(\mathbf{x}) = \int d\mathbf{x}'\; \psi_2(\mathbf{x}-\mathbf{x}')\, \psi_1(\mathbf{x}'), \tag{C.10}$$

the following relation holds for the Fourier transform:

$$\psi_3(\mathbf{p}) = (2\pi)^{d/2} \psi_2(\mathbf{p})\, \psi_1(\mathbf{p}). \tag{C.11}$$

Appendix D: Classical Mechanics

In classical physics a particle moves along a definite trajectory determined by the force exerted on it. For simplicity, we consider a single particle of mass m experiencing a conservative force, $F = -V'$, for which Newton's equation reads

$$m\ddot{x}_t = -V'(x_t, t). \tag{D.1}$$

Newton's equation is second order in time, so, to specify a solution uniquely, the position, $\mathbf{x}_t$, *and* the velocity, $\dot{\mathbf{x}}_t$, must both be specified at a certain time.

Introducing Lagrange's function, i.e. the kinetic energy *minus* the potential energy,

$$L(x_t, \dot{x}_t, t) \equiv \tfrac{1}{2}m\dot{x}_t^2 - V(x_t, t), \tag{D.2}$$

Newton's equation can be rewritten in the form

$$\frac{\partial L}{\partial x} - \frac{d}{dt}\left(\frac{\partial L}{\partial \dot{x}}\right) = 0. \tag{D.3}$$

Newton's equation can be specified in terms of a variational principle. In order to do so, consider the so-called classical action

$$S_{xtx't'}[x_t] \equiv \int_{t'}^{t} d\bar{t}\; L(x_{\bar{t}}, \dot{x}_{\bar{t}}, \bar{t}), \tag{D.4}$$

for an arbitrary path, $x_{\bar{t}}$, considered in the time interval from t' to t with associated initial, $x_{t'} = x'$, and final, $x_t = x$, points. The action is a so-called functional, i.e. it takes a function, here the path x_t, and attributes it a number.

Consider the difference in the value of the action for two nearby paths, x_t and $x_t + \delta x_t$, having the same initial and final points (therefore suppressing this information in the following)

$$\begin{aligned}
\delta S[x_t] &\equiv S[x_t + \delta x_t] - S[x_t] \\
&= \int_{t'}^{t} d\bar{t}\; \{\tfrac{1}{2}m(\dot{x}_{\bar{t}} + \delta\dot{x}_{\bar{t}})^2 - V(x_{\bar{t}} + \delta x_{\bar{t}}, \bar{t}) - \tfrac{1}{2}m\dot{x}_{\bar{t}}^2 + V(x_{\bar{t}}, \bar{t})\}.
\end{aligned} \tag{D.5}$$

Taylor-expanding the potential in the shift in the path gives, for the variation in the classical action to a small change in path,

$$\delta S[x_t] = \int_{t'}^{t} d\bar{t}\, \{m\dot{x}_{\bar{t}}\, \delta\dot{x}_{\bar{t}} - V'(x_{\bar{t}}, \bar{t})\, \delta x_{\bar{t}}\} + \mathcal{O}(\delta x_{\bar{t}}^2)$$

$$= \int_{t'}^{t} d\bar{t}\, \{-m\ddot{x}_{\bar{t}} - V'(x_{\bar{t}}, \bar{t})\}\, \delta x_{\bar{t}} + \mathcal{O}(\delta x_{\bar{t}}^2). \tag{D.6}$$

The last equation results from a partial integration, and the fact that there are no boundary terms since, by construction, there are assumed to be no variations of the starting and end points of the path, and $\mathcal{O}(\delta x_{\bar{t}}^2)$ signifies all the higher-order terms beyond linear order. If, for the path x_t, we choose the classical path, $x_t = x_t^{\text{cl}}$, the linear term vanishes, and we have obtained Hamilton's characterization of the classical motion of a particle: *the classical path is the one for which a small change in the path does not change the action*, i.e. there is no term proportional to the variation, δx_t, only terms of higher order. The classical path can thus be characterized as the one along which the action is an extremum, most often the one along which the kinetic energy *minus* the potential energy is a minimum, the principle of least action.

The form of Eq. (D.6) is

$$\delta S[x_t] = \int_{t'}^{t} d\bar{t}\, f(\bar{t})\, \delta x_{\bar{t}} + \mathcal{O}(\delta x_{\bar{t}}^2) \tag{D.7}$$

and the function $f(\bar{t})$ is called the functional derivative of the functional in question and denoted

$$f(\bar{t}) = \frac{\delta S[x_t]}{\delta x_{\bar{t}}} \tag{D.8}$$

and the variation of the action to a change in path can be expressed in the form

$$\delta S[x_t] = \int_{t'}^{t} d\bar{t}\, \frac{\delta S[x_t]}{\delta x_{\bar{t}}}\, \delta x_{\bar{t}} + \mathcal{O}(\delta x_{\bar{t}}^2). \tag{D.9}$$

Newton's equation can then, according to Eq. (D.6), be expressed as the vanishing of the functional derivative of the action at the classical path[1]

$$\left.\frac{\delta S}{\delta \mathbf{x}_t}\right|_{\mathbf{x}_t = \mathbf{x}_t^{\text{cl}}} = \mathbf{0}, \tag{D.10}$$

i.e. in terms of Hamilton's principle: the action is stationary for the classical path.

Finally, we show how to trade the information contained in Newton's second-order differential equation for two first-order differential equations. This is done by introducing Hamilton's function on phase space,

$$H(\mathbf{x}, \mathbf{p}, t) = \frac{\mathbf{p}^2}{2m} + V(\mathbf{x}, t), \tag{D.11}$$

[1] The derivation, seemingly performed in one spatial dimension, is by introducing the scalar product between vectors in Eq. (D.6), immediately seen to be identical for arbitrary dimensions.

and giving the variables dynamics through Hamilton's equations,

$$\dot{\mathbf{x}} = \frac{\partial H(\mathbf{x}, \mathbf{p}, t)}{\partial \mathbf{p}} \tag{D.12}$$

and

$$\dot{\mathbf{p}} = -\frac{\partial H(\mathbf{x}, \mathbf{p}, t)}{\partial \mathbf{x}}, \tag{D.13}$$

since, when $\mathbf{x}$ and $\mathbf{p}$ are interpreted as position and momentum, Eqs. (D.12) and (D.13) generate the same trajectory as Newton's equation for the initial condition $(\mathbf{x}_t, \mathbf{p}_t) = (\mathbf{x}_t, m\dot{\mathbf{x}}_t)$.

In phase space $(\mathbf{x}_t, \mathbf{p}_t)$, the dynamics is in terms of a first-order differential equation and therefore complete: specifying a point $(\mathbf{x}_0, \mathbf{p}_0)$, the future path in phase space is determined (and also the past path by time reversal symmetry, which then ensures that no two paths for particle evolution in phase space can meet each other).

For an isolated system, i.e. when the potential is independent of time, Hamilton's function is simply the energy of the particle, i.e. the only quantity that in general is conserved,

$$\frac{dH(\mathbf{x}, \mathbf{p})}{dt} = 0, \tag{D.14}$$

consisting of the kinetic energy *plus* the potential energy.

The Lagrange and Hamilton functions are generally related through a Legendre transformation,

$$H(\mathbf{x}_t, \mathbf{p}_t, t) = \dot{\mathbf{x}}_t \cdot \mathbf{p}_t - L(\mathbf{x}_t, \dot{\mathbf{x}}_t, t), \tag{D.15}$$

i.e.

$$\mathbf{p} = \frac{\partial L}{\partial \dot{\mathbf{x}}}, \tag{D.16}$$

giving the (canonical) momentum, $\mathbf{p}_t = m\dot{\mathbf{x}}_t$ (or the other way around, through Eq. (D.12)), and

$$\frac{dH(\mathbf{x}, \mathbf{p}, t)}{dt} = \frac{\partial H(\mathbf{x}, \mathbf{p}, t)}{\partial t} = -\frac{\partial L(\mathbf{x}, \dot{\mathbf{x}}, t)}{\partial t}, \tag{D.17}$$

where Hamilton's equations guarantee the validity of the first equality, and Eq. (D.15) the second.

Exercise D.1 The classical motion of a particle of mass m and charge e in electromagnetic fields, $\mathbf{E}(\mathbf{x}, t)$ and $\mathbf{B}(\mathbf{x}, t)$, is governed by Newton's equation as specified by the Lorentz force,

$$m\ddot{\mathbf{x}}_t = e\mathbf{E}(\mathbf{x}_t, t) + e\dot{\mathbf{x}}_t \times \mathbf{B}(\mathbf{x}_t, t). \tag{D.18}$$

Represent the fields, in concordance with Maxwell's equations, by the potentials, $\varphi(\mathbf{x}, t)$ and $\mathbf{A}(\mathbf{x}, t)$, scalar and vector potential,

$$\mathbf{E}(\mathbf{x}, t) = -\nabla\varphi(\mathbf{x}, t) - \frac{\partial \mathbf{A}(\mathbf{x}, t)}{\partial t}, \qquad \mathbf{B}(\mathbf{x}, t) = \nabla \times \mathbf{A}(\mathbf{x}, t). \tag{D.19}$$

Show that Hamilton's equations, employing the following Hamilton function,

$$H(\mathbf{x}_t, \mathbf{p}_t, t) = \frac{1}{2m}\left(\mathbf{p}_t - e\mathbf{A}(\mathbf{x}_t, t)\right)^2 + e\varphi(\mathbf{x}_t, t), \tag{D.20}$$

indeed reproduce Newton's equation, Eq. (D.18). In Hamilton's (and Lagrange's) formulation of classical mechanics, the electric and magnetic fields do not enter explicitly; instead, it is the scalar and vector potentials, the gauge fields. We note that classical mechanics is gauge-invariant, i.e. the equation of motion of the particle (as well as Maxwell's equations) are invariant (unchanged) with respect to the following adjustment of the fields, the gauge transformation:

$$\varphi(\mathbf{x}, t) \rightarrow \varphi(\mathbf{x}, t) + \frac{\partial \Lambda(\mathbf{x}, t)}{\partial t}, \qquad \mathbf{A}(\mathbf{x}, t) \rightarrow \mathbf{A}(\mathbf{x}, t) - \nabla_{\mathbf{x}}\Lambda(\mathbf{x}, t). \tag{D.21}$$

Solution

First, we consider the case where only a time-independent magnetic field is present. The first of Hamilton's equations, Eq. (D.12), gives

$$m\dot{\mathbf{x}}_t = \mathbf{p}_t - e\mathbf{A}(\mathbf{x}_t), \tag{D.22}$$

i.e. the kinetic momentum, $m\dot{\mathbf{x}}_t$, does not equal the canonical momentum. The second of Hamilton's equations, Eq. (D.13), gives

$$\dot{\mathbf{p}}_t = \frac{e}{m}(\mathbf{v}_t \cdot \nabla)\mathbf{A}(\mathbf{x}_t), \tag{D.23}$$

where $\mathbf{v}_t = \dot{\mathbf{x}}_t$ denotes the velocity of the particle. Combining the equations gives (suppressing the time dependence)

$$\begin{aligned} m\ddot{\mathbf{x}}_t = \dot{\mathbf{p}}_t &= e(\mathbf{v}_t \cdot \nabla)\mathbf{A}(\mathbf{x}_t) = e(\mathbf{v}_t \times \nabla \times \mathbf{A}(\mathbf{x}_t)) - e\mathbf{A}(\mathbf{x}_t)(\nabla \cdot \mathbf{v}_t) \\ &= e(\mathbf{v}_t \times \nabla \times \mathbf{A}(\mathbf{x}_t)) = e\mathbf{v}_t \times \mathbf{B}(\mathbf{x}_t), \end{aligned} \tag{D.24}$$

using well-known formulas from vector analysis. Repeat the derivation for the case of arbitrary electromagnetic fields.

Appendix E: Wave Function Properties

The wave function, being a probability amplitude *density*, is born a continuous function. Being a solution of the Schrödinger equation demands further smoothness. Consider the time-independent Schrödinger equation in one spatial dimension corresponding to energy E,

$$-\frac{\hbar^2}{2m}\frac{d^2\psi(x)}{dx^2} + V(x)\,\psi(x) = E\,\psi(x). \tag{E.1}$$

It demands a wave function to be twice differentiable, at least piecewise. Integrating the equation over the interval $[x - \Delta x, x + \Delta x]$ gives

$$\int\limits_{x-\Delta x}^{x+\Delta x} dx \left(-\frac{\hbar^2}{2m}\,\psi''(x) + V(x)\,\psi(x)\right) = E \int\limits_{x-\Delta x}^{x+\Delta x} dx\;\psi(x). \tag{E.2}$$

To lowest order in Δx the equation becomes

$$-\frac{\hbar^2}{2m}(\psi'(x+\Delta x) - \psi'(x-\Delta x)) + 2\Delta x V(x)\,\psi(x) = 2\Delta x\,E\,\psi(x). \tag{E.3}$$

Letting Δx approach zero shows that the derivative of the wave function is continuous (the pathological case of a singular potential is discussed shortly)

$$\lim_{\Delta x \to 0}(\psi'(x+\Delta x) - \psi'(x-\Delta x)) = 0. \tag{E.4}$$

Continuity of the derivative of the wave function implies continuity of the probability current density, as required by particle conservation.

If the potential has a finite jump in value at, say, position x, Eq. (E.1) is first integrated in the two continuous regions of the potential, giving (with $x^{\pm} \equiv x \pm 0$)

$$-\frac{\hbar^2}{2m}(\psi'(x+\Delta x) - \psi'(x^+)) + \Delta x V(x^+)\,\psi(x) = \Delta x\,E\,\psi(x) \tag{E.5}$$

and

$$-\frac{\hbar^2}{2m}(\psi'(x^-) - \psi'(x-\Delta x)) + \Delta x V(x^-)\psi(x) = \Delta x\,E\,\psi(x). \tag{E.6}$$

Subtract Eqs. (E.5) and (E.6) and divide by Δx, then upon letting Δx approach zero the second derivative is seen to have a jump determined by the jump in the potential,

$$\psi''(x^+) - \psi''(x^-) = \frac{2m}{\hbar^2}(V(x^+) - V(x^-))\psi(x). \tag{E.7}$$

Recall from Appendix B that the derivative of the step function is a delta function. A finite jump in the second derivative thus implies that the first derivative cannot have a discontinuity. Thus the derivative of the wave function is continuous,

$$\psi'(x+0) = \psi'(x-0), \tag{E.8}$$

even though the potential has a jump. The continuous first derivative of the wave function has a kink at the point where the potential is discontinuous. The discontinuity in the second derivative does not pose a problem, the second derivative having no requirement for a physical interpretation. A discontinuous potential is a convenient approximation to study, as stepwise constant potentials allow for exact solutions in terms of exponential and trigonometric functions. Since the probability current density is continuous, it leads to no pathologies.

Integrating formally the Schrödinger equation gives, for the first derivative of the wave function,

$$\frac{d\psi(x)}{dx} = \frac{2m}{\hbar^2} \int^{x} dx' \, (V(x') - E)\psi(x'), \tag{E.9}$$

also displaying that it is continuous even if the potential has a finite jump.

The above smoothness properties of wave functions are not particular to one spatial dimension. In three spatial dimensions, consider integrating the time-independent Schrödinger equation over an infinitesimal volume $\Delta\Omega$. The kinetic energy term is proportional to $\Delta\Omega$ (in view of the rest of the terms obviously being that). The Laplacian can be expressed as the divergence of the gradient, and using Gauss's theorem, the volume integral is converted into the corresponding surface integral, giving the equation

$$\mathcal{O}(\Delta\Omega) = \int_{\Delta\Omega} d\mathbf{x} \, \nabla \cdot \nabla\psi(\mathbf{x}) = \int_{S} d\mathbf{s} \cdot \nabla\psi(\mathbf{x}). \tag{E.10}$$

Choosing the volume $\Delta\Omega$ as a flat small (pill-box) volume, Eq. (E.10) becomes, in the limit of vanishing volume,

$$0 = d\mathbf{s} \cdot \nabla\psi(\mathbf{x}^+) - d\mathbf{s} \cdot \nabla\psi(\mathbf{x}^-). \tag{E.11}$$

The gradient of the wave function is a continuous vector field, as it takes the same value on infinitesimal close surfaces.

Consider two solutions, $\psi_1(x)$ and $\psi_2(x)$, to the time-independent Schrödinger equation (E.1), corresponding to the same energy value. Introducing the Wronskian, or Wronski determinant,

$$W(x) \equiv \psi_1(x)\psi_2'(x) - \psi_2(x)\psi_1'(x), \tag{E.12}$$

it follows from the time-independent Schrödinger equation that

$$\frac{dW(x)}{dx} = 0, \tag{E.13}$$

i.e. the Wronskian is a constant, W. If this constant is equal to zero, $W = 0$, the logarithmic derivatives of the two functions are equal,

$$\frac{d\ln(\psi_1(x))}{dx} = \frac{d\ln(\psi_2(x))}{dx}. \tag{E.14}$$

Integrating

$$\int_{x_1}^{x_2} d\ln\psi_1(x) = \int_{x_1}^{x_2} d\ln\psi_2(x) \tag{E.15}$$

gives

$$\ln\psi_1(x_2) - \ln\psi_1(x_1) = \ln\psi_2(x_2) - \ln\psi_2(x_1). \tag{E.16}$$

When varying, say, x_2, the equality can only be fulfilled if ψ_2 is proportional to ψ_1, $\psi_2 = c\,\psi_1$, i.e. the two solutions are dependent. For two independent solutions to the time-independent Schrödinger equation (E.1), the Wronskian is therefore a non-vanishing constant.

A most important departure from classical physics is that only discrete values of the energy are allowed quantum mechanically in confining potentials. To see this, consider a potential that vanishes far from the origin, $V(x \to \pm\infty) = 0$, and has only negative values with a lower bound, $V_{\min} < V(x) \leq 0$; the potential could be, for example, the shape of a smooth dip, i.e. an attractive potential, a potential well. A possible energy eigenvalue satisfies $E > V_{\min}$, as the average kinetic energy can only be non-negative (recall Exercise 1.5 on page 10). Assume that a solution with energy $E < 0$ exists. The time-independent Schrödinger equation

$$\psi''(x) = -\frac{2m}{\hbar^2}(E - V(x))\,\psi(x) \tag{E.17}$$

has qualitatively different solutions in spatial regions depending on whether $E - V(x)$ is positive or negative. Assume, for simplicity, that the potential only equals the assumed energy eigenvalue, E, for two spatial locations, $E - V(x_{l,r}) = 0$. Thus $E - V(x) > 0$ for $x_l < x < x_r$, and $E - V(x) < 0$ outside this interval.

Far enough away from the origin, $x \to \pm\infty$, the Schrödinger equation reduces, by assumption of the potential shape, to

$$\psi''(x) \simeq \frac{2m|E|}{\hbar^2}\,\psi(x). \tag{E.18}$$

In the spatial regions where $E < V(x)$, at large distances, the two independent solutions approach $\exp(\pm\kappa x)$, where $\kappa^2 = 2m|E|/\hbar^2$. One solution is decaying and the other diverging. The latter must be discarded, as it leads to the physically unwanted property of sending all the probability of the particle's whereabouts to infinity. A diverging function is not a proper wave function.[1] Since the other solution decays exponentially, a wave function corresponding to an energy eigenvalue $V_{\min} < E < 0$ is normalizable.

[1] Even superpositions of the stationary states corresponding to such divergent solutions cannot be made normalizable (as the expression in Eq. (2.20) for unlimited space shows).

Introducing the local de Broglie wave length, $\lambda(x) \equiv 2\pi\hbar/\sqrt{2m(E-V(x))}$, the time-independent Schrödinger equation can, in the middle spatial region where $E > V(x)$, be rewritten as

$$\psi''(x) = -\left(\frac{2\pi}{\lambda(x)}\right)^2 \psi(x), \qquad p(x) \equiv \frac{2\pi\hbar}{\lambda(x)} > 0, \tag{E.19}$$

and we have noted that $p(x) \equiv \sqrt{2m(E-V(x))}$ is the momentum that a classical particle at location x would have for given total energy E.

In the middle spatial region, the solution to the Schrödinger equation thus has an oscillatory behavior, since the prefactor of the wave function on the right-hand side of the equation is negative. For a smooth potential, the wave function oscillates with the local wave length $\lambda(x)$. These oscillations must be such that they join up smoothly with the values prescribed by the decaying solutions in the two outer regions. This can only happen if the wave oscillations are special. An E value any small magnitude away from the special value changes the wave length and the solution has lost its smoothness in the derivative at x_l and x_r, and is not a proper wave function. We therefore conclude that there are only allowed states with discrete energy eigenvalues in the E interval, $V_{\text{min}} < E < 0$.[2] The attractive property of the potential can lead to several discrete possible energy values with corresponding confined wave functions. Wave functions with the proper oscillation properties corresponding to these discrete energy values are, as shown, normalizable and are referred to as bound states, as the main probability for the particle is located in the well interior. Normalized energy eigenstates thus correspond to the discrete energy levels in the energy spectrum.[3]

The discreteness of energy eigenvalues of bound states in a confining potential such as the considered quantum well should be contrasted with tunneling through a barrier. Because the tunneling barrier is finite, both independent barrier solutions are physically acceptable since no divergence disqualification occurs; and with two independent solutions at our disposal, the smoothness requirement of the wave function can be met for the continuum of energy eigenvalues, as discussed in Section 2.1. Similarly, for the considered quantum well potential, the energy spectrum for $E > 0$, i.e. $E > V(x)$, is not subjected to the rejection of one of the independent mathematical solutions of the Schrödinger equation, as none are diverging, just oscillating. These wave functions are not normalizable but are at least bounded, the scattering states superposed to provide wave packet states as discussed in the study of tunneling.

The quantization of energy due to confined wave functions is not a special feature of one spatial dimension. In higher dimensions, the above analysis leads to an even stricter requirement since the smoothness of the wave function, in view of Eq. (E.11), should pertain to any spatial cross-section.

Next, consider the pathological case where the step in a potential is infinite and the potential stays infinite in a region, the so-called infinite wall region. If the potential is

[2] In the middle region $\psi''(x)/\psi(x) < 0$, whereas in the two other regions $\psi''(x)/\psi(x) > 0$. The delicacy of the matching is therefore due to a concave function in the middle region having to connect smoothly to convex functions.

[3] For a more quantitative demonstration of energy quantization, see the discussion in Section 6.1, where a definite potential is considered. Since a continuous function can be arbitrarily well approximated by stepwise constant pieces, a smooth confining potential gives the same qualitative features as step potentials.

infinite in a continuous region of space, then, in order for Eq. (E.3) to be fulfilled, there are two options. The first option is that it forces a solution, $\psi(x)$, of the equation also to be infinite in that region, which does not provide a meaningful wave function, as it leaves all probability in the wrong region. Consequently, the other option for the wave function to satisfy Eq. (E.3) is realized, *viz.* the wave function must vanish in the region where the potential is infinite. By continuity of the wave function, it thus vanishes at the boundary of an infinite wall. This conclusion is also directly obtained by letting the wall of the quantum well considered in Section 6.1 rise to infinity, resulting in the disappearance of the leakage of the wave function, as the binding energy is infinite. It is not possible to drive a quantum particle into a region of infinite potential. As a consequence, a particle cannot tunnel through such a region, leaving zero probability amplitude for the particle to be in the infinite wall region and, by linearity of the Schrödinger equation, vanishing probability amplitude beyond the infinite barrier. At an infinite wall boundary, the derivative of the wave function will not be continuous, as otherwise Eq. (E.5) would make the wave function vanish everywhere.

Exercise E.1 Consider the case of a particle of mass m in a box, i.e. the potential

$$V(x) = \begin{cases} \infty, & x < -a/2, \\ 0, & -a/2 \leq x \leq a/2, \\ \infty, & x > a/2. \end{cases} \tag{E.20}$$

Obtain the energies and energy eigenfunctions. Obtain the energies from the finite well solutions by taking the limit $V \to \infty$. Obtain the energy eigenfunctions similarly.

For calculational convenience, it can be efficient to consider other pathological cases of potentials, which at points are singular, i.e. have delta spikes. Consider the one-dimensional case of a single delta function potential located at x_0,

$$V(x) = w\,\delta(x - x_0). \tag{E.21}$$

Integrating the time-independent Schrödinger equation around the singular point of the potential, x_0, the second term on the left-hand side of Eq. (E.2) then becomes the finite term $\omega\psi(x_0)$, and Eq. (E.3) becomes

$$\psi'(x_0^+) - \psi'(x_0^-) = \frac{2m}{\hbar^2} w\psi(x_0), \tag{E.22}$$

stating that the first derivative of the wave function is not continuous at the singular point, i.e. the continuous wave function has a kink at the location of the delta function. Since the first derivative has a jump, the second derivative of the wave function becomes at x_0 the delta function

$$\psi''(x_0) = \frac{2m}{\hbar^2} w\delta(x_0), \tag{E.23}$$

with the strength equal to the jump (recall that the derivative of the step function is a delta function), and the two singular terms in the time-independent Schrödinger equation cancel each other so that the equation is fulfilled at the singular point.

As an example of a delta potential, consider the scattering properties of a delta function potential located at the origin, i.e. the potential in Eq. (E.21) with $x_0 = 0$. Assume that the particle is incoming from the left corresponding to the wave function (with $k \equiv k(E) = \sqrt{2mE}/\hbar$)

$$\psi_E(x) = \begin{cases} e^{ikx} + r\, e^{-ikx}, & x < 0, \\ t\, e^{ikx} & x > 0, \end{cases} \tag{E.24}$$

where $r \equiv r(E)$ and $t \equiv t(E)$ are the reflection and transmission amplitudes, respectively. Continuity of the wave function at $x = 0$ then gives the equation $1 + r = t$, and the discontinuity relation for the first derivative, Eq. (E.22), gives the condition

$$ikt - (ik - ikr) = \frac{2m}{\hbar^2} wt. \tag{E.25}$$

The tunneling probability for energy E, $T(E) = |t|^2$, becomes

$$T(E) = \frac{1}{|1 + imw/k\hbar^2|^2} = \frac{1}{1 + mw^2/(2\hbar^2 E)}. \tag{E.26}$$

Note that the transmission properties are independent of the sign of w; the delta *peak* transmits just like the delta *well*.

Calculating the probability current for $x > 0$ gives

$$S_>(x) = \frac{\hbar k}{m} |t|^2, \tag{E.27}$$

the transmitted current, and for $x < 0$,

$$S_<(x) = \frac{\hbar k}{2m}(1 - |r|^2 + r^* e^{2ikx} - re^{-2ikx}) + \text{c.c.}, \tag{E.28}$$

which with $r = |r|e^{-i\delta}$ becomes

$$S_<(x) = \frac{\hbar k}{2m}(1 - |r|^2 + 2i|r| \sin(2kx + \delta)) + \text{c.c.} = \frac{\hbar k}{m}(1 - |r|^2), \tag{E.29}$$

the incoming minus reflected probability current. Since an energy eigenstate is considered, the probability current density is independent of the location in the two regions, and seen to be continuous at the location of the delta function, $S_< = S_>$, since $|t|^2 = 1 - |r|^2$. The pathology of discontinuity in the first derivative of the wave function is thus not transferred to the physical quantity, the probability current. The two terms differing in the left and right derivative in the current according to Eq. (E.22) are purely imaginary, appearing with opposite sign and thereby canceling each other.

Exercise E.2 Check the correctness of the tunneling probability in Eq. (E.26) by noting that it should appear as the limiting expression for the square barrier, Eq. (2.82), in the limit $V \to \infty$ and $a \to 0$ in such a fashion that aV is kept constant at the value, $aV = w$.

Exercise E.3 Consider a particle with mass m in the delta potential

$$V(x) = -w\,\delta(x), \tag{E.30}$$

where $w > 0$. Does the potential have a bound state, $E < 0$?

Consider a particle in a time-independent potential $V(\mathbf{x})$. The potential and energy eigenvalues are real, and, since differentiation and complex conjugation are interchangeable operations, it follows that, if ψ_E is a solution of the time-independent Schrödinger equation with eigenvalue E, so is the complex conjugate function, ψ_E^*. Owing to the linearity of the time-independent Schrödinger equation, any linear combination of solutions corresponding to the same energy is also an energy eigenfunction corresponding to that energy. As a consequence of the Hamiltonian being real, $\hat{H}^* = \hat{H}$, energy eigenfunctions can thus, if warranted, be chosen as real functions

$$\psi_E^{(1)}(\mathbf{x}) = \psi_E(\mathbf{x}) + \psi_E^*(\mathbf{x}), \qquad \psi_E^{(2)}(\mathbf{x}) = -i(\psi_E(\mathbf{x}) - \psi_E^*(\mathbf{x})). \tag{E.31}$$

When the origin of space can be chosen such that the potential is inversion-symmetric, $V(-\mathbf{x}) = V(\mathbf{x})$, this symmetry can be reflected in the properties of the energy eigenfunctions. The time-independent Schrödinger equation

$$\left(-\frac{\hbar^2}{2m}\frac{\partial^2}{\partial \mathbf{x}^2} + V(\mathbf{x})\right)\psi_E(\mathbf{x}) = E\,\psi_E(\mathbf{x}) \tag{E.32}$$

is valid for any value of $\mathbf{x}$, so

$$\left(-\frac{\hbar^2}{2m}\frac{\partial^2}{\partial \mathbf{x}^2} + V(-\mathbf{x})\right)\psi_E(-\mathbf{x}) = E\,\psi_E(-\mathbf{x}), \tag{E.33}$$

which in view of the inversion symmetry, $V(-\mathbf{x}) = V(\mathbf{x})$, can be rewritten as

$$\left(-\frac{\hbar^2}{2m}\frac{\partial^2}{\partial \mathbf{x}^2} + V(\mathbf{x})\right)\psi_E(-\mathbf{x}) = E\,\psi_E(-\mathbf{x}), \tag{E.34}$$

i.e. if $\psi_E(\mathbf{x})$ is a solution with energy E, so is $\psi_E(-\mathbf{x})$. For an inversion-symmetric potential, the energy eigenfunctions can, due to the linearity of the time-independent Schrödinger equation, therefore be chosen as even or odd functions,

$$\psi_E^{(1)}(\mathbf{x}) = \psi_E(\mathbf{x}) + \psi_E(-\mathbf{x}), \qquad \psi_E^{(2)}(\mathbf{x}) = \psi_E(\mathbf{x}) - \psi_E(-\mathbf{x}). \tag{E.35}$$

If, for an energy E, there exists only one independent energy eigenfunction, this function is forced to be either even or odd; since an odd function cannot be proportional to an even function, one of the combinations must vanish, as otherwise the assumption of non-degeneracy is violated.

Appendix F: Transfer Matrix Properties

For an arbitrary barrier, a transfer matrix connects the amplitudes on the two free sides. The only difference compared to the discussion of the square barrier in Section 2.5 is that the matrix elements of the *barrier* matrices are the Wronski matrices of the two independent solutions in the barrier region,

$$\mathcal{K}(x_l^B) = \begin{pmatrix} \psi_1(x_l^B) & \psi_2(x_l^B) \\ \psi_1'(x_l^B) & \psi_2'(x_l^B) \end{pmatrix}, \qquad \mathcal{K}(x_r^B) = \begin{pmatrix} \psi_1(x_r^B) & \psi_2(x_r^B) \\ \psi_1'(x_r^B) & \psi_2'(x_r^B) \end{pmatrix}. \tag{F.1}$$

That the *barrier* matrices are invertible follows from their determinant being non-zero. The determinant

$$\det \mathcal{K}(x_l^B) = \psi_1(x_l^B)\,\psi_2'(x_l^B) - \psi_1'(x_l^B)\,\psi_2(x_l^B) \tag{F.2}$$

is the so-called Wronskian, which in Appendix E was shown to be a constant different from zero for independent solutions.

The relationships between the transfer matrix components, $T_{22} = T_{11}^*$ and $T_{21} = T_{12}^*$, are not specific to the square barrier, but a consequence of the time reversal invariance of the Schrödinger equation. If the function, $\psi_E(x)$, specified by the amplitude coefficients in the free regions as depicted in Figure F.1, is a solution of the time-independent Schrödinger equation with energy E, then, since the Hamiltonian is real, $\hat{H}^* = \hat{H}$, the complex conjugate of the function, $\psi_E^*(x)$, is also a solution with energy E,

$$\hat{H}\psi_E^*(x) = (\hat{H}\psi_E(x))^* = E\psi_E^*(x). \tag{F.3}$$

Complex conjugation, $(\exp(\pm ikx))^* = \exp(\mp ikx)$, makes, for example, $(A_l^L)^*$ the amplitude of the right-moving wave on the left in ψ_E^*, etc., i.e. complex conjugation interchanges the right- and left-moving indices ($r \leftrightarrow l$). The solution to the time-independent Schrödinger equation $\psi_E^*(x)$ corresponding to energy E is thus specified by the amplitudes as depicted in Figure F.2.

Matching for the solution $\psi_E^*(x)$, the amplitudes are complex-conjugated and their right- and left-moving indices are interchanged ($r \leftrightarrow l$) compared to the case of $\psi_E(x)$. We

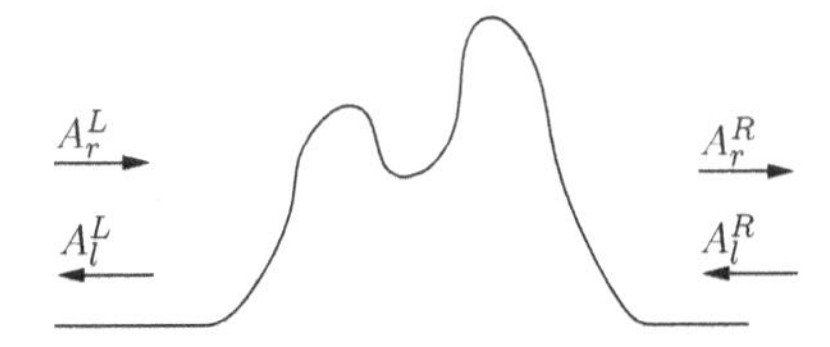

Figure F.1 Amplitudes specifying the solution $\psi_E(x)$.

Figure F.2 Amplitudes specifying the solution $\psi_E^*(x)$.

therefore get, according to the general relationship between left- and right-side amplitudes, Eq. (2.61), the relation between the complex-conjugated amplitudes:

$$\begin{pmatrix} A_l^R \\ A_r^R \end{pmatrix}^* = \mathcal{T}(x_r^B, x_l^B) \begin{pmatrix} A_l^L \\ A_r^L \end{pmatrix}^* . \tag{F.4}$$

Writing out the two equations as

$$(A_l^R)^* = T_{11}(A_l^L)^* + T_{12}(A_r^L)^*, \qquad (A_r^R)^* = T_{21}(A_l^L)^* + T_{22}(A_r^L)^* \tag{F.5}$$

and complex-conjugating them, whereby the complex conjugation is transferred to the transfer matrix elements, and changing to our original "r-upstairs and l-downstairs" notation, the relationship (F.4) is rewritten as

$$\begin{pmatrix} A_r^R \\ A_l^R \end{pmatrix} = \begin{pmatrix} T_{22}^* & T_{21}^* \\ T_{12}^* & T_{11}^* \end{pmatrix} \begin{pmatrix} A_r^L \\ A_l^L \end{pmatrix} . \tag{F.6}$$

Time reversal symmetry thus endows the transfer matrix with the property

$$\begin{pmatrix} T_{11} & T_{12} \\ T_{21} & T_{22} \end{pmatrix} = \begin{pmatrix} T_{22}^* & T_{21}^* \\ T_{12}^* & T_{11}^* \end{pmatrix} . \tag{F.7}$$

The displayed symmetry property of the transfer matrix can be expressed in matrix form as

$$\mathcal{T} = \sigma^x \mathcal{T}^* \sigma^x, \tag{F.8}$$

where we have introduced the Pauli matrix

$$\sigma^x = \begin{pmatrix} 0 & 1 \\ 1 & 0 \end{pmatrix}, \qquad \sigma^x \sigma^x = 1, \tag{F.9}$$

which sandwiches a matrix according to the rule

$$\sigma^x \begin{pmatrix} \cdot & \cdot \\ \cdot & \cdot \end{pmatrix} \sigma^x = \begin{pmatrix} \searrow & \swarrow \\ \nearrow & \nwarrow \end{pmatrix}, \tag{F.10}$$

i.e. interchange elements diagonally.

Next we derive the consequence of current conservation in a stationary state. Calculating the probability current for the stationary state in Eq. (2.6) on the left side of the barrier gives

$$S_L = \frac{\hbar k}{m}(|A_r^L|^2 - |A_l^L|^2). \tag{F.11}$$

The total current on the left side, the difference between the right- and left-going currents, is thus independent of the choice of position. Similarly, the probability current on the right side of the barrier is

$$S_R = \frac{\hbar k}{m}(|A_r^R|^2 - |A_l^R|^2). \tag{F.12}$$

For any stationary state, the continuity equation, Eq. (1.38), reduces to

$$\frac{dS(x)}{dx} = 0 \tag{F.13}$$

and the current is therefore the same for any position, in particular $S_R = S_L$ or[1]

$$|A_r^L|^2 - |A_l^L|^2 = |A_r^R|^2 - |A_l^R|^2. \tag{F.14}$$

Introducing the adjoint notation for mapping a column vector into a row vector,

$$\mathbf{A}^\dagger = \begin{pmatrix} A_r \\ A_l \end{pmatrix}^\dagger \equiv (A_r^*, A_l^*), \tag{F.15}$$

and noting that

$$\begin{pmatrix} A_r \\ -A_l \end{pmatrix} = \boldsymbol{\sigma}^z \begin{pmatrix} A_r \\ A_l \end{pmatrix}, \tag{F.16}$$

where

$$\boldsymbol{\sigma}^z = \begin{pmatrix} 1 & 0 \\ 0 & -1 \end{pmatrix}, \qquad \boldsymbol{\sigma}^z \boldsymbol{\sigma}^z = 1 \tag{F.17}$$

is the third Pauli matrix, Eq. (F.14) can be rewritten as

$$(\mathbf{A}^R)^\dagger \boldsymbol{\sigma}^z \mathbf{A}^R = |A_r^R|^2 - |A_l^R|^2 = |A_r^L|^2 - |A_l^L|^2 = (\mathbf{A}^L)^\dagger \boldsymbol{\sigma}^z \mathbf{A}^L. \tag{F.18}$$

For a matrix $\mathcal{T}$ operating on a column vector $\mathbf{A}$, the adjoint becomes

$$(\mathcal{T}\mathbf{A})^\dagger = \mathbf{A}^\dagger \mathcal{T}^\dagger, \tag{F.19}$$

where $\mathcal{T}^\dagger$ is the hermitian conjugate or adjoint matrix (transposed and complex-conjugated)

$$\mathcal{T}^\dagger = \begin{pmatrix} T_{11}^* & T_{21}^* \\ T_{12}^* & T_{22}^* \end{pmatrix} = (\mathcal{T}^{\mathrm{T}})^*. \tag{F.20}$$

Expressing the amplitudes on the right in terms of those on the left through the transfer matrix, Eq. (F.18) becomes

$$(\mathbf{A}^L)^\dagger \mathcal{T}^\dagger \boldsymbol{\sigma}^z \mathcal{T} \mathbf{A}^L = (\mathbf{A}^L)^\dagger \boldsymbol{\sigma}^z \mathbf{A}^L. \tag{F.21}$$

[1] We note that the current conservation condition, Eq. (F.14), for the scattering situation implies the transmission and reflection probability-conserving relation $T(E) + R(E) = 1$.

Since $\mathbf{A}^L$ is arbitrary, current conservation thus introduces, for any scattering potential, the *hyperbolic* constraint on the transmission matrix,

$$\mathcal{T}^\dagger \boldsymbol{\sigma}^z \mathcal{T} = \boldsymbol{\sigma}^z. \tag{F.22}$$

Multiplying in Eq. (F.22) with $\boldsymbol{\sigma}^z$ from the right, we get

$$\mathcal{T}^\dagger \boldsymbol{\sigma}^z \mathcal{T} \boldsymbol{\sigma}^z = \mathbf{1}, \tag{F.23}$$

and noting the sandwich property

$$\boldsymbol{\sigma}^z \begin{pmatrix} \cdot & \cdot \\ \cdot & \cdot \end{pmatrix} \boldsymbol{\sigma}^z = \begin{pmatrix} + & - \\ - & + \end{pmatrix}, \tag{F.24}$$

Eq. (F.23) becomes

$$\begin{pmatrix} 1 & 0 \\ 0 & 1 \end{pmatrix} = \begin{pmatrix} T_{11}^* & T_{21}^* \\ T_{12}^* & T_{22}^* \end{pmatrix} \begin{pmatrix} T_{11} & -T_{12} \\ -T_{21} & T_{22} \end{pmatrix} \tag{F.25}$$

or

$$\begin{pmatrix} 1 & 0 \\ 0 & 1 \end{pmatrix} = \begin{pmatrix} |T_{11}|^2 - |T_{21}|^2 & -T_{11}^* T_{12} + T_{21}^* T_{22} \\ T_{12}^* T_{11} - T_{21} T_{22}^* & -|T_{12}|^2 + |T_{22}|^2 \end{pmatrix}. \tag{F.26}$$

The off-diagonal elements thus vanish, in the way enforced by the time reversal symmetry constraint, Eq. (F.7), which also makes the diagonal elements equal to the determinant of the transfer matrix,

$$\mathbf{1} = (|T_{11}|^2 - |T_{21}|^2)\mathbf{1} \tag{F.27}$$

or equivalently

$$\det \mathcal{T} = 1, \tag{F.28}$$

i.e. the determinant of a transfer matrix quite generally equals one.

That the determinant of a transfer matrix equals one can also be arrived at using properties of the *barrier* matrices, $\mathcal{K}(x_{l,r})$. The determinant in Eq. (F.2) is the so-called Wronskian, which in Appendix E was shown to be a constant different from zero for independent solutions. Since the determinant of the inverse of a matrix is the inverse of the determinant, it follows from the matrix product form of the transfer matrix that its determinant equals one.

Exercise F.1 Consider a barrier potential with reflection symmetry, i.e. $V(-x) = V(x)$. Show that the transfer matrix then has the reflection symmetry property

$$\mathcal{T}(x_r^B, x_l^B) = \boldsymbol{\sigma}^x \mathcal{T}(x_l^B, x_r^B) \boldsymbol{\sigma}^x. \tag{F.29}$$

Show, by invoking the time reversal symmetry property, that the transfer matrix then has the property

$$\mathcal{T}^*(x_r^B, x_l^B) = \mathcal{T}(x_l^B, x_r^B). \tag{F.30}$$

Check that these properties are indeed satisfied in the case of the square barrier.

Exercise F.2 Show that, for the stationary state given in Eq. (2.6), the probability current in the barrier region is given by

$$S_B = \frac{2\hbar\kappa(E)}{m}\Im(A_l^B(E)A_r^{B^*}(E)). \tag{F.31}$$

Appendix G: Momentum

The use of position to describe the quantum state of a system, in terms of its wave function, $\psi(\mathbf{x})$, is complete and might be said to have a preferred practical status for intuitive reasons since phenomena (and experiments) are immediately described in space and time. However, to specify the laws of nature, i.e. the Hamiltonian, we have already in Section A.9 realized that we need to invoke more than simply the position coordinate, *viz.* the differential operator, which was seen to be related to the concept of momentum in classical mechanics according to the canonical quantization rule, Eq. (A.41). In the following this connection will be fully explored, resulting in a complementary description of a quantum system in terms of momentum probability amplitudes.

G.1 Time-of-Flight Experiment

Here we introduce what is meant by the *momentum* of a particle in quantum mechanics, and obtain a probability amplitude description of a physical system in terms of the momentum degree of freedom. To this end, consider a time-of-flight experiment. Suppose that, at a given time $t = 0$, we ascertain that a particle is in a definite region of space, say, by administering fast opening and closing of a set of shutters in an otherwise impenetrable wall to an approaching particle. A particle making it through the shutter at the initial time, $t = 0$, should be ascribed a wave function

$$\psi(\mathbf{x}, t = 0) = \psi_i(\mathbf{x}), \tag{G.1}$$

which is non-vanishing only in the region near the shutter, i.e. for $\mathbf{x} \sim \mathbf{0}$, as we choose our reference frame to have its origin in that region. At times subsequent to $t = 0$, the motion of the particle is free until at time t where particle position detectors are activated. Suppose the particle after the time interval t is detected to be in a volume element $\Delta\mathbf{x}$ around position $\mathbf{x}$, or as we shall say, the detector at that location *clicks*. We shall then *say* that the particle at time $t = 0$ had a *momentum* in the volume $\Delta\mathbf{p} \equiv (m/t)^3 \Delta\mathbf{x}$ around the momentum vector $\mathbf{p} \equiv m\mathbf{x}/t$.[1] Upon repeating the measurement (ascertaining the positions of identically prepared particles according to the described procedure), clicks will occur in other detectors. Which detector clicks is unpredictable and erratic, since if the detector at position

[1] Of course, a time-of-flight measurement is a *velocity* measurement, but we shall for later convenience phrase it in the equivalent terms of momentum.

x clicks it means that the particle interacted with this detector, i.e. arrived at it, and this event has according to quantum mechanics only ascribed the probability distribution $P(\mathbf{x}, t) = |\psi(\mathbf{x}, t)|^2$. Slowly but surely the erratic detector clicks build up this probability distribution, an example of which is shown in Figure G.1.

By definition, a corresponding distribution in initial momentum values is built, $P(\mathbf{p}, t = 0)\,\Delta\mathbf{p}$, since this histogram is the relative number of clicks in the detector at corresponding position $\mathbf{x} = \mathbf{p}t/m$ at measurement time t. The corresponding momentum measurement histogram thus emerges, as depicted in Figure G.2.

Thus, in terms of the definition of momentum, the two probability distributions are related according to

$$P_{\mathrm{p}}(\mathbf{p}, t = 0)\,\Delta\mathbf{p} = P_{\mathrm{x}}(\mathbf{x}, t)\,\Delta\mathbf{x} = |\psi(\mathbf{x}, t)|^2 \Delta\mathbf{x}, \tag{G.2}$$

where the subscripts on the probability distributions are introduced to indicate that they are different functions of their variables.

Using the fundamental dynamic equation, the relationship between position and momentum distributions can be established. Since the propagation to the detector is free, we can express the momentum probability distribution, Eq. (G.2), in terms of the initial wave function and the free propagator according to

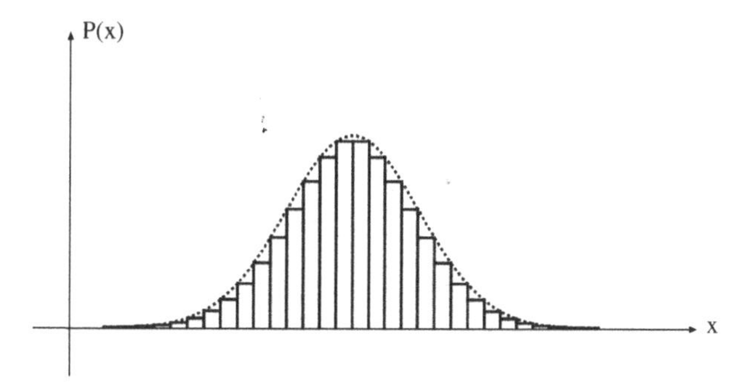

Figure G.1 A histogram of relative frequency detections of the particle locations.

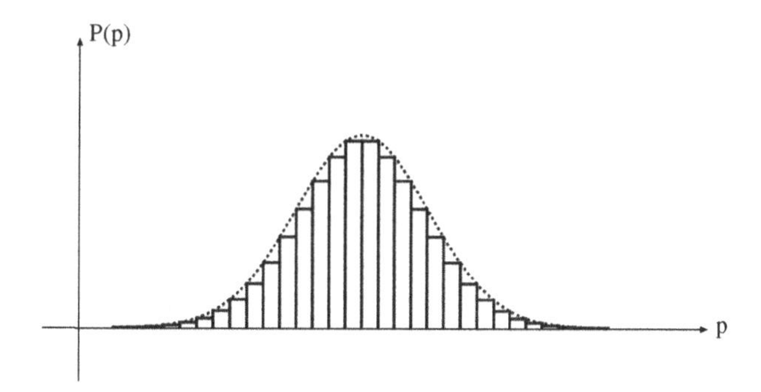

Figure G.2 Corresponding momentum histogram.

$$P_{\mathrm{p}}(\mathbf{p}, t=0)\, \Delta \mathbf{p} = \left| \int d\mathbf{x}'\; K_0(\mathbf{x}, t; \mathbf{x}', 0)\, \psi_i(\mathbf{x}') \right|^2 \Delta \mathbf{x}$$
$$= \left(\frac{m}{2\pi\hbar t}\right)^3 \left| \int d\mathbf{x}'\; \exp\left\{ \frac{im}{2\hbar} \frac{(\mathbf{x}-\mathbf{x}')^2}{t} \right\} \psi_i(\mathbf{x}') \right|^2 \Delta \mathbf{x}$$
$$= \left(\frac{1}{2\pi\hbar}\right)^3 \left| \int d\mathbf{x}'\; \exp\left\{ -\frac{i}{\hbar} \mathbf{x}' \cdot \left(\frac{m\mathbf{x}}{t} - \frac{m\mathbf{x}'}{2t} \right) \right\} \psi(\mathbf{x}', t=0) \right|^2 \Delta \mathbf{p}, \tag{G.3}$$

where, in the last equality, the relationship between position and momentum volumes has been invoked, the overall factor of one, $|\exp\{im\mathbf{x}^2/2\hbar t\}| = 1$, has been deleted, and the relationship between position and momentum volumes has been invoked. Since $\mathbf{x} = \mathbf{p}t/m$, the momentum probability distribution at time $t = 0$ is given by

$$P_{\mathrm{p}}(\mathbf{p}, t=0) = \frac{1}{(2\pi\hbar)^3} P_{\mathrm{x}}\left(\mathbf{x} = \frac{\mathbf{p}t}{m}, t\right)$$
$$= \frac{1}{(2\pi\hbar)^3} \left| \int d\mathbf{x}'\; \exp\left\{ -\frac{i}{\hbar} \mathbf{x}' \cdot \left(\mathbf{p} - \frac{m\mathbf{x}'}{2t} \right) \right\} \psi(\mathbf{x}', t=0) \right|^2. \tag{G.4}$$

The initial state, $\psi(\mathbf{x}', t=0)$, is localized near the shutter, with a linear extension of the wave function of size L_s, determined by the functioning of the shutter. The main contribution to the integral therefore comes from the region $|\mathbf{x}'| \leq L_s$. We therefore choose the time, t, of the measurement so that, for the region of momentum values of interest, we have $mL_s/t \ll \min\{|p_x|, |p_y|, |p_z|\}$. We can then neglect the quadratic term in $\mathbf{x}'$ in the exponent in Eq. (G.4), and obtain the momentum probability density at time $t = 0$,

$$P_{\mathrm{p}}(\mathbf{p}, t=0) = \frac{1}{(2\pi\hbar)^3} \left| \int d\mathbf{x}'\; e^{-(i/\hbar)\mathbf{x}'\cdot\mathbf{p}}\, \psi(\mathbf{x}', t=0) \right|^2. \tag{G.5}$$

Introducing the momentum probability *amplitude* according to the standard

$$P_{\mathrm{p}}(\mathbf{p}, t=0) = |\psi(\mathbf{p}, t=0)|^2, \tag{G.6}$$

we have, up to the usual arbitrary overall phase factor of a probability amplitude, that the position and momentum probability amplitudes are related through Fourier transformation,

$$\psi(\mathbf{p}, t=0) = \frac{1}{(2\pi\hbar)^{3/2}} \int d\mathbf{x}\; e^{-(i/\hbar)\mathbf{x}\cdot\mathbf{p}}\, \psi(\mathbf{x}, t=0). \tag{G.7}$$

Since the time t before activating the detectors is the choice of the experimenter, it can always be chosen long enough (for any given range of momenta of interest) so that the initial confinement due to the shutter arrangement is irrelevant. Thus considering such a general wave function, only localized due to it being normalized, leads to the conclusion that the position and momentum probability amplitudes are related through Fourier transformation,

$$\psi(\mathbf{p}, t) = \frac{1}{(2\pi\hbar)^{3/2}} \int d\mathbf{x}\; e^{-(i/\hbar)\mathbf{x}\cdot\mathbf{p}}\, \psi(\mathbf{x}, t). \tag{G.8}$$

As regards the wave function, $\psi(\mathbf{x})$, we speak of the position representation of the state, and as regards the momentum probability amplitude function, $\psi(\mathbf{p})$, we speak of the momentum representation of the state in question.

G.2 Momentum Representation

The momentum description of a quantum system is a complete description just like the position description with all analogous features, say, a momentum probability amplitude propagator. To obtain its relation to the position probability amplitude propagator, we express the momentum probability amplitude function in terms of the wave function,

$$\begin{aligned}\psi(\mathbf{p},t) &= \int \frac{d\mathbf{x}}{(2\pi\hbar)^{3/2}}\, e^{-(i/\hbar)\mathbf{x}\cdot\mathbf{p}}\, \psi(\mathbf{x},t) \\ &= \int \frac{d\mathbf{x}}{(2\pi\hbar)^{3/2}}\, e^{-(i/\hbar)\mathbf{x}\cdot\mathbf{p}} \int d\mathbf{x}'\, K(\mathbf{x},t;\mathbf{x}',t')\, \psi(\mathbf{x}',t'),\end{aligned} \tag{G.9}$$

where the last equality is obtained simply by using the fundamental equation specifying the dynamics, Eq. (A.12). Then introducing the inverse Fourier transform

$$\psi(\mathbf{x}',t') = \int \frac{d\mathbf{p}'}{(2\pi\hbar)^{3/2}}\, e^{(i/\hbar)\mathbf{p}'\cdot\mathbf{x}'}\, \psi(\mathbf{p}',t'), \tag{G.10}$$

Eq. (G.9) becomes

$$\psi(\mathbf{p},t) = \int d\mathbf{p}'\, K(\mathbf{p},t;\mathbf{p}',t')\, \psi(\mathbf{p}',t'), \tag{G.11}$$

where

$$K(\mathbf{p},t;\mathbf{p}',t') = \int \frac{d\mathbf{x}}{(2\pi\hbar)^{3/2}} \int \frac{d\mathbf{x}'}{(2\pi\hbar)^{3/2}}\, e^{-(i/\hbar)(\mathbf{x}\cdot\mathbf{p}-\mathbf{x}'\cdot\mathbf{p}')} K(\mathbf{x},t;\mathbf{x}',t'). \tag{G.12}$$

Interpreting Eq. (G.11) in terms of the multiplication rule and the superposition principle for momentum, $K(\mathbf{p},t;\mathbf{p}',t')$ is seen to be the conditional probability amplitude for the particle to have momentum $\mathbf{p}$ at time t *given* it had momentum $\mathbf{p}'$ at time t', i.e. it is the momentum probability amplitude propagator. In particular, we obtain from Eq. (G.11) at equal times that

$$K(\mathbf{p},t;\mathbf{p}',t) = \delta(\mathbf{p}-\mathbf{p}'), \tag{G.13}$$

i.e. a particle that definitely has the momentum $\mathbf{p}'$ is specified by the momentum probability amplitude function (up to a phase factor)[2]

$$\psi_{\mathbf{p}'}(\mathbf{p}) = \delta(\mathbf{p}-\mathbf{p}'). \tag{G.14}$$

[2] This could, of course, be written down immediately in analogy with the wave function of definite position, Eq. (A.14).

A particle that definitely has the momentum $\mathbf{p}$ is, according to Eqs. (G.14) and (G.10), described by the wave function

$$\psi_{\mathbf{p}}(\mathbf{x}) = \frac{1}{(2\pi\hbar)^{3/2}}\, e^{(i/\hbar)\mathbf{p}\cdot\mathbf{x}}, \tag{G.15}$$

i.e. a plane wave. A state of definite momentum $\mathbf{p}$ is described by a wave function that oscillates in space with the de Broglie wave length $\lambda = 2\pi\hbar/|\mathbf{p}|$, and its wave vector is related to the momentum of the particle according to the de Broglie relation, $\mathbf{p} = \hbar\mathbf{k}$.

We note that a particle in a state of definite momentum corresponds to equal (relative) probability for finding the particle anywhere in space. If the particle has a definite momentum, complete ignorance, indeterminacy, about its position reigns. We have struck upon the essential feature of quantum mechanics, complementarity: that all the physical attributes of a system, here the position and momentum of a particle, cannot simultaneously be ascribed with arbitrary accuracy. If the particle with definiteness has momentum $\mathbf{p}$, all position outcomes have equal probability (and vice versa).

Exercise G.1 Show that the wave functions of definite momentum, the plane waves, Eq. (G.15), are so-called *delta*-normalized since

$$\int d\mathbf{x}\; \psi^*_{\mathbf{p}}(\mathbf{x})\, \psi_{\mathbf{p}'}(\mathbf{x}) = \delta(\mathbf{p} - \mathbf{p}'). \tag{G.16}$$

Energy eigenfunctions corresponding to different energies are thus orthogonal.

Note similarly the identity

$$\int d\mathbf{p}\; \psi^*_{\mathbf{p}}(\mathbf{x}')\, \psi_{\mathbf{p}}(\mathbf{x}) = \delta(\mathbf{x} - \mathbf{x}'), \tag{G.17}$$

expressing the completeness of the plane waves.

If the particle is in a superposition of states of different momenta, i.e. its wave function is

$$\psi(\mathbf{x}) = \int d\mathbf{p}\; a_{\mathbf{p}}\, \psi_{\mathbf{p}}(\mathbf{x}) = \int \frac{d\mathbf{p}}{(2\pi\hbar)^3}\; a_{\mathbf{p}}\, e^{(i/\hbar)\mathbf{p}\cdot\mathbf{x}}, \tag{G.18}$$

then the probability that a momentum measurement will give the outcome $\mathbf{p}$ is, according to Eq. (G.8),

$$P(\mathbf{p}) = \frac{1}{(2\pi\hbar)^3} \left| \int d\mathbf{x}\; e^{-(i/\hbar)\mathbf{x}\cdot\mathbf{p}} \int d\mathbf{p}'\; a_{\mathbf{p}'}\, \psi_{\mathbf{p}'}(\mathbf{x}) \right|^2 = |a_{\mathbf{p}}|^2, \tag{G.19}$$

where the last equality follows from Eq. (G.16).

We can now interpret the Fourier transformation, Eq. (G.10),

$$\psi(\mathbf{x}, t) = \int d\mathbf{p}\; \psi(\mathbf{p}, t)\, \psi_{\mathbf{p}}(\mathbf{x}), \tag{G.20}$$

as the statement that any wave function can be expanded on the set of wave functions of definite momentum (with the expansion coefficients, $\psi(\mathbf{p}, t)$, given by Eq. (G.8)).

The momentum eigenfunctions form a so-called complete set of functions. We note that, according to Eq. (G.19), the absolute square of the expansion coefficients, $|\psi(\mathbf{p},t)|^2$, gives the probability that a momentum measurement at time t will give the outcome $\mathbf{p}$.

Exercise G.2 The possible momentum values are the set of real vectors. For example, the possible values of, say, the momentum along the x direction are all the real numbers, p_x. Introducing the notation for the differential operator,

$$\hat{\mathbf{p}} = \frac{\hbar}{i}\nabla_{\mathbf{x}}, \tag{G.21}$$

show that a state of definite momentum $\mathbf{p}$ is seen to satisfy the eigenvalue equation

$$\hat{\mathbf{p}}\,\psi_{\mathbf{p}}(\mathbf{x}) = \mathbf{p}\,\psi_{\mathbf{p}}(\mathbf{x}). \tag{G.22}$$

The eigenvalue equation represents the values of the physical quantity, momentum, as the eigenvalues of an operator whose eigenfunctions are the wave functions corresponding to the states of definite momentum, or so-called momentum eigenstates, specified by the eigenvalue in question. The differential operator, $\hat{\mathbf{p}}$, is called the momentum operator.

Having identified the momentum operator, we can now give a physical interpretation of the canonical quantization rule, Eq. (A.41). The Hamiltonian is obtained by substituting in Hamilton's function the momentum operator for the momentum, the *canonical quantization rule*.

Exercise G.3 Performing the Gaussian integrations in Eq. (G.12) for the case of a free particle, obtain for the free propagator in the momentum representation

$$K_0(\mathbf{p},t;\mathbf{p}',t') = e^{-(i/\hbar)\epsilon_{\mathbf{p}}(t-t')}\,\delta(\mathbf{p}-\mathbf{p}'). \tag{G.23}$$

The time evolution of a free particle in the momentum representation is thus described by a simple phase factor determined by the energy, $\epsilon_{\mathbf{p}} = \mathbf{p}^2/2m$, of the particle in the momentum state in question.

Exercise G.4 In view of Eqs. (G.23) and (G.11), the free evolution of the momentum probability amplitude is (this is, of course, the same conclusion, Eq. (1.21), we arrived at by solving the free particle Schrödinger equation in Section 1.2)

$$\psi(\mathbf{p},t) = e^{-(i/\hbar)\epsilon_{\mathbf{p}}(t-t')}\,\psi(\mathbf{p},t'). \tag{G.24}$$

The time evolution of a free particle in the momentum representation is thus described by a simple phase factor determined by the energy of the particle in the momentum state in question and elapsed time. A free particle state stays in the same momentum ray. The momentum probability distribution for a free particle is thus independent of time $|\psi(\mathbf{p},t)|^2 = |\psi(\mathbf{p},t')|^2$, or, as we shall say, the momentum is a constant of motion.

Returning to the time-of-flight measurement, according to Eq. (G.24) the time evolution of the Fourier transform of the wave function between the time of opening the shutter, $t = 0$, and the activation of the detectors is

$$\psi(\mathbf{p},t) = e^{-(i/\hbar)\epsilon_{\mathbf{p}}t}\, \psi_i(\mathbf{p},t=0). \tag{G.25}$$

Prior to the second position measurement in the time-of-flight experiment, the momentum probability distribution is thus identical to that after the operation of the shutter, the first position measurement, and the measured probability distribution equals that at the initial time.

In practice, it is not possible to prepare a particle in a state of definite momentum, and the main use of the plane waves is for formal considerations. The best one can do in this regard is to prepare a particle in a state where the momentum probability distribution is sharply peaked around a desired momentum value. The corresponding wave function is referred to as a wave packet, since a superposition of definite momentum states with values in volume $\Delta\mathbf{p}$ is a function localized in space, as the Fourier-transformed function has spatial width set by the lower bound $\Delta\mathbf{p}\,\Delta\mathbf{x} \geq (2\pi\hbar)^3$. When discussing particle motion in spatially varying potentials, it is beneficial to use spatially localized wave packets whose extension is small compared to the spatial variation of the potential, so that the potential can be assumed constant over its size.

Appendix H: Confined Particles

The time-independent Schrödinger equation for a free particle of mass m is

$$-\frac{\hbar^2}{2m}\nabla_{\mathbf{x}}^2\,\psi(\mathbf{x}) = E\,\psi(\mathbf{x}). \tag{H.1}$$

The general solution for the energy eigenvalue $E = \epsilon_{\mathbf{k}} = \hbar^2\mathbf{k}^2/2m$ is (recall Section 2.1.1)

$$\psi_{\epsilon_{\mathbf{k}}}(\mathbf{x}) = A_r\,e^{i\mathbf{k}\cdot\mathbf{x}} + A_l\,e^{-i\mathbf{k}\cdot\mathbf{x}}. \tag{H.2}$$

Except for the non-degenerate state of lowest energy, $E = 0$, the free particle energy states are infinitely degenerate; wave vectors on a sphere have the same energy. Each of the plane waves in Eq. (H.2) are eigenstates of momentum with eigenvalues $\pm\hbar\mathbf{k}$.

The electrons inside a material are confined; their wave functions vanish outside the material's somewhat blurry *surface*. The confinement potential at the edge of a material has, for example, the form shown in Figure H.1. The potential determines how electron wave functions decay into the vacuum side.

If we are interested in the bulk properties of a material, this tiny weight of the wave function is irrelevant compared to its bulk weight of probability almost equal to one. One can therefore assume that the probability is zero for the electron to be outside the box, corresponding to the potential being infinite outside the box. The wave function thus vanishes on the surfaces of the box, the so-called box boundary condition (recall the discussion in Appendix E), whereas the particle is free to move inside the box. The coefficients in Eq. (H.2) must thus be chosen such that the wave function vanishes at the boundaries, i.e. conspiring the two exponentials to become trigonometric functions. Choosing the origin in the middle of the box of volume $\Omega = L_xL_yL_z$, the energy eigenfunctions are (with $\mathbf{n} = (n_x, n_y, n_z)$)

$$\psi_{\mathbf{n}}(\mathbf{x}) = \prod_{\alpha=x,y,z}\sqrt{\frac{2}{L_\alpha}}\,\sin(k_{n_\alpha}(x_\alpha - L_\alpha/2)), \qquad k_{n_\alpha} = \frac{\pi}{L_\alpha}n_\alpha, \quad n_\alpha = 1,2,3,\ldots, \tag{H.3}$$

where the boundary condition restricts the wave vectors to the indicated values, and as a consequence the energies are quantized to the allowed values only,

$$E_{\mathbf{n}} = \frac{\pi^2\hbar^2}{2m}\left(\frac{n_x^2}{L_x^2} + \frac{n_y^2}{L_y^2} + \frac{n_z^2}{L_z^2}\right), \qquad n_x, n_y, n_z = 1,2,3,\ldots. \tag{H.4}$$

Since the energy eigenfunctions, Eq. (H.3), are real, the probability current density vanishes in these states. This is inconvenient when discussing transport properties (whereas the

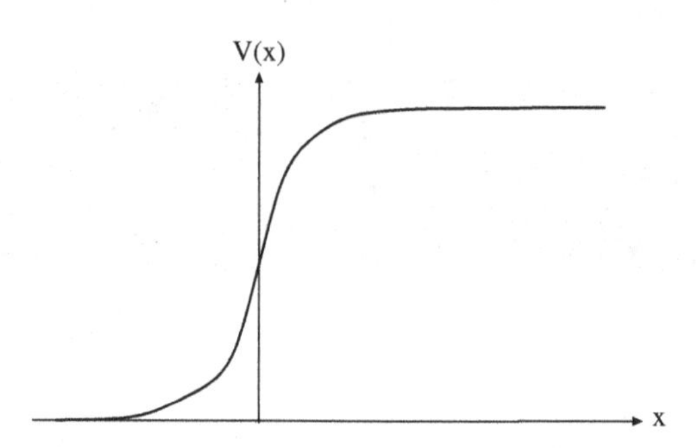

Figure H.1 Confining potential at a surface.

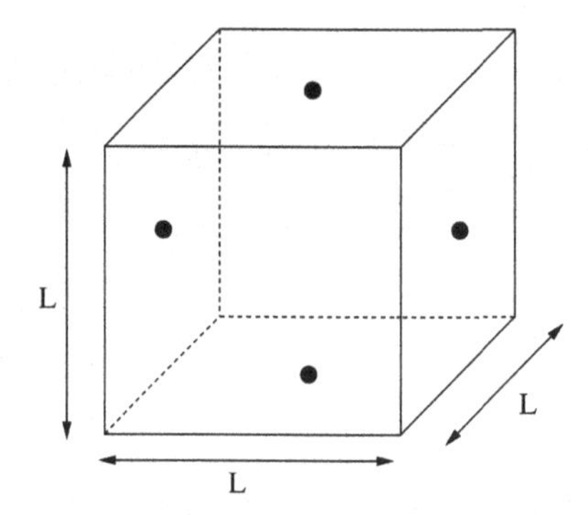

Figure H.2 Cubic box.

box boundary condition is suitable for considering thermodynamic properties), and instead we employ the plane waves

$$\psi_{\mathbf{p}}(\mathbf{x}) = \frac{1}{\sqrt{\Omega}}\, e^{(i/\hbar)\mathbf{p}\cdot\mathbf{x}}, \tag{H.5}$$

which have associated probability current density, $\mathbf{S} = \mathbf{p}/m\Omega$.

On these current-carrying states, we impose the periodic boundary condition, i.e. allowed wave functions are forced to take on the same value at points opposite each other on the pairwise opposing surfaces, examples of which are depicted by dots in Figure H.2.

The periodic boundary condition thus demands

$$\psi(\mathbf{x} + \mathbf{L}_\alpha) = \psi(\mathbf{x}), \qquad \mathbf{L}_\alpha = L\,\mathbf{e}_\alpha, \qquad \alpha = x, y, z, \tag{H.6}$$

where the triad of unit vectors $\mathbf{e}_\alpha$ defines in general the box of volume $\Omega = L_x L_y L_z$, i.e. the requirement

$$e^{(i/\hbar)\mathbf{p}\cdot\mathbf{L}_\alpha} = 1, \qquad \mathbf{p}_{\mathbf{n}} = 2\pi\hbar\,\tilde{\mathbf{n}}, \qquad \tilde{\mathbf{n}} = \left(\frac{n_x}{L_x}, \frac{n_y}{L_y}, \frac{n_z}{L_z}\right),$$
$$n_x, n_y, n_z = 0, \pm 1, \pm 2, \ldots, \tag{H.7}$$

and restricting the allowed momentum values as stated.

Any function that has the same values at opposing points on surfaces of a finite volume of space can be represented by its Fourier series (recall Appendix C). In other words, the plane waves, Eq. (H.5), restricted to the above discrete momentum values are a complete set of states.

The allowed momentum values reside on a lattice in momentum space, as depicted in Figure H.3. We note that there is one momentum state for each momentum volume of size $\Delta\mathbf{p}$, where

$$\Delta\mathbf{p} = \frac{(2\pi\hbar)^3}{\Omega}, \tag{H.8}$$

where for a cubic box the volume is L^3.

The periodic boundary condition is equivalent to a periodic repetition of the box. If the particle leaves the box at one side, say, it is in a wave packet state, the periodic boundary condition ensures that its *periodic-pode* enters at the *symmetric* point on the opposite surface of the box, as illustrated in Figure H.4. Particle conservation is thereby ensured. In mathematical terms, the periodic boundary condition makes the free particle Hamiltonian hermitian (and also the momentum operator), since the boundary terms in a partial integration in Eq. (1.11) cancel each other, the region of integration being the volume of the box.

Consider next an assembly of N non-interacting particles. The Hamiltonian is thus the sum of the Hamiltonians for the independent particles

$$\hat{H}_0 = \sum_{i=1}^{N} \frac{\hat{\mathbf{p}}_i^2}{2m_i} = -\sum_{i=1}^{N} \frac{\hbar^2}{2m_i}\Delta_i, \tag{H.9}$$

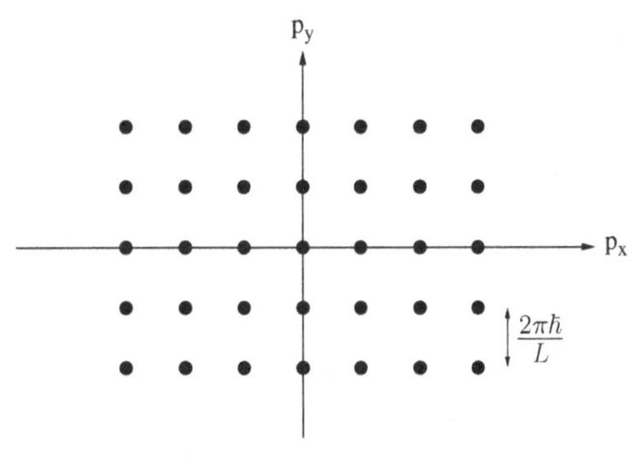

Figure H.3 For periodic boundary conditions, the allowed momentum quantum numbers reside on a lattice determined by the lattice constant $2\pi\hbar/L$.

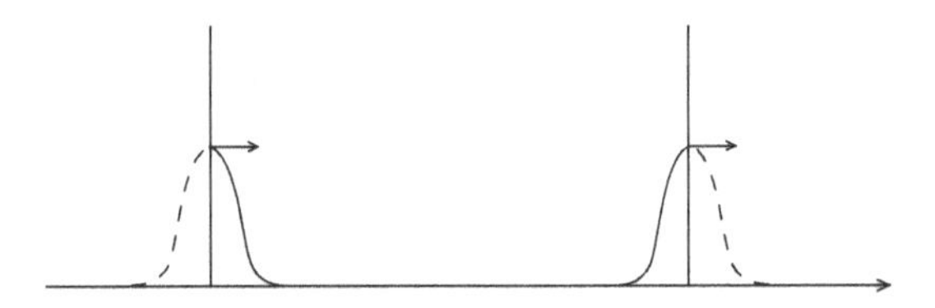

Figure H.4 With periodic boundary conditions, the particle re-enters as the periodic-pode when the particle leaves.

where m_i, $i = 1, \ldots, N$, are the masses of the particles. The time-independent Schrödinger equation

$$\left(-\sum_{i=1}^{N} \frac{\hbar^2}{2m_i} \Delta_i \right) \psi(\mathbf{x}_1, \ldots, \mathbf{x}_N) = E\, \psi(\mathbf{x}_1, \ldots, \mathbf{x}_N) \tag{H.10}$$

has energy eigenstates that are the product states of single-particle momentum eigenstates,

$$\psi_{\mathbf{p}_1,\ldots,\mathbf{p}_N}(\mathbf{x}_1, \ldots, \mathbf{x}_N) = \prod_{i=1}^{N} \psi_{\mathbf{p}_i}(\mathbf{x}_i) = \left(\frac{1}{\sqrt{\Omega}} \right)^N e^{(i/\hbar)\mathbf{p}_1 \cdot \mathbf{x}_1} e^{(i/\hbar)\mathbf{p}_2 \cdot \mathbf{x}_2} \ldots e^{(i/\hbar)\mathbf{p}_N \cdot \mathbf{x}_N}, \tag{H.11}$$

where the last equality sign assumes that the particles are confined in a box of volume Ω. These functions constitute a complete set for the N-particle system.

The basis states, Eq. (H.11), are simultaneous eigenstates of the energy and momentum operators for the N-particle system. The total momentum operator and the total momentum eigenvalue are

$$\hat{\mathbf{P}} = \sum_{i=1}^{N} \hat{\mathbf{p}}_i, \qquad \mathbf{P} = \sum_{i=1}^{N} \mathbf{p}_i \tag{H.12}$$

and the total energy value is

$$E(\mathbf{p}_1, \ldots, \mathbf{p}_N) = \sum_{i=1}^{N} \epsilon_i, \qquad \epsilon_i = \frac{\mathbf{p}_i^2}{2m_i}. \tag{H.13}$$

Assuming periodic boundary conditions, each momentum value $\mathbf{p}_i$ can take on the allowed values specified in Eq. (H.7), and an arbitrary N-particle wave function has its expansion in the basis states,

$$\psi(\mathbf{x}_1, \ldots, \mathbf{x}_N) = \sum_{\{\mathbf{p}_i\}_{i=1,\ldots,N}} c(\mathbf{p}_1, \mathbf{p}_2, \ldots, \mathbf{p}_N)\, \psi_{\mathbf{p}_1,\ldots,\mathbf{p}_N}(\mathbf{x}_1, \ldots, \mathbf{x}_N), \tag{H.14}$$

where the sum is over all possible N-tuples, $\{\mathbf{p}_i\}_{i=1,\ldots,N} \equiv (\mathbf{p}_1, \mathbf{p}_2, \ldots, \mathbf{p}_N)$, of momentum values.

The above description of an N-particle system is only correct if the particles are not identical (identical, at present, only implying that they have the same mass). The indistinguishability of identical particles has profound consequences for the allowed states of such a many-particle system, as described in Appendix I.

Exercise H.1 Consider two particles in the state described by the wave function $\psi(\mathbf{x}_1, \mathbf{x}_2, t)$. The probability for one of the particles to be at location $\mathbf{x}$ at time t is

$$P(\mathbf{x}, t) = \int d\mathbf{x}_2\, |\psi(\mathbf{x}, \mathbf{x}_2, t)|^2 + \int d\mathbf{x}_1\, |\psi(\mathbf{x}_1, \mathbf{x}, t)|^2, \tag{H.15}$$

which is the probability that the particle described by the first argument is at the location irrespective of the location of the other particle plus vice versa. If the wave function is normalized, the probability $P(\mathbf{x}, t)$ is normalized according to (recall Eq. (1.5))

$$\int d\mathbf{x}\, P(\mathbf{x}, t) = 2. \tag{H.16}$$

Use the Schrödinger equation to obtain the continuity equation and identify the probability current density.

Appendix I: Spin and Quantum Statistics

A particle has an internal quantum degree of freedom, its *spin* or *internal* angular momentum. For an electron, its spin can take on two values with respect to, say, the z direction. The electron is a spin-$1/2$ particle, $S = 1/2$, as angular momentum components in general have $2S + 1$ states. Indicating the two spin states by ± 1, the state where the particle is definitely in spin state s_z ($s_z = 1$ or $s_z = -1$) is

$$\chi_{s_z}(s_z') = \delta_{s_z, s_z'} = \begin{cases} 1, & s_z' = s_z, \\ 0, & s_z' \neq s_z. \end{cases} \tag{I.1}$$

For a complete specification, the internal degree of freedom of the particle must also be stated, and $\psi(\mathbf{x}, s_z)$ denotes the probability amplitude for the event of the particle at position $\mathbf{x}$ and with spin component s_z. When the position and spin degrees of freedom are independent of each other, $\psi(\mathbf{x}, s_z) = \psi(\mathbf{x})\,\chi(s_z)$, the probability for the event of the particle at position $\mathbf{x}$ and spin component s_z is

$$P(\mathbf{x}, s_z) = \psi^*(\mathbf{x})\chi^*(s_z)\,\psi(\mathbf{x})\chi(s_z) = |\psi(\mathbf{x})|^2\,|\chi(s_z)|^2. \tag{I.2}$$

Since the probability to be somewhere having some spin value equals one, we have

$$1 = \sum_{s_z=\pm 1} \int d\mathbf{x}\, P(\mathbf{x}, s_z) = \int d\mathbf{x}\, |\psi(\mathbf{x})|^2 \sum_{s_z=\pm 1} |\chi(s_z)|^2, \tag{I.3}$$

and we shall of course adopt the normalization

$$\sum_{s_z=\pm 1} |\chi(s_z)|^2 = 1, \tag{I.4}$$

the probabilities for the particle being in either of the spin states adding up to one, because observing the particle in either of the spin states is certain. For the case of interest, i.e. the basis state of a confined electron of a definite spin, its wave function is

$$\psi_{\mathbf{p},\pm 1}(\mathbf{x}, s_{z'}) = \delta_{\pm 1, s_{z'}} \frac{1}{\sqrt{\Omega}}\, e^{(i/\hbar)\mathbf{p}\cdot\mathbf{x}}. \tag{I.5}$$

Another quantum feature of nature is that the true identity of the same species of elementary particles is realized. Electrons are profoundly identical: there is nothing in nature to distinguish any two electrons – they are indistinguishable. Consider an assembly of N identical particles in a state described by the wave function $\psi(\mathbf{x}_1, s_1; \mathbf{x}_2, s_2; \ldots; \mathbf{x}_N, s_N)$, where s_i denotes either of the spin components. We use the convention that the particle referred to as particle *one* is described by the first argument of the wave function, and the particle referred to as particle *two* is described by the second argument, etc. The corresponding probability distribution for the whereabouts and spin components of the particles is then given by $|\psi(\mathbf{x}_1, s_1; \mathbf{x}_2, s_2; \ldots; \mathbf{x}_N, s_N)|^2$. Since the particles are indistinguishable, this probability is identical to the probability for the situation where we have interchanged the labeling of, say, the first two particles, or equivalently asking for the probability that the particle we refer to as the *second* particle is at position $\mathbf{x}_1$ and in spin state s_1 and the particle we refer to as the *first* particle is at position $\mathbf{x}_2$ and in spin state s_2. The probability distributions for the two situations must be the same,

$$|\psi(\mathbf{x}_1, s_1; \mathbf{x}_2, s_2; \ldots; \mathbf{x}_N, s_N)|^2 = |\psi(\mathbf{x}_2, s_2; \mathbf{x}_1, s_1; \ldots; \mathbf{x}_N, s_N)|^2, \tag{I.6}$$

as either configuration of particles is indistinguishable and therefore must be assigned the same probability, because otherwise the different probabilities would be a sign of distinguishability. The explicitly shown identity in Eq. (I.6) thus corresponds to the interchange of the spatial positions of particles one and two.

A wave function for identical particles must under the interchange of any pair of particles be either symmetric (in three spatial dimensions),

$$\psi(\mathbf{x}_1, s_1; \mathbf{x}_2, s_2; \ldots; \mathbf{x}_N, s_N) = \psi(\mathbf{x}_2, s_2; \mathbf{x}_1, s_1; \ldots; \mathbf{x}_N, s_N) \qquad \text{etc.,} \tag{I.7}$$

or antisymmetric,

$$\psi(\mathbf{x}_1, s_1; \mathbf{x}_2, s_2; \ldots; \mathbf{x}_N, s_N) = -\psi(\mathbf{x}_2, s_2; \mathbf{x}_1, s_1; \ldots; \mathbf{x}_N, s_N) \qquad \text{etc.} \tag{I.8}$$

Particles described by antisymmetric wave functions are called *fermions*, and particles described by symmetric wave functions are called *bosons*.

Since the electron has half-integer spin, according to the spin-statistics theorem it is a fermion, and the wave function for an assembly of electrons is antisymmetric under interchange of any two electrons.

Exercise I.1 Show that any two identical fermions cannot occupy the same state, as that would lead to a vanishing wave function. This is Pauli's exclusion principle, which has far-reaching consequences for the description of an assembly of identical fermions.

Appendix J: Statistical Mechanics

Statistical mechanics is the microscopic theory of the properties of macroscopic systems. The theory of thermal equilibrium states deals with the states typically reached by large bodies left to themselves, such as the water in a bath tub, which soon shows the same temperature on thermometers stuck in at different places. Since a macroscopic body consists of an astronomical number of particles, $N \sim 10^{22}$, a statistical description is needed. The *laws* of thermodynamics then emerge from statistical physics as empirical facts of certainty, *viz.* as the average assertions of the statistical theory where fluctuations in the system's properties are negligible, being macroscopically suppressed as being of order $1/N$.

A system in contact with a heat reservoir, a large body with a constant temperature, with which it can exchange energy, will not be in a definite energy state but will continually make transitions among different energy states. We are therefore interested in the probability, $P(E)$, that a certain energy state of the system is occupied. One can think of this probability as the average time spent by the system in the energy state in question. Alternatively, one can envisage a large number of identical systems, all in contact with a heat reservoir of the same temperature. The probability in question is then the probability that an arbitrary selection from the ensemble gives a system with the energy in question, the relative number of systems in the ensemble in that energy state.[1] The basic principle of equilibrium statistical mechanics states that the probability that a macroscopic system in thermal equilibrium at absolute temperature T is in the state with energy E is given by the Boltzmann factor

$$P(E) = Z^{-1}\, e^{-(E/kT)}, \tag{J.1}$$

where k is Boltzmann's constant (not a constant of nature, but the conventional converter between energy and temperature scales). As shown below, the factor $Z(T)$ secures that the probability distribution is properly normalized, and is called the partition function. The probability distribution, Eq. (J.1), can be inferred from the zeroth law of thermodynamics: the experimental fact (verified by sticking in thermometers) that two bodies having

[1] The two kinds of probability distributions are by ergodicity two realizations of the same probability distribution, the first realization being the one of interest if insisting on discussing a large but isolated single system, the latter, Gibbs' ensemble, being of calculational convenience. Proofs of ergodicity of isolated many-body systems are not easy, but ergodicity for simple realistic systems have been proved. If, on the other hand, one argues that a large system always has coupling to its environment, which is left unobserved, the corresponding reduced density matrix describing the system will reach a stationary state, i.e. all properties of the system are time-independent, and therefore is a function of the Hamiltonian of the system according to the von Neumann equation.

the same temperature will upon being brought into thermal contact remain in thermal equilibrium in a state with the same temperature as that of the two bodies when initially separated.

Let us, for simplicity, assume that the two systems are identical, say, two identical atomic gases enclosed in identical containers, so that their probability distributions for occupation of energy states are identical.[2] When separated and isolated, the two systems are independent and quantum mechanically described by the wave function that is a product of the wave functions for each of the individual systems, and the dynamics is governed by the Hamiltonian[3] $\hat{H}_{\rm IS} = \hat{H}_1 + \hat{H}_2$. In their thermal equilibrium state, each of the individual systems is attributed its probability distribution over its energies, $P_S(E)$. The probability, $P_{\rm IS}(E_1+E_2)$, for the two isolated systems to be in the state with total energy $E = E_1+E_2$, is then related to the probabilities that the two independent subsystems are in energy states E_1 and E_2 according to

$$P_{\rm IS}(E_1 + E_2) = P_S(E_1)\, P_S(E_2), \tag{J.2}$$

as the occupations of levels in two separate systems are statistically independent.

When the two systems are in contact, the Hamiltonian is modified to $\hat{H} = \hat{H}_1 + \hat{H}_2 + \hat{V}$, where $\hat{V}$ describes the additional interaction introduced by contact between the particles in the two bodies. For instance, assume that the particles are charged or interaction is mediated by the vibrations of atoms in the wall, leading to interaction between the atoms in the two gases bumping into the wall. However, since interaction is only effective between particles near the separating wall, this *surface* interaction has negligible effect on the energy levels of the total Hamiltonian compared to that of the bulk kinetic energy and possible, say, two-particle, interaction operative over the large volume of the system. In the thermodynamic limit of an infinite volume of the system, the term $\hat{V}$ can thus be dropped from the total Hamiltonian, leaving the thermal contact Hamiltonian equal to that of the isolated system case. The energy value of the combined system is thus constituted by the sum of energy values for the isolated system, $E = E_1 + E_2$. Just before contact, the probability for the total system to be in the states with the respective energies E_1 and E_2 is specified by Eq. (J.2). Immediately after contact, the two subsystems are in the same states as immediately before, and the probability, $P_{\rm CS}(E_1 + E_2)$, that the combined system in thermal contact is in energy state $E = E_1 + E_2$ thus equals that of the isolated case, $P_{\rm IS}(E_1 + E_2)$, and we have

$$P_{\rm CS}(E_1 + E_2) = P_S(E_1)\, P_S(E_2), \tag{J.3}$$

a relation valid for all values of the independent energy variables E_1 and E_2. The energy spectrum of any Hamiltonian is bounded from below, and since any constant added to a Hamiltonian leaves the physics invariant, we can assume that the ground-state energy

[2] Giving up on this assumption, as one should because the zeroth law of thermodynamics is universal, i.e. does not depend on the material structure of the bodies involved, would only lengthen the argument but lead to the same conclusion, as the following discussion shows. The conclusion valid for arbitrary systems in thermal contact is left as Exercise J.1.

[3] The two Hamiltonians are by assumption identical, the only difference being that they refer to the degrees of freedom in the two separate systems.

corresponds to energy value zero. Since in a macroscopic system the energy levels are infinitesimally close, the energy spectrum ranges in the thermodynamic limit continuously from zero to infinity. Setting $E_2 = 0$ in Eq. (J.3) gives

$$P_{CS}(E_1) = P_S(E_1)\, P_S(0). \tag{J.4}$$

This equation is valid for any energy value and we write

$$P_{CS}(E) = P_S(E)\, P_S(0). \tag{J.5}$$

Using this equation for the value $E = E_1 + E_2$ and Eq. (J.3) gives

$$P_S(E_1 + E_2)\, P_S(0) = P_{CS}(E_1 + E_2) = P_S(E_1)\, P_S(E_2), \tag{J.6}$$

and thereby for the distribution function $P_S(E)$ we obtain the equation

$$P_S(E_1 + E_2) = \frac{P_S(E_1)\, P_S(E_2)}{P_S(0)}. \tag{J.7}$$

This defines the function $P_S(E)$ to be the exponential function, or, if preferred, differentiating the equation with respect to E_2 and subsequently setting $E_2 = 0$ gives the differential equation

$$P_S'(E_1) = \frac{P_S'(0)}{P_S(0)}\, P_S(E_1). \tag{J.8}$$

This is another way of defining the exponential function and we have

$$P_S(E_1) = Z(T)\, e^{a(T)E_1}, \tag{J.9}$$

where $Z(T)$ is the normalization constant of the probability distribution, and $a(T)$ is negative as required by normalization. We therefore introduce the notation $\Theta(T) \equiv -1/a(T)$ and have

$$P_S(E) = Z(T)\, e^{-E/\Theta(T)}. \tag{J.10}$$

Exercise J.1 Show, by similar manipulations as above, that the formula for the probability of an energy state in the thermal equilibrium state, Eq. (J.10), is valid also for two systems that are not identical, having isolated probability distributions $P_1(E_1)$ and $P_2(E_2)$, respectively.

We can now identify Θ by using the result of the next appendix, realizing that the discussion there is valid without identifying Θ as kT, which is what we argue now. Consider a rarefied gas of particles. If Θ is larger than the chemical potential, $\Theta \gg \mu$, or rather $\exp(\mu/\Theta) \ll 1$, the fermion nature of the particles is irrelevant. The statistics reduces to Maxwell–Boltzmann statistics, for which one can derive the ideal gas law, thereby identifying Θ as the absolute temperature multiplied by Boltzmann's constant. We have hereby verified the basic principle of equilibrium statistical mechanics, Eq. (J.1).

Appendix K: Fermi–Dirac Distribution

Consider a gas of N identical fermions in thermal equilibrium at temperature T. The walls confining the particles vibrate due to its temperature, and as the particles bounce off the walls there is incessant energy exchange between the walls and the gas.[1] At each instant, electrons interacting with the walls are scattered into different energy states. Since a state of definite kinetic energy is also a state of definite momentum, the state of the gas can be described in momentum space where the occupations of different momentum states incessantly change. Owing to the exclusion principle, these dynamical processes are strongly restrained, as scattering can occur only into unoccupied states. As a result, the scattering activity will be limited to a thin (on the scale of the Fermi energy) layer around the Fermi surface, as the typical energy exchange is on the order of kT. In the thermal layer, the occupation and non-occupation of the states continually change, as depicted in Figure K.1.

The quantity of interest is the probability, $f(\mathbf{p}, s_z)$, that a single-particle state of energy $\epsilon_{\mathbf{p},s_z}$ is occupied for a Fermi gas in thermal equilibrium.[2] Owing to the quantum statistics of fermions, the exclusion principle, there are only two options as regards the occupation of a single-particle level: it is either occupied or empty. This fact allows for a simple quantum statistical argument to determine the thermal equilibrium distribution in the microcanonical ensemble considered here, a system with a fixed number of particles being considered.

In the following, we need not have in mind the Sommerfeld model of independent electrons in a box, but we can consider an arbitrary energy spectrum. For simplicity, we assume that none of the energy levels are degenerate, $\epsilon_1 < \epsilon_2 < \epsilon_3 < \ldots$.[3] An N-particle energy state is specified by stating the N-tuple $(\epsilon_{\lambda_1}, \epsilon_{\lambda_2}, \ldots, \epsilon_{\lambda_N})$ of occupied single-particle energy levels, where in accordance with Pauli's exclusion principle no level can appear twice. The energy of an N-particle state where the energy levels $\lambda_1, \lambda_2, \ldots, \lambda_N$ are occupied (and the rest of the levels are unoccupied) is

$$E_N(\{\epsilon_{\lambda_i}\}_{i=1,\ldots,N}) \equiv E_N(\epsilon_{\lambda_1}, \epsilon_{\lambda_2}, \ldots, \epsilon_{\lambda_N}) = \sum_{i=1}^{N} \epsilon_{\lambda_i}. \tag{K.1}$$

[1] For the electrons in a metal there is no literal *wall*, but at finite temperature the ions in the metal execute thermal vibrations and exchange energy with the electrons.

[2] This probability can be thought of either as the relative fraction of time the level is occupied, or, for an ensemble of identical systems, as the relative number of systems where the energy level is occupied (recall Appendix J).

[3] In the Sommerfeld model, this can be achieved by choosing the box to be non-cubic, $L_x \neq L_y \neq L_z$, and by having present a magnetic field to lift the spin degeneracy. The degenerate case then follows as a simple limiting case of letting the magnetic field disappear.

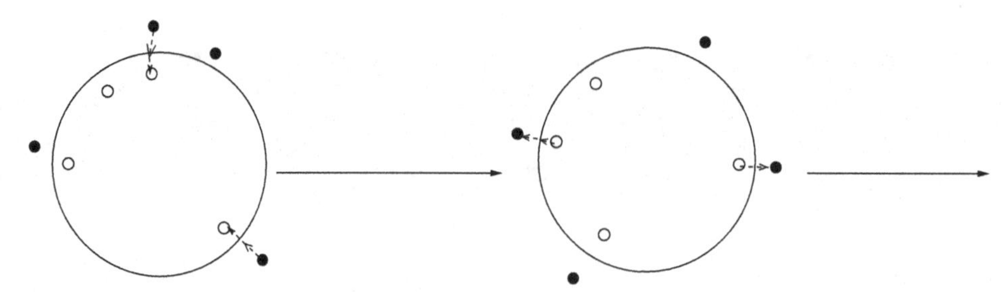

Figure K.1 Fermions interacting with vibrating walls change momentum states.

According to the basic principle of equilibrium statistical mechanics, the probability that in thermal equilibrium a particular energy state is realized is specified by the Boltzmann factor (recall Appendix J),

$$P_N(E_N(\{\epsilon_{\lambda_i}\}_{i=1,\ldots,N})) = Z^{-1} \exp\left\{-\frac{E_N(\{\epsilon_{\lambda_i}\}_{i=1,\ldots,N})}{kT}\right\} = Z^{-1} \exp\left\{-\frac{1}{kT}\sum_{i=1}^{N} \epsilon_{\lambda_i}\right\}. \tag{K.2}$$

The normalization factor for the probability distribution, the partition function Z, is determined by

$$\sum_{\{\epsilon_i\}_{i=1,\ldots,N}}{}' P_N(E_N(\{\epsilon_{\lambda_i}\}_{i=1,\ldots,N})) = 1, \tag{K.3}$$

giving

$$Z(T,V,N) = \sum_{\{\epsilon_{\lambda_i}\}_{i=1,\ldots,N}}{}' \exp\left\{-\frac{E_N(\{\epsilon_{\lambda_i}\}_{i=1,\ldots,N})}{kT}\right\} = \exp\left\{-\frac{1}{kT}\sum_{i=1}^{N}{}' \epsilon_{\lambda_i}\right\}, \tag{K.4}$$

where the summation is over N-tuples of energy configurations respecting the exclusion principle, i.e. the prime on the summation sign indicates that in an N-tuple all the energy values are different.

In terms of the free energy of the N-particle system[4]

$$F_N(T,V) = -kT \ln Z(T,V,N), \tag{K.5}$$

the probability for the occupation of an N-particle energy state takes the form

[4] A reader not familiar with *thermodynamics* and thereby the significance of the concept of free energy can without loss of understanding here take it as mere terminology. Statistical mechanics explains that the change in a system's free energy is the maximum work extractable. The case in question is for systems with fixed volume, Helmholtz *free energy*, $dF = -S\,dT + dW_m$, and the extractable work satisfies the inequality $L = -dW_m \leq -dF$ as $dQ = S\,dT$ is the heat generated in the process. For an isothermal process, $dT = 0$, the change in the free energy equals the work done on the system. The thermodynamic properties of a macroscopic system are obtainable from the partition function.

$$P_N(E_N(\{\epsilon_{\lambda_i}\}_{i=1,\dots,N})) = \exp\left\{\frac{F_N(T,V) - E_N(\{\epsilon_{\lambda_i}\}_{i=1,\dots,N})}{kT}\right\}$$
$$= \exp\left\{\frac{1}{kT}\left(F_N(T,V) - \sum_{i=1}^{N}\epsilon_{\lambda_i}\right)\right\}. \qquad \text{(K.6)}$$

The probability that the energy level λ is occupied, f_λ^N, is the number of such configurations divided by the total number of possible configurations, i.e. the probability of picking, out of all configurations, one of which the level λ is occupied, which is equal to the sum of probabilities for such configurations[5]

$$f_\lambda^N \equiv f^N(\epsilon_\lambda) = \sum_{\substack{\{\epsilon_{\lambda_i}\}_{i=1,\dots,N} \\ \exists\, \epsilon_{\lambda_i}:\, \epsilon_{\lambda_i}=\epsilon_\lambda}}{}^{\prime}\; P_N(E_N(\{\epsilon_{\lambda_i}\}_{i=1,\dots,N})). \qquad \text{(K.7)}$$

The discussion is valid for any number of particles, and the probability for the occupation of an energy state for an $(N+1)$-fermion system is, according to the general formula, Eq. (K.6), given by

$$P_{N+1}(E_{N+1}(\{\epsilon_{\lambda_i}\}_{i=1,\dots,N+1})) = \exp\left\{\frac{F_{N+1}(T,V) - E_{N+1}(\{\epsilon_{\lambda_i}\}_{i=1,\dots,N+1})}{kT}\right\}. \qquad \text{(K.8)}$$

The probability that state λ is occupied for the case of an $(N+1)$-fermion gas is accordingly

$$f_\lambda^{N+1} = \sum_{\substack{\{\epsilon_{\lambda_i}\}_{i=1,\dots,N+1} \\ \exists\, \epsilon_{\lambda_i}:\, \epsilon_{\lambda_i}=\epsilon_\lambda}}{}^{\prime}\; P_{N+1}(E_{N+1}(\{\epsilon_{\lambda_i}\}_{i=1,\dots,N+1}))$$
$$= \sum_{\substack{\{\epsilon_{\lambda_i}\}_{i=1,\dots,N+1} \\ \exists\, \epsilon_{\lambda_i}:\, \epsilon_{\lambda_i}=\epsilon_\lambda}}{}^{\prime}\; \exp\left\{\frac{F_{N+1}(T,V) - \sum_{i=1}^{N+1}\epsilon_{\lambda_i}}{kT}\right\}. \qquad \text{(K.9)}$$

The exclusion principle turns the normalization condition of the probability distribution into two distinct terms for any arbitrarily chosen level, ϵ_λ, splitting the configurations according to whether this single-particle level is occupied or not,

[5] If we considered probabilities for a finite outcome space, this identification of a subset outcome is trivial. Consider, for example, tossing one red and one blue dice. Then the probability that, say, red 3 is up, $P(3_r)$, is the number of configurations of the two dice with red up (which is 6) divided by the number of possible outcomes (which is 36), i.e. 6/36, which, of course, is equal to the sum of the probabilities of the configurations where red 3 is up, for the considered case the sum of six terms each with the configuration probability 1/36. Our situation is effectively a situation with a limited number of outcomes if, for a start, we limit our accuracy and assume that $\exp\{-100^{100}\}$ is effectively zero, cutting off all energy states with energy $\epsilon_\lambda > 100^{100}kT$. Letting the cut-off energy scale eventually approach infinity then captures the case of unlimited energy spectrum.

$$
\begin{aligned}
1 &= \sum_{\{\epsilon_i\}_{i=1,\ldots,N}}{}' P_N(E_N(\{\epsilon_{\lambda_i}\}_{i=1,\ldots,N})) \\
&= \sum_{\substack{\{\epsilon_{\lambda_i}\}_{i=1,\ldots,N} \\ \exists \epsilon_{\lambda_i}: \epsilon_{\lambda_i}=\epsilon_\lambda}}{}' P_N(E_N(\{\epsilon_{\lambda_i}\}_{i=1,\ldots,N})) + \sum_{\substack{\{\epsilon_{\lambda_i}\}_{i=1,\ldots,N} \\ \forall \epsilon_{\lambda_i}: \epsilon_{\lambda_i}\neq\epsilon_\lambda}}{}' P_N(E_N(\{\epsilon_{\lambda_i}\}_{i=1,\ldots,N}))
\end{aligned}
\tag{K.10}
$$

or

$$
f_\lambda^N = 1 - \sum_{\substack{\{\epsilon_{\lambda_i}\}_{i=1,\ldots,N} \\ \forall \epsilon_{\lambda_i}: \epsilon_{\lambda_i}\neq\epsilon_\lambda}}{}' P_N(E_N(\{\epsilon_{\lambda_i}\}_{i=1,\ldots,N})). \tag{K.11}
$$

Thus the probability that level λ is occupied is one minus the probability that level λ is not occupied. Instead of expressing the occupation number of a level, f_λ^N, in terms of all the configurations where the level in question is occupied, as in Eq. (K.7), by using the normalization of the distribution, Eq. (K.10), it can be expressed in terms of all the configurations where the level is *not* occupied.

The final observation needed is that the states of an $(N+1)$-fermion system where the energy level λ is occupied are in one-to-one correspondence with the states of the N-fermion system where the energy level λ is not occupied, as the depiction in Figure K.2 demonstrates, and the energies of the two states are related according to

$$
E_{N+1}(\{\epsilon_{\lambda_i}\}_{i=1,\ldots,N+1}) = E_N(\{\epsilon_{\lambda_i}\}_{i=1,\ldots,N}) + \epsilon_\lambda. \tag{K.12}
$$

Assuming that the energy level λ is occupied for the $(N+1)$-fermion state in question, its probability can be rewritten as

$$
P_{N+1}(E_{N+1}(\{\epsilon_{\lambda_i}\}_{i=1,\ldots,N+1})) = \exp\left\{\frac{1}{kT}\left(F_{N+1} - \epsilon_\lambda - \sum_{i=1}^{N}{}' \epsilon_{\lambda_i}\right)\right\}, \tag{K.13}
$$

where the prime indicates that the energy value ϵ_λ does not occur in the sum since it has been taken out explicitly and the labeling has been chosen such that $\epsilon_{\lambda_{N+1}} = \epsilon_\lambda$. The probability that energy level λ is occupied for the $(N+1)$-fermion system can thus be

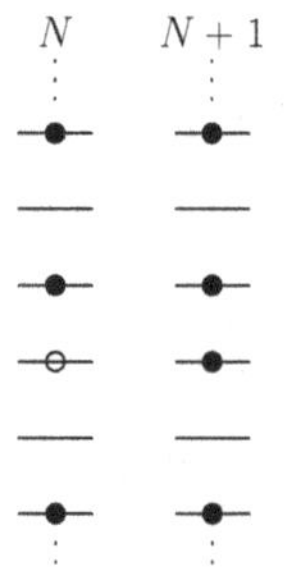

Figure K.2 The N-fermion configurations where the energy level ϵ_λ is not occupied is in one-to-one correspondence with all the $(N+1)$-fermion configurations where this energy level *is* occupied.

rewritten (by multiplying by $1 = e^{F_N} e^{-F_N}$) as

$$f_\lambda^{N+1} = \exp\left\{\frac{F_{N+1} - F_N - \epsilon_\lambda}{kT}\right\} \sum_{\substack{\{\epsilon_{\lambda_i}\}_{i=1,\dots,N} \\ \forall \epsilon_{\lambda_i}: \epsilon_{\lambda_i} \neq \epsilon_\lambda}}' \exp\left\{\frac{F_N(T,V) - \sum_{i=1}^{N}{}' \epsilon_{\lambda_i}}{kT}\right\}. \quad \text{(K.14)}$$

The summation is the same as that occurring in Eq. (K.11), giving

$$f_\lambda^{N+1} = e^{(F_{N+1} - F_N - \epsilon_\lambda)/kT} (1 - f_\lambda^N) \quad \text{(K.15)}$$

or equivalently

$$f_\lambda^N = 1 - e^{(\epsilon_\lambda + F_N - F_{N+1})/kT} f_\lambda^{N+1}, \quad \text{(K.16)}$$

an exact relation between the probabilities for occupation of level λ in an $(N+1)$- and N-fermion gas, respectively.

When N is large (in a piece of metal, the number of electrons is astronomically large, typically $N \sim 10^{22}$ per cubic centimeter), adding an extra particle, in accordance with the distribution (K.8), cannot change the occupation probability of a level significantly, and we expect (and shortly demonstrate) that

$$f_\lambda^{N+1} = f_\lambda^N + \mathcal{O}(1/N). \quad \text{(K.17)}$$

When the number of particles, N, is large, the probability for the occupation of an energy level, the mean occupation number, is therefore given by the Fermi function

$$f(\epsilon_\lambda) = \frac{1}{e^{(\epsilon_\lambda - \mu)/kT} + 1}, \quad \text{(K.18)}$$

where we have introduced the chemical potential

$$\mu = F_{N+1} - F_N, \quad \text{(K.19)}$$

the change in free energy due to adding a particle.

In the case of the Sommerfeld model, the probability that a momentum state $\mathbf{p}$, with either spin up or down, is occupied is

$$f(\epsilon_{\mathbf{p}}) = \frac{1}{e^{(\epsilon_{\mathbf{p}} - \mu)/kT} + 1}. \quad \text{(K.20)}$$

The Fermi function at zero temperature was encountered in Chapter 3. All levels up to the Fermi energy are filled, i.e. $f_{T=0}(\epsilon_{\mathbf{p}}) = \theta(\epsilon_F - \epsilon_{\mathbf{p}})$. Indeed, this is the zero temperature limit of Eq. (K.18), and identifying the chemical potential at zero temperature to be equal to the Fermi energy, we have $\mu(T=0) = \epsilon_F$. We now show that the temperature dependence of the chemical potential is negligible.

The chemical potential and the density of electrons are related according to (recall Eq. (3.70))

$$n = 2 \int_{-\infty}^{\infty} d\epsilon \, N(\epsilon) f(\epsilon), \quad \text{(K.21)}$$

where, since the density of states vanishes for negative energies, $N(\epsilon) = 0$ for $\epsilon < 0$, we can extend the lower integration limit to minus infinity.

Since the derivative of the Fermi function f' is peaked, we perform a partial integration, defining $\tilde{N}_0'(\epsilon) = N_0(\epsilon)$,

$$\int_{-\infty}^{\infty} d\epsilon\, N_0(\epsilon) f(\epsilon) = \Big[\tilde{N}_0(\epsilon) f(\epsilon)\Big]_{-\infty}^{\infty} - \int_{-\infty}^{\infty} d\epsilon\, \tilde{N}_0(\epsilon) f'(\epsilon)$$

$$= - \int_{-\infty}^{\infty} d\epsilon\, \tilde{N}_0(\epsilon) f'(\epsilon), \tag{K.22}$$

where the boundary terms vanish since we choose the integration constant according to

$$\tilde{N}_0(\epsilon) = \int_{-\infty}^{\epsilon} d\epsilon'\, N_0(\epsilon') = \int_{0}^{\epsilon} d\epsilon'\, N_0(\epsilon') = \frac{2}{3}\, \epsilon\, N_0(\epsilon). \tag{K.23}$$

Since $\tilde{N}_0(\epsilon)$ is a smooth function at the chemical potential, we Taylor-expand the density of states around the chemical potential and get

$$\int_{-\infty}^{\infty} d\epsilon\, \tilde{N}_0(\epsilon) f'(\epsilon) = \int_{-\infty}^{\infty} d\epsilon\, f'(\epsilon) \sum_{n=0}^{\infty} \frac{1}{n!} \tilde{N}_0^{(n)}(\mu)[\epsilon - \mu]^n. \tag{K.24}$$

Noting that the derivative of the Fermi function, Eq. (3.45), is an even function around the chemical potential, only even powers in the sum are non-zero. Noting the property

$$\int_{-\infty}^{\infty} d\epsilon \left(-\frac{\partial f}{\partial \epsilon}\right) = 1 \tag{K.25}$$

and making the shift of variable $x = (\epsilon - \mu)/kT$, we obtain the Sommerfeld expansion

$$\int_{-\infty}^{\infty} d\epsilon\, N_0(\epsilon) f(\epsilon) = \int_{-\infty}^{\mu} d\epsilon\, N_0(\epsilon) + \sum_{n=1}^{\infty} a_n N_0^{(2n-1)}(\mu)(kT)^{2n}, \tag{K.26}$$

where the numbers

$$a_n = -\frac{1}{(2n)!} \int_{-\infty}^{\infty} dx\, x^{2n} \frac{d}{dx}\left(\frac{1}{e^x + 1}\right) \tag{K.27}$$

are seen to be of order unity.

For a power-law function, $\tilde{N}_0(\epsilon) \propto \epsilon^{3/2}$,

$$\tilde{N}_0^{(n)}(\mu) \propto \frac{\tilde{N}_0(\mu)}{\mu^n} \tag{K.28}$$

and for the term of interest, $n = 1$,

$$\frac{\tilde{N}_0'(\mu)}{\mu} \propto \frac{\tilde{N}_0(\mu)}{\mu^2}. \tag{K.29}$$

The Sommerfeld expansion is thus an expansion in terms of the parameter $(kT/\mu)^2$ or, as we shortly realize, $(kT/\epsilon_F)^2$, and only the first term in this small parameter needs to be considered.

The Sommerfeld expansion, Eq. (K.26), for the density becomes

$$n = 2\int_0^{\mu} d\epsilon\; N(\epsilon) + 2a_1(kT)^2 N'(\mu) + \mathcal{O}((T/T_F)^4). \tag{K.30}$$

Since the number of particles is fixed, and at zero temperature given by

$$n = 2\int_0^{\epsilon_F} d\epsilon\; N(\epsilon), \tag{K.31}$$

we obtain

$$\int_{\mu}^{\epsilon_F} d\epsilon\; N(\epsilon) = a_1(kT)^2 N'(\mu) \tag{K.32}$$

and thereby

$$(\epsilon_F - \mu)N(\epsilon_m) = a_1(kT)^2 N'(\mu), \tag{K.33}$$

where ϵ_m is the mean value point, i.e. a value between μ and ϵ_F. Rearranging gives

$$\mu(T) = \epsilon_F\left(1 - \frac{N'(\mu)}{N(\epsilon_m)}\frac{(kT)^2}{\epsilon_F}\right). \tag{K.34}$$

Since the second term in parentheses is small at relevant temperatures, $T \ll T_F$, the chemical potential is close to the Fermi energy, $\mu(T) \simeq \epsilon_F$, and $N'(\mu)/N(\epsilon_m) \simeq N'(\epsilon_F)/N(\epsilon_F) = 1/(2\epsilon_F)$. The temperature dependence of the chemical potential is thus

$$\mu(T) = \epsilon_F - \frac{a_1}{2}\frac{kT^2}{\epsilon_F} = \epsilon_F\left(1 - \frac{\pi^2}{12}\left(\frac{T}{T_F}\right)^2\right), \tag{K.35}$$

where the last equality is obtained if one uses the irrelevant fact that $a_1 = \pi^2/6$. The chemical potential is thus pinned at the Fermi energy, $\mu(T) \simeq \epsilon_F$.

We now quantify the qualitative statements used to arrive at the expression for the Fermi–Dirac distribution. The relationship between the chemical potential and the number of electrons for the case of a gas of $N+1$ electrons reads

$$\frac{N+1}{V} = \frac{2}{V}\sum_{\mathbf{p}} \frac{1}{e^{(\epsilon_{\mathbf{p}} - \mu(N+1))/kT} + 1} \tag{K.36}$$

and subtracting the similar expression for the N-electron case we obtain

$$\frac{1}{V} = \frac{2}{V}\sum_{\mathbf{p}}\left(\frac{1}{e^{(\epsilon_{\mathbf{p}} - \mu(N+1))/kT} + 1} - \frac{1}{e^{(\epsilon_{\mathbf{p}} - \mu(N))/kT} + 1}\right). \tag{K.37}$$

We expect that adding one electron to an electron gas consisting of a huge number of electrons cannot change the chemical potential appreciably, i.e. $\mu(N+1) = \mu(N) + \Delta\mu$, where $\Delta\mu \ll \mu(N)$. Indeed, this is true, as $\mu \simeq \epsilon_F \propto (N/V)^{2/3}$. Taylor-expanding in Eq. (K.37) then gives

$$\frac{1}{V} = \frac{2}{V}\Delta\mu \sum_{\mathbf{p}} \left(-\frac{\partial f(\epsilon_{\mathbf{p}})}{\partial \epsilon_{\mathbf{p}}}\right) = 2\,\Delta\mu \int_0^\infty d\epsilon\, N_0(\epsilon) \left(-\frac{\partial f(\epsilon)}{\partial \epsilon}\right) \simeq 2N_0(\mu)\,\Delta\mu, \quad \text{(K.38)}$$

where we have used the facts that the derivative of the Fermi function, f', is peaked at the chemical potential, and that for relevant temperatures of an electron gas the temperature is always low, $kT \ll \mu$, thereby ensuring that $f(0) \simeq 1$. Since

$$N_0(\epsilon_F) = \frac{3n}{4\epsilon_F}, \quad \text{(K.39)}$$

the change in the chemical potential due to adding an extra particle is thus

$$\Delta\mu = \frac{1}{2N_0(\epsilon_F)V} = \frac{2}{3}\frac{\epsilon_F}{N}. \quad \text{(K.40)}$$

This is indeed completely negligible compared to ϵ_F for a gas containing $N \sim 10^{22}$ fermions.

Similarly, we obtain, using Eq. (3.45), that the difference in the occupancies of an energy level $\epsilon_{\mathbf{p}}$ in an $(N+1)$- and N-particle system is at most

$$\Delta f \equiv \max\left(f_{\epsilon_{\mathbf{p}}}^{N+1} - f_{\epsilon_{\mathbf{p}}}^{N}\right) = \max\left(-\frac{\partial f(\epsilon_{\mathbf{p}})}{\partial \epsilon_{\mathbf{p}}}\right)\Delta\mu = \frac{\Delta\mu}{4kT} = \frac{1}{6}\frac{\epsilon_F}{kT}\frac{1}{N}, \quad \text{(K.41)}$$

which is of order $1/N$ and therefore insignificant, thereby verifying Eq. (K.17).

Exercise K.1 The average energy of the Fermi gas at temperature T is given by

$$\bar{E}(T) = 2\sum_{\mathbf{p}} \epsilon_{\mathbf{p}} f(\mathbf{p}). \quad \text{(K.42)}$$

Use the Sommerfeld expansion to obtain for the specific heat the expression

$$c_V = \frac{1}{V}\left(\frac{\partial \bar{E}(T)}{\partial T}\right)_V = \frac{2\pi^2}{3}k^2 T N_0(\epsilon_F) + \mathcal{O}\left(\left(\frac{T}{T_F}\right)^3\right) = \frac{mk_F}{3\hbar^2}k^2 T. \quad \text{(K.43)}$$

The electronic contribution to the specific heat is thus proportional to the density of states at the Fermi energy or equivalently the (effective) mass of the fermion.

The physical interpretation of the electron gas contribution to the specific heat is simple: at *low* temperatures the specific heat is

$$c_V = \frac{\bar{E}(T) - \bar{E}(0)}{VT} \equiv \frac{\Delta E/V}{T} \quad \text{(K.44)}$$

and

$$\begin{aligned}\Delta E/V &= \#\text{ of excited electrons per unit volume} \times \text{their energy gain}\\ &\sim N(\epsilon_F)kT \times kT \end{aligned}\quad \text{(K.45)}$$

since, at finite temperature, $N(\epsilon_F)kT$ electrons per unit volume have been excited by the amount kT.

In metals there is also a lattice vibrational contribution to the specific heat, which has a cubic temperature dependence, T^3, and at sufficiently low temperatures only the electronic specific heat contribution that is linear in temperature survives, thereby making the density of states at the Fermi energy, or equivalently the electron mass, a measurable quantity.

Appendix L: Thermal Current Fluctuations

In thermal equilibrium, the physical quantities of a macroscopic body, such as current and voltage, fluctuate in time around their equilibrium values. Let us consider these fluctuations in an electron gas that is in thermal equilibrium at temperature T. Consider a metal wire of length L and cross-section S. The instantaneous electric current through any cross-section is

$$I(t) = S\frac{2e}{m}\frac{1}{\Omega}\sum_{\mathbf{p}} p_x f_{\mathbf{p}}(t), \tag{L.1}$$

where the x direction has been chosen perpendicular to the cross-section, and Ω is the volume of the system, $\Omega = LS$. The factor of 2 is due to the spin degeneracy of the energy states.

Owing to thermal agitation, the electrons are incessantly exchanging energy with their surroundings (in this case the ionic lattice degrees of freedom), and the actual occupation, $f_{\mathbf{p}}(t)$, of a momentum state $\mathbf{p}$ near the Fermi surface changes incessantly. (We neglect spin dynamics; as usual, spin only appears as a multiplicative factor of 2.) Owing to the exclusion principle, in some time intervals the value of occupation is one, the state is occupied, and in other time intervals the value is zero, the state is unoccupied. Averaged over a sufficiently long time, the average value emerges, $\overline{f_{\mathbf{p}}(t)} = f_{\mathbf{p}}$. To see this in the present model, let us keep track of occupation and non-occupation on average. In time intervals where $f_{\mathbf{p}}(t) = 1$, the deviation from the average value is $1 - f_{\mathbf{p}}$, and in time intervals where $f_{\mathbf{p}}(t) = 0$, the deviation from the average value is $0 - f_{\mathbf{p}}$. The value $f_{\mathbf{p}}(t) = 1$ occurs on average with the probability that the state is occupied, which in thermal equilibrium is the probability $f_{\mathbf{p}}$ ascribed by the Fermi function, and the value $f_{\mathbf{p}}(t) = 0$ occurs with the probability that the state is not occupied, which is given by the value $1 - f_{\mathbf{p}}$. The average deviation from the mean in state $\mathbf{p}$ is thus

$$\overline{f_{\mathbf{p}}(t) - f_{\mathbf{p}}} = (0 - f_{\mathbf{p}})(1 - f_{\mathbf{p}}) + (1 - f_{\mathbf{p}})f_{\mathbf{p}} = 0. \tag{L.2}$$

The average value thus emerges, $\overline{f_{\mathbf{p}}(t)} = f_{\mathbf{p}}$.

The average current therefore vanishes, as the current contribution from state $\mathbf{p}$ is canceled by the current contribution from state $-\mathbf{p}$, and the instantaneous current can therefore be rewritten as

$$I(t) = \frac{2e}{mL}\sum_{\mathbf{p}} p_x(f_{\mathbf{p}}(t) - f_{\mathbf{p}}). \tag{L.3}$$

The instantaneous value of the current squared is, according to Eq. (L.3),

$$I^2(t) = \left(\frac{2e}{mL}\right)^2 \sum_{\mathbf{p},\mathbf{p}'} p_x p'_x (f_{\mathbf{p}}(t) - f_{\mathbf{p}})(f_{\mathbf{p}'}(t) - f_{\mathbf{p}'}). \tag{L.4}$$

On average, the occupations of different momentum states are uncorrelated and statistically independent, and the average value of the fluctuations in the squared current is

$$\overline{I^2} = \left(\frac{2e}{mL}\right)^2 \sum_{\mathbf{p}} p_x^2 \overline{(f_{\mathbf{p}}(t) - f_{\mathbf{p}})^2}, \tag{L.5}$$

where we have already anticipated that, like all average values in an equilibrium state, it is time-independent. We can calculate the average of the population deviation squared of a momentum state as above. The fraction of time $f_{\mathbf{p}}$ is the state occupied and the deviation squared is $(1 - f_{\mathbf{p}})^2$, and the fraction of time $1 - f_{\mathbf{p}}$ is the state unoccupied and the deviation squared is $(0 - f_{\mathbf{p}})^2$. Therefore the mean squared occupation deviation is

$$\overline{(f_{\mathbf{p}}(t) - f_{\mathbf{p}})^2} = (0 - f_{\mathbf{p}})^2(1 - f_{\mathbf{p}}) + (1 - f_{\mathbf{p}})^2 f_{\mathbf{p}} = f_{\mathbf{p}}(1 - f_{\mathbf{p}}), \tag{L.6}$$

and the average value of the fluctuations in the squared current becomes

$$\begin{aligned}\overline{I^2} &= \left(\frac{2e}{mL}\right)^2 \sum_{\mathbf{p}} p_x^2 (1 - f_{\mathbf{p}}) f_{\mathbf{p}} \\ &= \Omega kT \left(\frac{2e}{mL}\right)^2 \frac{1}{3} \int_0^\infty d\epsilon_{\mathbf{p}}\, N_0(\epsilon_{\mathbf{p}})\, 2m\epsilon_{\mathbf{p}} \left(-\frac{\partial f(\epsilon_{\mathbf{p}})}{\partial \epsilon_{\mathbf{p}}}\right).\end{aligned} \tag{L.7}$$

In the last equality, use has been made of the identity valid for the Fermi function (recall Eq. (3.45))

$$(1 - f_{\mathbf{p}}) f_{\mathbf{p}} = -kT \frac{\partial f(\epsilon_{\mathbf{p}})}{\partial \epsilon_{\mathbf{p}}} \tag{L.8}$$

and the integral is evaluated by the standard method of Chapter 3, giving

$$\overline{I^2} = \frac{A}{L} \frac{2ne^2}{m} kT \tag{L.9}$$

and in the Drude model, according to Eq. (4.116),

$$\overline{I^2} = \frac{kT}{\tau R}, \tag{L.10}$$

where the impurity mean free time is typically $\tau = 10^{-13}$ s. At room temperature, $kT_{300\mathrm{K}} \simeq 25\,\mathrm{meV} \simeq 4 \times 10^{-21}$ J, we thus have $\overline{I^2} \simeq 4 \times 10^{-8}\,\mathrm{J\,s}/R$. Compared to an average current, $I = V/R$, the relative fluctuation is

$$\frac{\sqrt{\overline{I^2}}}{I} \simeq \frac{2 \times 10^{-4}}{\sqrt{R}} \frac{R}{V} = 2 \times 10^{-4} \frac{\sqrt{R}}{V}. \tag{L.11}$$

For relevant resistances and voltages, the fluctuations are thus small.

Appendix M: Gaussian Wave Packets

Consider the normalized Gaussian wave packet

$$\psi_{\mathbf{x}_0\mathbf{p}_0}(\mathbf{x}) = \left(\frac{1}{2\pi a^2}\right)^{3/4} \exp\left\{-\frac{(\mathbf{x}-\mathbf{x}_0)^2}{4a^2} + \frac{i}{\hbar}\,\mathbf{p}_0\cdot\mathbf{x}\right\}, \tag{M.1}$$

where a is a real number and $\mathbf{x}_0$ and $\mathbf{p}_0$ are real vectors. The function is seen to be centered around the position value $\mathbf{x}_0$ and has a width of size a. The option of having different widths in the different Cartesian directions is a trivial but not needed generalization, and we refer to the choice in Eq. (M.1) as the symmetric case. The real part of the Gaussian function is depicted in Figure M.1, in the one-dimensional case or in arbitrary dimension a cross-sectional view along the direction of $\mathbf{p}_0$. The imaginary part would look similar, just shifted a quarter of a wave length, having peaks where the real part has troughs and vice versa, since it is just the shift from the cosine to the sine function.

The Gaussian function, Eq. (M.1), is a product of two functions, the momentum eigenfunction, the plane wave corresponding to momentum value $\mathbf{p}_0$, and an amplitude,

$$\psi_{\mathbf{x}_0\mathbf{p}_0}(\mathbf{x}) = A_{\mathbf{x}_0}(\mathbf{x})\, e^{(i/\hbar)\mathbf{p}_0\cdot\mathbf{x}}, \tag{M.2}$$

where the Gaussian envelope

$$A_{\mathbf{x}_0}(\mathbf{x}) = \left(\frac{1}{2\pi a^2}\right)^{3/4} \exp\left\{-\frac{(\mathbf{x}-\mathbf{x}_0)^2}{4a^2}\right\} \tag{M.3}$$

gives the spatial extent of the Gaussian wave packet, the function (and image) marked by dashes in Figure M.1. The de Broglie wave length, $\lambda_0 = 2\pi\hbar/|\mathbf{p}_0|$, of the plane wave is indicated in Figure M.1.

The average position for a particle in the Gaussian state,

$$\langle\mathbf{x}\rangle = \langle\mathbf{x}\rangle_{\mathbf{x}_0\mathbf{p}_0} = \int d\mathbf{x}\,\mathbf{x}\,|\psi_{\mathbf{x}_0\mathbf{p}_0}(\mathbf{x})|^2 = \left(\frac{1}{2\pi a^2}\right)^{3/2} \int d\mathbf{x}\,\mathbf{x}\,e^{-(\mathbf{x}-\mathbf{x}_0)^2/2a^2}, \tag{M.4}$$

is by a shift of integration variable and normalization of the Gaussian function immediately seen to equal the real vector $\mathbf{x}_0$, $\langle\mathbf{x}\rangle = \mathbf{x}_0$.

Evaluating the average value of the position squared by first doing the Gaussian integral in the two directions other than the α direction gives

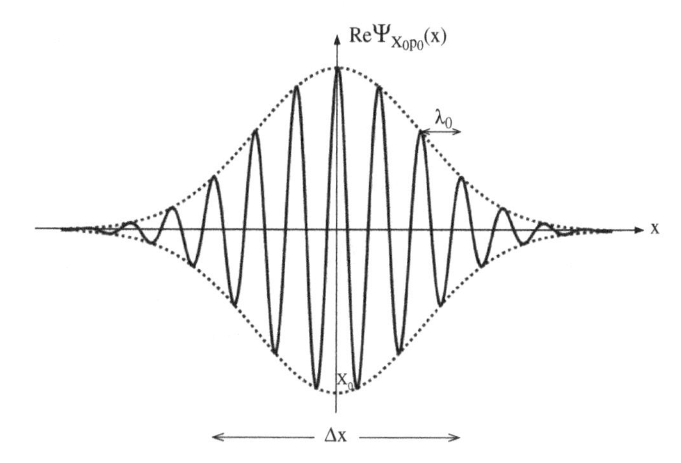

Figure M.1 Real part of the Gaussian wave packet.

$$\int d\mathbf{x}\, x_\alpha^2\, |\psi_{\mathbf{x}_0\mathbf{p}_0}(\mathbf{x})|^2 = \left(\frac{1}{2\pi a^2}\right)^{3/2} \left(\sqrt{2\pi a^2}\right)^2 \int_{-\infty}^{\infty} dx_\alpha\, x_\alpha^2\, e^{-(x_\alpha - x_{0_\alpha})^2/2a^2}. \quad \text{(M.5)}$$

Express $x_\alpha^2 = -x_{0_\alpha}^2 + 2x_{0_\alpha}x_\alpha + (x_\alpha - x_{0_\alpha})^2$, obtaining, using Eq. (M.4) for the linear x_α term,

$$\int d\mathbf{x}\, x_\alpha^2\, |\psi_{\mathbf{x}_0\mathbf{p}_0}(\mathbf{x})|^2 = \left(\frac{1}{2\pi a^2}\right)^{1/2} \Bigg(- x_{0_\alpha}^2 \sqrt{2\pi a^2} + 2x_{0_\alpha}x_{0_\alpha}\sqrt{2\pi a^2} + \int_{-\infty}^{\infty} dx_\alpha\, (x_\alpha - x_{0_\alpha})^2\, e^{-(x_\alpha - x_{0_\alpha})^2/2a^2} \Bigg). \quad \text{(M.6)}$$

Shifting the integration variable in the last term, the integral is evaluated by noting

$$\int_{-\infty}^{\infty} dx_\alpha\, x_\alpha^2\, e^{-cx_\alpha^2} = -\frac{d}{dc} \int_{-\infty}^{\infty} dx_\alpha\, e^{-cx_\alpha^2} = -\frac{d}{dc}\left(\frac{\pi}{c}\right)^{1/2} = \frac{\sqrt{\pi}}{2}(c)^{-3/2} \quad \text{(M.7)}$$

and the average value of the position squared becomes

$$\int d\mathbf{x}\, x_\alpha^2\, |\psi_{\mathbf{x}_0\mathbf{p}_0}(\mathbf{x})|^2 = x_{0_\alpha}^2 + \left(\frac{1}{2\pi a^2}\right)^{1/2} \frac{\sqrt{\pi}}{2} \left(\frac{1}{2a^2}\right)^{-3/2}. \quad \text{(M.8)}$$

The α direction position variance is thus seen to be given by the parameter a,

$$\langle \delta x_\alpha^2 \rangle_{\mathbf{x}_0\mathbf{p}_0} \equiv \langle (x_\alpha - x_{0_\alpha})^2 \rangle_{\mathbf{x}_0\mathbf{p}_0} = \langle x_\alpha^2 \rangle_{\mathbf{x}_0\mathbf{p}_0} - \langle x_\alpha \rangle_{\mathbf{x}_0\mathbf{p}_0}^2 = \langle x_\alpha^2 \rangle_{\mathbf{x}_0\mathbf{p}_0} - x_{0_\alpha}^2 = a^2, \quad \text{(M.9)}$$

equal for the three Cartesian directions for the chosen symmetric case, $\delta x \equiv \langle \delta x_\alpha^2 \rangle_{\mathbf{x}_0\mathbf{p}_0} = a$.

The momentum probability amplitude corresponding to the Gaussian wave packet,

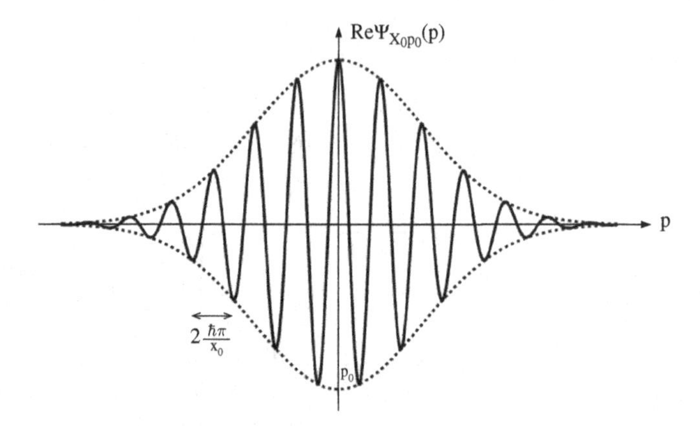

Figure M.2 Real part of the momentum probability amplitude of a Gaussian state.

Eq. (M.1), is easily obtained, as the Fourier transform is a Gaussian integral, giving in three spatial dimensions[1]

$$\psi_{\mathbf{x}_0\mathbf{p}_0}(\mathbf{p}) = \left(\frac{2\delta x^2}{\pi\hbar^2}\right)^{3/4} \exp\left\{-\frac{(\mathbf{p}-\mathbf{p}_0)^2\delta x^2}{\hbar^2} - \frac{i}{\hbar}(\mathbf{p}-\mathbf{p}_0)\cdot\mathbf{x}_0\right\}, \tag{M.10}$$

i.e. a Gaussian function centered around the momentum value $\mathbf{p}_0$. Superposing plane waves with a Gaussian weight thus gives the Gaussian function.

The real part of the momentum probability amplitude of a Gaussian state is depicted in Figure M.2.

As in the position case, one realizes immediately from Eq. (M.10) that $\mathbf{p}_0$ equals the average momentum,

$$\langle\mathbf{p}\rangle = \langle\mathbf{p}\rangle_{\mathbf{x}_0\mathbf{p}_0} = \int d\mathbf{p}\,\mathbf{p}\,|\psi_{\mathbf{x}_0\mathbf{p}_0}(\mathbf{p})|^2 = \mathbf{p}_0, \tag{M.11}$$

and the Gaussian wave function can be expressed as

$$\psi_{\mathbf{x}_0\mathbf{p}_0}(\mathbf{x}) = \left(\frac{1}{2\pi\delta x^2}\right)^{3/4} \exp\left\{-\frac{(\mathbf{x}-\langle\mathbf{x}\rangle)^2}{4\delta x^2} + \frac{i}{\hbar}\langle\mathbf{p}\rangle\cdot\mathbf{x}\right\}. \tag{M.12}$$

The Gaussian function is thus completely specified in terms of its two lowest moments, average value and variance.

Evaluating the average value of the momentum squared by first doing the Gaussian integral in the two other directions than the α direction gives

$$\int d\mathbf{p}\,p_\alpha^2\,|\psi_{\mathbf{x}_0\mathbf{p}_0}(\mathbf{p})|^2 = \left(\frac{2\delta x^2}{\pi\hbar^2}\right)^{3/2}\left(\sqrt{\frac{\pi\hbar^2}{2\delta x^2}}\right)^2 \int_{-\infty}^{\infty} dp_\alpha\,p_\alpha^2\,e^{-(2\delta x^2/\hbar^2)(p_\alpha - p_{0\alpha})^2}. \tag{M.13}$$

[1] First do the one-dimensional case, and then appeal to the multiplicative nature of Fourier transformation in additional dimensions as $e^{a+b} = e^a e^b$. Then we can immediately write down the result in any spatial dimension as expressed by the scalar product notation.

Express $p_\alpha^2 = (p_\alpha - p_{0_\alpha})^2 - p_{0_\alpha}^2 + 2p_{0_\alpha}p_\alpha$, giving

$$\int d\mathbf{p}\; p_\alpha^2\, |\psi_{\mathbf{x}_0\mathbf{p}_0}(\mathbf{p})|^2 = \left(\frac{2\delta x^2}{\pi\hbar^2}\right)^{1/2} \left(-p_{0_\alpha}^2 \sqrt{\frac{\pi\hbar^2}{2\delta x^2}} + 2p_{0_\alpha}p_{0_\alpha}\sqrt{\frac{\pi\hbar^2}{2\delta x^2}} + \int_{-\infty}^{\infty} dp_\alpha\, (p_\alpha - p_{0_\alpha})^2\, e^{-(2\delta x^2/\hbar^2)(p_\alpha - p_{0_\alpha})^2} \right). \quad \text{(M.14)}$$

The integral is evaluated according to Eq. (M.7), and the α direction momentum variance is seen to be given by

$$\langle \delta p_\alpha^2 \rangle_{\mathbf{x}_0\mathbf{p}_0} \equiv \langle p_\alpha^2 \rangle_{\mathbf{x}_0\mathbf{p}_0} - \langle p_\alpha \rangle_{\mathbf{x}_0\mathbf{p}_0}^2 = \langle p_\alpha^2 \rangle_{\mathbf{x}_0\mathbf{p}_0} - p_{0_\alpha}^2 = \frac{\hbar^2}{4\langle \delta x_\alpha^2 \rangle_{\mathbf{x}_0\mathbf{p}_0}}. \quad \text{(M.15)}$$

For the symmetric Gaussian wave packet, Eq. (M.1), the momentum variance is equal in the different Cartesian directions, $\delta p^2 = \langle \delta p_\alpha^2 \rangle_{\mathbf{x}_0\mathbf{p}_0}$, and the momentum probability amplitude can be expressed in terms of its two lowest moments,

$$\psi_{\mathbf{x}_0\mathbf{p}_0}(\mathbf{p}) = \left(\frac{1}{2\pi\delta p^2}\right)^{3/4} \exp\left\{ -\frac{(\mathbf{p} - \langle \mathbf{p} \rangle)^2}{4\delta p^2} - \frac{i}{\hbar}(\mathbf{p} - \langle \mathbf{p} \rangle) \cdot \langle \mathbf{x} \rangle \right\}. \quad \text{(M.16)}$$

In the Gaussian state, the position and momentum variances satisfy the lower bound in Heisenberg's uncertainty relation,

$$\delta p\, \delta x = \frac{\hbar}{2}, \quad \text{(M.17)}$$

in fact the Gaussian functions are characterized by this property.

Exercise M.1 Show that operating with the free particle Hamiltonian on the Gaussian function gives

$$-\frac{\hbar^2}{2m}\nabla_{\mathbf{x}}^2\, \psi_{\mathbf{x}_0\mathbf{p}_0}(\mathbf{x}) = \left(\frac{3\hbar^2}{4m\delta x^2} - \frac{\hbar^2}{2m}\left(-\frac{(\mathbf{x} - \mathbf{x}_0)}{2\delta x^2} + \frac{i\mathbf{p}_0}{\hbar} \right)^2 \right) \psi_{\mathbf{x}_0\mathbf{p}_0}(\mathbf{x}). \quad \text{(M.18)}$$

Show, by evaluating the integral, that the average kinetic energy of a particle in the Gaussian wave packet state is, with $\epsilon(\mathbf{p}_0) = \mathbf{p}_0^2/2m$,

$$-\frac{\hbar^2}{2m}\int d\mathbf{x}\; \psi^*_{\mathbf{x}_0\mathbf{p}_0}(\mathbf{x})\nabla_{\mathbf{x}}^2\, \psi_{\mathbf{x}_0\mathbf{p}_0}(\mathbf{x}) = \left(\frac{3\hbar^2}{8m\delta x^2} + \epsilon(\mathbf{p}_0) \right), \quad \text{(M.19)}$$

which, expressed in terms of the momentum variance of the state, becomes the expression in Eq. (4.5). The result, Eq. (4.5), also follows immediately from Eq. (M.15), as the average kinetic energy is related to the momentum variance, since

$$\int d\mathbf{p}\; \epsilon(\mathbf{p})\, |\psi_{\mathbf{x}_0\mathbf{p}_0}(\mathbf{p})|^2 = \frac{1}{2m}\langle \mathbf{p}^2 \rangle_{\mathbf{x}_0\mathbf{p}_0} = \frac{1}{2m}(\mathbf{p}_0^2 + 3\delta p^2). \quad \text{(M.20)}$$

Owing to the spatial localization of the wave packet, the average energy gets the additional contribution $\delta p^2/2m$ for each Cartesian direction as dictated by Heisenberg's uncertainty principle.

Of importance is that the Gaussian functions are a complete set as they satisfy the closure relation

$$\int_{-\infty}^{\infty} \frac{d\mathbf{x}_0\, d\mathbf{p}_0}{(2\pi\hbar)^3}\, \psi_{\mathbf{x}_0,\mathbf{p}_0}(\mathbf{x})\, \psi^*_{\mathbf{x}_0,\mathbf{p}_0}(\mathbf{x}') = \delta(\mathbf{x}-\mathbf{x}'). \tag{M.21}$$

They can therefore be used as a basis set for describing arbitrary particle states. The completeness relation, Eq. (M.21), is easily demonstrated, as the left-hand side, according to Eq. (M.1), has the form

$$\int_{-\infty}^{\infty} \frac{d\mathbf{x}_0\, d\mathbf{p}_0}{(2\pi\hbar)^3}\, \psi^*_{\mathbf{x}_0,\mathbf{p}_0}(\mathbf{x})\, \psi_{\mathbf{x}_0,\mathbf{p}_0}(\mathbf{x}')$$
$$= \frac{e^{-(\mathbf{x}^2+\mathbf{x}'^2)/4\delta x^2}}{(\sqrt{2\pi\delta x^2}\,)^3} \int_{-\infty}^{\infty} d\mathbf{x}_0\, \exp\left\{-\frac{1}{2\delta x^2}\,\mathbf{x}_0^2 + \frac{(\mathbf{x}+\mathbf{x}')\cdot\mathbf{x}_0}{2\delta x^2}\right\}$$
$$\times \int_{-\infty}^{\infty} \frac{d\mathbf{p}_0}{(2\pi\hbar)^3}\, \exp\left\{-\frac{i}{\hbar}\,\mathbf{p}_0\cdot(\mathbf{x}-\mathbf{x}')\right\}. \tag{M.22}$$

The $\mathbf{p}_0$ integration gives the delta function $\delta(\mathbf{x}-\mathbf{x}')$, allowing us to set $\mathbf{x}' = \mathbf{x}$ in the rest of the expression. The remaining Gaussian $\mathbf{x}_0$ integration then cancels the prefactor and Eq. (M.21) is obtained. We note that the closure relation, Eq. (M.21), is invariant to the interchange $\mathbf{x} \leftrightarrow \mathbf{x}'$, or, equivalently, complex conjugation leaves the right-hand side unchanged. With the closure relation at hand, any function can be represented as a superposition of Gaussian functions,

$$\psi(\mathbf{x}) = \int_{-\infty}^{\infty} \frac{d\mathbf{x}_0\, d\mathbf{p}_0}{(2\pi\hbar)^3}\, C(\mathbf{x}_0,\mathbf{p}_0)\, \psi_{\mathbf{x}_0,\mathbf{p}_0}(\mathbf{x}), \tag{M.23}$$

with expansion coefficients

$$C(\mathbf{x}_0,\mathbf{p}_0) = \int d\mathbf{x}\, \psi^*_{\mathbf{x}_0,\mathbf{p}_0}(\mathbf{x})\, \psi(\mathbf{x}) \tag{M.24}$$

and, though integration is over all space, it is in practice only over the region near $\mathbf{x}$, respectively $\mathbf{x}_0$, in the two integrals.

The overlap of Gaussian functions is easily calculated, as the integral is Gaussian,

$$\begin{aligned}
&\langle \mathbf{x}_0, \mathbf{p}_0 \mid \mathbf{x}'_0, \mathbf{p}'_0 \rangle \\
&\equiv \int d\mathbf{x}\, \psi^*_{\mathbf{x}_0,\mathbf{p}_0}(\mathbf{x})\, \psi_{\mathbf{x}'_0,\mathbf{p}'_0}(\mathbf{x})\, \psi(\mathbf{x}) \\
&= \left(\frac{1}{2\pi\delta x^2}\right)^{3/2} \int d\mathbf{x}\, \exp\left\{-\frac{(\mathbf{x}-\mathbf{x}_0)^2}{4\delta x^2} - \frac{(\mathbf{x}-\mathbf{x}'_0)^2}{4\delta x^2}\right\} \exp\left\{-\frac{i}{\hbar}(\mathbf{p}_0 - \mathbf{p}'_0)\cdot \mathbf{x}\right\} \\
&= \exp\left\{-\frac{(\mathbf{x}_0-\mathbf{x}'_0)^2}{8\delta x^2} - \frac{(\mathbf{p}_0-\mathbf{p}'_0)^2}{8\delta p^2} - \frac{i}{2\hbar}(\mathbf{p}_0 - \mathbf{p}'_0)\cdot(\mathbf{x}_0 + \mathbf{x}'_0)\right\}. \qquad \text{(M.25)}
\end{aligned}$$

The Gaussian functions are thus not orthogonal, because

$$\langle \mathbf{x}_0, \mathbf{p}_0 \mid \mathbf{x}'_0, \mathbf{p}'_0 \rangle \neq \delta(\mathbf{x}_0 - \mathbf{x}'_0)\, \delta(\mathbf{p}_0 - \mathbf{p}'_0), \qquad \text{(M.26)}$$

i.e. the scalar product does not vanish for different Gaussian states.

Any Gaussian function, say, the one labeled by $(\mathbf{x}_0, \mathbf{p}_0)$, has an expansion on the complete set of Gaussian functions,

$$\psi_{\mathbf{x}_0,\mathbf{p}_0}(\mathbf{x}) = \int_{-\infty}^{\infty} \frac{d\mathbf{x}'_0\, d\mathbf{p}'_0}{(2\pi\hbar)^3} \langle \mathbf{x}'_0, \mathbf{p}'_0 \mid \mathbf{x}_0, \mathbf{p}_0 \rangle\, \psi_{\mathbf{x}'_0,\mathbf{p}'_0}(\mathbf{x}), \qquad \text{(M.27)}$$

and the expansion coefficients, Eq. (M.25), are non-vanishing, though most of them are exponentially small. Expressing the integral as a sum over infinitesimal phase space volumes gives (n labeling the volumes in phase space)

$$\psi_{\mathbf{x}_0,\mathbf{p}_0}(\mathbf{x}) = \frac{\Delta\mathbf{x}_0 \Delta\mathbf{p}_0}{(2\pi\hbar)^3}\, \psi_{\mathbf{x}_0,\mathbf{p}_0}(\mathbf{x}) + \sum_{n\neq 0} \frac{\Delta\mathbf{x}'_n \Delta\mathbf{p}'_n}{(2\pi\hbar)^3} \langle \mathbf{x}'_0, \mathbf{p}'_0 \mid \mathbf{x}_n, \mathbf{p}_n \rangle\, \psi_{\mathbf{x}'_n,\mathbf{p}'_n}(\mathbf{x}). \qquad \text{(M.28)}$$

The Gaussian functions are thus not linearly independent. The basis function $\psi_{\mathbf{x}_0,\mathbf{p}_0}$ can be expressed in terms of the other Gaussian functions. Thus not all the basis functions are needed to make a complete set – the set of Gaussian functions is said to be over-complete. To rid the Gaussian basis of its over-completeness and just have a complete set on which any function has an expansion, we must figure out how to prune the over-complete set. Choosing the spatial variable to be the same, the free propagator, Eq. (1.27), specifies the amplitude for the event of the particle at a later time to be at the initial location

$$\left(\frac{m}{2\pi i\hbar t}\right)^{3/2} = K_0(\mathbf{x}, t; \mathbf{x}, 0) = \int_{-\infty}^{\infty} \frac{d\mathbf{p}}{(2\pi\hbar)^3}\, e^{-(i/\hbar)\epsilon(\mathbf{p})t}, \qquad \text{(M.29)}$$

where the last equality follows from the inverse Fourier transformation of Eqs. (G.12) and (G.23) or, of course, by direct evaluation of the Gaussian momentum integral. Changing from momentum to energy integration in the integral in Eq. (M.29) gives, according to Eq. (3.23),

$$K_0(\mathbf{x}, t; \mathbf{x}, 0) = \int_0^{\infty} d\epsilon(\mathbf{p})\, N(\epsilon(\mathbf{p}))\, e^{-(i/\hbar)\epsilon(\mathbf{p})t}, \qquad \text{(M.30)}$$

where

$$N(E) = \frac{\sqrt{2m^3E}}{2\pi^2\hbar^3} \tag{M.31}$$

is the density of states, the number of energy levels per unit energy per unit volume.

Integrating the space-independent function over a volume of size Ω gives

$$\Omega K_0(\mathbf{x},t;\mathbf{x},0) = \int_\Omega d\mathbf{x}\, K_0(\mathbf{x},t;\mathbf{x},0) = \int_\Omega \frac{d\mathbf{x}\,d\mathbf{p}}{(2\pi\hbar)^3}\, e^{-(i/\hbar)\epsilon(\mathbf{p})t} \tag{M.32}$$

or, in view of Eq. (M.30), a relation between phase space and energy integrals,

$$\int_\Omega \frac{d\mathbf{x}\,d\mathbf{p}}{(2\pi\hbar)^3}\, e^{-(i/\hbar)\epsilon(\mathbf{p})t} = \Omega \int_0^\infty dE\; N(E)\, e^{-(i/\hbar)Et}. \tag{M.33}$$

Writing the integral as a sum over infinitesimal energy intervals gives

$$\Omega \int_0^\infty dE\, N(E)\, e^{-(i/\hbar)Et} = \sum_n \Delta E_n \Omega N(E_n)\, e^{-(i/\hbar)E_n t}. \tag{M.34}$$

Choosing ΔE_n such that $1 = \Delta E_n\, \Omega N(E_n)$, the energy intervals contain exactly one state, and are indeed infinitesimal in the large volume limit. Writing also the phase space integral as a sum (n labeling the volumes in phase space $(\mathbf{x},\mathbf{p})$)

$$\int_\Omega \frac{d\mathbf{x}\,d\mathbf{p}}{(2\pi\hbar)^3}\, e^{-(i/\hbar)\epsilon(\mathbf{p})t} = \sum_n \frac{\Delta\mathbf{x}_n \Delta\mathbf{p}_n}{(2\pi\hbar)^3}\, e^{-(i/\hbar)\epsilon(\mathbf{p}_n)t}, \tag{M.35}$$

the relation in Eq. (M.33) has, in view of Eq. (M.34), been transformed into

$$\sum_n e^{-(i/\hbar)E_n t} = \sum_n \frac{\Delta\mathbf{x}_n \Delta\mathbf{p}_n}{(2\pi\hbar)^3}\, e^{-(i/\hbar)\epsilon(\mathbf{p}_n)t}. \tag{M.36}$$

Choosing labeling so that $E_n = \epsilon(\mathbf{p}_n)$, the identity in Eq. (M.36) thus demands that $\Delta\mathbf{x}_n\Delta\mathbf{p}_n = (2\pi\hbar)^3$, i.e. when counting allowed Gaussian states, each phase space volume $\Delta\mathbf{x}_n\Delta\mathbf{p}_n$ of size $(2\pi\hbar)^3$ should contain only one quantum state (per spin). We should thus prune the set of Gaussian basis functions to have values in phase space only on, say, a cubic lattice where each unit cell around a lattice point $(\mathbf{x}_0, \mathbf{p}_0)$ has volume $\Delta\,\mathbf{x}_c\Delta\mathbf{p}_c = (2\pi\hbar)^3$. Phase space summations are thus, according to Eq. (M.33), convertible into energy integration according to

$$\frac{1}{\Omega}\sum_c \frac{\Delta\mathbf{x}_c\Delta\mathbf{p}_c}{(2\pi\hbar)^3} \;\leftrightarrow\; \int d\epsilon\; N(\epsilon). \tag{M.37}$$

The Gaussian states on the lattice can be used as a complete basis set for the description of arbitrary particle states according to the closure relation[2]

$$\sum_c \frac{\Delta\mathbf{x}_c\Delta\mathbf{p}_c}{(2\pi\hbar)^3}\,\psi^*_{\mathbf{x}_c,\mathbf{p}_c}(\mathbf{x})\,\psi_{\mathbf{x}_c,\mathbf{p}_c}(\mathbf{x}') = \delta(\mathbf{x}-\mathbf{x}'). \tag{M.38}$$

Therefore, quantum states of a particle can be expressed as a superposition of Gaussian wave packet states

$$\psi(\mathbf{x}) = \sum_c C(\mathbf{x}_c, \mathbf{p}_c)\,\psi_{\mathbf{x}_c,\mathbf{p}_c}(\mathbf{x}), \tag{M.39}$$

where the summation is over all the lattice points corresponding to the labels $(\mathbf{x}_c, \mathbf{p}_c)$ residing on a lattice in $(\mathbf{x}, \mathbf{p})$ space, the phase space, and where each phase space cell around the lattice point has volume

$$\Delta\mathbf{x}_c\Delta\mathbf{p}_c = (2\pi\hbar)^3. \tag{M.40}$$

The exact choice of location of the lattice sites is not important when interest is only in particle states whose wave functions are smooth on the scale $\Delta\mathbf{x}_c$.

Exercise M.2 Consider the antisymmetric function in two variables built by Gaussian functions

$$\psi_{\mathbf{x}_1\mathbf{p}_1,\mathbf{x}_2\mathbf{p}_2}(\mathbf{x}'_1,\mathbf{x}'_2) = \frac{1}{\sqrt{2}}(\psi_{\mathbf{x}_1\mathbf{p}_1}(\mathbf{x}'_1)\,\psi_{\mathbf{x}_2\mathbf{p}_2}(\mathbf{x}'_2) - \psi_{\mathbf{x}_2\mathbf{p}_2}(\mathbf{x}'_1)\,\psi_{\mathbf{x}_1\mathbf{p}_1}(\mathbf{x}'_2)). \tag{M.41}$$

The two cell indices should be different, as otherwise the wave function vanishes, i.e. $(\mathbf{x}_1, \mathbf{p}_1) \neq (\mathbf{x}_2, \mathbf{p}_2)$.

Show that the average of the kinetic energy operator according to Eq. (M.19) becomes, with $\hat{h}_i \equiv -(\hbar^2/2m)\nabla^2_{\mathbf{x}'_i}$, $i = 1, 2$,

$$\begin{aligned}
&\langle\mathbf{x}_1\mathbf{p}_1, \mathbf{x}_2\mathbf{p}_2|\hat{h}_1 + \hat{h}_2|\mathbf{x}_1\mathbf{p}_1, \mathbf{x}_2\mathbf{p}_2\rangle \\
&\equiv \int d\mathbf{x}'_1\, d\mathbf{x}'_2\, \psi^*_{\mathbf{x}_1\mathbf{p}_1,\mathbf{x}_2\mathbf{p}_2}(\mathbf{x}'_1,\mathbf{x}'_2)(\hat{h}_1 + \hat{h}_2)\psi_{\mathbf{x}_1\mathbf{p}_1,\mathbf{x}_2\mathbf{p}_2}(\mathbf{x}'_1,\mathbf{x}'_2) \\
&= \sum_{i=1,2}\left(\epsilon(\mathbf{p}_i) + \frac{\delta p^2}{2m}\right) - \frac{1}{2}\int d\mathbf{x}'_1\, \psi^*_{\mathbf{x}_1\mathbf{p}_1}(\mathbf{x}'_1)\hat{h}_1\psi_{\mathbf{x}_2\mathbf{p}_2}(\mathbf{x}'_1)\int d\mathbf{x}'_2\, \psi^*_{\mathbf{x}_2\mathbf{p}_2}(\mathbf{x}'_2)\psi_{\mathbf{x}_1\mathbf{p}_1}(\mathbf{x}'_2) \\
&\quad -\frac{1}{2}\int d\mathbf{x}'_2\, \psi^*_{\mathbf{x}_1\mathbf{p}_1}(\mathbf{x}'_2)\hat{h}_2\psi_{\mathbf{x}_2\mathbf{p}_2}(\mathbf{x}'_2)\int d\mathbf{x}'_1\, \psi^*_{\mathbf{x}_2\mathbf{p}_2}(\mathbf{x}'_1)\psi_{\mathbf{x}_1\mathbf{p}_1}(\mathbf{x}'_1) \\
&\quad -\frac{1}{2}\int d\mathbf{x}'_1\, \psi^*_{\mathbf{x}_2\mathbf{p}_2}(\mathbf{x}'_1)\hat{h}_1\psi_{\mathbf{x}_1\mathbf{p}_1}(\mathbf{x}'_1)\int d\mathbf{x}'_2\, \psi^*_{\mathbf{x}_1\mathbf{p}_1}(\mathbf{x}'_2)\psi_{\mathbf{x}_2\mathbf{p}_2}(\mathbf{x}'_2)
\end{aligned}$$

[2] For a harmonic oscillator, it is well known that the following statement can be rigorously proved: the Gaussian lattice states are still complete when removing one state, but incomplete when removing two states.

$$-\frac{1}{2}\int d\mathbf{x}_2'\ \psi^*_{\mathbf{x}_2\mathbf{p}_2}(\mathbf{x}_2')\hat{h}_2\psi_{\mathbf{x}_1\mathbf{p}_1}(\mathbf{x}_2')\int d\mathbf{x}_1'\ \psi^*_{\mathbf{x}_1\mathbf{p}_1}(\mathbf{x}_1')\psi_{\mathbf{x}_2\mathbf{p}_2}(\mathbf{x}_1')$$

$$=\sum_{i=1,2}\left(\epsilon(\mathbf{p}_i)+\frac{\delta p^2}{2m}\right)$$

$$-\left(\int d\mathbf{x}_1'\ \psi^*_{\mathbf{x}_1\mathbf{p}_1}(\mathbf{x}_1')\hat{h}_1\psi_{\mathbf{x}_2\mathbf{p}_2}(\mathbf{x}_1')\int d\mathbf{x}_2'\ \psi^*_{\mathbf{x}_2\mathbf{p}_2}(\mathbf{x}_2')\psi_{\mathbf{x}_1\mathbf{p}_1}(\mathbf{x}_2')+\text{c.c.}\right)$$

$$=\sum_{i=1,2}\left(\epsilon(\mathbf{p}_i)+\frac{\delta p^2}{2m}\right)-2\,\Re(\langle\mathbf{x}_1\mathbf{p}_1|\hat{h}|\mathbf{x}_2\mathbf{p}_2\rangle\langle\mathbf{x}_2\mathbf{p}_2\mid\mathbf{x}_1\mathbf{p}_1\rangle), \tag{M.42}$$

i.e. equal to the sum of the single-particle energies and a term that is the product of Hamiltonian and wave function overlaps.

The state in Eq. (M.41) is not normalized, because

$$\begin{aligned}
&\langle\mathbf{x}_1\mathbf{p}_1,\mathbf{x}_2\mathbf{p}_2\mid\mathbf{x}_1\mathbf{p}_1,\mathbf{x}_2\mathbf{p}_2\rangle\\
&\equiv\int d\mathbf{x}_1'\,d\mathbf{x}_2'\ |\psi^*_{\mathbf{x}_1\mathbf{p}_1,\mathbf{x}_2\mathbf{p}_2}(\mathbf{x}_1',\ \mathbf{x}_2')|^2\\
&=\frac{1}{2}\int d\mathbf{x}_1'\,d\mathbf{x}_2'\ \Big[|\psi_{\mathbf{x}_1\mathbf{p}_1}(\mathbf{x}_1')|^2\,|\psi_{\mathbf{x}_2\mathbf{p}_2}(\mathbf{x}_2')|^2+(\mathbf{x}_1'\leftrightarrow\mathbf{x}_2')\text{-term}\\
&\quad-(\psi_{\mathbf{x}_1\mathbf{p}_1}(\mathbf{x}_1')^*\,\psi_{\mathbf{x}_2\mathbf{p}_2}(\mathbf{x}_1')\psi_{\mathbf{x}_2\mathbf{p}_2}(\mathbf{x}_2')^*\psi_{\mathbf{x}_1\mathbf{p}_1}(\mathbf{x}_2')+\text{c.c.})\Big]\\
&=\frac{1}{2}\Big[1+1-2\,\Re(\langle\mathbf{x}_1\mathbf{p}_1\mid\mathbf{x}_2\mathbf{p}_2\rangle\,\langle\mathbf{x}_2\mathbf{p}_2\mid\mathbf{x}_1\mathbf{p}_1\rangle)\Big]\\
&=1-|\langle\mathbf{x}_1\mathbf{p}_1\mid\mathbf{x}_2\mathbf{p}_2\rangle|^2\\
&=1-e^{-(\mathbf{x}_1-\mathbf{x}_2)^2/4\delta x^2}e^{-(\mathbf{p}_1-\mathbf{p}_2)^2/4\delta x^2}\\
&=\det(\langle\mathbf{x}_i\mathbf{p}_i\mid\mathbf{x}_j\mathbf{p}_j\rangle).
\end{aligned}\tag{M.43}$$

The state in Eq. (M.41) for normalization thus requires the factor

$$\begin{aligned}
\mathcal{N}_{\mathbf{x}_1\mathbf{p}_1,\mathbf{x}_2\mathbf{p}_2}&=\left(1-e^{-(\mathbf{x}_1-\mathbf{x}_2)^2/4\delta x^2}e^{-(\mathbf{p}_1-\mathbf{p}_2)^2/4\delta x^2}\right)^{-1/2}\\
&=\left(\det(\langle\mathbf{x}_i\mathbf{p}_i\mid\mathbf{x}_j\mathbf{p}_j\rangle)\right)^{-1/2}.
\end{aligned}\tag{M.44}$$

Exercise M.3 Show that, for the antisymmetrized three-particle state, the average of the kinetic energy operator is

$$\langle\mathbf{x}_1\mathbf{p}_1,\mathbf{x}_2\mathbf{p}_2,\mathbf{x}_3\mathbf{p}_3|\hat{h}_1+\hat{h}_2+\hat{h}_3|\mathbf{x}_1\mathbf{p}_1,\mathbf{x}_2\mathbf{p}_2,\mathbf{x}_3\mathbf{p}_3\rangle$$

$$=\left(\epsilon(\mathbf{p}_1)+\frac{\delta p^2}{2m}\right)(1-|\langle\mathbf{x}_2\mathbf{p}_2|\mathbf{x}_3\mathbf{p}_3\rangle|^2)$$

$$+\left(\epsilon(\mathbf{p}_2)+\frac{\delta p^2}{2m}\right)(1-|\langle\mathbf{x}_1\mathbf{p}_1|\mathbf{x}_3\mathbf{p}_3\rangle|^2)+\left(\epsilon(\mathbf{p}_3)+\frac{\delta p^2}{2m}\right)(1-|\langle\mathbf{x}_1\mathbf{p}_1|\mathbf{x}_2\mathbf{p}_2\rangle|^2)$$
$$+2\,\Re\Big[\langle\mathbf{x}_3\mathbf{p}_3|\hat{h}|\mathbf{x}_1\mathbf{p}_1\rangle(\langle\mathbf{x}_1\mathbf{p}_1|\mathbf{x}_2\mathbf{p}_2\rangle\langle\mathbf{x}_2\mathbf{p}_2|\mathbf{x}_3\mathbf{p}_3\rangle-\langle\mathbf{x}_1\mathbf{p}_1|\mathbf{x}_3\mathbf{p}_3\rangle)\Big]$$
$$+2\,\Re\Big[\langle\mathbf{x}_2\mathbf{p}_2|\hat{h}|\mathbf{x}_3\mathbf{p}_3\rangle(\langle\mathbf{x}_1\mathbf{p}_1|\mathbf{x}_2\mathbf{p}_2\rangle\langle\mathbf{x}_3\mathbf{p}_3|\mathbf{x}_1\mathbf{p}_1\rangle-\langle\mathbf{x}_3\mathbf{p}_3|\mathbf{x}_2\mathbf{p}_2\rangle)\Big]$$
$$+2\,\Re\Big[\langle\mathbf{x}_1\mathbf{p}_1|\hat{h}|\mathbf{x}_2\mathbf{p}_2\rangle(\langle\mathbf{x}_2\mathbf{p}_2|\mathbf{x}_3\mathbf{p}_3\rangle\langle\mathbf{x}_3\mathbf{p}_3|\mathbf{x}_1\mathbf{p}_1\rangle-\langle\mathbf{x}_2\mathbf{p}_2|\mathbf{x}_1\mathbf{p}_1\rangle)\Big]. \tag{M.45}$$

Consider dressing the particles with spin

$$\psi_{\mathbf{x}_1\mathbf{p}_1s_1^z,\mathbf{x}_2\mathbf{p}_2s_2^z}(\mathbf{x}_1',s_1^{z\prime},\mathbf{x}_2',s_2^{z\prime})=\frac{\mathcal{N}_{s_1^zs_2^z}}{\sqrt{2}}\Big[\psi_{\mathbf{x}_1\mathbf{p}_1}(\mathbf{x}_1')\,\chi_{s_1^z}(s_1^{z\prime})\,\psi_{\mathbf{x}_2\mathbf{p}_2}(\mathbf{x}_2')\,\chi_{s_2^z}(s_2^{z\prime})$$
$$-\,\psi_{\mathbf{x}_2\mathbf{p}_2}(\mathbf{x}_1')\,\chi_{s_2^z}(s_2^{z\prime})\,\psi_{\mathbf{x}_1\mathbf{p}_1}(\mathbf{x}_2')\,\chi_{s_1^z}(s_1^{z\prime})\Big]. \tag{M.46}$$

The scalar product, the normalization integral, is then

$$\langle\mathbf{x}_1\mathbf{p}_1s_1^z,\mathbf{x}_2\mathbf{p}_2s_2^z\mid\mathbf{x}_1\mathbf{p}_1s_1^z,\mathbf{x}_2\mathbf{p}_2s_2^z\rangle$$
$$\equiv\sum_{s_1^{z\prime}s_2^{z\prime}}\int d\mathbf{x}_1'\,d\mathbf{x}_2'\;\psi^*_{\mathbf{x}_1\mathbf{p}_1s_1^z,\mathbf{x}_2\mathbf{p}_2s_2^z}(\mathbf{x}_1',s_1^{z\prime},\mathbf{x}_2',s_2^{z\prime})\psi_{\mathbf{x}_1\mathbf{p}_1s_1^z,\mathbf{x}_2\mathbf{p}_2s_2^z}(\mathbf{x}_1',s_1^{z\prime},\mathbf{x}_2',s_2^{z\prime})$$
$$=\frac{\mathcal{N}^2_{s_1^zs_2^z}}{2}\sum_{s_1^{z\prime}s_2^{z\prime}}\int d\mathbf{x}_1'\,d\mathbf{x}_2'$$
$$\times\Big[\psi^*_{\mathbf{x}_1\mathbf{p}_1s_1^z}(\mathbf{x}_1',s_1^{z\prime})\psi^*_{\mathbf{x}_2\mathbf{p}_2s_2^z}(\mathbf{x}_2',s_2^{z\prime})-\psi^*_{\mathbf{x}_2\mathbf{p}_2s_2^z}(\mathbf{x}_1',s_1^{z\prime})\psi^*_{\mathbf{x}_1\mathbf{p}_1s_1^z}(\mathbf{x}_2',s_2^{z\prime})\Big]$$
$$\times\Big[\psi_{\mathbf{x}_1\mathbf{p}_1s_1^z}(\mathbf{x}_1',s_1^{z\prime})\psi_{\mathbf{x}_2\mathbf{p}_2s_2^z}(\mathbf{x}_2',s_2^{z\prime})-\psi_{\mathbf{x}_2\mathbf{p}_2s_2^z}(\mathbf{x}_1',s_1^{z\prime})\psi_{\mathbf{x}_1\mathbf{p}_1s_1^z}(\mathbf{x}_2',s_2^{z\prime})\Big]$$
$$=\frac{\mathcal{N}^2_{s_1^zs_2^z}}{2}\int d\mathbf{x}_1'\,d\mathbf{x}_2'\;\Big\{|\psi_{\mathbf{x}_1\mathbf{p}_1}(\mathbf{x}_1')|^2\,|\psi_{\mathbf{x}_2\mathbf{p}_2}(\mathbf{x}_2')|^2+(\mathbf{x}_1'\leftrightarrow\mathbf{x}_2')\text{-term}$$
$$-\sum_{s_1^{z\prime}s_2^{z\prime}}\Big[\psi^*_{\mathbf{x}_1\mathbf{p}_1}(\mathbf{x}_1')\,\psi_{\mathbf{x}_2\mathbf{p}_2}(\mathbf{x}_1')\psi^*_{\mathbf{x}_2\mathbf{p}_2}(\mathbf{x}_2')\psi_{\mathbf{x}_1\mathbf{p}_1}(\mathbf{x}_2')\,\delta_{s_1^zs_1^{z\prime}}\,\delta_{s_2^zs_1^{z\prime}}\,\delta_{s_2^zs_2^{z\prime}}\,\delta_{s_1^zs_2^{z\prime}}+\text{c.c.}\Big]\Big\}$$
$$=\frac{\mathcal{N}^2_{s_1^zs_2^z}}{2}\Big[1+1-(\,|\langle\mathbf{x}_1\mathbf{p}_1|\mathbf{x}_2\mathbf{p}_2\rangle|^2\,\delta_{s_1^zs_2^z}+\text{c.c.})\Big]$$
$$=\mathcal{N}^2_{s_1^zs_2^z}\left(1-|\langle\mathbf{x}_1\mathbf{p}_1|\mathbf{x}_2\mathbf{p}_2\rangle|^2\,\delta_{s_1^zs_2^z}\right). \tag{M.47}$$

The normalization factor is thus dependent, in addition, on the spin overlap, which by orthonormality is either zero or one,

$$\mathcal{N}_{s_1^zs_2^z}=\left(1-|\langle\mathbf{x}_1\mathbf{p}_1|\mathbf{x}_2\mathbf{p}_2\rangle|^2\,\delta_{s_1^zs_2^z}\right)^{-1/2}, \tag{M.48}$$

leaving the state simply normalized if the spins are different, the non-orthogonality of the Gaussian states being ineffective in that case.

The probability that one of the particles is at location $\mathbf{x}$ irrespective of its spin value and location and the spin of the other particle is

$$P(\mathbf{x}) = \left(\frac{1}{2} \sum_{s_1^{z\prime} s_2^{z\prime}} \int d\mathbf{x}_2' \, |\psi_{\mathbf{x}_1\mathbf{p}_1 s_1^z, \mathbf{x}_2\mathbf{p}_2 s_2^z}(\mathbf{x}, s_1^{z\prime}, \mathbf{x}_2', s_2^{z\prime})|^2 \right.$$
$$\left. + \frac{1}{2} \sum_{s_1^{z\prime} s_2^{z\prime}} \int d\mathbf{x}_1' \, |\psi_{\mathbf{x}_1\mathbf{p}_1 s_1^z, \mathbf{x}_2\mathbf{p}_2 s_2^z}(\mathbf{x}_1', s_1^{z\prime}, \mathbf{x}, s_2^{z\prime})|^2 \right)$$
$$= \sum_{s_1^{z\prime} s_2^{z\prime}} \int d\mathbf{x}_2' \, |\psi_{\mathbf{x}_1\mathbf{p}_1 s_1^z, \mathbf{x}_2\mathbf{p}_2 s_2^z}(\mathbf{x}, s_1^{z\prime}, \mathbf{x}_2', s_2^{z\prime})|^2. \tag{M.49}$$

Reading off from Eq. (M.47) gives

$$P(\mathbf{x}) = \frac{\mathcal{N}^2_{s_1^z s_2^z}}{2} \Big[|\psi_{\mathbf{x}_1\mathbf{p}_1}(\mathbf{x})|^2 + |\psi_{\mathbf{x}_2\mathbf{p}_2}(\mathbf{x})|^2$$
$$- \left(\psi^*_{\mathbf{x}_1\mathbf{p}_1}(\mathbf{x}) \, \psi_{\mathbf{x}_2\mathbf{p}_2}(\mathbf{x}) \langle \mathbf{x}_2\mathbf{p}_2 | \mathbf{x}_1\mathbf{p}_1 \rangle \delta_{s_1^z s_2^z} + \text{c.c.} \right) \Big] \tag{M.50}$$

and thereby

$$\int d\mathbf{x} \, P(\mathbf{x}) = \frac{\mathcal{N}^2_{s_1^z s_2^z}}{2} \left(2 - 2 \, |\langle \mathbf{x}_2\mathbf{p}_2 | \mathbf{x}_1\mathbf{p}_1 \rangle|^2 \delta_{s_1^z s_2^z} \right) = 1. \tag{M.51}$$

The average density of the particles is then

$$n(\mathbf{x}) = 2P(\mathbf{x}), \qquad \int d\mathbf{x} \, n(\mathbf{x}) = 2. \tag{M.52}$$

Exercise M.4 By noting that

$$\langle \mathbf{x}_1\mathbf{p}_1 | \mathbf{x}^n | \mathbf{x}_2\mathbf{p}_2 \rangle \equiv \int d\mathbf{x} \, \psi^*_{\mathbf{x}_1\mathbf{p}_1}(\mathbf{x}) \, \mathbf{x}^n \psi_{\mathbf{x}_2\mathbf{p}_2}(\mathbf{x})$$
$$= \left(\frac{\hbar}{i} \frac{\partial}{\partial \mathbf{p}_2} \right)^n \langle \mathbf{x}_1\mathbf{p}_1 | \mathbf{x}_2\mathbf{p}_2 \rangle, \tag{M.53}$$

show that

$$\langle \mathbf{x}_1\mathbf{p}_1 | \mathbf{x} | \mathbf{x}_2\mathbf{p}_2 \rangle = \left(\frac{\mathbf{x}_1 + \mathbf{x}_2}{2} + \frac{\hbar}{i} \frac{\mathbf{p}_1 - \mathbf{p}_2}{4\delta p^2} \right) \langle \mathbf{x}_1\mathbf{p}_1 | \mathbf{x}_2\mathbf{p}_2 \rangle \tag{M.54}$$

and

$$\langle \mathbf{x}_1\mathbf{p}_1 | \mathbf{x}^2 | \mathbf{x}_2\mathbf{p}_2 \rangle = \left(\frac{(\mathbf{x}_1 - \mathbf{x}_2)^2}{4} + \frac{\hbar(\mathbf{x}_1 - \mathbf{x}_2) \cdot (\mathbf{p}_1 - \mathbf{p}_2)}{4i \, \delta p^2} + \frac{\hbar^2}{4\delta p^2} \right.$$
$$\left. - \frac{\hbar^2 (\mathbf{p}_1 - \mathbf{p}_2)^2}{16\delta p^4} \right) \langle \mathbf{x}_1\mathbf{p}_1 | \mathbf{x}_2\mathbf{p}_2 \rangle, \tag{M.55}$$

and using

$$\langle \mathbf{x}_1\mathbf{p}_1|\hat{h}|\mathbf{x}_2\mathbf{p}_2\rangle = \langle \mathbf{x}_2\mathbf{p}_2|\hat{h}|\mathbf{x}_1\mathbf{p}_1\rangle^* \tag{M.56}$$

show that

$$\langle \mathbf{x}_1\mathbf{p}_1|\hat{h}|\mathbf{x}_2\mathbf{p}_2\rangle = \left(\epsilon(\mathbf{p}_1) + \epsilon(\mathbf{p}_2) + \frac{i\hbar(\mathbf{p}_1 + \mathbf{p}_2)\cdot(\mathbf{x}_1 - \mathbf{x}_2)^2}{8m\delta x^2} - \frac{\hbar^2(\mathbf{x}_1 - \mathbf{x}_2)^2}{16m\delta x^4} + \frac{(\mathbf{p}_1 - \mathbf{p}_2)^2}{4m}\right)\frac{\langle \mathbf{x}_1\mathbf{p}_1|\mathbf{x}_2\mathbf{p}_2\rangle}{2}. \tag{M.57}$$

To rid the Gaussian Bloch wave packets of their over-completeness (recall Eq. (8.88)) they must, like the free Gaussian states, be pruned. Consider

$$\frac{1}{\Omega}\int\limits_{\Omega,\mathrm{Bz}_1}\frac{d\mathbf{x}\,d\mathbf{k}}{(2\pi)^3}\,e^{-(i/\hbar)\epsilon_n(\mathbf{k})t} = \int\limits_{\mathrm{Bz}_1}\frac{d\mathbf{k}}{(2\pi)^3}\,e^{-(i/\hbar)\epsilon_n(\mathbf{k})t} = \int\limits_{\epsilon_{\min}}^{\epsilon_{\max}} d\epsilon\; N_n(\epsilon)\,e^{-(i/\hbar)\epsilon t}, \tag{M.58}$$

where the last equality follows from Eq. (9.75), and thereby

$$\sum_m \Delta\epsilon_m \Omega N_n(\epsilon_m)\,e^{-(i/\hbar)\epsilon_m t} = \sum_c \frac{\Delta\mathbf{x}_c\Delta\mathbf{k}_c}{(2\pi)^3}\,e^{-(i/\hbar)\epsilon_n(\mathbf{k}_c)t}. \tag{M.59}$$

Choosing $\Delta\epsilon_m$ such that $\Delta\epsilon_m\Omega N_n(\epsilon_m)$ equals one, the energy intervals contain exactly one state, and are indeed infinitesimal in the large volume limit. Choosing labeling so that $\epsilon_m = \epsilon_n(\mathbf{k}_c)$, the identity (M.59) thus demands that $\Delta\mathbf{x}_c\Delta\mathbf{k}_c = (2\pi)^3$. The Gaussian Bloch wave packet states are thus pruned to reside on a lattice in $(\mathbf{x}, \mathbf{k})$ space where each volume of size $(2\pi)^3$ contains one quantum state (per spin).

Appendix N: Wave Packet Dynamics

The quantum dynamics of Gaussian states is easily assessed for motion in homogeneous fields. Consider, first, the free evolution of the Gaussian wave packet that at time $t = 0$ is centered around position $x = 0$ and momentum p, the initial wave function[1]

$$\psi_{0p}(x, t = 0) = \left(\frac{1}{2\pi\delta x^2}\right)^{1/4} \exp\left\{-\frac{x^2}{4\delta x^2} + \frac{i}{\hbar}px\right\}, \tag{N.1}$$

already encountered in Exercise 1.3 on page 6. Instead of, as there, employing the spatial propagator, one can, in view of Eq. (G.23), also first (trivially) propagate the momentum probability amplitude, Eq. (M.10), and then Fourier-transform to real space. Either way, one obtains Eq. (1.30), which can be rewritten, with $x_0(t) = pt/m$ and $\epsilon(p) = p^2/2m$, as

$$\begin{aligned}\psi_{0p}(x,t) &= \left(\frac{\delta x^2}{2\pi\delta x_t^4}\right)^{1/4} \exp\left\{\frac{im}{2\hbar t}x^2\right\} \exp\left\{-i\frac{m\delta x^4(x - x_0(t))^2}{2\hbar t|\delta x_t^2|^2}\right\} \\ &\quad\times \exp\left\{-\frac{\delta x^2(x - x_0(t))^2}{4|\delta x_t^2|^2}\right\} \\ &= \left(\frac{\delta x^2}{2\pi\delta x_t^4}\right)^{1/4} \exp\left\{\frac{i}{\hbar}px - \frac{i}{\hbar}\epsilon(p)t\right\} \exp\left\{-\frac{\delta x^2(x - x_0(t))^2}{4|\delta x_t^2|^2}\right\} \\ &\quad\times \exp\left\{\left(\frac{imx^2}{2\hbar t} - \frac{i}{\hbar}px + \frac{i}{\hbar}\epsilon(p)t\right)\left(1 - \frac{\delta x^4}{|\delta x_t^2|^2}\right)\right\},\end{aligned} \tag{N.2}$$

where

$$\delta x_t^2 = \delta x^2\left(1 + \frac{i\hbar t}{2m\delta x^2}\right) = \delta x^2\left(1 + \frac{2i\,\delta p^2}{m\hbar}t\right), \tag{N.3}$$

and therefore

$$|\delta x_t^2| = \delta x^2\left(1 + \left(\frac{\hbar t}{2m\delta x^2}\right)^2\right)^{1/2}. \tag{N.4}$$

The last rewriting in Eq. (N.2) expresses the wave function in terms of the free particle wave function corresponding to energy $\epsilon(p)$ multiplied by a Gaussian envelope centered at $x_0(t)$ and an intricate phase factor, thereby making contact with the expression in Eq. (4.85). For short times, where wave packet spreading can be neglected, the wave function is the free particle solution corresponding to energy $\epsilon(p)$ multiplied by a Gaussian envelope.

[1] The one-dimensional case is considered, first, and generalization to arbitrary dimensions is done straightforwardly later.

The center of the envelope function moves with the velocity

$$\frac{dx_0(t)}{dt} = \frac{p}{m}, \tag{N.5}$$

the group velocity of the superposition of plane waves, which equals the velocity of a free particle of momentum p according to classical mechanics, and therefore equals the derivative of the kinetic energy corresponding to momentum p,

$$v_p \equiv \frac{dx_0(t)}{dt} = \frac{p}{m} = \frac{\partial\epsilon(p)}{\partial p}, \qquad \epsilon(p) = \frac{p^2}{2m}. \tag{N.6}$$

The absolute square of the wave function at time t is

$$\begin{aligned} |\psi_{0p}(x,t)|^2 &= \left(\frac{\delta x^2}{2\pi|\delta x_t|^4}\right)^{1/2} \exp\left\{-\frac{m\delta x^2(x-x_0(t))^2}{2\hbar t}\left(\frac{i}{\delta x_t^2} - \frac{i}{(\delta x_t^2)^*}\right)\right\} \\ &= \left(\frac{\delta x^2}{2\pi|\delta x_t^2|^2}\right)^{1/2} \exp\left\{-\frac{m\delta x^2(x-x_0(t))^2}{2\hbar t}\frac{\hbar t/m}{|\delta x_t^2|^2}\right\}, \end{aligned} \tag{N.7}$$

i.e. the probability density at time t is

$$P_{0p}(x,t) = |\psi_{0p}(x,t)|^2 = \sqrt{\frac{\delta x^2}{2\pi|\delta x_t^2|^2}} \exp\left\{-\frac{\delta x^2(x-x_0(t))^2}{2|\delta x_t^2|^2}\right\}, \tag{N.8}$$

which is a Gaussian function centered at $x_0(t)$, exhibiting diminishing peak value and a corresponding broadening as governed by the constraint of normalization. Except for the quantum feature of wave packet spreading, the dynamics is classical. The spatial spreading of the Gaussian wave packet is due to the uncertainty in momentum, $\delta p = \hbar/(2\,\delta x)$, in the initial state, the momentum probability amplitude function also of Gaussian shape (see Eq. (M.10)).

The probability current in the state in Eq. (N.2) is (recall Eq. (1.30))

$$S_{0p}(x,t) = \frac{\frac{x}{t}\left(\frac{\hbar t}{2m\delta x^2}\right)^2 + \frac{x_0(t)}{t}}{1+\left(\frac{\hbar t}{2m\delta x^2}\right)^2} \frac{\exp\left\{-\frac{[x-x_0(t)]^2}{2\delta x^2\{1+[\hbar t/(2m\delta x^2)]^2\}}\right\}}{\left\{2\pi\delta x^2\left[1+\left(\frac{\hbar t}{2m\delta x^2}\right)^2\right]\right\}^{1/2}}. \tag{N.9}$$

The Gaussian envelope makes the expression only non-vanishing when x is close to $x_0(t)$, $x \sim x_0(t)$, and, if spreading is negligible, $\hbar t/(2m\delta x^2) \ll 1$, the probability current reduces to

$$S_{0p}(x,t) \simeq \frac{x_0(t)}{t}\sqrt{\frac{1}{2\pi\delta x^2}} \exp\left\{-\frac{[x-x_0(t)]^2}{2\delta x^2}\right\}, \tag{N.10}$$

i.e. to the expression in Eq. (4.90).

If, in the expression for the probability density, Eq. (N.8), first $\hbar$ is set to zero, removing wave packet spreading, and then the initial spreading δx^2 is set to zero, it becomes

$$P_{0p}^{\text{clas}}(x,t) = \delta(x - x_0(t)), \tag{N.11}$$

i.e. the classical trajectory expressed in the language of probability densities: the probability for the particle to be anywhere but on the classical trajectory is zero.

The average position of the particle at time t is immediately obtained (recall the similar calculation in Appendix M) as

$$\langle x(t)\rangle \equiv \int_{-\infty}^{\infty} dx\, x\, P_{0p}(x,t) = x_0(t) = \frac{p}{m}\,t. \tag{N.12}$$

The position variance at time t,

$$\Delta x_t^2 \equiv \langle (x - \langle x(t)\rangle)^2\rangle_t = \langle x^2\rangle_t - \langle x(t)\rangle^2 = \langle x^2\rangle_t - x_0^2(t), \tag{N.13}$$

equals

$$\Delta x_t^2 = \int_{-\infty}^{\infty} dx\, x^2\, P_{0p}(x,t) - x_0^2(t) \tag{N.14}$$

and becomes (recall the similar calculation in Appendix M)

$$\Delta x_t^2 = \delta x^2 \left(1 + \left(\frac{\hbar t}{2m\delta x^2}\right)^2\right) \tag{N.15}$$

or equivalently

$$\Delta x_t^2 = \frac{|\delta x_t^2|^2}{\delta x^2}. \tag{N.16}$$

The Gaussian probability distribution, Eq. (N.8), is thus completely characterized by the average position and variance,

$$P_{0p}(x,t) = \sqrt{\frac{1}{2\pi\Delta x_t^2}}\, \exp\left\{-\frac{(x - \langle x(t)\rangle)^2}{2\Delta x_t^2}\right\}. \tag{N.17}$$

Exercise N.1 If the initial state is located at x_0,

$$\psi_{x_0p_0}(x, t=0) = \left(\frac{1}{2\pi\delta x^2}\right)^{1/4} \exp\left\{-\frac{(x-x_0)^2}{4\delta x^2} + \frac{i}{\hbar}p_0 x\right\}, \tag{N.18}$$

show that the Gaussian state at a later time is

$$\psi_{x_0p_0}(x,t) = \left(\frac{\delta x^2}{2\pi\delta x_t^2}\right)^{1/4} \exp\left\{\frac{im}{2\hbar t}(x^2 - x_0^2 - 2x_0(x + x_0(t)))\right\}$$
$$\times \exp\left\{-\frac{im\delta x^4}{2\hbar t|\delta x_t^2|^2}(x - x_0(t))^2\right\} \exp\left\{-\frac{\delta x^2}{2\,|\delta x_t^2|^2}(x - x_0(t))^2\right\}, \tag{N.19}$$

i.e. an intricate phase factor but the real envelope function one could have guessed at as now $x_0(t) = x_0 + p_0 t/m$.

Exercise N.2 Show that the momentum variance of a free particle stays constant,

$$\Delta p_t^2 \equiv \langle (p - \langle p(t) \rangle_t)^2 \rangle_t = \hbar^2/4\delta x^2 = \Delta p_{t=0}^2 \equiv \delta p^2 \tag{N.20}$$

and

$$\Delta x_t^2 \, \Delta p_t^2 = \frac{|\delta x_t^2|^2}{\delta x^2} \, \Delta p_{t=0}^2 = \frac{\hbar^2}{4} \left(\frac{|\delta x_t^2|}{\delta x^2} \right)^2 = \frac{\hbar^2}{4} \left(1 + \left(\frac{\hbar t}{2m\delta x^2} \right)^2 \right). \tag{N.21}$$

The free particle Gaussian wave packet evolution in three spatial dimensions amounts to the same calculation for the two other Cartesian coordinates and (assuming as usual the same spreading in all directions) gives

$$\psi_{0\mathbf{p}}(\mathbf{x}, t) = \left(\frac{\delta x^2}{2\pi \delta x_t^4} \right)^{3/4} \exp\left\{ \frac{im}{2\hbar t} \mathbf{x}^2 \right\} \exp\left\{ -\frac{im\delta x^2}{2\hbar t} \frac{(\mathbf{x} - \mathbf{x}_0(t))^2}{\delta x_t^2} \right\}, \tag{N.22}$$

where

$$\mathbf{x}_0(t) = \frac{\mathbf{p}}{m} t = \mathbf{v}_\mathbf{p} \, t, \qquad \mathbf{v}_\mathbf{p} = \frac{d\mathbf{x}_0(t)}{dt} = \frac{\partial \epsilon(\mathbf{p})}{\partial \mathbf{p}}. \tag{N.23}$$

Next we turn to the motion of a Gaussian wave packet in the presence of a constant force, F, i.e. the potential $V(x) = -Fx$, considering the one-dimensional case first. Differentiating the function[2]

$$K_\mathrm{F}(x, t; x', t') = K_0(x, t; x', t') \exp\left\{ \frac{i}{2\hbar} (t - t') \left(F(x + x') - \frac{F^2 (t - t')^2}{12m} \right) \right\}, \tag{N.24}$$

where K_0 is the free propagator, it is seen to satisfy the equation

$$i\hbar \frac{\partial K_\mathrm{F}(x, t; x', t')}{\partial t} = H_\mathrm{F} K_\mathrm{F}(x, t; x', t'), \tag{N.25}$$

where

$$H_\mathrm{F} = -\frac{\hbar^2}{2m} \frac{\partial^2}{\partial x^2} - Fx, \tag{N.26}$$

i.e. K_F is the propagator for a particle experiencing a constant force, as it clearly satisfies the initial condition (A.13). The propagator is thus a Gaussian function.

Assuming the state of the particle at time $t = 0$ is the Gaussian wave packet specified by average position $x = 0$ and momentum value p, i.e. the function given in Eq. (N.1), the wave function at a later time is

$$\psi_\mathrm{F}(x, t) = \int dx' \; K_\mathrm{F}(x, t; x', t') \, \psi_{0p}(x', t'). \tag{N.27}$$

[2] Not stooping to this short cut, but taking a constructive way of obtaining the propagator, one starts from the path integral expression for the propagator, which then, for the considered case, leads to the stated result by a Gaussian integration; see for example reference [1].

The integration is Gaussian, giving (the result also obtained in a different way at the end of this section)

$$\psi_{\mathrm{F}}(x,t) = \left(\frac{\delta x^2}{2\pi\delta x_t^4}\right)^{1/4} \exp\left\{-\frac{im\delta x^2}{2\hbar t}\frac{(x - x_{\mathrm{F}}(t))^2}{\delta x_t^2}\right\}$$

$$\times \exp\left\{\frac{im}{2\hbar t}\left(x^2 + \frac{Ft^2}{m}x - \frac{F^2t^4}{12m^2}\right)\right\}, \tag{N.28}$$

where the center of the envelope moves as a classical particle in the constant force, since

$$x_{\mathrm{F}}(t) = pt/m + Ft^2/2m, \tag{N.29}$$

i.e. $m\ddot{x}_{\mathrm{F}}(t) = F$, and δx_t^2 is given by Eq. (1.31). The spreading of the wave packet is not influenced by the presence of the constant force, remaining the same as that for a free particle.

The probability density is Gaussian

$$P_{\mathrm{F}}(x,t) = |\psi_{\mathrm{F}}(x,t)|^2 = \sqrt{\frac{1}{2\pi\Delta x_t^2}} \exp\left\{-\frac{(x - x_{\mathrm{F}}(t))^2}{2\Delta x_t^2}\right\}, \tag{N.30}$$

where Δx_t^2 is given by Eq. (1.33), Since the average position is

$$\langle x(t)\rangle = x_{\mathrm{F}}(t), \tag{N.31}$$

the position probability distribution is expressed by the position average value and the variance

$$P_{\mathrm{F}}(x,t) = \sqrt{\frac{1}{2\pi\Delta x_t^2}} \exp\left\{-\frac{(x - \langle x(t)\rangle)^2}{2\Delta x_t^2}\right\}. \tag{N.32}$$

Exercise N.3 Perform the Gaussian integration to obtain the momentum probability amplitude function for the wave function in Eq. (N.28),

$$\psi_{\mathrm{F}}(p',t) = \int_{-\infty}^{\infty} \frac{dx}{(2\pi\hbar)^{1/2}}\, e^{-(i/\hbar)xp'}\, \psi_{\mathrm{F}}(x,t), \tag{N.33}$$

and thereby the momentum probability distribution,

$$P_{\mathrm{F}}(p',t) = |\psi_{\mathrm{F}}(p',t)|^2 = \left(\frac{1}{2\pi\delta p^2}\right)^{1/2} e^{-(p'-(p+Ft))^2/2\delta p^2}. \tag{N.34}$$

Show that the average momentum is

$$\langle p(t)\rangle = p + Ft \tag{N.35}$$

and thereby that the momentum probability distribution

$$P_{\mathrm{F}}(p',t) = |\psi_{\mathrm{F}}(p',t)|^2 = \left(\frac{1}{2\pi\delta p^2}\right)^{1/2} e^{-(p'-\langle p(t)\rangle)^2/2\delta p^2} \tag{N.36}$$

is completely characterized by the average value and the conserved momentum variance. We note that the dynamics is described purely parametrically as

$$P_{\mathrm{F}}(p',t) = P_{0p(t)}(p',t=0), \qquad p(t) \equiv p + Ft. \tag{N.37}$$

Exercise N.4 Show that the momentum propagator for a particle in a constant force is

$$K_{\mathrm{F}}(p,t:p',0) = e^{-i(p^3-p'^3)/6m\hbar F}\,\delta(p-p'-Ft). \tag{N.38}$$

The result in three spatial dimensions amounts to the same calculation for the two other Cartesian coordinates and the evolution in the presence of a force is (assuming as usual the same spreading in all directions)

$$\psi_{0\mathbf{p}}(\mathbf{x},t) = \left(\frac{\Delta x^2}{2\pi\delta x_t^4}\right)^{3/4} \exp\left\{-\frac{im\Delta x^2}{2\hbar t}\frac{(\mathbf{x}-\mathbf{x}_{\mathrm{F}}(t))^2}{\delta x_t^2}\right\}$$
$$\times \exp\left\{\frac{im}{2\hbar t}\left(\mathbf{x}^2 + \frac{\mathbf{F}\cdot\mathbf{x}}{m}t^2 - \frac{\mathbf{F}^2 t^4}{12m^2}\right)\right\}. \tag{N.39}$$

The average of the position for the Gaussian wave packet equals the position of a classical particle in a constant force,

$$\langle \mathbf{x}(t)\rangle = \mathbf{x}_{\mathrm{F}}(t), \tag{N.40}$$

whose dynamics is dictated by Newton's equation

$$m\,\frac{d^2\mathbf{x}_{\mathrm{F}}(t)}{dt^2} = \mathbf{F}. \tag{N.41}$$

It is instructive to obtain, for a particle experiencing a constant force, the wave function that initially is a plane wave. The Schrödinger equation for a particle in a constant force F,

$$i\hbar\frac{\partial\psi(x,t)}{\partial t} = \left(-\frac{\hbar^2}{2m}\frac{\partial^2}{\partial x^2} - Fx\right)\psi(x,t), \tag{N.42}$$

is seen by direct differentiation to have the solution

$$\psi_p(x,t) = \exp\left\{\frac{i}{\hbar}(p+Ft)x\right\}\,\exp\left\{-\frac{i}{\hbar}\int_0^t dt'\,\frac{1}{2m}(p+Ft')^2\right\}, \tag{N.43}$$

which at time $t=0$ is the plane wave

$$\psi_p(x,t=0) = e^{(i/\hbar)px}, \tag{N.44}$$

the free particle energy eigenfunction corresponding to energy $\epsilon(p) = p^2/2m$. Introduce

$$p(t) \equiv p + Ft, \qquad \epsilon(p(t)) = \frac{(p+Ft)^2}{2m} = \frac{1}{2}mv^2(t), \qquad v(t) = \frac{p(t)}{m}, \tag{N.45}$$

which is the momentum (velocity) and energy evolution dictated by Newton's equation for a particle experiencing a constant force. The time dependence of the solution can thus be completely parameterized by $p(t)$ as

$$\psi_p(x,t) = \exp\left\{\frac{i}{\hbar}p(t)x\right\} \exp\left\{-\frac{i}{\hbar}\int_0^t dt'\, \epsilon(p(t'))\right\} \equiv \psi_{p(t)}(x). \tag{N.46}$$

The parameter, $p(t)$, characterizing the state thus traverses the free particle energy parabola with a rate of change

$$\frac{d\epsilon(p(t))}{dt} = \frac{d\epsilon(p(t))}{dp(t)}\frac{dp(t)}{dt} = \frac{p(t)}{m}F = v(t)\,F \tag{N.47}$$

equal to the classical work per unit time done by the force.

Exercise N.5 Show that the superposition (notation defined above)

$$\psi(x,t) = \int_{-\infty}^{\infty} dp\; g(p)\, \exp\left\{\frac{i}{\hbar}p(t)x\right\} \exp\left\{-\frac{i}{\hbar}\int_0^t dt'\, \epsilon(p(t'))\right\} \tag{N.48}$$

is the solution to the Schrödinger equation for a particle in a constant force.

Choosing

$$g(p) = N \exp\left\{-\frac{(p-p_0)^2}{4\delta p^2} - \frac{i}{\hbar}(p-p_0)x_0\right\}, \tag{N.49}$$

where N is a normalization constant, show that the superposition at time $t = 0$ is equal to the Gaussian function localized at (x_0, p_0), Eq. (M.1),

$$\psi(x, t=0) = \psi_{x_0 p_0}(x). \tag{N.50}$$

Calculate the wave function at a later time.

Appendix O: Screening by Symmetry Method

Instead of solving the differential equation in Section 4.6.2 by Fourier analysis, the solution can also be found by elementary means and a symmetry consideration. The mean field potential must be spherically symmetric, $\phi(\mathbf{r}) = \phi(|\mathbf{r}|) = \phi(r)$, since the adjustment of the electronic charge distribution responding to a point charge must be spherically symmetric because the Sommerfeld model is isotropic. The Laplacian operates on a spherically symmetric function according to

$$\Delta_{\mathbf{r}}\, \phi(|\mathbf{r}|) = \frac{1}{r}\frac{\partial^2}{\partial r^2}\,(r\phi(r)) = \frac{1}{r^2}\frac{\partial}{\partial r}\left(r^2 \frac{\partial}{\partial r}\phi(r)\right), \tag{O.1}$$

where the last rewriting is most suitable for our purpose. Integrating Eq. (4.61) over the sphere of radius r_0 centered at the impurity is then simple in spherical coordinates, and gives (dividing by the 4π from the angular integration)

$$\int_0^{r_0} dr\, \frac{\partial}{\partial r}\left(r^2 \frac{\partial \phi(r)}{\partial r}\right) = \kappa_s^2 \int_0^{r_0} dr\, r^2 \phi(r) - \frac{e}{4\pi\epsilon_0}, \tag{O.2}$$

where the last term originates from the delta function term. The integral on the left-hand side is trivially evaluated,

$$\int_0^{r_0} dr\, \frac{\partial}{\partial r}\left(r^2 \frac{\partial \phi(r)}{\partial r}\right) = \left[r^2 \phi'(r)\right]_0^{r_0} = \left[rF'(r)\right]_0^{r_0} - \int_0^{r_0} dr\, F'(r), \tag{O.3}$$

the second equality being a rewriting, as $F(r) \equiv r\phi(r)$, and Eq. (O.2) becomes

$$\left[rF'(r)\right]_0^{r_0} - \int_0^{r_0} dr\, F'(r) = \kappa_s^2 \int_0^{r_0} dr\, rF(r) - \frac{e}{4\pi\epsilon_0}. \tag{O.4}$$

Differentiating with respect to r_0 gives

$$F''(r_0) = \kappa_s^2\, F(r_0), \tag{O.5}$$

which has the general solution, renaming $r_0 \to r$,

$$F(r) = A\, e^{-\kappa_s r} + B\, e^{\kappa_s r}, \tag{O.6}$$

and thereby for the mean field potential

$$\phi(r) = A\, \frac{e^{-\kappa_s r}}{r} + B\, \frac{e^{\kappa_s r}}{r}. \tag{O.7}$$

Calculating the total (plus and minus) charge, $Q(r_0)$, inside the sphere of radius r_0 around the impurity gives

$$Q(r_0) = \int_0^{r_0} d\mathbf{r}\, (e\delta(\mathbf{r}) - e\delta n(\mathbf{r})) = e - 2e^2 N_0 \int_0^{r_0} d\mathbf{r}\, \phi(\mathbf{r}), \tag{O.8}$$

where the last equality follows from Eq. (4.60). The physical boundary condition of the problem is that $Q(r_0)$ must vanish for large values of r_0, since the system is charge-neutral, and the term proportional to B violates this condition, in fact leads to the unphysical result of infinite total charge, and therefore $B = 0$.[1] Evaluating the integral determines the constant

$$0 = Q(r_0 \to \infty) = e - \frac{2e^2 N_0 4\pi A}{\kappa_s^2}, \qquad A = \frac{e}{4\pi}, \tag{O.9}$$

thereby obtaining the result, Eq. (4.71), for the screened potential. Close to the impurity, an electron experiences the unscreened impurity charge, its bare Coulomb potential.

[1] The electric field cannot grow without limit far from the impurity; the potential must vanish far from a neutral mixture of plus and minus charges.

Appendix P: Commutation and Common Eigenfunctions

The existence of common eigenfunctions of two operators is equivalent to the two operators commuting, $\hat{A}\hat{B} = \hat{B}\hat{A}$. Our interest will be the case encountered in Section 7.2, the Hamiltonian for a particle in a one-dimensional periodic potential and the commuting crystal translation operator, $[\hat{T}_a, \hat{H}] = 0$. The time-independent Schrödinger equation for a particle in a one-dimensional potential is a linear differential equation of *second* order and therefore has, for each possible eigenvalue, ϵ, in general at most two independent solutions, say, $\tilde{\psi}_1$ and $\tilde{\psi}_2$, which are eigenfunctions of the Hamiltonian, $\hat{H}\tilde{\psi}_i = \epsilon\,\tilde{\psi}_i$. Their independence means that they are not proportional, $\tilde{\psi}_1 \neq c\,\tilde{\psi}_2$, and thus they describe two different physical states with the same energy. If $\tilde{\psi}_i$, $i = 1, 2$, are energy eigenfunctions with eigenvalue ϵ, it follows from the commutation of $\hat{H}$ and $\hat{T}_a$ (recall Eq. (7.6)) that so is the translated function $\hat{T}_a\tilde{\psi}_i$. Any energy eigenfunction can be expressed in terms of the two independent solutions, so that

$$\hat{T}_a\tilde{\psi}_1 = b_1\tilde{\psi}_1 + b_2\tilde{\psi}_2, \qquad \hat{T}_a\tilde{\psi}_2 = c_1\tilde{\psi}_1 + c_2\tilde{\psi}_2. \tag{P.1}$$

If the coefficients b_2 and c_1 accidentally vanish, $b_2 = 0$ and $c_1 = 0$, we have also struck upon two independent eigenfunctions of the translation operator $\hat{T}_a$, corresponding to eigenvalues b_1 and c_2, respectively. In general, however, the common eigenfunctions of two commuting operators must be constructed, the case corresponding to b_2 and c_1 non-vanishing, $b_2 \neq 0$ and $c_1 \neq 0$.

Consider operating on a superposition and use Eq. (P.1),

$$\begin{aligned}\hat{T}_a(d_1\tilde{\psi}_1 + d_2\tilde{\psi}_2) &= d_1(b_1\tilde{\psi}_1 + b_2\tilde{\psi}_2) + d_2(c_1\tilde{\psi}_1 + c_2\tilde{\psi}_2)\\ &= (d_1b_1 + d_2c_1)\,\tilde{\psi}_1 + (d_1b_2 + d_2c_2)\,\tilde{\psi}_2.\end{aligned} \tag{P.2}$$

Choosing $d_1 = (\lambda_1 - c_2)d_2/b_2$, the coefficient of $\tilde{\psi}_2$ becomes equal to $\lambda_1 d_2$. In the superposition, Eq. (P.2), d_2 is assumed non-zero, $d_2 \neq 0$, as otherwise the considered function would vanish, because now $d_1 \propto d_2$, and thus cannot be a wave function describing a particle. The so-far arbitrary number λ_1 can now be chosen such that

$$\lambda_1 d_1 = d_1b_1 + d_2c_1, \tag{P.3}$$

whereby the coefficient of $\tilde{\psi}_1$ in Eq. (P.2) becomes $\lambda_1 d_1$, and the constructed energy eigenfunction solution, $\psi_1 \equiv d_1\tilde{\psi}_1 + d_2\tilde{\psi}_2$, is thus, with this choice of the parameters d_1 and λ_1, also an eigenfunction of the translation operator, because Eq. (P.2) reduces to

$$\hat{T}_a\psi_1 = \lambda_1\psi_1. \tag{P.4}$$

Inserting the expression for d_1 into the requirement (P.3), then, after multiplication by b_2 and canceling an overall factor of d_2, the requirement on λ_1 is seen to be equivalent to requiring λ_1 to be a solution of the second-order equation

$$\lambda_1^2 - (b_1 + c_2)\lambda_1 + (b_1 c_2 - b_2 c_1) = 0. \tag{P.5}$$

For a superposition $\psi_2 = e_1\tilde{\psi}_1 + e_2\tilde{\psi}_2$, we similarly have

$$\begin{aligned} \hat{T}_a(e_1\tilde{\psi}_1 + e_2\tilde{\psi}_2) &= e_1(b_1\tilde{\psi}_1 + b_2\tilde{\psi}_2) + e_2(c_1\tilde{\psi}_1 + c_2\tilde{\psi}_2) \\ &= (e_1 b_1 + e_2 c_1)\tilde{\psi}_1 + (e_1 b_2 + e_2 c_2)\tilde{\psi}_2. \end{aligned} \tag{P.6}$$

Choosing the coefficient $e_1 = (\lambda_2 - c_2)e_2/b_2$, the coefficient of $\tilde{\psi}_2$ becomes equal to $\lambda_2 e_2$. The coefficient of $\tilde{\psi}_1$ becomes

$$e_1 b_1 + c_1 e_2 = \left(\frac{(\lambda_2 - c_2)b_1}{b_2} + c_1\right) e_2. \tag{P.7}$$

Subtracting this expression from $\lambda_2 e_1$ and inserting the expression for e_1 gives

$$\begin{aligned} \lambda_2 e_1 - (e_1 b_1 + c_1 e_2) &= e_2 \left(\frac{1}{b_2}\lambda_2^2 - \left(\frac{c_2}{b_2} + \frac{b_1}{b_2}\right)\lambda_2 + \left(\frac{c_2 b_1}{b_2} - c_1\right)\right) \\ &= \frac{e_2}{b_2}\left[\lambda_2^2 - (c_2 + b_1)\lambda_2 + (c_2 b_1 - c_1 b_2)\right]. \end{aligned} \tag{P.8}$$

Choosing λ_2, in general, as the other solution of Eq. (P.5), then Eq. (P.8) becomes $e_1 b_1 + e_2 c_1 = \lambda_2 e_1$, and ψ_2 thereby becomes the eigenfunction of the translation operator with eigenvalue λ_2,

$$\hat{T}_a \psi_2 = \lambda_2 \psi_2. \tag{P.9}$$

The constructed solutions, ψ_1 and ψ_2, are independent solutions of the time-independent Schrödinger equation, since, if $\psi_1 \propto \psi_2$, i.e. $d_1\tilde{\psi}_1 + d_2\tilde{\psi}_2 \propto e_1\tilde{\psi}_1 + e_2\tilde{\psi}_2$, this would immediately lead to $\tilde{\psi}_1 \propto \tilde{\psi}_2$, in contradiction with *their* assumed independence. As a consequence of the two operators commuting, they thus have common eigenfunctions (the inverse statement, of course, trivially true). Two independent energy eigenfunctions for a particle in a periodic potential can therefore, according to Eqs. (P.9) and (P.4), be chosen which are also eigenfunctions of the translation operator.[1]

[1] In the above example, one of the operators, the Hamiltonian, has two-fold degenerate eigenstates. The above analysis can be straightforwardly generalized to the case of N-fold degeneracy, and sets of more than two commuting operators.

Appendix Q: Interband Coupling

Before evaluating the interband coupling strength, the property in Eq. (8.63) is established. Differentiating Eq. (7.82) with respect to the Bloch number gives

$$\hat{H}(k)\,\frac{\partial u_{n'k}(x)}{\partial k} + \left(\frac{\partial \hat{H}(k)}{\partial k}\right) u_{n'k}(x) = \epsilon_{n'}(k)\,\frac{\partial u_{n'k}(x)}{\partial k} + u_{n'k}(x)\,\frac{\partial \epsilon_{n'}(k)}{\partial k} \tag{Q.1}$$

and taking the scalar product with u_{nk} gives

$$\int_0^a dx\; u^*_{nk}(x)\left(\hat{H}(k)\,\frac{\partial u_{n'k}(x)}{\partial k} + \frac{\hbar}{m}\left(\frac{\hbar}{i}\frac{\partial}{\partial x} + \hbar k\right) u_{n'k}(x)\right)$$
$$= \epsilon_{n'}(k)\int_0^a dx\; u^*_{nk}(x)\,\frac{\partial u_{n'k}(x)}{\partial k} + \frac{\partial \epsilon_n(k)}{\partial k}\,\delta_{n,n'}, \tag{Q.2}$$

which upon using the hermitian property of $\hat{H}(k)$ becomes

$$(\epsilon_n(k) - \epsilon_{n'}(k))\int_0^a dx\; u^*_{nk}(x)\,\frac{\partial u_{n'k}(x)}{\partial k}$$
$$= -\frac{\hbar^2}{im}\int_0^a dx\; u^*_{nk}(x)\,\frac{\partial u_{n'k}(x)}{\partial x} + \left(\frac{\partial \epsilon_n(k)}{\partial k} - \frac{\hbar^2 k}{m}\right)\delta_{n,n'}. \tag{Q.3}$$

For different bands, $n \neq n'$, Eq. (Q.3) thus becomes Eq. (8.63).

To be specific regarding the two-band model considered in Section 8.6, the lower band is assumed to have maximum at $k = 0$, whereby the upper band has a minimum at the zone center (the Bragg point case is seen to be treated analogously). The set of functions $\{u_{n0}\}_{n=1,2,3,\dots}$ is complete and any periodic function can be expanded in this basis. In practice, only those states nearest in energy need to be taken into account, and in the two-band model we therefore have

$$u_{nk}(x) = c_1^{(n)}(k)\,u_{10}(x) + c_2^{(n)}(k)\,u_{20}(x), \qquad n = 1, 2. \tag{Q.4}$$

Then the so-called **k** . **p** method is used to take us beyond the quadratic energy dispersion. The *Hamiltonian* for the u function, Eq. (7.82), is rewritten as

$$\hat{H}(k) = \hat{H}(0) + \frac{\hbar^2 k^2}{2m} + \frac{\hbar k}{m}\hat{p}, \qquad \hat{p} \equiv \frac{\hbar}{i}\frac{\partial}{\partial x}, \tag{Q.5}$$

and the eigenvalue equation $\hat{H}(k)\,u_{nk} = \epsilon_n(k)\,u_{nk}$ thus takes the form

$$\left(\epsilon_1(0) + \frac{\hbar^2k^2}{2m} + \frac{\hbar k}{m}\hat{p}\right) c_1^{(n)}(k)\,u_{10} + \left(\epsilon_2(0) + \frac{\hbar^2k^2}{2m} + \frac{\hbar k}{m}\hat{p}\right) c_2^{(n)}(k)\,u_{20}$$
$$= \epsilon_n(k)[c_1^{(n)}(k)\,u_{10} + c_2^{(n)}(k)\,u_{20}]. \tag{Q.6}$$

Taking scalar products with u_{10} and u_{20}, and noting that, in view of Eqs. (8.22) and (8.23),

$$p_{11} \equiv \langle u_{10}|\hat{p}|u_{10}\rangle = m\bar{v}_{10} = \frac{m}{\hbar}\frac{\partial\epsilon_1(k)}{\partial k}\bigg|_{k=0} = 0, \tag{Q.7}$$

and similarly $p_{22} = 0$, turns the Hamiltonian in the two-band model into the 2×2 matrix that gives the eigenvalue equation

$$\begin{pmatrix} \epsilon_2(0) + \hbar^2k^2/2m & (\hbar k/m)p_{21} \\ (\hbar k/m)p_{12} & \epsilon_1(0) + \hbar^2k^2/2m \end{pmatrix} \begin{pmatrix} c_2^{(n)}(k) \\ c_1^{(n)}(k) \end{pmatrix} = \epsilon_n(k) \begin{pmatrix} c_2^{(n)}(k) \\ c_1^{(n)}(k) \end{pmatrix}, \tag{Q.8}$$

where

$$p_{21} \equiv \langle u_{20}|\hat{p}|u_{10}\rangle = p_{12}^*. \tag{Q.9}$$

Since u_{10} and u_{20} can be chosen as real functions, because the current vanishes in these states, p_{21} can be assumed purely imaginary. The secular equation is

$$\begin{vmatrix} -\epsilon_n(k) + \Delta + \hbar^2k^2/2m & (\hbar k/m)p_{21} \\ (\hbar k/m)p_{12} & -\epsilon_n(k) + \hbar^2k^2/2m \end{vmatrix} = 0, \tag{Q.10}$$

where, for ease of algebra, the energy is now measured relative to the top of the lower band, $\epsilon_1(0)$, and $\Delta = \epsilon_2(0) - \epsilon_1(0)$ is the energy gap. The quadratic equation has the two roots

$$\epsilon_n(k) = \frac{\Delta}{2} + \frac{\hbar^2k^2}{2m} + (-1)^n\,\frac{d(k)}{2}, \qquad n = 1, 2, \tag{Q.11}$$

where

$$d(k) = \sqrt{\Delta^2 + \frac{4\hbar^2|p_{12}|^2}{m^2}k^2}. \tag{Q.12}$$

The band masses become

$$\frac{1}{m_n} = (-1)^n\,\frac{1}{\hbar^2}\frac{\partial^2\epsilon_n(k=0)}{\partial k^2} = \frac{2\,|p_{12}|^2}{m^2\Delta} + (-1)^n\frac{1}{m}, \qquad n = 1, 2, \tag{Q.13}$$

where parameters are assumed such that both band masses are positive, i.e. $|p_{12}|^2 > m\Delta/2$, in accordance with the kind of band structure of interest, *viz.* where the lower band has maximum at the zone center, a typical situation of a valence and conduction band of an insulator or semiconductor. Introducing

$$\frac{1}{m_0} \equiv \frac{1}{m_1} + \frac{1}{m_2} \tag{Q.14}$$

gives

$$|p_{12}| = \frac{m}{2}\sqrt{\frac{\Delta}{m_0}}, \qquad d(k) = \sqrt{\Delta^2 + \frac{\hbar^2\Delta}{m_0}k^2}, \tag{Q.15}$$

and we note that $m_0 < m/2$.

According to Eq. (Q.11), the band energies can be analytically continued to complex k values, giving branch points where $d(q) = 0$, i.e. at $q_\pm = \pm i\sqrt{m_0\Delta}/\hbar$, as determined by the square root function.

Possible stationary points for the function $\chi(k)$, Eq. (8.60), correspond to the analytically continued energy band functions being equal,

$$\epsilon_2(q) = \epsilon_1(q), \quad \text{i.e.} \quad d(q) = 0, \tag{Q.16}$$

and gives, according to Eq. (Q.15), locations on the imaginary axis[1]

$$q_\pm = \pm i\sqrt{\frac{m_0\Delta}{\hbar^2}}, \tag{Q.17}$$

i.e. the stationary points of $\chi(k)$ coincide with the branch points of the energy. If $\Delta \ll 2\pi^2\hbar^2/m_0a^2$, then q is located close to zero on the scale of π/a.

What is needed is $\gamma_{12}(k)$ near a stationary point. According to Eq. (Q.8), the coefficients are related according to (with the adopted convention of zero energy)

$$c_1^{(n)}(k) = \frac{(\hbar p_{12}/m)k}{\epsilon_n(k) - \hbar^2k^2/2m}\, c_2^{(n)}(k), \tag{Q.18}$$

i.e.

$$c_1^{(n)}(k) = \begin{cases} \dfrac{(2\hbar p_{12}/m)k}{\Delta + d(k)}\, c_2^{(2)}(k), & n = 2, \\[2ex] \dfrac{(2\hbar p_{12}/m)k}{\Delta - d(k)}\, c_2^{(1)}(k), & n = 1. \end{cases} \tag{Q.19}$$

The two normalized functions, Eq. (Q.4), diagonalizing the two-band Hamiltonian are therefore

$$u_{2k}(x) = \frac{1}{\sqrt{2d(k)}}\left(\sqrt{d(k) + \Delta}\, u_{20}(x) + \frac{(2\hbar p_{12}/m)k}{\sqrt{d(k) + \Delta}}\, u_{10}(x)\right) \tag{Q.20}$$

and

$$u_{1k}(x) = \frac{1}{\sqrt{2d(k)}}\left(\sqrt{d(k) - \Delta}\, u_{20}(x) - \frac{(2\hbar p_{12}/m)k}{\sqrt{d(k) - \Delta}}\, u_{10}(x)\right). \tag{Q.21}$$

These functions can be analytically continued into the complex plane, being *energy* eigenfunctions corresponding to energies for the complex k values, complex-valued energies.

Employing the fact that p_{12} is purely imaginary, $p_{12} = i|p_{12}|s(p_{12})$, where s is short for the sign function, $s(p_{12}) \equiv \text{sign}(p_{12})$, we can, according to Eq. (Q.15), rewrite $2\hbar p_{12}k/m = is(p_{12})s(k)\sqrt{d^2(k) - \Delta^2}$, and

$$u_{2k}(x) = \frac{1}{\sqrt{2d(k)}}\left(\sqrt{d(k) + \Delta}\, u_{20}(x) + is(p_{12})s(k)\sqrt{d(k) - \Delta}\, u_{10}(x)\right) \tag{Q.22}$$

[1] For a three-dimensional cubic crystal with electric field along the k_x direction, the only difference would be $q \to \pm i\sqrt{m_r\Delta/\hbar^2 + k_y^2 + k_z^2}$, i.e. the additional real Bloch vector components along the y and z directions.

and

$$u_{1k}(x) = \frac{1}{\sqrt{2d(k)}}\left(\sqrt{d(k)-\Delta}\,u_{20}(x) - is(p_{12})s(k)\sqrt{d(k)+\Delta}\,u_{10}(x)\right) \tag{Q.23}$$

and we note that $u_{n-k} = u^*_{nk}$.

Evaluating the derivative gives

$$\frac{d}{dk}\left(\frac{\sqrt{d(k)\pm\Delta}}{\sqrt{2d(k)}}\right) = \frac{\mp\Delta}{(2d(k))^{3/2}\sqrt{d(k)\pm\Delta}}\,d'(k), \qquad d'(k) = \frac{\hbar^2\Delta}{m_0}\frac{k}{d(k)}, \tag{Q.24}$$

and the quantity of interest becomes

$$\frac{\partial u_{1k}(x)}{\partial k} = \frac{\Delta d'(k)}{(2d)^{3/2}}\left(\frac{1}{\sqrt{d-\Delta}}\,u_{20}(x) + is(p_{12})s(k)\frac{1}{\sqrt{d+\Delta}}\,u_{10}(x)\right). \tag{Q.25}$$

The interband coupling

$$\gamma_{21}(k) = \int_0^a dx\; u^*_{2k}(x)\,\frac{\partial u_{1k}(x)}{\partial k} = \int_0^a dx\; u_{2-k}(x)\,\frac{\partial u_{1k}(x)}{\partial k} \tag{Q.26}$$

then becomes

$$\gamma_{21}(k) = \frac{\Delta d'(k)}{(2d)^{3/2}}\int_0^a dx\left(\sqrt{d+\Delta}\,u_{20}(x) - is(p_{12})s(k)\sqrt{d-\Delta}\,u_{10}(x)\right)$$
$$\times\left(\frac{1}{\sqrt{d-\Delta}}\,u_{20}(x) + is(p_{12})s(k)\frac{1}{\sqrt{d+\Delta}}\,u_{10}(x)\right), \tag{Q.27}$$

and thereby

$$\gamma_{21}(k) = \frac{\Delta d'(k)}{(2d(k))^2}\left(\frac{\sqrt{d(k)+\Delta}}{\sqrt{d(k)-\Delta}} + \frac{\sqrt{d(k)-\Delta}}{\sqrt{d(k)+\Delta}}\right)$$
$$= \frac{\Delta d'(k)}{2d(k)\sqrt{(d(k))^2-\Delta^2}} = \sqrt{\frac{\hbar^2\Delta}{4m_0}}\,\frac{\Delta}{(d(k))^2}\,\frac{k}{\sqrt{k^2}}. \tag{Q.28}$$

Introducing $q \equiv +i\sqrt{m_0\Delta/\hbar^2}$ and noting that $d^2(k) = (k^2-q^2)\hbar^2\Delta/m_0$, the interband coupling takes the form

$$\gamma_{21}(k) = \frac{\sqrt{m_0\Delta}}{4\hbar q}\frac{k}{\sqrt{k^2}}\left(\frac{1}{k-q} - \frac{1}{k+q}\right). \tag{Q.29}$$

Near the stationary point, $k \simeq q$, $\gamma_{21}(k)$ thus exhibits the simple pole specified in Eq. (8.64).

Appendix R: Common Crystal Structures

The geometrically simplest kind of three-dimensional crystal to envision would be identical ions occupying the points of a simple cubic lattice, a monatomic Bravais lattice. The corresponding Bravais lattice would then be *identical* to the cubic lattice in question, and as basis vectors we could, for example, choose the three orthogonal vectors connecting an ion with its three nearest neighbors. However, this is not how an assembly of identical atoms crystallizes at ambient conditions (except for the alpha phase of polonium). This has to do with the electrostatic energetics of stacking spheres with the size of the filled shell part of an atom's electronic configuration (the ions), and the distribution in space of its valence electrons. The alkali metals, for example, have body-centered cubic (bcc) crystal structure, where, in addition to the corners of the cubes of a simple cubic lattice, also the centers of the cubes are occupied by the monovalent alkali ions. A possible choice of basis vectors for this structure could be the vectors connecting a corner ion with two nearest corner ions and the ion at the center closest to these (that the bcc lattice indeed is a Bravais lattice, i.e. looks equivalent from its corners and centers, see Exercise R.1).[1]

Semiconductors have typically covalent bonding. Semiconductors such as silicon or germanium have the diamond structure, as do almost all the important semiconductors. Silicon and germanium atoms have four valence electrons, so, as discussed in Section 6.6, they arrange themselves so that around any atom are four nearest neighbors, which together make up a tetrahedron in space.[2]

Exercise R.1 Draw enough sites of a bcc lattice to realize that it can be viewed as constituted by two interpenetrating simple cubic Bravais lattices, one making up the "corners" and the other making up the "centers". Realizing that "corners" and "centers" can be interchanged, they are indeed equivalent points and the bcc lattice is indeed a Bravais lattice.

Assume that the two simple cubic lattices have the lattice constant a, i.e. are generated by the primitive vectors $a\hat{\mathbf{x}}$, $a\hat{\mathbf{y}}$ and $a\hat{\mathbf{z}}$, and show that

[1] In order to understand the content of the chapters, a detailed knowledge of crystal lattice structures is not needed, and the exercises are just meant for a reader with inclination for knowing the various types of structures of interest.

[2] Crystallographically, this is a face-centered cubic (fcc) crystal with eight atoms in the unit cell. There are three important directions in such a crystal. The directions of the cube edges are called (100)-directions; the directions from the center of the cube to the middle of each of the 12 edges are called (110)-directions; and the directions from the center to the eight cube vertices are called the (111)-directions. The atoms are differently densely packed in the various planes.

$$\mathbf{a}_1 = a\hat{\mathbf{x}}, \qquad \mathbf{a}_2 = a\hat{\mathbf{y}}, \qquad \mathbf{a}_3 = \frac{a}{2}(\hat{\mathbf{x}} + \hat{\mathbf{y}} + \hat{\mathbf{z}}) \tag{R.1}$$

is a possible choice of primitive vectors for the Bravais lattice describing the bcc lattice.

Show that

$$\mathbf{a}_1 = \frac{a}{2}(\hat{\mathbf{y}} + \hat{\mathbf{z}} - \hat{\mathbf{x}}), \qquad \mathbf{a}_2 = \frac{a}{2}(\hat{\mathbf{z}} + \hat{\mathbf{x}} - \hat{\mathbf{y}}), \qquad \mathbf{a}_3 = \frac{a}{2}(\hat{\mathbf{x}} + \hat{\mathbf{y}} - \hat{\mathbf{z}}) \tag{R.2}$$

is another possible choice of primitive vectors. Notice the cyclic permutation symmetry of these expressions.

Another common form into which solids crystallize is an fcc crystal. The fcc Bravais lattice is obtained from the simple cubic lattice by adding the center points of the six faces of each cube. That this lattice indeed is a Bravais lattice, i.e. looks equivalent from its corners and face center points, is realized by the following consideration. Draw a cube and mark its corner and face center points. Then draw lines from the face center points to the face center points of adjacent cubes, thereby forming a cube equal in size to the original cube. One then realizes that the corner points of the original cube now appear as face center points of the newly formed cube, indeed *corner* and *center* points are equivalent.

Exercise R.2 Show that

$$\mathbf{a}_1 = \frac{a}{2}(\hat{\mathbf{y}} + \hat{\mathbf{z}}), \qquad \mathbf{a}_2 = \frac{a}{2}(\hat{\mathbf{z}} + \hat{\mathbf{x}}), \qquad \mathbf{a}_3 = \frac{a}{2}(\hat{\mathbf{x}} + \hat{\mathbf{y}}) \tag{R.3}$$

is a possible symmetric choice of primitive vectors for the fcc Bravais lattice, and notice that they connect a corner point to three nearest face center points.

Exercise R.3 Show that the bcc and fcc lattices are reciprocal to each other, i.e. the reciprocal lattice for the bcc lattice is the fcc lattice, and vice versa.

Rock salt has its sodium and chloride ions sitting on alternating points of a simple cubic lattice, i.e. each ion has six of the other kind of ions as nearest neighbors. The ions occupy the sites of a simple cubic lattice but obviously lack its translation symmetry since the ions are different. The two different sets of ions occupy the sites of two interpenetrating fcc lattices, which is seen by drawing some crystal sites. The crystal structure of this compound can therefore be described as an fcc Bravais lattice with a basis consisting of a chloride ion at the origin, $\mathbf{0}$, and a sodium ion at the center, $(\hat{\mathbf{x}} + \hat{\mathbf{y}} + \hat{\mathbf{z}})a/2$, of the simple cubic cell defining one of the fcc lattices.

Exercise R.4 Show that both the bcc and fcc Bravais lattices can be described as a simple cubic Bravais lattice spanned by the three corner-to-corner vectors $a\hat{\mathbf{x}}$, $a\hat{\mathbf{y}}$ and $a\hat{\mathbf{z}}$ with a basis: in the bcc case, with a two-point basis, one point at a cubic corner, $\hat{\mathbf{0}}$, and the other at the point $a(\hat{\mathbf{x}} + \hat{\mathbf{y}} + \hat{\mathbf{z}})/2$; and in the fcc case, with a four-point basis at the points $\hat{\mathbf{0}}$, $a(\hat{\mathbf{x}} + \hat{\mathbf{y}})/2$, $a(\hat{\mathbf{y}} + \hat{\mathbf{z}})/2$ and $a(\hat{\mathbf{z}} + \hat{\mathbf{x}})/2$.

Appendix S: Effective Mass Approximation

Often, only the band structure near band edges is of interest in semiconductors. For example, these are the states of interest in a doped semiconductor, since they are the states populated in the conduction bands and empty in the valence bands and thus responsible for carrying electric currents. Typically the band edges correspond to points at the zone center or on the boundary of the Brillouin zone, and since these often are points of high symmetry, the energy there takes on extremal values according to the analysis of Section 9.7.

S.1 Band Edge as Potential

Assuming, for example, an extremum at the zone center, the band structure close to $\mathbf{k} = \mathbf{0}$ is quadratic,

$$\epsilon_n(\mathbf{k}) = \epsilon_n(\mathbf{0}) + \frac{\hbar^2}{2} \sum_{\alpha\beta} k_\alpha \mathcal{M}^{-1}_{\alpha\beta} k_\beta + \cdots , \tag{S.1}$$

where

$$\mathcal{M}^{-1}_{\alpha\beta} \equiv \left. \frac{\partial^2 \epsilon_n(\mathbf{k})}{\partial k_\alpha \, \partial k_\beta} \right|_{\mathbf{k}=\mathbf{0}} \tag{S.2}$$

is called the inverse effective mass tensor. Since the inverse effective mass tensor is seen to be symmetric, i.e. hermitian and real, a set of principal axes in $\mathbf{k}$-space can be chosen making it diagonal (any real hermitian matrix can be diagonalized by an orthogonal matrix),

$$\mathcal{M}_n^{-1} = \begin{pmatrix} 1/m_x^* & 0 & 0 \\ 0 & 1/m_y^* & 0 \\ 0 & 0 & 1/m_z^* \end{pmatrix} , \tag{S.3}$$

and in the principal axis coordinate system the energy spectrum is parabolic

$$\epsilon_n(\mathbf{k}) \simeq \epsilon_n(\mathbf{0}) + \frac{\hbar^2}{2} \left(\frac{k_x^2}{m_x^*} + \frac{k_y^2}{m_y^*} + \frac{k_z^2}{m_z^*} \right) \tag{S.4}$$

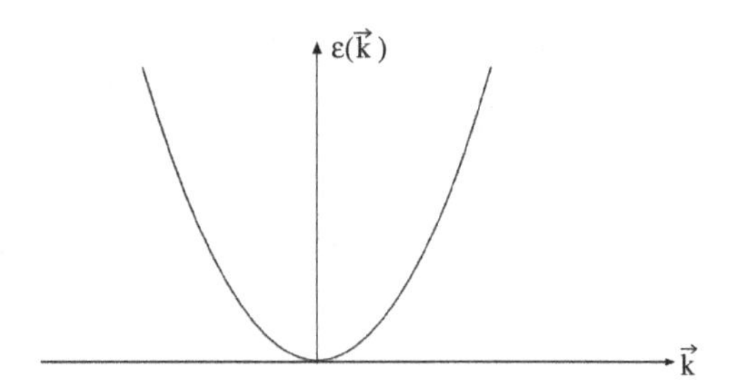

Figure S.1 Quadratic or parabolic energy spectrum.

and depicted in Figure S.1. However, in contrast to that of a free particle, the curvature can be different in different directions, as described by the different effective band masses, and motion of an electron in different directions can thus differ. In semiconductors, the effective masses due to band structure effects typically differ by one or two orders of magnitude either lighter or heavier than the electron mass.

The energies close to the zone center are thus equivalently determined by the equation

$$\left(\frac{\hbar^2}{2m_x^*}\frac{\partial^2}{\partial x^2}+\frac{\hbar^2}{2m_y^*}\frac{\partial^2}{\partial y^2}+\frac{\hbar^2}{2m_z^*}\frac{\partial^2}{\partial z^2}+\epsilon_n(\mathbf{0})\right)e^{i\mathbf{k}\cdot\mathbf{x}}=\epsilon_n(\mathbf{k})\,e^{i\,\mathbf{k}\cdot\mathbf{x}} \tag{S.5}$$

and the electron moves in the crystal as if in vacuum except that its mass has changed from its free electron value, and is different for motion in different directions. In addition, the electron is exposed to a constant potential, specified by the band edge energy, a feature irrelevant for a homogeneous semiconductor, but crucial for understanding the properties of heterostructures.

Qualitatively, the functioning of a so-called heterostructure, a material consisting of layers of different semiconductors, can now be understood. Suppose a crystal B is grown on top of a crystal A. An electron then experiences different periodic potentials in the two materials. Under certain conditions, discussed in detail in Chapter 12, the wave function equation for such a structure reduces to an effective mass equation, with the sole modification that the band edge potentials and effective masses differ in the two materials. The different band gaps in the different materials provide a step potential, a new tool for controlling carriers, since, for example, conduction electrons can be forced into the narrow-band-gap region. For its own sake and in particular for its use in Chapter 12, we delve into the effective mass approximation and the so-called $\mathbf{k}\cdot\mathbf{p}$ method,

S.2 The k · p method

Band structure information, such as effective mass values, can be and often are simply taken from experiments. However, it is enlightening to obtain some features from analytical results.[1] The equation determining the energy spectrum, $\epsilon_n(\mathbf{k})$, and the periodic part of the Bloch function, Eq. (9.36), can be written as

[1] *Ab initio* calculations of band structures are nowadays possible by number crunching using a computer.

$$\left\{\frac{\hat{\mathbf{p}}^2}{2m} + V(\mathbf{r}) + \frac{\hbar^2\mathbf{k}^2}{2m} + \frac{\hbar\mathbf{k}}{m}\cdot\hat{\mathbf{p}}\right\} u_{n\mathbf{k}}(\mathbf{r}) = \epsilon_n(\mathbf{k})\, u_{n\mathbf{k}}(\mathbf{r}) \tag{S.6}$$

and has the form

$$\hat{H}_{\mathbf{k}}\, u_{n\mathbf{k}}(\mathbf{r}) = \epsilon_n(\mathbf{k})\, u_{n\mathbf{k}}(\mathbf{r}), \tag{S.7}$$

where

$$\hat{H}_{\mathbf{k}} = \hat{H}_0 + \hat{W}(\mathbf{k}) \tag{S.8}$$

with

$$\hat{H}_0 \equiv \frac{\hat{\mathbf{p}}^2}{2m} + V(\mathbf{r}) \tag{S.9}$$

and

$$\hat{W}(\mathbf{k}) \equiv \frac{\hbar^2\mathbf{k}^2}{2m} + \frac{\hbar\mathbf{k}}{m}\cdot\hat{\mathbf{p}}. \tag{S.10}$$

We note that

$$\hat{W}(\mathbf{0}) = 0. \tag{S.11}$$

In addition to satisfying Eq. (S.6), the functions $u_{n\mathbf{k}}$ are, according to Bloch's theorem, required to satisfy the periodic boundary condition, Eq. (9.33).

As an example, suppose that we are interested in the band structure of band n near the zone center, $\mathbf{k} = \mathbf{0}$. These energy eigenvalues, near the band edge, can be calculated in perturbation theory by treating $\hat{W}(\mathbf{k})$ as the perturbation. The unperturbed energies and eigenfunction are specified by the band edge energy,

$$\hat{H}_0 u_{n\mathbf{0}}(\mathbf{r}) = \epsilon_n(\mathbf{0})\, u_{n\mathbf{0}}(\mathbf{r}), \qquad n = 1, 2, 3, \ldots. \tag{S.12}$$

Since the energy is assumed to have an extremum at the zone center, the first-order term in $\mathbf{k}$ vanishes, being proportional to

$$\langle n, \mathbf{0}|\mathbf{k}\cdot\hat{\mathbf{p}}|n, \mathbf{0}\rangle = m\mathbf{k}\cdot\mathbf{v}_n(\mathbf{0}) = m\mathbf{k}\cdot\frac{\partial\epsilon_n(\mathbf{0})}{\partial\mathbf{k}} = 0. \tag{S.13}$$

The second-order expression is therefore needed, giving, assuming for simplicity that the nth band is non-degenerate,

$$\epsilon_n(\mathbf{k}) \simeq \epsilon_n(\mathbf{0}) + \frac{\hbar^2\mathbf{k}^2}{2m} + \sum_{n'\,(\neq n)} \frac{\left|\langle n, \mathbf{0}|(\hbar\mathbf{k}/m)\cdot\hat{\mathbf{p}}|n', \mathbf{0}\rangle\right|^2}{\epsilon_n(\mathbf{0}) - \epsilon_{n'}(\mathbf{0})}. \tag{S.14}$$

Equivalently

$$\epsilon_n(\mathbf{k}) \simeq \epsilon_n(\mathbf{0}) + \frac{\hbar^2}{2}\mathbf{k}\cdot\boldsymbol{\mathcal{M}}_n^{-1}\mathbf{k}, \tag{S.15}$$

where the inverse effective mass tensor (for the nth band) is given by

$$[\boldsymbol{\mathcal{M}}_n^{-1}]_{\alpha\beta} = \frac{1}{m}\delta_{\alpha\beta} + \frac{1}{m^2}\sum_{n'\,(\neq n)} \frac{p^{\alpha}_{nn'}p^{\beta}_{n'n} + p^{\alpha}_{n'n}p^{\beta}_{nn'}}{\epsilon_n(\mathbf{0}) - \epsilon_{n'}(\mathbf{0})}, \tag{S.16}$$

where

$$p^{\alpha}_{nn'} = \langle n, \mathbf{0} | \hat{p}_{\alpha} | n', \mathbf{0} \rangle = \int_V d\mathbf{r}\; \psi^*_{n\mathbf{0}}(\mathbf{r})\, \hat{p}_{\alpha}\, \psi_{n'\mathbf{0}}(\mathbf{r}) = \int_V d\mathbf{r}\; u^*_{n\mathbf{0}}(\mathbf{r})\, \hat{p}_{\alpha}\, u_{n'\mathbf{0}}(\mathbf{r}) \tag{S.17}$$

is the matrix element of the momentum operator between the Bloch states in question. The inverse effective mass tensor is seen to be a symmetric matrix and can thus be diagonalized.

We note the general features that, due to the sign of the denominator in Eq. (S.14), contributions from higher energy bands make the effective mass heavier and tend to flatten the band, whereas contributions from lower bands have the opposite effect. In either case, the effect of any given band is thus to repel the band in question, the general feature of level repulsion. The need to often include several bands makes the determination of the effective mass tensor complicated, and in practice one takes into account only a few bands that are close in energy, as the energy denominator makes the contributions from far-away bands negligible.

Our interest is to solve Eq. (S.7), and we recall that, for any $\mathbf{k}$, the functions $\{u_{n\mathbf{k}}(\mathbf{r})\}_{n=1,2,\ldots}$ constitute a complete orthogonal set of functions for the space of functions satisfying Eq. (S.6) and the periodic boundary condition (9.33). Any eigenfunction of $\hat{H}_{\mathbf{k}}$ can therefore be expressed as a superposition of the eigenfunctions of, say, $\hat{H}_{\mathbf{k}_0}$,

$$u_{n\mathbf{k}}(\mathbf{r}) = \sum_{n'} a_{n'}(\mathbf{k})\, u_{n'\mathbf{k}_0}(\mathbf{r}), \tag{S.18}$$

where the summation is over all the bands, and the parametric dependence on the Bloch vector of the expansion coefficients is indicated. In the following, we assume for generality that the band edges are located at $\mathbf{k}_0$. Inserting into Eq. (S.7), and taking the scalar product with $u_{n_1\mathbf{k}_0}$, gives an equation for each band index n_1,

$$0 = a_{n_1}(\mathbf{k}) \left(\epsilon_{n_1}(\mathbf{k}_0) - \epsilon_n(\mathbf{k}) + \frac{\hbar^2}{2m}(\mathbf{k}^2 - \mathbf{k}_0^2) \right) + \sum_{n'} a_{n'}(\mathbf{k}) \frac{\hbar}{m}(\mathbf{k} - \mathbf{k}_0) \cdot \langle n_1 \mathbf{k}_0 | \hat{\mathbf{p}} | n' \mathbf{k}_0 \rangle. \tag{S.19}$$

The exact energies $\epsilon_n(\mathbf{k})$ are thus determined by the secular equation, the determinant of the coefficients being equal to zero, and specified in terms of all the $\epsilon_n(\mathbf{k}_0)$ values, $n = 1, 2, 3, \ldots$, and the momentum operator matrix elements. In practice, only a few bands need to be taken into account.

In Appendix Q, the $\mathbf{k} \cdot \mathbf{p}$ method is used to calculate the interband coupling in the two-band model.

Appendix T: Integral Doubling Formula

Consider an integral over a volume, Ω, where the integrand is the product of a periodic function, $p(\mathbf{r})$, having the periodicity property of a lattice, $p(\mathbf{r}+\mathbf{R_n}) = p(\mathbf{r})$, and a function, $S(\mathbf{r})$, slowly varying on the scale of the unit cell. We now establish that the integral approximately equals the product of the integrals of the two functions divided by the volume,

$$\int_{\Omega} d\mathbf{r}\; S(\mathbf{r})\, p(\mathbf{r}) \simeq \frac{1}{\Omega}\int_{\Omega} d\mathbf{r}\; p(\mathbf{r}) \int_{\Omega} d\mathbf{r}\; S(\mathbf{r}), \tag{T.1}$$

where the volume is assumed to conform with the periodicity of the periodic function, i.e. equals a number of lattice unit cells.

The periodic function has a Fourier expansion

$$p(\mathbf{r}) = \sum_{\mathbf{K}} p_{\mathbf{K}}\, e^{i\mathbf{K}\cdot\mathbf{r}}, \tag{T.2}$$

where the Fourier coefficients are given by

$$p_{\mathbf{K}} = \frac{1}{\Omega}\int_{\Omega} d\mathbf{r}\; e^{-i\mathbf{r}\cdot\mathbf{K}} p(\mathbf{r}). \tag{T.3}$$

The set of $\mathbf{K}$ vectors are the reciprocal lattice vectors, i.e. $e^{i\mathbf{K}\cdot\mathbf{R_n}} = 1$ for all lattice and reciprocal lattice vectors (recall Section 9.3).

Any location in the volume can be reached by first going to the nearest unit cell center and from there to the point in question, $\mathbf{r} = \mathbf{R}_n + \mathbf{r}'$, as depicted in Figure 9.7. Adding up the contributions to the integral from each unit cell gives

$$\int_{\Omega} d\mathbf{r}\; S(\mathbf{r})\, p(\mathbf{r}) = \sum_{\mathbf{K}} p_{\mathbf{K}} \sum_{\mathbf{R_n}}{}' \int_{v} d\mathbf{r}'\; S(\mathbf{R_n}+\mathbf{r}')\, e^{i\mathbf{K}\cdot\mathbf{r}'}, \tag{T.4}$$

where the prime on the lattice site summation indicates that the sum is only over lattice sites within the volume, and v denotes the volume of the unit cell, $\Omega = Nv$, where N is the number of unit cells.

Using the fact that, for the slowly varying function, its variation within a unit cell can be neglected,

$$\int_{\Omega} d\mathbf{r}\; S(\mathbf{r})\, p(\mathbf{r}) \simeq \sum_{\mathbf{K}} p_{\mathbf{K}} \sum_{\mathbf{R_n}}{}' S(\mathbf{R_n}) \int_{v} d\mathbf{r}\; e^{i\mathbf{K}\cdot\mathbf{r}}. \tag{T.5}$$

The approximation has now been quantified: the correction term is on the order of the ratio of the lattice constant to the scale of variation of the slowly varying function. Since $\mathbf{K}$

is a vector that respects the symmetry of the unit cell, the integration of the exponential function produces the Kronecker function, $v\,\delta_{\mathbf{K},\mathbf{0}}$ (recall Eq. (9.42)), which makes the $\mathbf{K}$ summation collapse to only the value $\mathbf{K} = \mathbf{0}$, and we have

$$\begin{aligned}\int_{\Omega} d\mathbf{r}\; S(\mathbf{r})\, p(\mathbf{r}) &\simeq v p_{\mathbf{K}=\mathbf{0}} \sum_{\mathbf{R_n}}{}' S(\mathbf{R_n}) \\ &\simeq \frac{1}{\Omega} \int_{\Omega} d\mathbf{r}\; p(\mathbf{r}) \int_{\Omega} d\mathbf{r}\; S(\mathbf{r}).\end{aligned} \tag{T.6}$$

References

[1] J. Rammer (1998) *Quantum Transport Theory*, Frontiers in Physics, vol. 99. Reading, MA: Perseus Books (paperback edition 2004).

[2] J. Zak (1988) Comment on "Time evolution of Bloch electrons in a homogeneous electric field". *Phys. Rev.* B **38**, 6322.
J. B. Krieger and G. J. Iafrate (1988) Reply to "Comment on 'Time evolution of Bloch electrons in a homogeneous electric field'". *Phys. Rev.* B **38**, 6324.

[3] J. Schwinger and D. S. Saxon (1968) *Discontinuities in Waveguides: Notes on Lectures by Julian Schwinger*. New York: Gordon and Breach.

[4] B. J. van Wees, H. van Houten, C. W. J. Beenakker, J. G. Williamson, L. P. Kouwenhoven, D. van der Marel and C. T. Foxon (1988) Quantized conductance of point contacts in a two-dimensional electron gas. *Phys. Rev. Lett.* **60**, 848.

[5] F. Ott and J. Lunney (1998) Quantum conduction: a step-by-step guide. *Europhysics News*, **29**, no. 1, January/February, pp. 13–16.

[6] N. Garcia and J. L. Costa-Krämer (1996) Quantum-level phenomena in nanowires, *Europhysics News*, **27**, no. 3, May/June, pp. 89–91.

[7] E. Buks, R. Schuster, M. Heiblum, D. Mahalu and V. Umansky (1998) Dephasing in electron interference by a 'which-path' detector. *Nature (London)* **391**, 871.

[8] D. Sprinzak, E. Buks, M. Heiblum and H. Shtrikman (2000) Controlled dephasing of electrons via a phase sensitive detector. *Phys. Rev. Lett.* **84**, 5820.

Index